Principles and Applications of Thermal Analysis

Principles and Applications of Thermal Analysis

Edited by

Paul Gabbott

© 2008 by Blackwell Publishing Ltd

Blackwell Publishing editorial offices:
Blackwell Publishing Ltd, 9600 Garsington Road, Oxford OX4 2DQ, UK
  Tel: +44 (0)1865 776868
Blackwell Publishing Professional, 2121 State Avenue, Ames, Iowa 50014-8300, USA
  Tel: +1 515 292 0140
Blackwell Publishing Asia Pty Ltd, 550 Swanston Street, Carlton, Victoria 3053, Australia
  Tel: +61 (0)3 8359 1011

The right of the Author to be identified as the Author of this Work has been asserted in accordance with the Copyright, Designs and Patents Act 1988.

All rights reserved. No part of this publication may be reproduced, stored in a retrieval system, or transmitted, in any form or by any means, electronic, mechanical, photocopying, recording or otherwise, except as permitted by the UK Copyright, Designs and Patents Act 1988, without the prior permission of the publisher.

Designations used by companies to distinguish their products are often claimed as trademarks. All brand names and product names used in this book are trade names, service marks, trademarks or registered trademarks of their respective owners. The Publisher is not associated with any product or vendor mentioned in this book.

This publication is designed to provide accurate and authoritative information in regard to the subject matter covered. It is sold on the understanding that the Publisher is not engaged in rendering professional services. If professional advice or other expert assistance is required, the services of a competent professional should be sought.

First published 2008 by Blackwell Publishing Ltd

ISBN-13: 978-1-4051-3171-1

Library of Congress Cataloging-in-Publication Data

Applications of thermal analysis / edited by Paul Gabbott. – 1st ed.
    p. ; cm.
  Includes bibliographical references and index.
  ISBN-13: 978-1-4051-3171-1 (hardback : alk. paper)
 1. Thermal analysis.   2. Colorimetric analysis.   3. Thermal analysis–Industrial applications.   I. Gabbott, Paul.

QD117.T4A67 2007
543′.26–dc22
                                    2007029148

Set in Minion 10/12pt by Aptara Inc., New Delhi, India
Printed and bound in Singapore by Markono Print Media Pte Ltd

The publisher's policy is to use permanent paper from mills that operate a sustainable forestry policy, and which has been manufactured from pulp processed using acid-free and elementary chlorine-free practices. Furthermore, the publisher ensures that the text paper and cover board used have met acceptable environmental accreditation standards.

For further information on Blackwell Publishing, visit our website:
www.blackwellpublishing.com

Contents

Abbreviations   xv

List of Contributors   xvi

1   A Practical Introduction to Differential Scanning Calorimetry   *Paul Gabbott*   1
    1.1   Introduction   2
    1.2   Principles of DSC and types of measurements made   2
       1.2.1   A definition of DSC   2
       1.2.2   Heat flow measurements   3
       1.2.3   Specific heat (C_p)   3
       1.2.4   Enthalpy   5
       1.2.5   Derivative curves   5
    1.3   Practical issues   6
       1.3.1   Encapsulation   6
       1.3.2   Temperature range   8
       1.3.3   Scan rate   8
       1.3.4   Sample size   10
       1.3.5   Purge gas   10
       1.3.6   Sub-ambient operation   11
       1.3.7   General practical points   11
       1.3.8   Preparing power compensation systems for use   11
    1.4   Calibration   12
       1.4.1   Why calibrate   12
       1.4.2   When to calibrate   12
       1.4.3   Checking performance   13
       1.4.4   Parameters to be calibrated   13
       1.4.5   Heat flow calibration   13
       1.4.6   Temperature calibration   15
       1.4.7   Temperature control (furnace) calibration   16
       1.4.8   Choice of standards   16

		1.4.9	Factors affecting calibration	16
		1.4.10	Final comments	17
	1.5	Interpretation of data		17
		1.5.1	The instrumental transient	17
		1.5.2	Melting	18
		1.5.3	The glass transition	22
		1.5.4	Factors affecting T_g	24
		1.5.5	Calculating and assigning T_g	25
		1.5.6	Enthalpic relaxation	26
		1.5.7	T_g on cooling	30
		1.5.8	Methods of obtaining amorphous material	31
		1.5.9	Reactions	34
		1.5.10	Guidelines for interpreting data	40
	1.6	Oscillatory temperature profiles		42
		1.6.1	Modulated temperature methods	42
		1.6.2	Stepwise methods	44
	1.7	DSC design		46
		1.7.1	Power compensation DSC	46
		1.7.2	Heat flux DSC	47
		1.7.3	Differential thermal analysis DTA	48
		1.7.4	Differential photocalorimetry DPC	48
		1.7.5	High-pressure cells	49
	Appendix: standard DCS methods			49
	References			50
2	Fast Scanning DSC *Paul Gabbott*			51
	2.1	Introduction		52
	2.2	Proof of performance		52
		2.2.1	Effect of high scan rates on standards	52
		2.2.2	Definition of HyperDSC^{TM}	54
		2.2.3	The initial transient	54
		2.2.4	Fast cooling rates	54
	2.3	Benefits of fast scanning rates		57
		2.3.1	Sensitivity	57
		2.3.2	Measurement of sample properties without unwanted annealing effects	57
		2.3.3	Separate overlapping events based on different kinetics	59
		2.3.4	Speed of analysis	59
	2.4	Application to polymers		61
		2.4.1	Melting and crystallisation processes	61
		2.4.2	Comparative studies	64
		2.4.3	Forensic studies	65
		2.4.4	Effect of heating rate on the sensitivity of the glass transition	67
		2.4.5	Effect of heating rate on the temperature of the glass transition	68
		2.4.6	Effect of heating rate on T_g of annealed materials (and enthalpic relaxation phenomena)	72

2.5		Application to pharmaceuticals	76
	2.5.1	Purity of polymorphic form	76
	2.5.2	Identifying polymorphs	78
	2.5.3	Determination of amorphous content of materials	79
	2.5.4	Measurements of solubility	81
2.6		Application to water-based solutions and the effect of moisture	82
	2.6.1	Measurement of T_g in frozen solutions and suspensions	82
	2.6.2	Material affected by moisture	83
2.7		Practical aspects of scanning at fast rates	83
	2.7.1	Purge gas	83
	2.7.2	Sample pans	84
	2.7.3	Sample size	85
	2.7.4	Scan rate	85
	2.7.5	Instrumental settings	85
	2.7.6	Cleanliness	85
	2.7.7	Getting started	86
References			86

3 Thermogravimetric Analysis   *Rod Bottom*   87
   3.1 Introduction   88
   3.2 Design and measuring principle   89
      3.2.1 Buoyancy correction   90
   3.3 Sample preparation   92
   3.4 Performing measurements   93
      3.4.1 Influence of heating rate   93
      3.4.2 Influence of crucible   94
      3.4.3 Influence of furnace atmosphere   95
      3.4.4 Influence of residual oxygen in inert atmosphere   95
      3.4.5 Influence of reduced pressure   96
      3.4.6 Influence of humidity control   97
      3.4.7 Special points in connection with automatic sample changers   97
      3.4.8 Inhomogeneous samples and samples with very small changes in mass   98
   3.5 Interpreting TGA curves   98
      3.5.1 Chemical reactions   99
      3.5.2 Gravimetric effects on melting   101
      3.5.3 Other gravimetric effects   101
      3.5.4 Identifying artefacts   103
      3.5.5 Final comments on the interpretation of TGA curves   104
   3.6 Quantitative evaluation of TGA data   104
      3.6.1 Horizontal or tangential step evaluation   104
      3.6.2 Determination of content   106
      3.6.3 The empirical content   107
      3.6.4 Reaction conversion, α 110
   3.7 Stoichiometric considerations   111
   3.8 Typical application: rubber analysis   111

3.9		Analysis overview	112
3.10		Calibration and adjustment	112
	3.10.1	Standard TGA methods	113
3.11		Evolved gas analysis	114
	3.11.1	Brief introduction to mass spectrometry	115
	3.11.2	Brief introduction to Fourier transform infrared spectrometry	115
	3.11.3	Examples	117
Reference			118

4 Principles and Applications of Mechanical Thermal Analysis *John Duncan* 119

4.1		Thermal analysis using mechanical property measurement	120
	4.1.1	Introduction	120
	4.1.2	Viscoelastic behaviour	121
	4.1.3	The glass transition, T_g	123
	4.1.4	Sub-T_g relaxations	124
4.2		Theoretical considerations	125
	4.2.1	Principles of DMA	125
	4.2.2	Moduli and damping factor	127
	4.2.3	Dynamic mechanical parameters	127
4.3		Practical considerations	128
	4.3.1	Usage of DMA instruments	128
	4.3.2	Choosing the best geometry	129
	4.3.3	Considerations for each mode of geometry	133
	4.3.4	Static force control	134
	4.3.5	Consideration of applied strain and strain field	134
	4.3.6	Other important factors	135
	4.3.7	The first experiment – what to do?	136
	4.3.8	Thermal scanning experiments	137
	4.3.9	Isothermal experiments	137
	4.3.10	Strain scanning experiments	138
	4.3.11	Frequency scanning experiments	138
	4.3.12	Step-isotherm experiments	138
	4.3.13	Creep – recovery tests	138
	4.3.14	Determination of the glass transition temperature, T_g	139
4.4		Instrument details and calibration	140
	4.4.1	Instrument drives and transducers	140
	4.4.2	Force and displacement calibration	141
	4.4.3	Temperature calibration	141
	4.4.4	Effect of heating rate	142
	4.4.5	Modulus determination	142
4.5		Example data	143
	4.5.1	Amorphous polymers	143
	4.5.2	Semi-crystalline polymers	145
	4.5.3	Example of α and β activation energy calculations using PMMA	147
	4.5.4	Glass transition, T_g, measurements	150

		4.5.5	Measurements on powder samples	152
		4.5.6	Effect of moisture on samples	155
	4.6	Thermomechanical analysis		156
		4.6.1	Introduction	156
		4.6.2	Calibration procedures	157
		4.6.3	TMA usage	157
	Appendix: sample geometry constants			162
	References			163
5	Applications of Thermal Analysis in Electrical Cable Manufacture *John A. Bevis*			164
	5.1	Introduction		165
	5.2	Differential scanning calorimetry		165
		5.2.1	Oxidation studies (OIT test)	165
		5.2.2	Thermal history studies	166
		5.2.3	Cross-linking processes	168
		5.2.4	Investigation of unknowns	171
		5.2.5	Rapid scanning with DSC	172
	5.3	Thermomechanical analysis		175
		5.3.1	Investigation of extrusion defects	176
		5.3.2	Cross-linking	177
		5.3.3	Material identification	177
		5.3.4	Extrusion studies	178
		5.3.5	Fire-retardant mineral insulations	179
	5.4	Thermogravimetric analysis		180
		5.4.1	Practical comments	180
		5.4.2	Investigation of composition	181
		5.4.3	Carbon content	185
		5.4.4	Rapid scanning with TGA	186
	5.5	Combined studies		188
	5.6	Concluding remarks		189
	References			189
6	Application to Thermoplastics and Rubbers *Martin J. Forrest*			190
	6.1	Introduction		191
	6.2	Thermogravimetric analysis		192
		6.2.1	Background	192
		6.2.2	Determination of additives	194
		6.2.3	Compositional analysis	199
		6.2.4	Thermal stability determinations	206
		6.2.5	High resolution TGA and modulated TGA	208
		6.2.6	Hyphenated TGA techniques and evolved gas analysis	210
	6.3	Dynamic mechanical analysis		212
		6.3.1	Background	212
		6.3.2	Determination of polymer transitions and investigations into molecular structure	215

		6.3.3	Characterisation of curing and cure state studies	218
		6.3.4	Characterisation of polymer blends and the effect of additives on physical properties	220
		6.3.5	Ageing, degradation and creep studies	224
		6.3.6	Thermal mechanical analysis	227
	6.4	Differential scanning calorimetry		228
		6.4.1	Background	228
		6.4.2	Crystallinity studies and the characterisation of polymer blends	231
		6.4.3	Glass transition and the factors that influence it	236
		6.4.4	Ageing and degradation	238
		6.4.5	Curing and cross-linking	241
		6.4.6	Blowing agents	243
		6.4.7	Modulated DSC	244
		6.4.8	HyperDSC™	245
		6.4.9	Microthermal analysis	246
	6.5	Other thermal analysis techniques used to characterise thermoplastics and rubbers		247
		6.5.1	Dielectric analysis	247
		6.5.2	Differential photocalorimetry (DPC)	248
		6.5.3	Thermally stimulated current (TSC)	248
		6.5.4	Thermal conductivity analysis (TCA)	249
	6.6	Conclusion		249
	References			250
7	Thermal Analysis of Biomaterials *Showan N. Nazhat*			256
	Abbreviations			257
	7.1	Biomaterials		257
		7.1.1	Introduction	257
	7.2	Material classes of biomaterials		260
		7.2.1	Metals and alloys	260
		7.2.2	Ceramics and glasses	260
		7.2.3	Polymers	261
		7.2.4	Composites	262
	7.3	The significance of thermal analysis in biomaterials		262
	7.4	Examples of applications using dynamic mechanical analysis (DMA) in the development and characterisation of biomaterials		263
		7.4.1	Particulate and/or fibre-filled polymer composites as bone substitutes	264
		7.4.2	Absorption and hydrolysis of polymers and composites	270
		7.4.3	Porous foams for tissue engineering scaffolds	271
	7.5	Examples of applications using DSC in the development and characterisation of biomaterials		275
		7.5.1	Thermal history in particulate-filled degradable composites and foams	275
		7.5.2	Plasticisation effect of solvents	276
		7.5.3	Thermal stability and degradation	278
		7.5.4	Setting behaviour of inorganic cements	278

7.6	Examples of applications using differential thermal analysis/thermogravimetric analysis (DTA/TGA) in the development and characterisation of biomaterials		280
	7.6.1	Bioactive glasses	280
	7.6.2	Thermal stability of bioactive composites	282
7.7	Summary		283
References			283

8 Thermal Analysis of Pharmaceuticals   *Mark Saunders*   286

8.1	Introduction		287
8.2	Determining the melting behaviour of crystalline solids		288
	8.2.1	Evaluating the melting point transition	288
	8.2.2	Melting point determination for identification of samples	289
8.3	Polymorphism		290
	8.3.1	Significance of pharmaceutical polymorphism	291
	8.3.2	Thermodynamic and kinetic aspects of polymorphism: enantiotropy and monotropy	292
	8.3.3	Characterisation of polymorphs by DSC	293
	8.3.4	Determining polymorphic purity by DSC	297
	8.3.5	Interpretation of DSC thermograms of samples exhibiting polymorphism	302
8.4	Solvates and hydrates (pseudo-polymorphism)		303
	8.4.1	Factors influencing DSC curves of hydrates and solvates	304
	8.4.2	Types of desolvation/dehydration	305
8.5	Evolved gas analysis		308
8.6	Amorphous content		310
	8.6.1	Introduction	310
	8.6.2	Characterisation of amorphous solids: the glass transition temperature	311
	8.6.3	Quantification of amorphous content using DSC	313
8.7	Purity determination using DSC		315
	8.7.1	Types of impurities	315
	8.7.2	DSC purity method	316
	8.7.3	Practical issues and potential interferences	317
8.8	Excipient compatibility		320
	8.8.1	Excipient compatibility screening using DSC	321
	8.8.2	Excipient compatibility analysis using isothermal calorimetry	321
8.9	Microcalorimetry		323
	8.9.1	Introduction	323
	8.9.2	Principles of isothermal microcalorimetry	324
	8.9.3	High-sensitivity DSC	324
	8.9.4	Pharmaceutical applications of isothermal microcalorimetry	325
References			327

9 Thermal Methods in the Study of Foods and Food Ingredients
   Bill MacNaughtan, Imad A. Farhat   330
   9.1   Introduction   331
   9.2   Starch   332
         9.2.1   Starch structure   332
         9.2.2   Order in the granule   335
         9.2.3   The glass transition   336
         9.2.4   Extrusion and expansion   337
         9.2.5   Mechanical properties   338
         9.2.6   Starch retrogradation   338
         9.2.7   Effect of sugars   339
         9.2.8   Lipid–amylose complexes   339
         9.2.9   Multiple amorphous phases: polyamorphism   339
         9.2.10 Foods   341
   9.3   Sugars   343
         9.3.1   Physical properties   343
         9.3.2   Sugar glasses   345
         9.3.3   Sugar crystallisation   345
         9.3.4   Effect of ions on crystallisation   347
         9.3.5   Effect of ions on the glass transition temperature   348
         9.3.6   Ageing   349
         9.3.7   The Maillard reaction   350
         9.3.8   Mechanical properties   350
         9.3.9   Foods   350
   9.4   Fats   353
         9.4.1   Solid/liquid ratio   353
         9.4.2   Phase diagrams   355
         9.4.3   Polymorphic forms and structure   356
         9.4.4   Kinetic information   356
         9.4.5   Non-isothermal methods   359
         9.4.6   DSC at high scanning rates   361
         9.4.7   Mechanical measurements   362
         9.4.8   Lipid oxidation and the oxidation induction
                 time test   362
         9.4.9   Foods   363
   9.5   Proteins   365
         9.5.1   Protein denaturation and gelation   365
         9.5.2   Differential scanning microcalorimetry   366
         9.5.3   Aggregation   369
         9.5.4   Glass transition in proteins   371
         9.5.5   Ageing in proteins   371
         9.5.6   Gelatin in a high-sugar environment   373
         9.5.7   Mechanical properties of proteins   373
         9.5.8   Foods   373
         9.5.9   TA applied to other areas   377
         9.5.10 Interactions with polysaccharides and other materials   377

9.6	Hydrocolloids		378
	9.6.1	Definitions	378
	9.6.2	Structures in solution	379
	9.6.3	Solvent effects	381
	9.6.4	The rheological T_g	381
	9.6.5	Glassy behaviour in hydrocolloid/high-sugar systems	382
	9.6.6	Foods	386
9.7	Frozen systems		387
	9.7.1	Bound water?	387
	9.7.2	State diagram	388
	9.7.3	Mechanical properties of frozen sugar solutions	390
	9.7.4	Separation of nucleation and growth components of crystallisation	391
	9.7.5	Foods	392
	9.7.6	Cryopreservation	394
9.8	Thermodynamics and reaction rates		395
	9.8.1	Studies on mixing	395
	9.8.2	Isothermal titration calorimetry	396
	9.8.3	Reaction rates	398
	9.8.4	Thermogravimetric analysis	399
	9.8.5	Sample controlled TA	401
References			402

10 Thermal Analysis of Inorganic Compound Glasses and Glass-Ceramics
David Furniss, Angela B. Seddon — 410

10.1	Introduction		411
10.2	Background glass science		411
	10.2.1	Nature of glasses	411
	10.2.2	Crystallisation of glasses	413
	10.2.3	Liquid–liquid phase separation	416
	10.2.4	Viscosity of the supercooled, glass-forming liquid	416
10.3	Differential thermal analysis		418
	10.3.1	General comments	418
	10.3.2	Experimental issues	420
	10.3.3	DTA case studies	421
10.4	Differential scanning calorimetry		426
	10.4.1	General comments	426
	10.4.2	Experimental issues	427
	10.4.3	DSC case studies	427
	10.4.4	Modulated Differential Scanning Calorimetry case studies	432
10.5	Thermomechanical Analysis		432
	10.5.1	General comments	432
	10.5.2	Experimental issues	433
	10.5.3	Linear thermal expansion coefficient (α) and dilatometric softening point (M_g)	433

　　　　　10.5.4　Temperature coefficient of viscosity: introduction　　436
　　　　　10.5.5　TMA indentation viscometry　　438
　　　　　10.5.6　TMA parallel-plate viscometry　　443
　　10.6　Final comments　　447
　　References　　448

Appendix　　450

Glossary　　453

Further Reading　　458

Web Resources　　458

Index　　459

Abbreviations

$C_{g'}$	Concentration of maximally freeze-concentrated glass
CPMAS NMR	Cross-polarisation magic angle spinning nuclear magnetic resonance
DMA	Dynamic mechanical analysis/analyser
DSC	Differential scanning calorimetry/calorimeter
DSM/NC	Differential scanning micro/nanocalorimetry
DTA	Differential thermal analysis/analyser
DTG	Derivative thermogravimetric curves
DVS	Dynamic vapour sorption
EGA	Evolved gas analysis
FTIR	Fourier transform infrared spectrometer
ITC	Isothermal titration calorimetry
MS	Mass spectrometer
RAF	Rigid amorphous fraction
RVA	Rapid viscoanalyser
SAXS	Small angle X-ray diffraction
SEC	Size exclusion chromatography
T_g	Glass transition temperature
T_g'	Glass transition temperature of the maximally freeze-concentrated glass
TEM	Transmission electron microscopy
TGA	Thermogravimetric analysis/analyser

TG-FTIR	Thermogravimetric analyser coupled to a Fourier transform infrared spectrometer
TG-MS	Thermogravimetric analyser coupled to a mass spectrometer
TG-GC-FTIR	TGA analyser employing a Chromatography Stage and Fourier Transform Infrared spectroscopy for evolved gas analysis
TMA	Thermomechanical analysis/analyser
TNM	Tool Narayanaswamy Moynihan
WAXS	Wide-angle X-ray diffraction

Contributors

John A. Bevis
Prysmian Cables and Systems Limited
Chickenhall Lane
Eastleigh
Hampshire
SO50 9YU

Rod Bottom
Mettler-Toledo Ltd
64 Boston Road
Beaumont Leys
Leicester, LE4 1AW

John Duncan
Triton Technology Limited
3 The Courtyard, Main Street
Keyworth, Nottinghamshire
NG12 5AW

Imad A. Farhat
Firmenich S.A.
7 Rue de la Bergère
CH-1217 Meyrin
Switzerland

Martin J. Forrest
Principal Consultant – Polymer Analysis
Rapra Technology Limited
Shawbury
Shrewsbury
SY4 4NR

David Furniss
Materials Engineering
Wolfson Building
University Park
Nottingham
NG7 2RD

Paul Gabbott
PETA Solutions
PO Box 188
Beaconsfield
Bucks HP9 2GB

Bill MacNaughtan
Nottingham University
Dept of Food Sciences
Sutton Bonnington
Loughborough
Leicestershire, LE12 5RD

Showan N. Nazhat
Department of Mining, Metals and
   Materials Engineering
McGill University
MH Wong Building
3610 University Street
Montreal
Quebec
Canada
H3A 2B2

Mark Saunders
29 Radnor Road
Earley
Reading
RG6 7NP

Angela Seddon
Materials Engineering
Wolfson Building
University Park
Nottingham
NG7 2RD

Chapter 1
A Practical Introduction to Differential Scanning Calorimetry

Paul Gabbott

Contents

1.1	Introduction	2
1.2	Principles of DSC and types of measurements made	2
	1.2.1 A definition of DSC	2
	1.2.2 Heat flow measurements	3
	1.2.3 Specific heat (C_p)	3
	1.2.4 Enthalpy	5
	1.2.5 Derivative curves	5
1.3	Practical issues	6
	1.3.1 Encapsulation	6
	1.3.2 Temperature range	8
	1.3.3 Scan rate	8
	1.3.4 Sample size	10
	1.3.5 Purge gas	10
	1.3.6 Sub-ambient operation	11
	1.3.7 General practical points	11
	1.3.8 Preparing power compensation systems for use	11
1.4	Calibration	12
	1.4.1 Why calibrate	12
	1.4.2 When to calibrate	12
	1.4.3 Checking performance	13
	1.4.4 Parameters to be calibrated	13
	1.4.5 Heat flow calibration	13
	1.4.6 Temperature calibration	15
	1.4.7 Temperature control (furnace) calibration	16
	1.4.8 Choice of standards	16
	1.4.9 Factors affecting calibration	16
	1.4.10 Final comments	17
1.5	Interpretation of data	17
	1.5.1 The instrumental transient	17
	1.5.2 Melting	18

	1.5.3	The glass transition	22
	1.5.4	Factors affecting T_g	24
	1.5.5	Calculating and assigning T_g	25
	1.5.6	Enthalpic relaxation	26
	1.5.7	T_g on cooling	30
	1.5.8	Methods of obtaining amorphous material	31
	1.5.9	Reactions	34
	1.5.10	Guidelines for interpreting data	40
1.6	Oscillatory temperature profiles		42
	1.6.1	Modulated temperature methods	42
	1.6.2	Stepwise methods	44
1.7	DSC design		46
	1.7.1	Power compensation DSC	46
	1.7.2	Heat flux DSC	47
	1.7.3	Differential thermal analysis DTA	48
	1.7.4	Differential photocalorimetry DPC	48
	1.7.5	High-pressure cells	49
Appendix: standard DSC methods		49
References			49

1.1 Introduction

Differential scanning calorimetry (DSC) is the most widely used of the thermal techniques available to the analyst and provides a fast and easy to use method of obtaining a wealth of information about a material, whatever the end use envisaged. It has found use in many wide-ranging applications including polymers and plastics, foods and pharmaceuticals, glasses and ceramics, proteins and life science materials; in fact virtually any material, allowing the analyst to quickly measure the basic properties of the material. Many of the application areas are dealt with in greater depth within the chapters of this book, and the principles involved extend to many other materials that may not be mentioned specifically. It is in fact a fascinating technique and the purpose of this introduction is to provide an insight into this method of measurement, to provide the necessary practical guidance a new user will need to go about making measurements, and to give understanding about the information that can be obtained and how to interpret the data.

1.2 Principles of DSC and types of measurements made

1.2.1 A definition of DSC

A DSC analyser measures the energy changes that occur as a sample is heated, cooled or held isothermally, together with the temperature at which these changes occur.

The energy changes enable the user to find and measure the transitions that occur in the sample quantitatively, and to note the temperature where they occur, and so to characterise a material for melting processes, measurement of glass transitions and a range of more complex events. One of the big advantages of DSC is that samples are very easily encapsulated, usually with little or no preparation, ready to be placed in the DSC, so that measurements can be quickly and easily made. For further details of instrumental design and basic equations, refer to Section 1.7 at the end of this chapter.

1.2.2 Heat flow measurements

The main property that is measured by DSC is heat flow, the flow of energy into or out of the sample as a function of temperature or time, and usually shown in units of mW on the y-axis. Since a mW is a mJ/s this is literally the flow of energy in unit time. The actual value of heat flow measured depends upon the effect of the reference and is not absolute. What matters is that a stable instrumental response or baseline is produced against which any changes can be measured. The starting point of the curve on the y-axis may be chosen as one of the starting parameters, and it should be set at or close to zero.

Two different conventions exist for the display of the heat flow curve: one shows endotherms in the downward direction, the other upward. The operator has a choice with most software packages. Traditionally, with heat flux systems (Section 1.7.2) endotherms are shown as going down, since endothermic transitions result in a negative temperature differential, whilst with power compensation systems (Section 1.7.1) they are shown as going up since with this principle endothermic transitions result in an increase in power supplied to the sample. In this chapter data are shown with endotherms up.

The value of measuring energy flow is that it enables the analyst to identify the range of different transitions that may occur in the sample as it is heated or cooled; the main transitions are described in Section 1.5.

1.2.3 Specific heat (C_p)

The specific heat (heat capacity, C_p) of a material can be determined quantitatively using DSC and is designated C_p since values are obtained at constant pressure. Traditionally, this is done by subtracting a baseline from the heat flow curve in the manner described below, but values may also be obtained using *modulated temperature* techniques, Section 1.6. The subtracted curve referenced against a standard gives a quantitative value of C_p, Figure 1.1. The accuracy that can be obtained depends upon the instrument and method in use.

In practice the traditional standard test method (see Appendix on p49) provides a fairly rapid method for determination of C_p and many manufacturers provide software specifically designed to comply with this. Three runs are required, each consisting of an isothermal period, temperature ramp and final isotherm. This method is applied identically to the succeeding runs:

1. First run: a baseline with uncrimped empty pans placed in the furnace.
2. Second run: as above but adding a reference (typically sapphire) to the sample pan.
3. Third run: replace the reference with your sample.

The three curves are brought up on the screen, isothermals matched, data subtracted and referenced against the standard. Most software packages will do this automatically, and if the

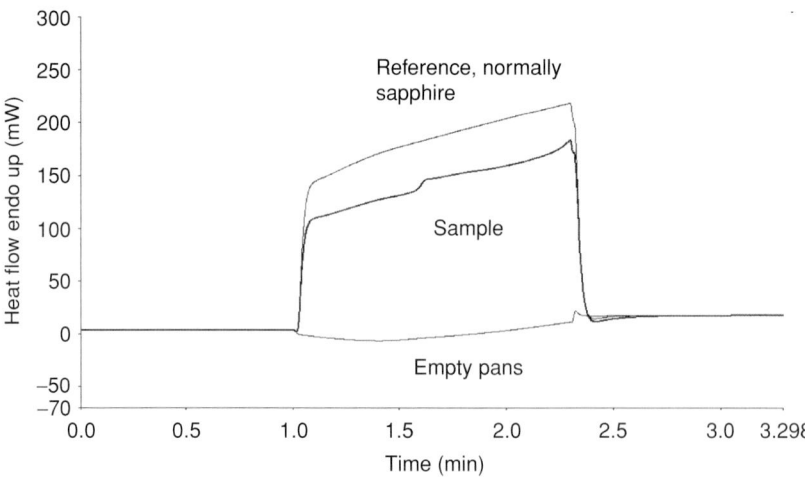

Figure 1.1 Heat capacity of PET obtained using fast scanning techniques showing the three traces required for subtraction. The height of the sample compared to the empty pan is divided by the scan rate and the mass of sample to obtain a value for C_p. This is referenced against a known standard such as sapphire for accuracy. If small heating steps of, for example, 1°C are used the area under the curve can be used to calculate C_p. This calculation is employed as an option in stepwise heating methods.

differing weight and heat capacity of sample pans are taken into account then the baseline and reference runs may be used for subsequent samples, provided the DSC is stable. In fact, because the procedure is based on a subtraction technique between measurements made at different times, any drift will cause error. The DSC must be very stable and in practice it is best not to use an instrument at the extremes of its temperature range where stability may be compromised. The standard most often used is sapphire, and the mass used should be similar to the sample; in any event the sample should not be a great deal larger or errors will be increased. This method relies on the measurement of the heat flow of the sample compared to that of an empty pan. Whilst there may be a number of factors which dictate the scan rate of choice it should be noted that faster scan rates result in increased values of heat flow giving increased accuracy of measurement, and this also minimises the time of the run and potential drift of the analyser. It has been reported that fast scan rates used by fast scan DSC (Chapter 2) can give extremely accurate data [1].

A similar principle is employed in stepwise heating methods where the temperature may be raised by only a fraction of a degree between a series of isotherms. This is reported to give a very accurate value for C_p because of the series of short temperature intervals.

Specific heat data can be of value in its own right since this information is required by chemists and chemical engineers when scaling up reactions or production processes, it provides information for mathematical models, and is required for accurate kinetic and other advanced calculations. It can also help with curve interpretation since the slope of the curve is fixed and absolute, and small exothermic or endothermic events identified. Overall, it gives more information than the heat flow trace because values are absolute, but it does take more time, something often in short supply in industry.

1.2.4 Enthalpy

The enthalpy of a material is energy required to heat the material to a given temperature and is obtained by integrating the heat capacity curve. Again many software packages provide for the integration of the C_p curve to provide an enthalpy curve. Enthalpy curves are sometimes used for calculations, for example when calculating fictive temperature (see Glossary), and can help in understanding why transitions have the shape they do. In the cases where amorphous and crystalline polymer materials exhibit significantly different enthalpies, the measurement of enthalpy can allow an estimate of crystallinity over a range of temperatures as the polymer is heated [2].

1.2.5 Derivative curves

Derivative curves are easily obtained from the heat flow curve via a mathematical algorithm and aid with interpretation of the data. Typically they can help define calculation limits, and can aid with the resolution of data, particularly where overlapping peaks are concerned. The first derivative curve is useful for examining stepwise transitions such as the glass transition, and is very useful for thermogravimetric analysis (TGA) studies (Chapter 3) where weight loss produces a step. The second derivative of a peak is more easily interpreted than the first derivative. In this case the data are inverted, but any shoulders in the original data will resolve into separate peaks in the second derivative curve. It is particularly useful for examining melting processes to help identify shoulders in the peak shape due to multiple events. An example is shown in Figure 1.2. The second derivative produces a maximum or minimum for each inflection of the original curve. Shoulders in the original curve show up as peaks in the second derivative. The shoulder in this example is quite clear, but often the second derivative can pick out multiple events when the original data are much less clear.

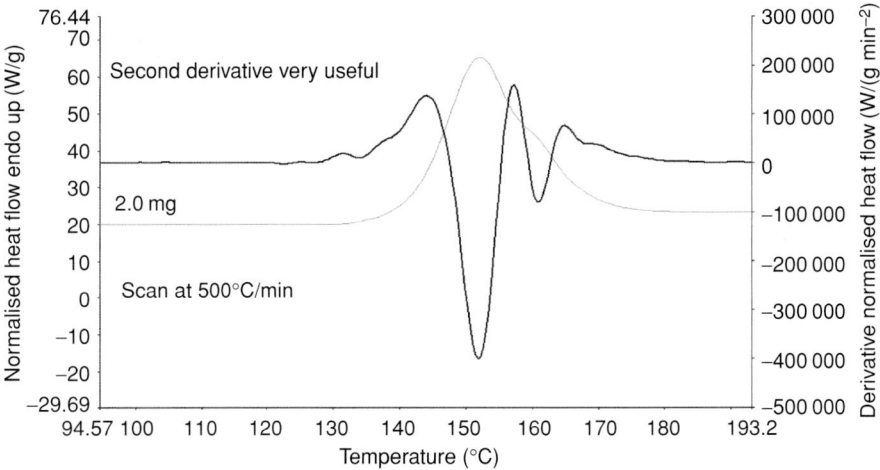

Figure 1.2 Indomethacin form 2 scanned at 500°C/min. The shoulder on the melt resolves into a separate peak in the second derivative, which shows a doublet pointing downwards.

The higher the derivative level the greater is the noise that is generated, so good quality data are needed for higher derivative studies. Curve fitting or smoothing techniques used on the heat flow curve before generating a derivative can also be very useful. In general, sharp events and inflections produce the best derivative curves. Studies at high rate also produce very good derivative curves since the rates of change are increased.

1.3 Practical issues

1.3.1 Encapsulation

Encapsulation is necessary to prevent contamination of the analyser, and to make sure that the sample is contained and in good thermal contact with the furnace. One of the most annoying sights after a DSC run is to find a pan burst with the contents all over the furnace, or unexpected peaks and bumps in the DSC trace with no real cause or meaning. Such errors can be due to poor sample encapsulation, which is one of the most important areas to be considered in order to obtain good data. Most manufacturers provide a range of sample pans for different purposes, with a range of different sizes, pan materials, and associated temperature and pressure ranges. Consideration should be given to the following issues when choosing pans and encapsulating your samples.

1.3.1.1 Temperature range

Make sure that the pan is specified for the desired temperature range and will not melt during a scan; remember that aluminium must not be used above 600°C. Pan materials such as gold (mp 1063°C), platinum or alumina can be used at higher temperatures.

1.3.1.2 Pressure build-up and pan deformation

Pressure build-up in an inappropriately chosen pan is a frequent cause of difficulties. It is important to determine whether the sample needs to be run in a hermetically sealed pan or not; dry samples unlikely to evolve significant amounts of volatiles below decomposition do not need to be sealed. Yet if run in a hermetically sealed pan then pressure will build up inside the pan as it is heated. Many hermetically sealed aluminium pans cannot withstand high internal pressure and will deform (not necessarily visible) and result in potential artefacts in the trace as heat transfer to the sample changes. Ultimately sample leakage and bursting can occur, which usually results in contamination of the analyser. The best solution for such systems is to work with crimped pans that do not seal, or use lids with holes in. If not available then it may be best to pierce the pan lid before encapsulation so that pressure does not build up. Sometimes one hole will block with sample (particularly if a hole is made after loading the sample) and back pressure will force sample out of the pan giving more artefacts, so it is better to have more than one hole in the lid.

In the case of foods and other water-containing materials hermetic sealing is important since volatile loss will mask other transitions and will dry out the sample. It is then necessary to use a pan system capable of withstanding the anticipated pressure. The maximum operating temperature of a sealed pan may vary but may be below 100°C. Some thicknesses of aluminium and type of pan offer slightly higher temperature ranges but consideration should be given to 'O' ring sealed stainless steel pans which can hold pressures above 20 bar. These have been designed for use with biological materials or for systems involving condensation reactions, e.g. phenolics.

In the case of samples taken to very high temperature or pressure, capsules which hold internal pressures up to 150 bar should be used. Disposable high-pressure capsules which avoid the need for cleaning are available. For hazard evaluation, gold-plated high-pressure pans should be used, since these should be inert towards the sample.

1.3.1.3 Reactions with the pan

Samples that react with a pan can cause serious damage to an analyser since they may also react with the furnace beneath. Solder pastes and inorganic salts are typical of the type of samples where care must be taken. If in doubt check it out separately from the analyser, and then choose a pan type which is inert. Sometimes the effect of catalysis is of interest and copper pans may be used to provide a catalytic effect. Aluminium pans are normally made of very high-purity metal to prevent unwanted catalytic effects.

1.3.1.4 Pan cleanliness

Most pans can be used as received but sometimes a batch may be found to be slightly contaminated, possibly with a trace of machine oils used in pan manufacture. If so, pans can be cleaned by volatilising off the oil. Heating to 300 °C should be more than adequate for this. If using a hot plate, do not heat a lot of pans all together or they may stick together. The use of clean pans is also important for very sensitive work with fast scan DSC.

1.3.1.5 Pans with very small holes

Some pan and lid systems have been developed with very small holes, typically about 50 μm diameter that have found use with hydrate and solvate loss. They can be used with ordinary samples to relieve internal pressure, so long as the hole does not block, but are intended to improve resolution of mass loss from a sample. As volatiles do not tend to escape through such a small hole until the volatile partial pressure exceeds atmospheric pressure boiling point can be determined.

1.3.1.6 Liquid samples

Liquid samples must be placed in sealed pans of a type that can withstand any internal pressure build-up. Do not overfill the pan or contaminate the sealing surfaces, which will prevent sealing and cause leakage. When sealing a liquid, bring the dies together gently to avoid splashing the sample.

1.3.1.7 Sample contact

Samples need to be in good thermal contact with the pan. Liquids and powders, when pressed down, give good thermal contact, other samples should be cut with a flat surface that can be placed against the base of the pan. Avoid grinding materials unless you are sure it will not change their properties. If possible, films should not be layered to prevent multiple effects from the same transition, though this may be the only way to get enough samples. If so, take care to make sure that the films are pressed well together. Low-density samples provide poor heat transfer so should be compressed. Some crimping processes do this automatically; with others it may be of value to compress a sample between two pan bases or using a flat pan lid as an insert. Take care not to deform a pan and discard any pans that are obviously deformed before use. Sometimes pans of slightly thicker aluminium can give better heat transfer because they retain a flatter base.

1.3.1.8 Spillage

A frequent cause of contamination is from sample attaching to the outside of the pan. Check and remove any contaminant sticking to the outside of the pan, particularly the base of the pan. A soft brush is good for removing powders.

1.3.1.9 Use without pans

Occasionally, very stable samples may be run without pans, for example stable metals over a low temperature range. These benefit from increased thermal transfer to the sample. Thermally conductive pastes have also been used on occasion to try to improve thermal transfer, but such measures require great care, and the use of helium or other highly conductive purge gas may be a better alternative.

1.3.2 Temperature range

The starting temperature should be well below the beginning of the first transition that you want to measure in order to see it clearly following a period of flat baseline. This should take into account the period of the initial transient where the scan rate is not yet fully controlled and the baseline is not stable. With ambient DSC systems the starting temperature is often around 30°C. The upper temperature should be below the decomposition temperature of the sample. Decomposing a material in a DSC normally gives rise to a very noisy drifting response and the evolved volatiles will contaminate the system. It is good practice to establish the decomposition temperature first using a TGA analyser if available.

1.3.3 Scan rate

Traditionally, the most common scan rate used by thermal analysts is 10°C/min, but with commercially available instruments rates can be varied between 0.001 and 500°C/min, often to significant advantage. Choice of scan rate can affect the following areas:

1. *Sensitivity.* The faster the scan rate the greater the sensitivity; see the heat flow equation, Section 1.7. If a sample with a small transition is received where great care is thought necessary, there is no point in scanning slowly since the transition is likely to be missed altogether. Whilst the energy of any given transition is fixed and should therefore be the same whatever the scan rate, the fact is that the faster the scan rate,

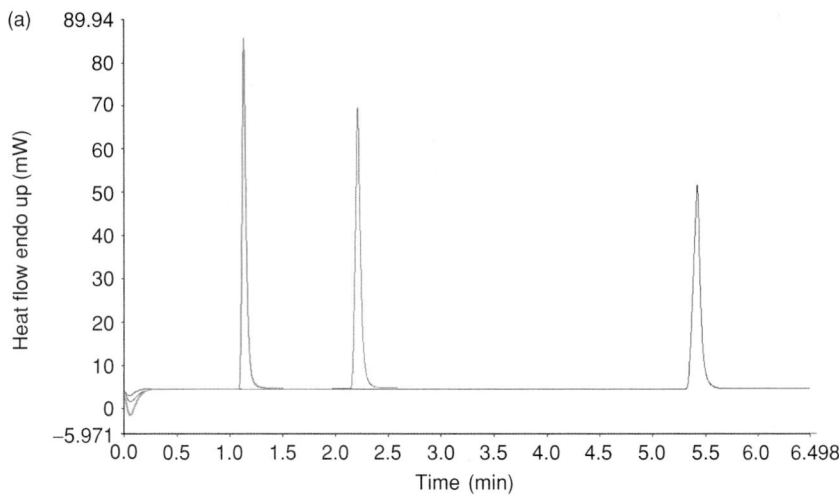

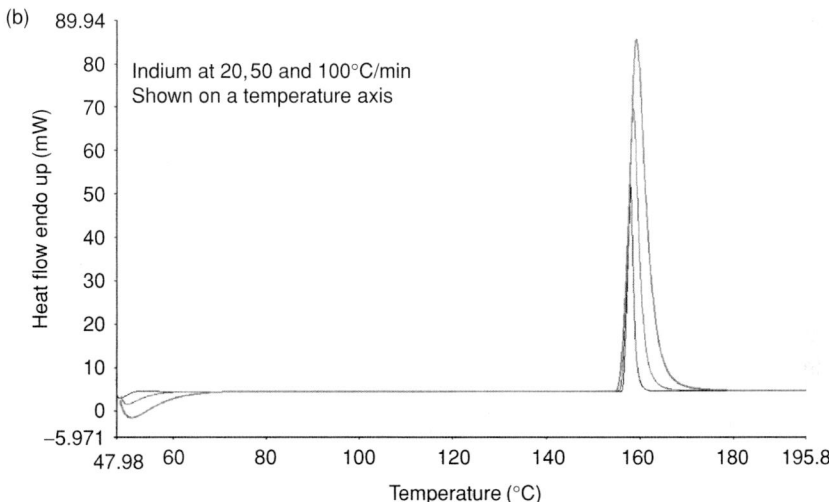

Figure 1.3 The effect of increasing scan rate on indium. Part (a) (upper curve) is shown with the x-axis in time; part (b) (lower curve) is shown with the x-axis in temperature. The same energy flows faster in a shorter time period at the faster rates, giving larger peaks.

the bigger the transition appears. Increasing scan rate increases sensitivity. The reason for this is that a DSC measures the *flow* of energy and during a fast scan the flow of energy increases, though over a shorter time period. A slow scan results in lower flows of energy over a longer time period, but since DSC data are usually shown with the x-axis as temperature, it simply looks like a transition is bigger at the faster rates; see Figures 1.3a and 1.3b. Hence an increased scan rate leads to an increase in sensitivity, so do not use slow rates for small difficult-to-find transitions, unless otherwise unavoidable.

2. *Temperature calibration.* Calibrations need to be performed prior to analysis to ensure that the scan rate of use is calibrated. Different instruments use different approaches to calibration and manufacturer's instructions should be followed.
3. *Resolution.* Because of thermal gradients across a sample the faster the scan rate the lower the resolution, and the slower the scan rate the sharper the resolution. Thermal gradients can be reduced by reducing sample size and improving thermal contact with the pan by good encapsulation.
4. *Transition kinetics.* Slow events such as a cure reaction may not complete if scanned quickly and may be displaced to a higher temperature where they can occur more rapidly. The kinetics of an event may need to be considered when choosing a scan rate.
5. *Time of analysis.* Speed of analysis is an issue in many businesses, and higher rates speed up throughput.

1.3.4 Sample size

The amount of sample used will vary according to the sample and application. In general, make sure that excessive amounts of sample are not added to the pan. Sample collapse when melting or softening can give rise to noise during or after the transition. Reducing sample size minimises the chances of this happening. Often just a few milligrams or less is sufficient, typically 1–3 mg for pharmaceutical materials, though for very weak transitions and for accurate heat of fusion measurements more sample may be needed. Note that when considering accuracy of energy measurements the accuracy of the balance should be taken into account. For most work a five-figure balance is needed, and six figures (capable of measuring to the microgram level) for more accurate heats of fusion. For polymers slightly larger samples of 10 mg are often used, but again this depends on the sample and the transition of interest. See the following chapters for practical discussion of individual application areas. As a general comment, if the sample is to be melted then choose a lower weight, though for accurate heat of fusion values typically 5 mg is needed in order to measure the weight accurately, possibly higher if the balance accuracy is not very great. Choose a pan that can cope with sample size needed. Note that one of the major causes of contamination is also from using too large a sample size.

1.3.5 Purge gas

Purge gases are used to control the sample environment, purge volatiles from the system and prevent contamination. They reduce noise by preventing internal convection currents, prevent ice formation in sub-ambient systems, and can provide an active atmosphere. They are recommended for use and essential for some systems. The most common purge gas is nitrogen, which provides a generally inert atmosphere and prevents sample oxidation and is probably the default choice. Air and oxygen are sometimes used for oxidative tests such as OIT (oxidative induction time) measurements. Helium is used for work at very low temperatures where nitrogen and oxygen would condense, and is recommended for fast scan DSC studies; see Chapter 2. Other gases such as argon have been used and this can be helpful when operating at higher temperatures, typically above 600°C. This gas has a lower thermal conductivity, which results in lower heat loses but at lower temperatures results in decreased resolution and sensitivity. Air, nitrogen and oxygen have similar thermal conductivities and instrument calibration is unaffected if they are switched. The use of helium or argon requires separate calibrations to be performed. For totally inert performance use high-purity gases together with copper or steel gas lines, and give time for the sample area to be fully purged of air before beginning a run.

Gas flows are controlled either by mass flow controllers or by using a pressure regulator and restrictor. If you want to check for gas flows, do not put the exit line into a beaker of water. Suck back can occur, destroying the furnace, and bubble formation causes noise. A soap bubble flow meter can be used if needed, but any device used to measure the normally low purge gas flows should not in itself cause a restriction.

1.3.6 Sub-ambient operation

A variety of cooling systems are available for most instruments and these are now mostly automatic in operation. Issues of filling liquid nitrogen systems and of availability mean that refrigerated coolers (intracoolers) are often preferred and some systems can operate at below $-100°C$. Intracoolers are preferred for use with autosamplers since there is no risk of the coolant running out.

1.3.7 General practical points

Cleanliness is next to godliness as far as DSC use is concerned. Contamination will occur if a pan is placed on a dirty surface and then transferred to the furnace, so keep the analyser and the working area clean and tidy, and discard used pans to avoid mixing them up. DSC furnaces should be kept clean; refer to manufacturers instructions for the most appropriate cleaning method. Avoid abrasives and be aware of the dangers of using flammable solvents. Take note of specific instructions for any particular type of furnace.

- Do not heat aluminium pans above 600°C.
- Do not operate furnaces in an oxidising atmosphere above their recommended limits.
- Do not use excessive force when cleaning a furnace.
- Discard used samples immediately after use before they become mixed up with fresh samples.

1.3.8 Preparing power compensation systems for use

With power compensation DSC analysers there are two small furnaces and some specific practices to be aware of.

Before starting and before turning on any sub-ambient accessories, inspect the furnaces to make sure that they are clean and the surrounding block is dry and free from condensation. Clean furnaces with a swab and suitable solvent, and complete this with a furnace burn-off in air at about 600°C or above. For this purpose, the furnace should be opened to allow air access and volatiles to escape easily. Any pans should be removed. The furnaces are constructed of 95% platinum, 5% iridium, which will not oxidise within the temperature range of the analyser. Though organics will burn off, metal and inorganic content may burn in, so make sure that the furnaces are as clean as possible before a burn-off. The platinum furnace covers should also be clean; inspect the underside to make sure. The furnace covers should not be swapped about between sample and reference as they will have slightly different heat capacities which will affect instrument response run to run. They should fit easily, otherwise they may be incorrectly fitted. They can be reformed if needed using a reforming tool. Incorrectly fitting covers will lead to significant slope on the baseline. If a sample has leaked and stuck on the pan or a cover to the furnace, it is often helpful to heat the furnace to a point at which the sample softens and its adhesive properties are lost. The sample and pan can then be removed and the furnace cleaned.

The underside of the swing-away or rotating cover should be inspected from time to time to make sure that it is clean; particularly if sublimation is suspected. If the guard rings beside the furnaces are contaminated these can be removed and cleaned. They should be replaced with the slot facing the purge gas exit port, which is situated centrally behind the furnaces.

Turn on the purge gas and cooling system and allow them to stabilize; this applies to sub-ambient or ambient systems where water circulation is used. Stable background temperatures and purge gas flow provide a controlled environment from which stable and reproducible performance can be obtained. Dry purge gases must be used for the instrument air shield if operating below ambient. If the system has been left for any period of time it is good to condition the analyser by heating to 400–500°C before beginning your work.

1.4 Calibration

1.4.1 Why calibrate

There is potential for slight inaccuracy of measurements in all DSC analysers because sensors, no matter how good, are not actually embedded in the sample, and the sensors themselves are also a potential variable. Therefore in order to ensure accuracy and repeatability of data, a system must be calibrated and checked under the conditions of use. It is important to understand that calibration itself is just a tool or process designed by the manufacturer to adjust the analyser to give a slightly different temperature or energy response. The accuracy and acceptability of the system can only be judged by separately checking the system against accepted standards. A good overview of currently available standards for DSC with comments on accuracy and procedure is given in [3]. See also the details in this section.

1.4.2 When to calibrate

When the system is out of accepted specifications! It is important to distinguish between the process of calibration and the process of checking specification. If a system is brand new, has undergone some type of service, or is to be used under new conditions, it may be in need of complete calibration, but a system in regular use should be checked regularly and calibrated when it is shown to be out of specification. Many systems, particularly in the pharmaceutical industry, will be used under GLP (good laboratory practices) or other regulations and have established guidelines as to when and how often checks should be made.

1.4.3 Checking performance

In many industries frequency and method of checking, together with accepted limits, will be determined by the standard operating procedure (SOP) adopted and may well be done on a daily basis. Regular performance checks are common sense. If a system is only checked once every 6 months and then found to be in error, 6 months work is suspect.

The most common procedure is to run an indium standard under the normal test conditions and measure the heat of fusion value and melting onset temperature. These values are then compared with literature values and a check made against accepted limits. For many industries limits of ±0.5°C for temperature or 1% for heat of fusion may be accepted, though tighter limits of ±0.3°C and 0.1% may also be adopted. The choice of limits depends on how accurate you need to be. Indium is the easiest standard to use because of its stability and relatively low melting point of 156.6°C, which means it can often be reused, provided it is not heated above 180°C.

1.4.4 Parameters to be calibrated

Areas subject to calibration are as follows:

(a) *Heat flow or energy.* This is normally performed using the heat of fusion of a known standard such as indium. As an alternative, heat flow may be calibrated directly using a standard of known specific heat.
(b) *Temperature recorded by analyser.* This is performed using the melting point of known standards.
(c) *Temperature control of the analyser.* Sometimes called furnace calibration.

Use the same conditions for calibration as will be used for subsequent tests. If a range of different conditions are to be used then a range of different calibrations must be performed, one for each different set of conditions. Some analysers/software make allowance for changes in scan rate, so simplifying calibration procedures, but the effectiveness of calibration should in all cases be checked and verified for each set of conditions and scan rates used.

Some instruments may require set default conditions to be restored before measurements are made. The manufacturers' instructions for calibration should in all cases be followed even if they deviate from the general principles outlined here. Use standard values obtained with the actual standards since different batches of standards may have different values.

1.4.5 Heat flow calibration

The y-axis of a DSC trace is in units of heat flow; this needs to be calibrated. This is usually performed via a heat of fusion measurement as shown in Figure 1.4. The area under the

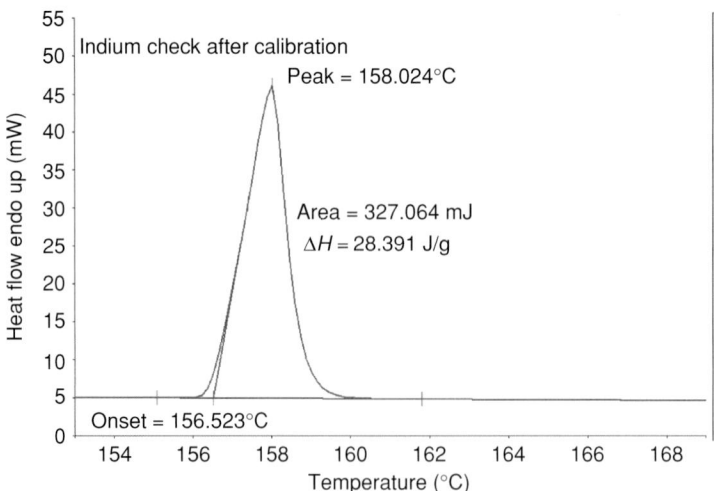

Figure 1.4 Indium run as a check after a calibration procedure has been completed showing the onset calculation for melting point, and area calculation for heat of fusion. This particular curve would benefit from a higher data point collection rate to remove the stepping effect of the data and improve overall accuracy.

melting curve is used since this reflects the heat flow as a function of temperature or time. Some software packages may also make provision for the y-axis itself to be calibrated and this is done against a heat capacity standard such as sapphire.

1.4.5.1 Procedure

An appropriate standard is chosen and heated under the same conditions (e.g. scan rate, purge gas and pan type) as the tests to be performed. Measure the heat of fusion. This information is then entered into the calibration section of the software. Indium is probably the best standard to use but others may be available in order to check a wider temperature range. The weight of standard should be sufficient to be accurate, typically 5–10 mg and weighed with a six-figure balance for best accuracy. Inaccuracy of weight measurement will limit the accuracy of heat of fusion measurements. An inert purge gas should be used for calibration to minimise potential oxidation. Nitrogen can be used even if samples are to be subsequently run in air or oxygen since all these gases have similar thermal properties and can be used interchangeably as far as calibration is concerned. A clean melting profile is required, without irregularities such as spikes or a kink in the leading edge of the melt caused by sample collapse and flow. For this reason, the sample should be flattened when placed in the pan; most standard samples are sufficiently soft and they may be flattened using the flat part of tweezers before placing in the pan. It is also good practice to use indium on the reheat after it has been melted once, since this will improve the thermal performance. Indium may be reused provided it has not been overheated (see Section 1.4.8); other standards should be discarded after use, though metals such as lead, tin or zinc, which show irregularities on initial heating, may be used on the reheat since this is better than taking data from a poor trace. In all cases stop the run once melting is complete, and if a standard is taken well above its melting point do not reuse it even if the initial heat is unsuitable. Heat of fusion values are more noticeably affected than temperature when a standard deteriorates.

If the y-axis scale is to be calibrated directly then the approach used for C_p measurement (see Section 1.2.3) should be employed. Values obtained for a C_p standard are then entered against standard values.

1.4.6 Temperature calibration

A typical temperature onset measurement is also shown in Figure 1.4. Note that melting point is determined from the onset of melt, not the peak maximum. Normally, at least two standards are needed to adequately calibrate for temperature, and ideally they should span the temperature range of interest, though if working in an ambient environment two widely spaced standards such as indium and lead or zinc are often sufficient. Temperature response is normally linear, so measurements outside of the standards range are normally accurate, but if in doubt then check against a standard. It is the verification process which is the critical aspect of calibration, not the actual procedure used with any specific system. For sub-ambient work a sub-ambient standard is advisable. Organic liquids do not make good reference materials but they are often the only available materials. Obtain as pure a material as possible. Cyclohexane is quite useful having two transitions, a crystal transition at $-87\,°C$ and melt at $6.5\,°C$. Table 1.1 shows a list of standards.

Table 1.1 Commonly used standards and reference materials

Standard	Melting point (°C)	Heat of fusion (J/g)
Indium	156.6	28.42
Tin	231.9	
Lead	327.5	
Zinc	419.5	108.26
K_2SO_4	585.0	
$K_2Cr_2O_7$	670.5	

Substance	Transition	Transition temperature (°C)
Cyclopentane	Crystal	−151.16
Cyclopentane	Crystal	−135.06
Cyclohexane	Crystal	−87.06
Cyclohexane	Melt	6.54
n-Heptane	Melt	−90.56
n-Octane	Melt	−56.76
n-Decane	Melt	−29.66
n-Dodecane	Melt	−9.65
n-Octadecane	Melt	28.24

1.4.6.1 Procedure

Use the same conditions as for the subsequent tests that you want to perform. Measure the onset values and enter them into the calibration software. Sample weights are not so critical if temperature measurement only is being made and typically a few milligrams should be used. The onset value of indium melt from the heat of fusion test is usually employed to save repeating the process.

1.4.7 Temperature control (furnace) calibration

In principle, calibration of the control of the analyser requires the same approach as temperature calibration: the information from temperature control should be compared with that of various standards. However, to save repeating the whole process again software often uses an automatic procedure to match the information already obtained with the control routines. Some parameters for this, e.g. the temperature range, may need to be chosen.

1.4.8 Choice of standards

Table 1.1 lists details of commonly used reference materials. Materials should be used once and then discarded. However, indium is one significant exception to this rule, since it can be reheated many times, provided it is not overheated. Indium fusion values repeat to very high precision (0.1%) provided it is not heated above 180°C. To ensure complete recrystallisation between tests it should be cooled to 50°C or below after each test. Since indium can be obtained to high purity, is very stable, and has a very useful melting point, it makes an ideal standard for most analysis and can be obtained as a certified reference material (CRM) (see below). Other metals listed should be discarded after use.

For temperature ranges where no metal transitions are appropriate other materials may be used as reference materials. These may not be certified materials but may be the best available. Use fresh materials of as high purity as possible.

1.4.8.1 Certified reference materials

These materials have a certificate giving values obtained after the material has been tested by a range of certified laboratories. This is not necessarily a certificate of purity but they are regarded as the ultimate in reference materials. They can be obtained from Lab of the Government Chemist [4]. Other materials may, for example, possess a certificate of high purity allowing use of theoretical melting and heat of fusion values. These types of materials are termed reference materials and are also widely accepted.

1.4.9 Factors affecting calibration

A number of factors are known to affect the response of a system and if varied may require different calibration settings. These include the following:

- Instrument set-up and stability
- Use of cooling accessories
- Scan rate
- Purge gas and flow rate
- Pan type

Varying any of these factors can affect calibration, though effects may be more significant for one instrument than another. Initially, ensure that the instrument is set up properly, all services turned on and stable. Most analysers will contain some analogue circuitry which will cause slight drift as they warm up, therefore make sure that instruments have been switched on for a while (typically at least an hour) before calibration. There may be other instrument settings in specific analysers that affect calibration, so users should have a good familiarity with the analyser being used before proceeding with calibration. Use of cooling systems can cause temperature values to shift whilst they are cooling down, so make sure that these are switched on and stable. If using different scan rates, particularly the very fast rates employed by fast scan DSC, you should ensure that the analyser is suitably calibrated for the scan rate of use. Purge gases such as air, oxygen and nitrogen have similar thermal effects and can be used interchangeably, provided scan rates are not altered. The higher conductivity of helium or lower conductivity of argon can have very significant effects on calibration and therefore systems must be calibrated with these gases, if they are to be used. Typically, helium is used at low temperatures or with high scan rates, whilst argon may give better performance at high temperatures. Differing pan types generally do not have too much effect but if thermal contact is changed, for example by the differing thickness or material type of a pan base, then calibration can be affected. If in doubt, then check.

1.4.10 Final comments

Calibration is a process of check and adjustment. It is not complete until you have checked the analyser afterwards to ensure that values obtained are acceptable; see Figure 1.4. Verification ensures that the analyser is in good order, not the calibration procedure itself. A good overview of the calibration procedure together with recognised standards available is given in [3]. Note: Always follow manufacturers' recommended procedures when calibrating a system.

1.5 Interpretation of data

1.5.1 The instrumental transient

When the run begins, it takes a short period of time for energy to be transmitted to both sample and reference in order to produce the required heating rate, so there is always a slight period of instability at the start of a run before a stable heating (or cooling) rate is established. It often appears as an endothermic step but may vary run to run. This period is termed the *transient*. In Figures 1.3a and 1.3b where a complete trace can be seen the transient shows as the instability at the start of the run before a flat baseline is established.

The time taken by a transient will vary from instrument to instrument but will always be present. Values can be as low as a few seconds to well in excess of a minute, varying slightly with the mass in the furnace. After this period the baseline may show a slight slope. The heat flow curve reflects changes in heat capacity of the sample, compared to the reference, and this may result in a slight positive or negative slope. It is good to have a period of straight baseline before the first transition, so start well below the temperature of the first transition to allow for this and the transient. A transient will also occur when a scan is completed, for example when switching to an isotherm, and can on occasions mask measurements made during the isotherm. If this occurs it may be possible to minimise the effects by subtraction. For example, when performing an isothermal cure reaction the post-cured sample can be rerun under identical conditions and the transient portion can then be used for subtraction from the original trace.

1.5.2 Melting

A crystalline melt is one of the most commonly measured transitions, and appears as a peak on the heat flow trace (see Figure 1.4 for the melt of a single crystal such as indium and Figure 1.6 for the broader melt of a polymeric material). It is described as a first-order transition having a discontinuous step in the first-order derivative of the Gibbs free energy equation [5].

For first-order transitions

$$dG = -S\,dT + V\,dP$$

where $S = -(dG/dT)_P$ and $V = (dG/dP)_T$ discontinuities (steps) are observed in entropy, enthalpy or the volume. The first derivative of a step is a peak, so the heat flow curve, which is the first derivative of the enthalpy curve, shows a peak from a melt transition.

Melting is an endothermic process since the sample absorbs energy in order to melt. Integrating the peak area gives the heat of fusion ΔH_f and this can be a simple process but often requires careful choice of cursor positions for the integration.

1.5.2.1 Single crystal melt

In principle, melting of a single crystal should not produce a symmetrical peak shape, though on occasion it may look so. When melting begins, the sample will essentially remain at the melting temperature whilst solid and liquid are in equilibrium and melting progresses. This results in a peak with a straight leading edge whose slope reflects the rate of energy transfer to the sample. The peak maximum represents the end of the equilibrium melt region and the trace then drops back to the baseline, the energy under the tail of the peak resulting from energy needed to heat the liquid sample to the temperature of the furnace. The resulting peak shape is not symmetrical and at slow rates looks more triangular (Figure 1.5). In a heat flux system it is the temperature difference between sample and reference that creates the signal that is reported as the heat flow. In a power compensation system, a temperature difference still occurs since solid and liquid in equilibrium will remain at the melting point until melting is completed. From this it will be seen that the melting point of a single crystal

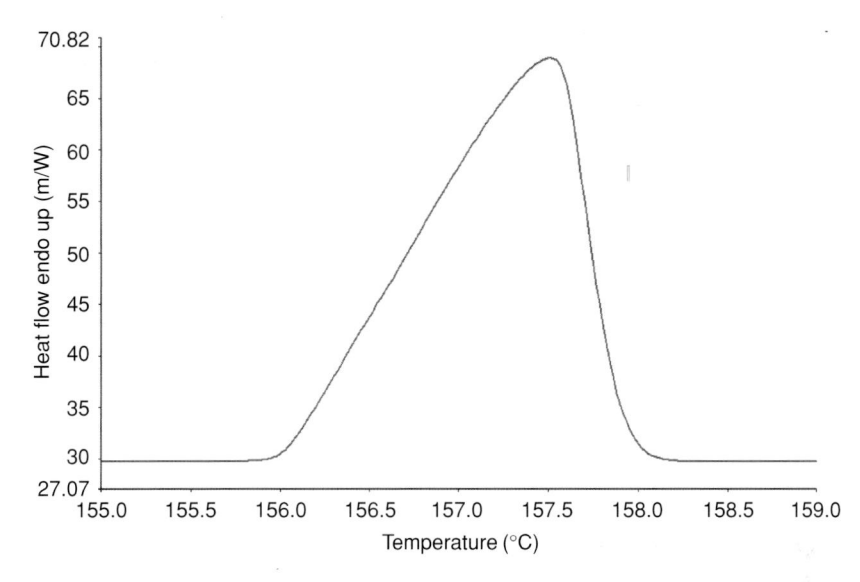

Figure 1.5 Indium heated at 5°C/min showing an almost triangular melting profile typical of single crystal melt at lower scan rates. The slope of the leading edge of the melt of a pure material such as indium gives a value for the thermal resistance constant R_0.

sample is reflected by the onset temperature, which is where melting begins, and not the peak maximum value. The peak maximum value and the peak height value will be influenced by sample weight and possibly by encapsulation procedure, since poor thermal contact will result in a broader peak with lower peak height. The two primary and most accurate pieces of information drawn from single crystal melt are therefore the heat of fusion (melting area) and melting point (onset temperature) Figure 1.4.

1.5.2.2 Thermal resistance constant (R_0 value)

This is used in a number of calculations. In theory a 100% pure material should melt with an infinitely sharp and infinitely narrow peak if there were no thermal gradients across a sample (sometimes called thermal lags). However, because it takes time for energy to flow into a sample the peak is broadened and reduced in height. The slope of the leading edge of the melt profile of such a material (e.g. a pure indium standard) should therefore provide a measure of the maximum possible rate of absorption of energy into a material, or put another way, it is a measure of the thermal resistance to energy absorption. A value of R_0 measured under the appropriate scan rate, pan type, and other experimental conditions is used to correct kinetic and purity calculations amongst others. It is easily measured, but note the units specified by any given calculation.

1.5.2.3 Broad melting processes

Many materials do not contain single crystals but a range of crystals of varying stability that melt over a broad temperature range. Typical of these are polymeric systems, including

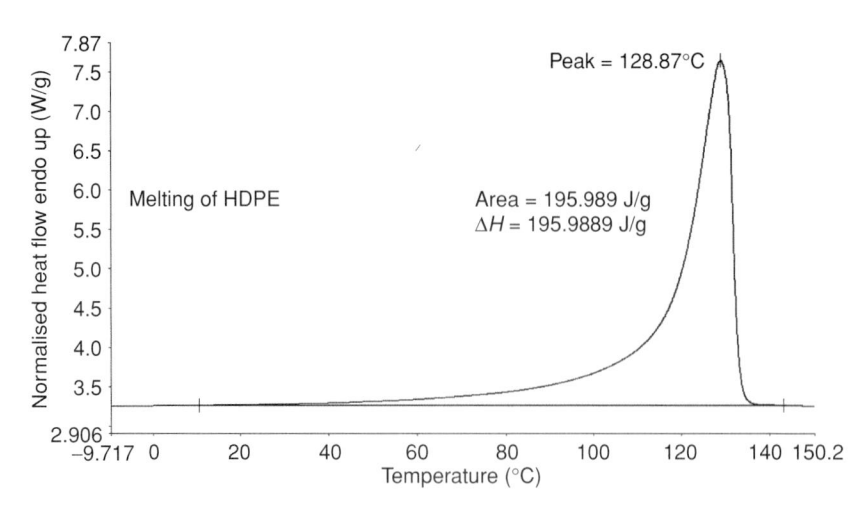

Figure 1.6 Melting of high-density polyethylene showing calculation of the heat of fusion from the area under the curve, and melting point measured from the peak maximum, unlike single crystal melt. The accuracy of the heat of fusion needs to be assessed in conjunction with the sample weight and the balance used and is unlikely to extend beyond a decimal point.

thermoplastics, foods and biological materials. An example of a broad polymer melt is given in Figure 1.6. Trying to calculate an onset temperature for such a process is fairly difficult and possibly meaningless, since melting is too broad and gradual. The peak maximum temperature of a broad melt is the most meaningful and therefore used in practice. For many materials it represents the temperature where melting is complete, though it will be subject to variation with differing run parameters such as sample weight, so it is good practice when comparing samples to use the same weight and conditions for each analysis.

1.5.2.4 Choice of peak integration baselines

Calculation limits need careful selection, particularly of broad peaks where the gradual start to melting may be hard to distinguish, and small changes can have a significant effect on the values obtained. The run may need to be started well below the melting region in order to distinguish the flat baseline from the gradual beginning of melt. Significantly expanding the data to see just the baseline area and where the transition begins can be very helpful, but it is down to the operator to decide the limits.

On some occasions, the heat capacity of a sample after melt may be noticeably different from that of the sample before melt, resulting in a peak with an obvious step underneath. In this situation, the actual baseline for peak integration is usually unknown but a number of different approaches may be made to the peak area calculation. The choice of a straight linear baseline is probably the most reproducible provided care is given to the selection of cursor positions. Sigmoidal baseline calculations provide another approach where the baseline extends linearly from before and after the melt and following a suitable mathematical algorithm provides a 'sigmoidal' shaped baseline for the calculation; see Figure 1.7. In practice these calculations look quite convincing, but as with all extrapolated calculations, they

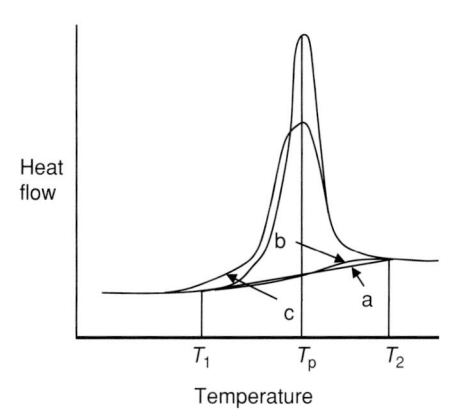

Figure 1.7 DSC peak and two possible baseline constructions (a and b). Also shown is the effect of impurity (curve c) which reduces the melting point and broadens the melting profile.

tend to be less reproducible. On other occasions, it may be preferred to extend horizontal baselines from the right or left, particularly if isothermal conditions have been used. Modern software now gives a wide choice of parameters for a range of different requirements, including partial area calculations for broad and multi-peaked processes.

1.5.2.5 Effect of impurities

When comparing melting peaks it is important to remember the effect of impurities, which may vary between samples. This is particularly important in foods and pharmaceuticals where impurities both broaden the melting process and lower the onset temperature, as shown in Figure 1.7. Water can act as an impurity in materials such as starch [6], so effect of impurities should be considered when comparing melting processes. Examination of the peak shape can also allow an estimate of purity; see Section 8.2.4.

1.5.2.6 Peak separation

On some occasions peaks may not be fully separated. To calculate the correct heat of fusion, it is best to separate the different events if possible. Because the actual peak shape does not follow any particular mathematical distribution it is not generally possible to perform an accurate mathematical de-convolution procedure, although where a particular peak shape has been well defined this may be a possible approach. The main approach to this situation is to try to improve resolution by reducing sample weight and by reducing scan rate. Employing helium as a purge gas has also been found to be beneficial in some systems, since its higher thermal conductivity leads to smaller thermal gradients. Depending on the kinetics of the differing events an increased scan rate may also improve separation. Bear in mind that if small samples are used very accurate weight measurement is needed for accurate results. If peaks cannot be separated, and some processes genuinely overlap, then partial area calculations provide a route to estimate the relative energies involved in the different parts of the transition.

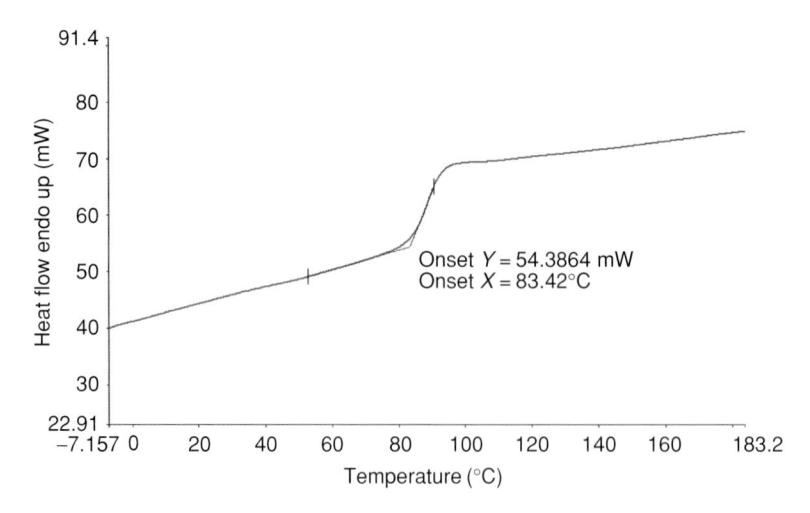

Figure 1.8 T_g of PET showing the step in the heat flow trace observed at T_g together with an onset calculation.

1.5.3 The glass transition

The glass transition (T_g) is the temperature assigned to a region above which amorphous (non-crystalline) materials are fluid or rubbery and below which they are immobile and rigid, simply frozen in a disordered, non-crystalline state. Strictly speaking, this frozen disordered state is not described as solid, a term which applies to the crystalline state, but materials in this state are hard and often brittle. However it is a frozen liquid state, where molecular movement can occur albeit over very long time periods. Material in this frozen liquid state is defined as being a glass, and when above T_g is defined as being in a fluid or rubbery state. Materials in their amorphous state exhibit very substantial changes in physical properties as they pass through the glass transition and it is one of the most commonly measured transitions using thermal methods; see [7] for a review. Figure 1.8 shows the typical step shape of a T_g as measured by DSC. It is exhibited only by amorphous (non-crystalline) structure; fully crystalline materials do not show this transition, and it represents the boundary between the hard (glassy) and soft (rubbery) state.

It is described as a second-order transition, since it has a discontinuity in the second-order derivative of the Gibbs free energy equation with respect to temperature [5].

For second-order transitions

$$-(d^2G/dT^2)_P = -(dS/dT)_P = Cp/T$$
$$(d^2G/dP^2)_T = -(dV/dP)_T = -\beta V$$
$$d/dT(dG/dP)_T = -(dV/dT)_P = \alpha V$$

Therefore discontinuities would be observed in heat capacity (heat flow), isothermal compressibility and constant pressure expansivity; and such steps, over a finite temperature range, are indeed observed at the glass transition.

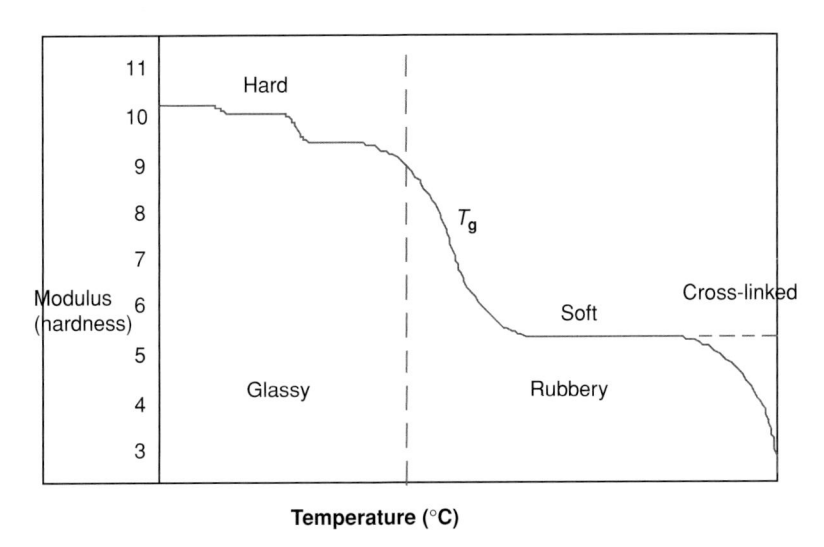

Figure 1.9 Modulus (hardness) of an amorphous polymeric material as a function of temperature showing the dramatic change in hardness that occurs in the glass transition region. Small lower temperature events are termed β- and γ-transitions.

The glass transition is quite a complex event and has been explained in different ways. (See explanations given in Chapter 4 for mechanical issues and Chapter 10 as related to glasses and ceramics.) One approach involves consideration of the energy involved in movement. At very low temperatures, molecules have insufficient energy to move and materials are immobile and very hard. When gradually heated, a typical long-chain molecule will exhibit small transitions where rotations and vibrations of side groups or side chains can begin, and in some cases crank shaft type motions of the main chain itself. Eventually, a temperature is reached where one complete molecule of the structure can move independently of another resulting in significant softening of the material and this is designated the glass transition. Figure 1.9 shows a diagram of the modulus (hardness) of a material as it is heated through T_g. Above T_g fully amorphous materials are soft and fluid unless bonded together in some way. The T_g is the most significant of the transitions shown in Figure 1.9; transitions observed at lower temperatures are termed β- and γ-transitions respectively, and they too have their effects on materials properties, particularly toughness, since they provide a mechanism for the dispersion of energy from high-frequency impact.

After a molecule has passed through T_g it has greater ability to move in various molecular motions, described as an increase in degrees of freedom of movement. For a given molecule these various movements may not be well defined, but they represent additional ways of absorbing energy, which means that an increase in specific heat is observed. As a second-order transition this appears as a step in the heat flow curve, which reflects changes in specific heat, and as an inflection in the enthalpy curve; see Figure 1.10.

The significance of the T_g will depend upon the amorphous content. Some materials are fully amorphous and therefore T_g will have a very significant influence on material properties. For example, fully amorphous thermoplastics will become fluid above T_g, whereas partially crystalline materials will remain solid until the melting point is reached. Fully crystalline

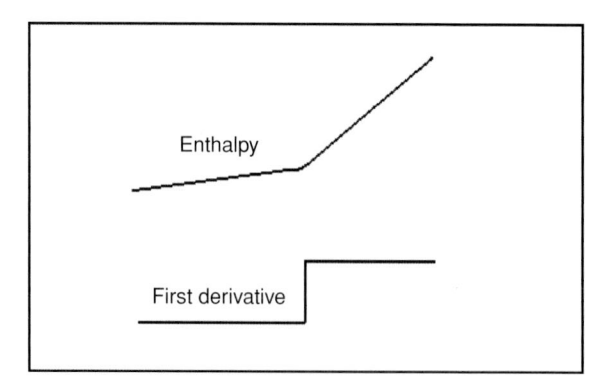

Figure 1.10 Diagram of an enthalpy curve as a sample is heated through T_g. A heat flow curve shows a broad step at T_g since it measures energy flow, which is a derivative of the enthalpy curve.

materials will not show a T_g, however, materials that are semi-crystalline will exhibit both a glass transition and a melt. The height of the T_g transition is dependent upon amorphous content and can provide a measure of it; further discussion of this can be found in Section 2.5.3. Partially, crystalline materials may also contain amorphous material which is not totally free to move because of its close proximity to crystals of the material, and some long-chain individual molecules may be part amorphous while a different section may be part of a crystalline structure. In Chapter 10 these have been referred to as tie points. Since such molecules are not totally free to move and may not undergo the glass transition this material has been termed the rigid amorphous fraction or RAF [8]. See Section 9.2.9 for further discussion of RAF, and possible multiple amorphous phases – polyamorphism.

Materials in the glassy state are in general stable, and neither react or decompose very quickly because the molecules are immobile over short time periods. This feature is used to advantage in freeze-drying where materials are specifically produced in a glassy state to render them stable and yet easily used; see Section 1.5.8. Other materials such as epoxy resins need to be heated above their T_g in order for them to react and cross-link.

1.5.4 Factors affecting T_g

The glass transition temperature is influenced by many parameters including effects of additives, for example plasticisers which are added to reduce the temperature of T_g; see Figure 1.11. One of the most significant plasticisers is moisture. This is well known to affect the T_g of composites, pharmaceuticals and many other materials. It can lead to loss of strength or stability at a given temperature. Cornflakes are a typical example of a material which is hard and brittle when dry at room temperature, yet with the addition of water (milk) the T_g is reduced and the cornflakes become very soft and pliable. Reactions, such as a cure reaction, which increase the molecular weight of a material will increase the temperature of T_g, since molecules need more energy and a higher temperature to move independently of one another.

Another factor is the flexibility of the chemical bonds in the molecule itself. Materials with flexible bonds such as Si–O or CH_2–CH_2 have low-temperature glass transitions whereas more rigid structures have higher temperature transitions.

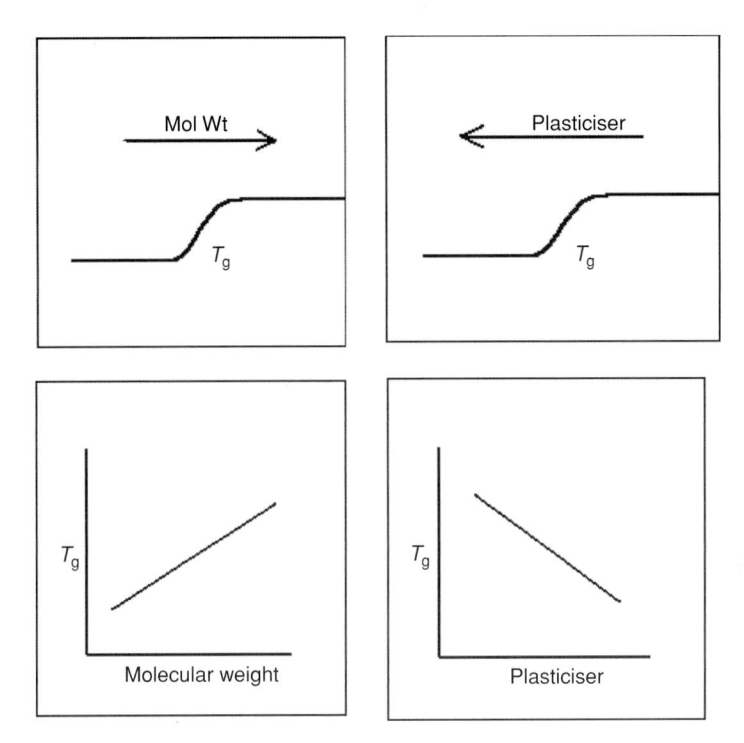

A Practical Introduction to Differential Scanning Calorimetry

Figure 1.11 Diagrams showing the effect of plasticiser or molecular weight on the glass transition temperature. Increasing molecular weight will increase T_g whereas the inclusion of plasticiser will reduce T_g.

1.5.5 Calculating and assigning T_g

The glass transition is not a single point and occurs over a region in temperature, in some cases many tens of degrees, therefore care is required when quoting a value for T_g. A choice can be made from the following calculations:

Onset	The intersection of tangents from the baseline and initial slope of the step; see Figure 1.12.
Midpoint	The middle of the step measured as half the step height; see Figure 1.12.
Endpoint	The intersection of tangents from the step and final baseline; see Figure 1.12.
Inflection point	The inflection point of the step.
Fictive temperature	This is a specific value and comes from the definition of the fictive temperature as being the temperature of the frozen glassy state at which a material is in equilibrium (see Section 1.5.6). The calculation is taken from the enthalpy curve and is the intersection of tangents taken from above and below the T_g. The same value can be obtained by drawing tangents to the integral of the heat flow trace.

In practice, one of the most often used values is the midpoint calculation but there is no rule that says that this or any other value should be used, so the method of calculation

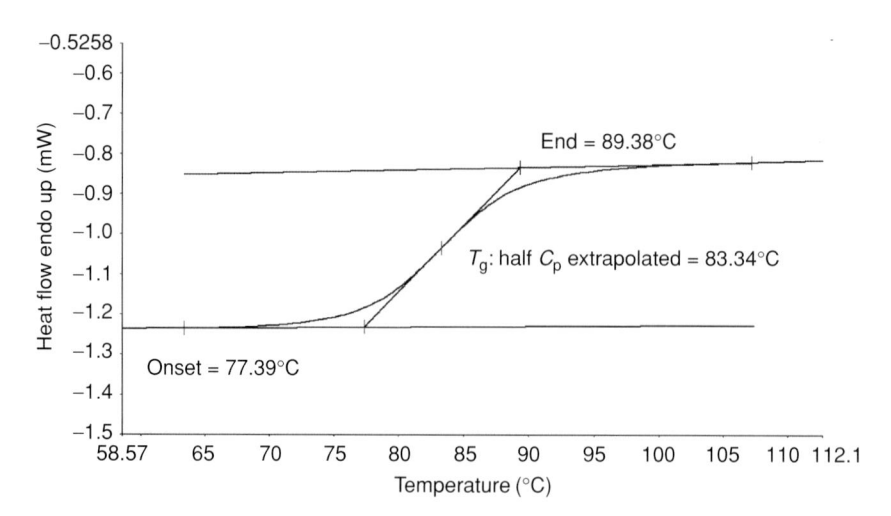

Figure 1.12 Different calculations of T_g. The onset, midpoint and endpoint are shown. The method of calculation is best quoted with the T_g value.

should be appropriate to the investigation at hand and should be quoted with the result. If a standard method is being followed then the procedure of calculation should be defined by the method, though the type of calculation used should always be quoted to avoid confusion.

1.5.6 Enthalpic relaxation

Sometimes a peak can occur on top of the glass transition which can affect the calculation of T_g. Large effects which entirely swamp the small glass transition event can sometimes be found, and these are possibly due to relaxation of mechanically induced stresses resulting from the production of materials such as films or fibres where mechanical pulling is used to orient molecules in one particular direction. Heating such a material results in a peak which is formed as previously stressed molecules are able to rearrange in the T_g region, absorbing energy in the process. If this occurs then a sample can be reheated to observe the glass transition, though care may be needed to ensure that the material is otherwise unchanged in the process.

Enthalpic relaxation phenomena usually give rise to smaller peaks (see Figure 1.13) and often result when heating a material which has been cooled slowly through T_g or held just below the T_g for a period of time. This will allow the material to anneal (relax) into a lower enthalpic state; see Figure 1.14. A lower glass transition temperature is associated with a lower enthalpic state, since the glass transition of a material is not a fixed thermodynamic event, and the temperature of transition is dependent to some extent upon how it has been cooled, and the energy state into which it has been annealed.

On heating a polymer which has been annealed into a low enthalpic state a small peak may appear almost like an overshoot, referred to as an enthalpic relaxation event. If heated very slowly this peak may not appear and a true value of T_g is then obtained, but at normal

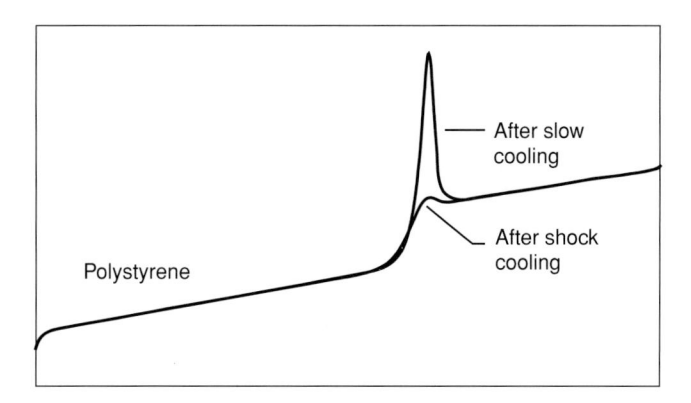

[Figure showing DSC curves for polystyrene with labels "After slow cooling", "After shock cooling", and "Polystyrene"]

Figure 1.13 Polystyrene heated at 10°C/min. This shows an example of a marked enthalpic relaxation peak which appears on top of the T_g on heating after the sample has been cooled slowly.

scan rates of 10 or 20°C/min a slight overshoot of the true T_g of an annealed material is often observed. The lower dashed line in Figure 1.14 shows an example. At this point, the material will pass into an energy state which is below the equilibrium energy state for the material, shown as the fixed line in Figure 1.14. When the material passes through T_g and the molecules become mobile they will also absorb energy in order to regain equilibrium, giving rise to a

[Figure showing enthalpy vs thermodynamic temperature with labels "Fictive temperature T_f or enthalpy H", "Cool", "Heat", "$T = T_f$", and "$C_{p\,norm} = dT_f/dT$"]

Thermodynamic temperature T

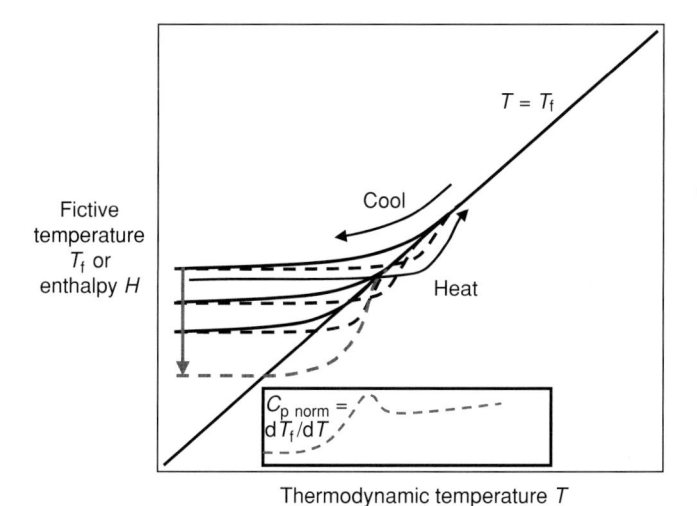

Figure 1.14 The enthalpy curve of a material in the T_g region. If cooled slowly through T_g or held isothermally (annealed) just below T_g a material can relax into a lower energy state as indicated. Heating an annealed material at normal heating rates is likely to produce a slight 'overshoot' of the true T_g onset resulting in a step as well as an inflection in the enthalpy trace. The resulting heat flow trace shown in the box is a step with a peak on top. The equilibrium line for the material is the solid line from bottom left to top right. It can be seen from the lower dotted line that an annealed sample heated at sufficient rate enters a non-equilibrium state of low energy before going through T_g at an elevated temperature and absorbing energy to regain equilibrium.

small endotherm. In consequence, a small enthalpic relaxation peak associated with the glass transition indicates that measurement has been shifted slightly from the true position. Since different T_g values are associated with the same material depending upon how it has been annealed, care needs to be taken not only with the measurement of the T_g but also with the conditioning of a material before measurement if a T_g value is to be quoted. The enthalpic relaxation peak will also affect the onset and midpoint calculations, since the shape of the transition is altered. If midpoint T_g calculations are to be performed it is normal to put the end cursor after the peak.

The true T_g value of such a material is then lower than that indicated by the normal DSC step calculation. Fast cooling is likely to prevent annealing and so remove any relaxation phenomena (one guideline is to first cool the polymer through T_g at the same rate or faster than it is to be heated), and this may be one preferred route prior to measurement; another is to use an enthalpic or fictive T_g calculations (see Section 1.5.5). This involves integration of the heat flow curve to give the enthalpy curve. A 'true' value of T_g can then be obtained using an extrapolated onset calculation of the enthalpy curve from above and below the T_g region; effectively, this seeks to reproduce the curve shown in Figure 1.14, though the results are likely to be less reproducible because it is an extrapolated calculation. This type of calculation can be done automatically with some software packages. However, it is worth noting that the T_g is neither a fixed point, nor does it occur at a fixed temperature as does a melt; it is dependent upon the thermal conditioning of the material. A very slow cooling rate, or slow heating rate, can permit annealing of a sample as can a modulated temperature method, so resulting in a lower T_g value. The change in T_g often associated with changing scan rate may in reality be due to annealing of a material.

The term 'ageing' or 'ageing peaks' has also been applied to the phenomenon of enthalpic relaxation, since it has been seen as an effect following storage. Many materials are not deliberately annealed but as they are stored the temperatures experienced over time may permit annealing, so the term ageing has been applied. It is observed in many amorphous food substances; starch, sugars and, less clearly, proteins are reported as being the prime examples (see Chapter 9).

See Section 10.2 and Figure 10.1 for further discussion of enthalpic relaxation of materials (glasses).

Enthalpic relaxation behaviour can be formalised in the TNM (Tool Narayanaswamy and Moynihan) description of ageing [9].

The equations describing this behaviour are as follows:

$$\tau_0 = A \exp[x\Delta H/RT + (1-x)\Delta H/RT_f]$$
$$T_f = T_0 + dT'\{1 - \exp[-(dT''/q\tau_0)^\beta]\} \quad \text{and} \quad Cp(\text{normalised}) = dT_f/dT$$

Where x and β are the non-linearity and non-exponential parameters respectively, A is the pre-exponential factor, ΔH is the activation enthalpy, τ_0 is the time constant and T_f is the fictive temperature. The effects of changes in some of these parameters on the overshoot in heat capacity can be seen in Table 1.2.

Moynihan gives an excellent description of the principles behind the TNM model. While the approach may at first appear complex, the behaviour of these equations is actually

Table 1.2 The effect of changes in the particular parameter on the ageing trace. $T-T_\infty$ can replace the absolute temperature T if a WLF type treatment of the data is used

Parameter	A	ΔH	x	β	T_∞
Effect on trace	Position	Position and sensitivity	Peak development	Transition width	Compression

comparatively simple and represents an interplay between time and temperature effects. The whole formalism is based on superposition theory, which states that the current state of the material is the sum of all the responses to previous temperature steps, in the case of a temperature profile. It is vital that the system starts from an initial equilibrium temperature. The effects of different rates of heating and cooling, annealing steps where the time alone is allowed to increase and so relax the whole system, and multiple annealing steps can be included. The latter is particularly relevant to food products involving multiple processing at different temperatures and storage for different times at different temperatures. All these temperature histories can be simply accounted for by summing all the different steps involved according to superposition theory. Figure 1.15 shows a theoretical temperature profile and the corresponding heat capacity trace generated using superposition theory. The memory effect in these materials is particularly interesting; note the multiple peaks in Figure 1.15. Further discussion of this subject and the effects of scan rate in particular can be found in Section 2.4.6.

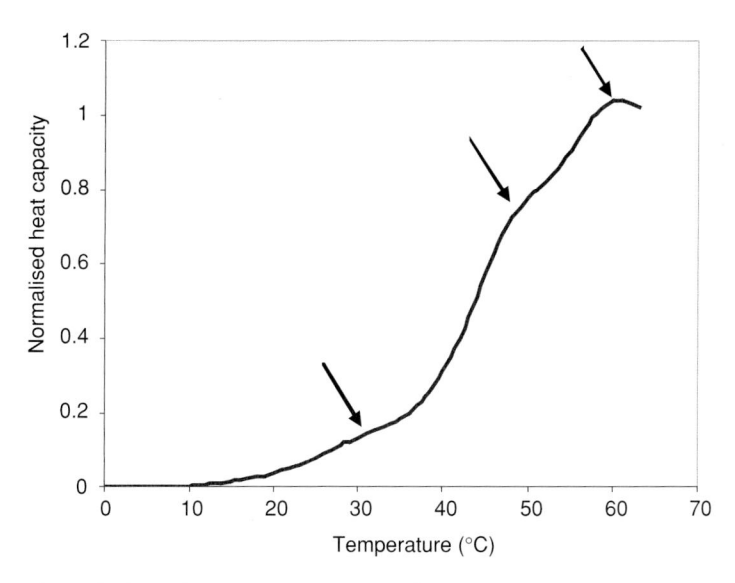

Figure 1.15 The multiple peaks generated using the following TNM parameters. Temp high = 63°C, Temp low = −5°C, Temp anneal1 = 32°C, Temp anneal2 = 5°C, time anneal1 = 4800, time anneal2 = 480,000, all cooling rates = 20°C/min, heating rate = 10°C/min. Parameters were $x = 0.632$, $\beta = 0.355$, $A = 6.57 \times 10^{-65}$, $\Delta H = 97.1$ kJ/mol.

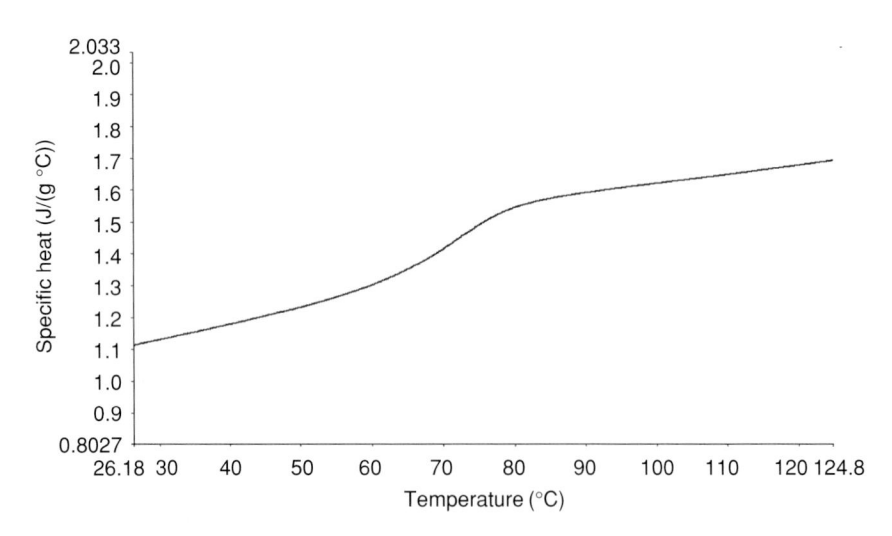

Figure 1.16 The heat capacity trace of PET cooling at 200°C/min using fast scan DSC. The T_g is clearly shown as a step similar to that found on heating.

1.5.7 T_g on cooling

If a sample is cooled through the T_g then a step in the DSC signal is again observed; see Figure 1.16. This shows the measurement of *heat capacity* on cooling, a step similar to that found on heating. However, the *heat flow* curve of T_g is quite different and apparently shows a step going in the other direction, as if there is an increase in heat flow on cooling as well as heating; see Figure 1.17. This is because the flow of energy on cooling is actually a negative

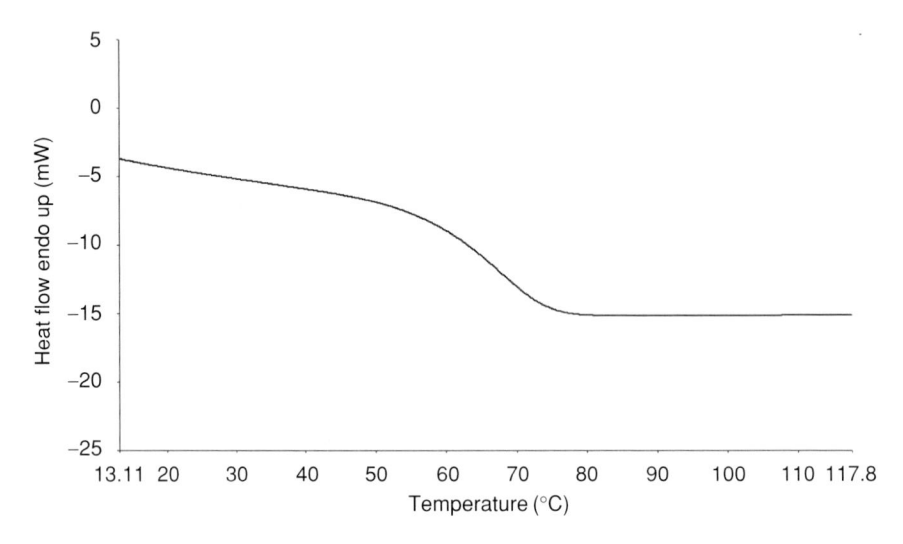

Figure 1.17 Heat flow trace of PET cooled at 150°C/min showing the T_g which appears as an apparent step increase in heat flow as cooling proceeds, but in reality is a step decrease in negative heat flow.

heat flow, not always apparent because the y-axis is not absolute, and as the sample goes through T_g a step decrease in heat flow is measured, which appears as a step towards zero. Thus heat flow curves on heating and cooling should appear more like a mirror image of each other reflected against the zero heat flow line.

1.5.8 Methods of obtaining amorphous material

1.5.8.1 Rapid cooling

Many materials can exhibit a T_g if they are produced in a way that prevents crystallisation. One such method is rapid cooling at a rate that is sufficiently fast that the material does not have time for crystallisation. Even water if cooled sufficiently rapidly can exist in a glassy state with a T_g which has been reported in the region of $-136°C$ (see Table 9.3). Bear in mind that the value recorded will vary depending upon whether onset, midpoint, inflection point, etc. are being quoted (See Section 1.5.5). Depending upon the method of production not only polymers but also many small molecules can and do exist in a glassy state, and many mainly crystalline materials may well possess a small amount of amorphous material, particularly if the material has been ground to reduce particulate size. It is often useful if trying to find a T_g, to cool a material rapidly from the melt before reheating, so as to maximise the amorphous content and increase the size of the glass transition.

1.5.8.2 Spray-drying

One way of obtaining amorphous material is to dry rapidly from solution. In solution, the solute is in disordered state and if dried very quickly it does not have time to recrystallise, and so gives an amorphous product. The process of spray-drying is often used with pharmaceuticals and can therefore give an amorphous product, though some crystallisation may occur. Resulting products need to be kept dry to prevent the T_g from being reduced by plasticisation with water or the material may recrystallise.

1.5.8.3 Freeze-drying and phase diagrams

The term freeze-drying has become familiar to many people as foods that can be found on supermarket shelves, such as coffee, have been produced in a freeze-dried state. In fact, many foods and pharmaceuticals are freeze-dried and the aim is to produce a material which is stable and does not decompose, yet which can be rapidly reconstituted with the addition of water into a useable, desirable form. Since materials are effectively stable in the glassy state, freeze-drying is useful because it produces a glassy amorphous material with a T_g above room temperature so that it is stable and can be stored on the shelf, but is very porous and can be dissolved very easily. DSC is a valuable tool for measuring melting and freezing points to construct phase diagrams and to measure the T_g of freeze-dried materials, and the process is described below in some detail.

An appreciation of phase diagrams greatly aids the understanding of the principles of freeze-drying and the literature on mixtures and melting of materials such as fats. Figure 1.18a represents a trivial phase diagram and shows the condition of perfect mixtures of

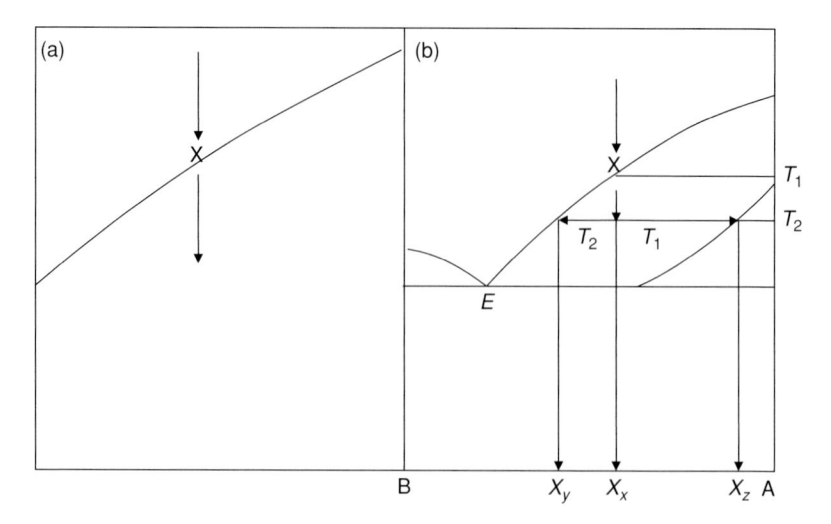

Figure 1.18 A phase diagram of a perfect mixture (a) and a eutectic mixture (b).

components in the liquid and solid phases. As the temperature is reduced the liquid solidifies at point x to a solid of exactly the same composition. In Figure 1.18b, if a mixture of ratio A/B is cooled from above the upper phase boundary, commonly referred to as the solidus line, the first thermal event occurs when this is crossed at position x. In fact this type of cooling method, coupled with accurate temperature measurement, was originally used to construct phase diagrams. At point x, temperature T_1, a phase enriched with A begins to crystallise out and the composition of the remaining material follows the curve towards point E. At any temperature, say T_2, the weight of the enriched phase and the remaining mixture can be calculated according to the length of the tie lines T_1 and T_2.

Using conservation of mass

$$W = W_y + W_z$$

and conservation of component A

$$XW = X_y W_y + X_z W_z$$

therefore

$$W_y/W_z = \{X_z - X_x\}/\{X_x - X_y\} = T_1/T_2$$

Here W_n is the weight of phase n and X_n is the mole fraction of component A in phase n. This process continues until the remaining mixture reaches the point E where a solid mixture of composition E (eutectic) crystallises out. Naturally, the same arguments hold if the eutectic point is approached from the opposite side where B is the crystallising component.

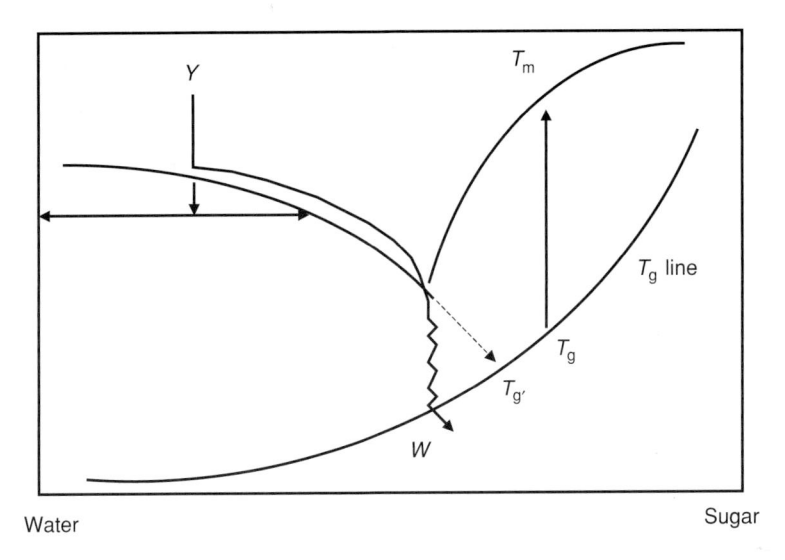

Figure 1.19 Supplemented state diagram for a sugar–water mixture.

Phase diagrams similar to the above, but where a pure compound crystallises out, apply to compounds such as benzene and naphthalene. However, if we consider a different system where A is assumed to be sugar and B water, an interesting observation is made.

By similar arguments and starting from position Y in Figure 1.19, pure ice would be expected to crystallise out until a point is reached where pure sugar and pure water would crystallise separately to form a solid mixture at point T_g''. However, the viscosity of the material is such that equilibrium behaviour is prevented and a material is formed having a raised concentration of water (point W in Figure 1.19). The material formed is a non-equilibrium uniform glass. The sample has been cooled at a rate where equilibrium can no longer be maintained, and it is commonly held that any material showing this behaviour will form a glass. The moisture content of a glass can subsequently be altered by relative humidity adjustment and the transition temperature T_g studied as a function of the concentration of the components. There are several equations, which describe the resultant T_g as a function of the concentration of the components. Perhaps the most common is the Gordon–Taylor equation [10]

$$T_g = \{x_1 T_{g1} + K x_2 T_{g2}\}/\{x_1 + K x_2\}$$

This can be seen to be equivalent to the Couchman–Karasz equation where k is equal to the ratio of the specific heats instead of $\rho_1 \Delta\alpha_2/\rho_2 \Delta\alpha_1$ as in the original paper [11]. P and $\Delta\alpha$ are the density and thermal expansivity respectively:

$$\ln T_g = \sum x_i \Delta C_{pi} \ln T_{gi} \Big/ \sum x_i \Delta C_{pi}$$

This equation can also be expressed in a linear approximation.

The Ten Brinke modification of the Couchman–Karasz equation was proposed to describe the lowering of the glass transition temperature of a cross-linked polymer network by a low molecular weight diluent [12].

This results in the following equation:

$$T_g = -\Delta C_{p2}/\Delta C_{p1}(T_{g1} - T_{g2})x_2$$

All these equations describe to a greater or lesser degree of accuracy the variation of the glass transition temperature with composition.

It follows that a material which is to be prepared by freeze-drying is first cooled, allowing the formation of pure water crystals and increasingly concentrated solution of the solid which eventually enters into the region of the glassy state. Pressure is then reduced and the pure water crystals sublimed away, a process which requires the input of energy. The process can be controlled so that the fraction of water remaining is low, and the consequent T_g of the final material is above ambient. This allows the material to be stored in a stable dry state which can be easily reconstituted by the addition of water.

It is important to keep these materials dry because an ingress of moisture into the material will lead to a reduction in T_g and in the resulting stability of the material. It is therefore important to know the T_g of these materials. One difficulty encountered is that the material may dry out if being heated slowly, so changing the measured value. An alternative route is to measure the T_g of the dried material and then with knowledge of the moisture content back-calculate the T_g of the material using the Gordon–Taylor equation. High scan rates can also be used to measure the T_g of moisture-containing materials before they have chance to dry out; see Section 2.3.3.

When heating amorphous materials through the T_g region the expected T_g step can be followed by recrystallisation as the system seeks to reestablish its equilibrium state. Some materials, particularly low-density freeze-dried materials, are prone to physical collapse once heated through T_g and this can also result in significant transitions in the T_g region or following T_g. These are usually not very reproducible because they result from sample movement, and faster scan rates can benefit the analysis since the T_g can be measured before the sample has had time to move.

1.5.8.4 Non-Arrhenius behaviour

When considering glasses, it would be surprising if the issue of non-Arrhenius behaviour did not arise. It is found in rheological studies that the effective zero temperature where mobility ceases should be greater than zero kelvin, which is the implicit assumption in the Arrhenius equation. Certain rheological plots can be linearised by assuming a value of $T_\infty = T_g - 50$. It is this term which appears in the WLF equation (see Chapter 4), and is thought to be the limiting underlying glass transition temperature at infinite time. By adding 50°C to this value, the rheological glass transition can be given a more real-world value but is effectively a reference value. A similar procedure has been carried out in annealing studies [13].

1.5.9 Reactions

The most common reaction that is measured is the cure reaction found in rubbers and composites and which appears as a broad exothermic peak; see Figure 1.20. In general, any

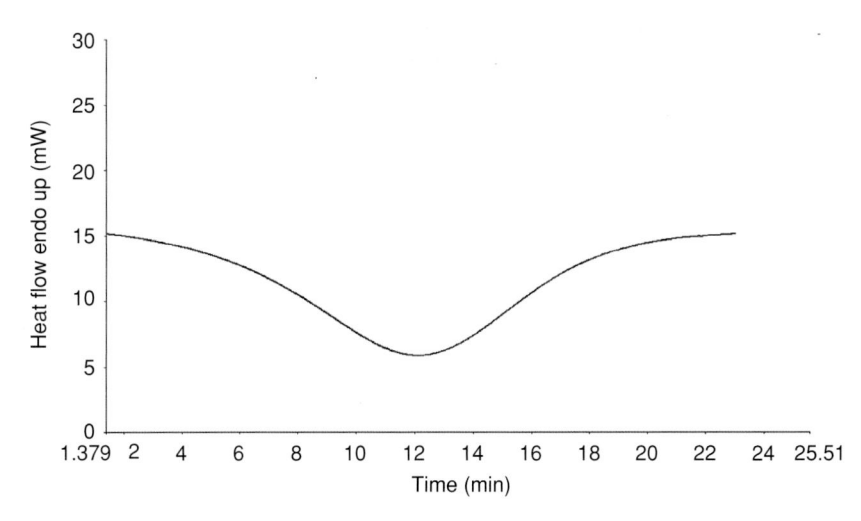

Figure 1.20 A cure reaction measured by DSC showing a broad exothermic peak. Slow scan rates (or isothermals) are normally preferred for such measurements to allow time for the reaction to occur prior to decomposition.

exothermic response comes from some type of reaction, be it decomposition, crystallisation or chemical interaction.

1.5.9.1 Cure reactions

Cure reactions in composites and other materials are normally slow events compared to typical scan times, and in order to measure them slow scan rates need be used so that the cure reaction can be measured fully before decomposition temperatures are reached. The alternative is to heat to an isothermal temperature where time can be given for the reaction to come to completion. This is also helpful when making calculations since a flat line extrapolated from the reacted portion of the sample provides an accurate baseline for peak integration. Note that subsequent heating to a higher isothermal temperature may permit further reaction to occur, since the extent of reaction may be limited by increasing viscosity and the material becoming hard. (Effectively, the T_g increases towards the isothermal temperature and upon reaching it the material becomes a stable glassy structure.) Often a T_g can be observed prior to a cure reaction, in a pre-mixed system the material will be stable below T_g so it needs to be heated above this temperature for reaction to begin. Modulated methods have also found value for the measurement of cure since the underlying heat capacity of the material can be measured as a function of increasing temperature, during the cure reaction. This can provide a more accurate peak baseline for the measurement of energy during a temperature scan.

From a slow scan, the onset and energy of reaction can be measured which can be of value in itself and can also give some idea of the potential hazard involved. These curves can be further investigated using kinetic analysis. Data can be fit to a variety of mathematical models, yielding activation energy and other information. Arrhenius-type models are the most common and are often available as part of the instrument software. Whilst data can be obtained from a single scan, the accuracy of the data is open to question since the software will generally be able to fit the data and produce values for activation energy, order of reaction and rate constant, whether accurate or not. Data obtained from isothermal reactions is to be preferred. This is taken from a series of isothermal curves run at different temperatures, and results can be compared from different isothermal values. If data from a temperature scan are used they should first be checked against isothermal data to assess their accuracy.

In any kinetics work the control of the sample temperature is important, and if isothermal measurements are taken from an analyser with heat flux principle it should be noted that the heat flow measurement is obtained from the deviation of the sample temperature from the cell temperature. It is the furnace temperature that is controlled, not the sample temperature, so true isothermal measurements cannot be made using this approach. Once the basic kinetic parameters have been established predictions can be made as a function of temperature, time, and extent of reaction, though the accuracy depends upon how well the data fit to the model used.

1.5.9.2 Arrhenius kinetics

Most kinetics start with the assumption that the rate of reaction is a function of degree of conversion and sample temperature.

The degree of conversion is given by

$$d\alpha/dt = k(T) f(\alpha)$$

where $k(T)$ is described by the Arrhenius expression

$$k(T) = Z \exp(-E_a/RT)$$

Here Z is the pre-exponential constant, E_a is the activation energy of the reaction, R is the universal gas constant, and T is the absolute temperature in kelvin.

Combining the above equations in logarithmic form gives

$$\ln[d\alpha/dt] = Zf(\alpha) - E_a/RT$$

A plot of $\ln[d\alpha/dt]$ versus the reciprocal of temperature should then give a straight line of slope $-E_a/R$.

This is the so-called Arrhenius plot. Since the slope is dependent upon activation energy, plots of these types are often displayed within kinetics software to allow the operator to check the consistency of the data. This is of particular value where isothermal data are obtained at a series of different isothermal temperatures. A similar expression can be derived for kinetics based on a series of different scan rates, and results in an Arrhenius plot of scan rate versus reciprocal temperature. These plots are then checked at a series of desired conversions.

Once obtained, the kinetic parameters can be used to predict reaction and storage times at times and temperatures not so conveniently measured, and may also give an indication of hazard.

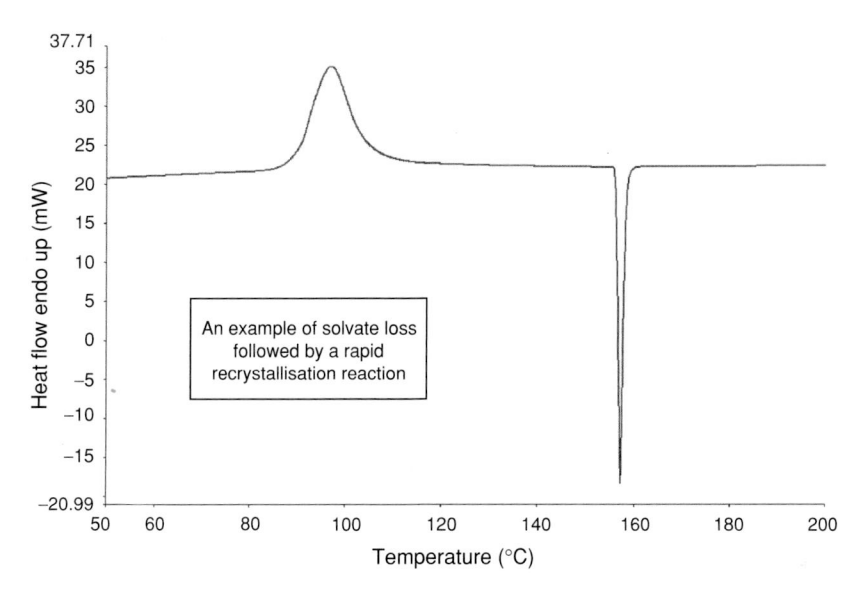

Figure 1.21 An example of solvate loss, the initial endothermic peak, followed by rapid recrystallisation of the destabilised crystalline structure.

1.5.9.3 Recrystallisation reactions

Many pharmaceuticals and foods are prone to polymorphism, where exist a number of distinct crystal structures, and recrystallisation between polymorphs can occur. These are usually fairly rapid events compared to the cure reaction, particularly if resulting from desolvation or dehydration where a solvent or water molecule has been ejected from a crystal leaving it unstable. A typical example is shown in Figure 1.21.

Loss of solvate and hydrate are both observed as endothermic events resulting from the loss of mass from a pan; in a sealed high-pressure pan no transition may be observed. The loss of moisture from a sample can therefore be a significant issue since it can obscure smaller events occurring in the same region, for example the glass transition in a polyamide or a freeze-dried material such as heparin (Figure 2.7).

Thermoplastic polymers are also prone to recrystallisation processes, particularly after rapid cooling employed in many manufacturing processes. Sometimes these are very evident as with PET (Figure 2.5), though sometimes the overall annealing effect does not show an overall exothermic event. The kinetics of recrystallisation processes are usually described by the Avrami equation discussed below.

1.5.9.4 The Avrami equation

The Avrami equation is crucial in the description of crystallisation and other processes. The equation has been applied to areas as diverse as corrosion, reaction kinetics and the growth of micro-organisms. If applied correctly, it can give information on the type of nucleation (homogeneous or heterogeneous) as well as the geometry of crystallisation, for example whether crystals are growing spherically (in three dimensions), or linearly with a needle-shaped form.

Crystallisation can be thought of as the expanding waves when raindrops fall on a pond. The probability that n circles will pass over a point P is given by the Poisson distribution $\lambda^n \exp(-E)/n!$, where E is the expected value.

To obtain E, the contribution of each annulus of width dr centred at a distance r from P is calculated and integrated for all values of r from 0 to vt, where v is the radial velocity of the wave. The integration is permissible because expectation is additive. Nuclei at distances exceeding vt will not reach P. The annulus has an area $2\pi r \, dr$ and during a time $(t - r/v)$, a point within the annulus will be capable of sending out a circle that will cross P before time t. If Ω is the 2D nucleation rate then

$$dE = \Omega(t - r/v)2\pi r \, dr \quad \text{and} \quad E = 2\pi\Omega \int_0^{vt} (tr - r^2/v) \, dr = 1/3 \pi \Omega v^2 t^3$$

and the chance that no circle passes over point P is

$$P_0 = \lambda^0 \exp(-E)/0!$$

or

$$\alpha = \exp(-\pi \Omega v^2 t^3)$$

The crucial point is that the formula is derived for the chance that P escapes being covered by a circle, i.e. $n = 0$. In other words, this expression is for the uncrystallised material.

If the nuclei are pre-existing, the number of circles is represented by a nucleation density ω and the number present in an area dA is $\omega \, dA = \omega r \, dr$. The function to be integrated contains no time terms, such as $(t - r/v)$ as all the nuclei are present at the start and none are added. Therefore,

$$E = 2\pi\omega \int_0^{vt} r \, dr = \pi \omega v^2 t^2 \quad \text{and} \quad \alpha = \exp(-\pi \omega v^2 t^2)$$

These two previous functions have the forms

$$1 - \exp(-kt^3) \quad \text{and} \quad 1 - \exp(-kt^2)$$

Note in particular the value for n.

Similarly, for expanding spheres from nuclei appearing sporadically in time, the annulus is now a spherical shell and the differential of expectation is

$$dE = \Omega'(t - r/v)4\pi r^2 \, dr$$

Table 1.3 The different exponents and expressions for the rate constant in the Avrami equation for different crystallisation morphology

Mechanism of crystal growth	Sporadic nucleation (primary) n	K	Predetermined nucleation (seeded) n	k
Polyhedral	4	$\pi G^3 I/3$	3	$4\pi G^3 I'/3$
Plate-like	3	$\pi d G^2 I/3$	2	$\pi d G^2 I'$
Linear	2	$\pi d^2 G I/r$	1	$\pi d^2 G I'/2$

The Avrami equation here is $\varphi = 1 - \exp(-kt)^n$.

where Ω' is the volume or 3D nucleation rate. This gives a higher power of n in the expression for the crystallising fraction $(1 - \alpha)$

$$1 - \exp(-kt^4)$$

This can be summed up in Table 1.3, which shows that fits of these sigmoidal curves to crystallisation data give information on the rate constant as well as the geometry.

The Avrami equation is frequently applied directly to isothermal DSC data where the heat output of the crystallisation process is thought to describe exactly the crystallisation. It can also be applied in modified form to non-isothermal data such as the crystallisation of PET on cooling. It should be noted that there is a certain ambiguity in the exponent value. For example, a value of 3 can mean either 2D nucleation from nuclei appearing randomly in time, or 3D nucleation from pre-existing nuclei.

The issue of induction time is important. Frequently, in order to obtain sensible values for the exponent n, it is necessary to subtract the induction time, or the time taken for the formation of a stable nucleus. The equation is then

$$\text{Rate} = 1 - \exp[-k(t - t_{\text{ind}})^n]$$

Similarly, the value of the rate constant is frequently put inside the brackets in order to have constant dimensions. The Avrami equation has found increasing application in the foods area, for instance, in the crystallisation of fats as well as more traditional polymer areas.

1.5.9.5 Explosives and decomposition reactions

Explosives are a particular class of exothermic reaction where the size of the exotherm is very large. Using small sample sizes of 1–2 mg in the DSC allows the reaction to be controlled and measurements can be made without damage to the analyser. The DSC can be considered as a screen for hazardous materials; usually high-pressure pans are used to encapsulate the sample, and a kinetic evaluation produced. If potential hazard is indicated then samples of larger size can be run on other equipment, for example isothermal/adiabatic calorimetry, to obtain more exact information.

1.5.9.6 Oxidative Induction Time (OIT test)

In general, sample decomposition should be avoided when using a DSC. They are usually noisy, dirty interactions of exothermic decomposition energy and of endothermic volatile loss and often result in the contamination of the analyser. However, one type of test that does measure decomposition is OIT test, used frequently with oils, polymers and foods. Often production and manufacturing conditions are quite extreme compared to storage and usage conditions, and it is necessary to know whether a material can stand these production conditions. The OIT test is an isothermal test which involves heating the material to the desired test temperature under inert conditions (make sure that the cell is purged of all oxygen before heating). Once holding isothermally at the test temperature the atmosphere is switched to an oxygen atmosphere, and the run continued until the material begins to exotherm, which indicates the start of oxidative decomposition. The run is then stopped. The time from the gas switch to the onset of decomposition is calculated and is known as the oxidative induction time. It is usual to employ automatic gas switching for this test and software will usually employ an OIT calculation which automatically calculates from the gas switching time. Oxygen is normally specified for this test rather than air and it is normal to use a small sample size, e.g. 1–2 mg, and use a completely open pan. When using oils the surface area is all that matters and sample sizes may be smaller; however, the loss of volatiles from oils and similar materials will affect the results. It is for this reason that external pressure cells have been developed and typically operate to about 40 bar, though some may go higher. These suppress the loss of volatiles from open pans and allow measurements to be made with volatile materials. A good example of the use of the OIT test is given in Section 5.2.1.

1.5.10 Guidelines for interpreting data

First of all optimise the data. Data should be expanded to a scale where transitions of interest can be clearly seen in the context of the trace. It may be helpful to slope the data so that areas of flat response are shown as flat. The eye can interpret more easily from the horizontal. The process of sloping simply pivots the data graphically; it is not a curve fitting or smoothing process so has no effect on transitions or calculations performed. Sometimes transitions are missed simply because they are very small compared to a major transition. Inspect the whole trace carefully if looking for small events. Remember, if a transition cannot be repeated it is unlikely to be real. If in doubt repeat the analysis.

In addition, find out as much about the sample as possible, it is difficult to set up a good method or to give clear interpretations of a trace when working without full information. The sort of information commonly needed is an idea of melting point or decomposition temperature, but more fundamentally it is better to have some idea of what transitions to look for and why. For example:

1. What type of sample is it?
2. What type of transitions can it undergo?
3. What would any changes appear as?
4. What is the temperature range of interest?
5. What data are available from complimentary techniques, e.g. TGA?
6. Has there been any previous analysis?

Then examine the data:

1. Is the event an endotherm or exotherm?
2. Is the event repeatable on a fresh sample or on a reheat?
3. What happens on cooling?
4. Is the event the same in a sealed and unsealed pan?
5. Is the transition sharp or gradual, large or small?
6. Does the event look real? Thermal events are not normally excessively sharp.

Examine the sample pan and look for sample leakage, decomposition or bursting which may have influenced the trace. In some cases, it may be useful to reweigh the pan to see if sample has volatilised and escaped.

The following general comments may help with the assignment of peaks:

Large well-defined endotherms

1. Pure crystal melting
2. Liquid crystal
3. Solvate/hydrate loss
4. Stress relaxation

Shallow broad endotherms

1. Moisture loss
2. Residual solvent/monomer loss
3. Sublimation
4. Range of molecular weights melting

Large well-defined exotherms

1. Crystallisation in/from the liquid state
2. Degradation of highly unstable material
3. Crystal structure rearrangement

Shallow and broad exotherms

1. Cure reaction
2. Degradation/oxidation

Calculations performed may require careful placement of cursors so that extrapolated tangents and baselines give sensible intercepts. Few software packages offer any automatic calculations, so individual interpretation is required. Often the derivative can help, and with peaks it can be useful to expand the trace to a high level to see where the transition actually began, reverting to a smaller scale once the calculation is performed.

1.5.10.1 *Baselines, subtraction, and averaging*

Ideally, the baseline recorded from a DSC with no pans should be flat. A truly flat response is not expected when there is a sample in a DSC since the heat flow data should reflect the heat capacity of the sample, which will increase as a function of temperature. In practice, if the thermal mass of the sample and reference are not exactly matched a positive or negative slope may arise; often more clearly seen with power compensation systems which measure differential heat capacity directly. The slope of a DSC curve is not an issue, unless extreme, since data can be resloped as required using data handling at the end of a run. Difficulties arise when the baseline is curved or contains anomalies as a result of contamination. It may be that simple cleaning is required but if the instrument itself cannot be adjusted then it may be preferable to run an instrument baseline and subtract it from the data. It should be established first that the baseline is totally stable and reproducible; if it is not then the cause of any instability should be identified.

Where a baseline is to be subtracted it should be run under the same conditions as the sample, and software may permit real-time or post-run subtraction. Real time is preferred since the data set is then permanently saved, though it is helpful if both the subtracted data and un-subtracted data are available for view.

Subtraction routines can also be useful to view differences between materials. A lot of thermal analysis work involves comparison of one material with another. Subtracting curves allows differences to be seen in a more exact way.

When a transition is very small it is possible to improve the sensitivity of the analysis by an averaging process. Just as with techniques such as FTIR (Fourier transform infrared spectroscopy) where the signal improves with the number of scans, repeated DSC scans of the same material can be averaged to improve the final response, though at slow scan rates this can be time consuming. Algorithms for this may be included in the instrument software.

1.6 Oscillatory temperature profiles

When a cyclic temperature profile is applied to a sample the heat flow signal will oscillate as a result of the temperature program, and the size of the oscillation will be a function of the heat capacity of the sample. Therefore, the amplitude of the heat flow signal allows a heat capacity value to be obtained. This is similar to DMA (see Chapter 4) where the amplitude of the oscillation allows a modulus value to be obtained. Whilst other methods already exist to provide heat capacity, the value of this method is that the heat capacity measurement is separated from other potentially overlapping events, such as reactions or stress relaxations and can also be obtained with increased sensitivity compared to the slow linear scan rates of traditional DSC.

The sinusoidal approach by Reading [14, 15] introduced the terminology of *reversing heat flow* for what is essentially the heat capacity trace, *total heat flow* for the average of the modulated heat flow trace, which is the conventional DSC trace, and *non-reversing heat flow* for the kinetic response. Since the T_g is observed as a step in the heat flow trace the reversing heat flow signal has been used to make measurements of T_g. This curve should show events that are truly reversing in the sense that the same event can be observed upon reheating or recooling.

1.6.1 Modulated temperature methods

In many DSC traces, there is evidence of a 'kinetic signal' being present in addition to the usual changes in heat capacity such as those due to the glass transition and melting. This is

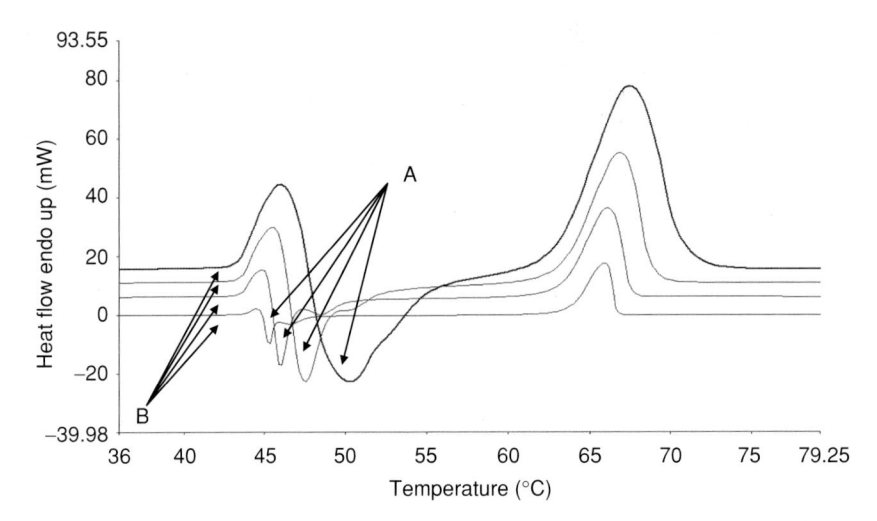

Figure 1.22 The recrystallisation event in this fat sample is shifted as a function of scan rate whereas the melting events remain unaffected.

often hinted at by the kinetic signal being sensitive to the scanning rate while other signals appear to be relatively unaffected (Figure 1.22). Essentially, the kinetics of slow events such as reactions and recrystallisation events are influenced by the scan rate and so shifted to higher temperatures as a function of increasing scan rate. Modulated temperature DSC (MTDSC) techniques allow the reversing heat flow (heat capacity) signal which is measured over the short time period (the period of the modulation) to be separated from slower transitions that occur over a much longer time period. Thus the glass transition can be separated from relaxation, recrystallisation and other events that might obscure it and make the measurements clearer. Relaxation and other non-reversing events should then appear on the non-reversing curve. With appropriate parameters the sensitivity of measurement can be enhanced by the increased rate of scanning that occurs during part of the cycle, which results in an increased heat flow signal. Hence difficult-to-find glass transitions can become more obvious when using modulated methods. MTDSC can also be used in quasi-isothermal mode such that no overall temperature ramp is induced and allows measurement of C_p values at a single point. This can allow improved resolution of events where a change in C_p is observed. During melting, energy is continually absorbed by a material and a steady state is not maintained, making analysis of the data more complex.

For many applications a qualitative analysis of the resulting data will be all that is necessary, for example giving values for the glass transition temperature. If a fully quantitative analysis is required it is important to decide on the limits of applicability. Schawe and Hohne [16, 17] have given clear guidelines as to when the reversing and non-reversing signals may be accurately separated from each other. They have stated that only in the case where the reaction is fast, such that the system is always in equilibrium relative to the temperature change and the enthalpy is independent of temperature, will the reversing and non-reversing signals be reliably separated. In addition, the amplitude of temperature modulation must be sufficiently small that the term bT_a in the temperature-dependent expansion of the reaction rate can be neglected. This is equivalent to a reaction, which although clearly time dependent has a weak temperature dependence and so will be in equilibrium for small temperature steps. This can be summed up in the following equations.

The degree of reaction can be expanded as

$$\alpha(T) = a + bT + cT^2 + \cdots$$

where α is the degree of reaction, a, b and c are constants, and T is the temperature. The scanning rate is

$$\beta = \beta_0 + \omega_0 T_a \cos \omega_0 t$$

where β_0 is the linear rate of temperature increase, and ω_0 and T_a are the frequency and the amplitude of modulation respectively.

The heat flow into the sample $= C \times$ scanning rate, where C is a generalised heat capacity

$$C = C_p + H_r \, d\alpha/dT$$

Therefore the heat flow into the sample is

$$(C_p + H_r \, d\alpha/dT) \times (\beta_0 + \omega_0 T_a \cos \omega_0 t)$$

T_a can be chosen such that the effect of modulation on the reaction is small and so a linear approximation to the effect of the temperature profile on the reaction can be used.

Therefore, the heat flow into the sample is $[Cp\beta_0 + Hr\,d\alpha/dT\,\beta_0] + Cp\omega_0 T_a \cos \omega_0 t$ where H_r is the enthalpy of reaction. It can be seen that in this case the reversing signal can be obtained from the second term in the periodic heat flow, as there is no complication of other contributions such as the energy of reaction appearing in this term. The conventional DSC trace can then be used as a background for subtraction to obtain the energy of the kinetic transition, provided the frequency dependence of the reversing signal is weak relative to the modulation frequency. This is a rather stringent test, even the well-known cold crystallisation of PET fails to completely meet these criteria, except at the very lowest of modulation amplitudes and underlying heating rates.

If the modulation amplitude is higher, such that T_a cannot be neglected or it causes the reaction to behave non-linearly, i.e. requires higher terms in the expansion for $\alpha(T)$, then a simple separation can no longer be made. Further, if the signal is delayed in time from the stimulus, in this case the temperature rate, then the treatment of the data cannot be analysed in the previous way and needs to be treated in a linear-response framework along the lines of rheology and dielectric data to obtain in-phase and out-of-phase components. Interpretation of data can then become very complex and difficult to relate to events in the material.

1.6.2 Stepwise methods

Using a stepwise approach, the method of calculating thermodynamic (reversing) and isokinetic baseline (non-reversing) signals is different from the methods given in the previous

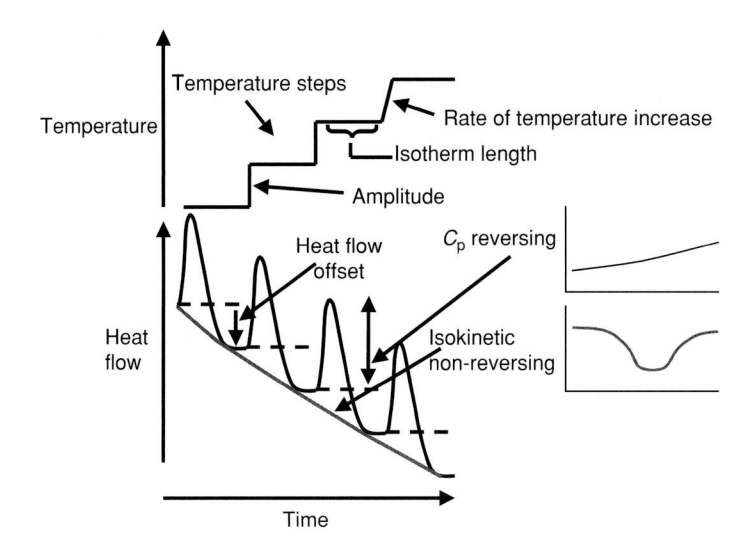

Figure 1.23 A diagram of the temperature profile used in step-based methods and the heat flow signal that results. Heat capacity is calculated either from peak height as in the classical method, or peak area of the heat flow trace. The kinetic response is obtained from the isothermal portion.

section [18] and the above equations do not apply directly. As Figure 1.23 shows, the 'non-reversing' signal can more correctly be labelled the 'slow' component, and is the power offset at the end of the isotherm, whilst the reversing trace is calculated from the amount of heat required to raise the sample temperature by a given amount – a conventional heat capacity measurement. This can be achieved using either amplitude of heat flow, or the area under the curve. In order to achieve a reasonable (if only approximate) separation into reversing or rapid and non-reversing or slow signals it is important that the isotherm length is chosen correctly. If it is too short, then an artificially high signal can appear in the non-reversing trace. In general, the isothermal period should be long enough for the system to reach stability, typically 30–60 s. Moreover, if the slow signal varies with time or the isothermal profile changes with temperature, as has been shown for the rapid transitions between polymorphic forms such as in tripalmitin or chocolate, then there will be unavoidable sources of error. Additionally, if the amplitude of the step is too large, the peaks will be widened due to averaging so reducing resolution. To a first approximation, the slope of the step is unimportant in determining the heat capacity; at least in this form of analysis.

One example of information obtained from step-based methods is shown in Figure 1.24. The heat capacity of the material is obtained, separated from the crystallisation and relaxation phenomena which appear in the kinetic trace, leading to a clearer measurement of the glass transition. This is typical of the type of information available from modulated methods in general.

There is also a step-based equivalent technique founded on linear-response theory. This has been presented by Schick and co-workers [19] and is formally equivalent to the real and imaginary heat capacities generated by modulation at a single frequency. In principle, it is therefore possible to apply a Fourier transform to step-scan type data and break down the overall response into heat flow responses to a series of sinusoidal temperature modulations.

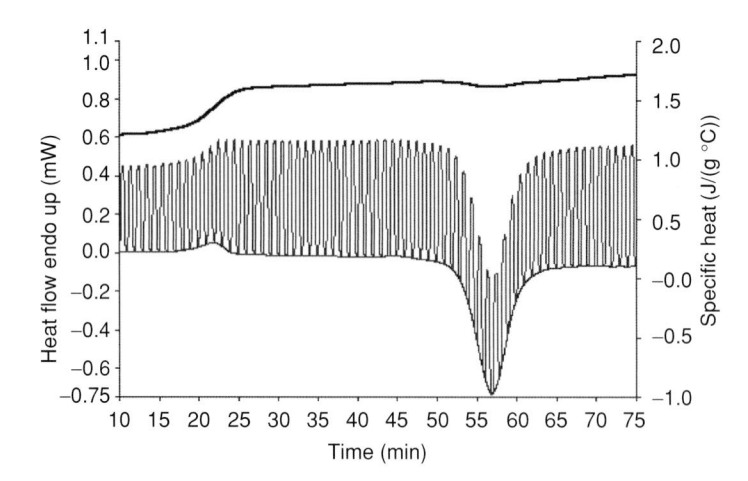

Figure 1.24 Step-scan trace of PET shown as a function of time. The step increase in C_p identifies the glass transition, whilst a slight reduction also occurs as the sample recrystallises. The IsoK curve matches the lower envelope of the heat flow data. Quantitative data of these types require a baseline subtraction in accord with the requirements of the C_p method. Qualitative data showing the events can be obtained if this is not performed.

In this case, the slope of the initial step is important as this contains the high-frequency information. Any phase differences observed between stimulus and response and not caused by the DSC instrument itself will invalidate the simple reversing/non-reversing methods, but are coped with in linear-response theory.

1.7 DSC design

1.7.1 Power compensation DSC

Power compensation DSC has at its heart two small identical furnaces, one for the sample and one for the reference (normally an empty pan), the reference being the right-hand furnace; Figure 1.25. These are both heated at a pre-programmed heating (or cooling) rate and power compensations are applied to either furnace as required to maintain this rate. In the resulting DSC trace the difference in energy flowing into the sample furnace is compared to the inert reference and plotted as a function of temperature or time. This design measures flow of energy directly in mW or J/s.

The fundamental equation of DSC is

$$\text{DSC signal (W/g)} = \text{Heat Capacity (J/(K g))} \times \text{Scanning Rate (K/s)}$$
$$dH/dt = dH/dT \times dT/dt$$

Therefore the raw heat flow signal can be viewed as a form of heat capacity. In practice, it reflects the changes occurring in heat capacity, and the absolute value is obtained when the

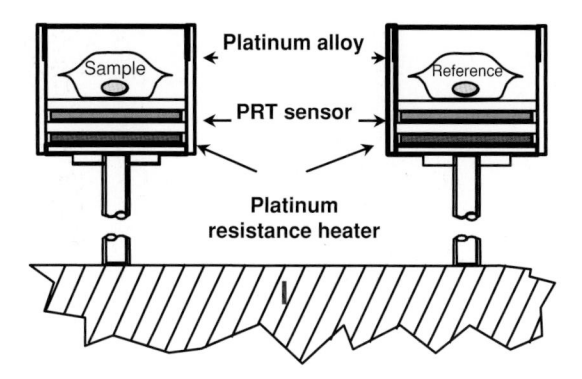

Figure 1.25 Diagram of a power compensation DSC. In this system, both the sample and reference furnaces are heated at a programmed heating/cooling rate. In order to maintain this rate when transitions occur in the sample, a power compensation circuit increases or reduces power to either furnace as required in order to maintain the heating rate. The power compensation circuit therefore reflects the energy changes occurring in the sample and is presented on the screen as a function of temperature or time. This technique measures energy changes directly.

method used takes into account the contribution of the empty pans and reference together with the scan rate.

The small furnaces of this system can be heated or cooled at very low rates to very high rates and are ideal for a range of different techniques, particularly fast scan DSC. Power-compensated DSC also permits true isothermal operation, since under constant temperature conditions both the sample and furnace are held isothermally. The temperature range of use is from liquid nitrogen temperatures to around 730°C.

1.7.2 Heat flux DSC

Heat flux DSC is of a single furnace design with a temperature sensor (or multiple sensors) for each of the sample and reference pans located within the same furnace; see Figure 1.26. Sample and reference pans are placed in their required positions and the furnace heated at the pre-programmed heating (or cooling) rate. When transitions in the sample are encountered a temperature difference is created between sample and reference. On continued heating beyond the transition this difference in temperature decreases as the system reaches equilibrium in accordance with the time constant τ (see Glossary) of the system. It is the difference in temperature or Δt signal that is the basic parameter measured. Modern analysers are carefully calibrated so that the Δt signal is converted to a heat flow equivalent and this is displayed as a function of temperature or time. The reason a difference in temperature is created is easily understood if melting is considered. When melting of a single crystal occurs the resulting mixture of solid and liquid remains at the melting point until melting is complete, so the temperature of the sample will fall behind that of the reference. Typical heat flux DSC analysers can be used from liquid nitrogen temperatures to a maximum of around 700°C similar to power compensation DSC, though modern

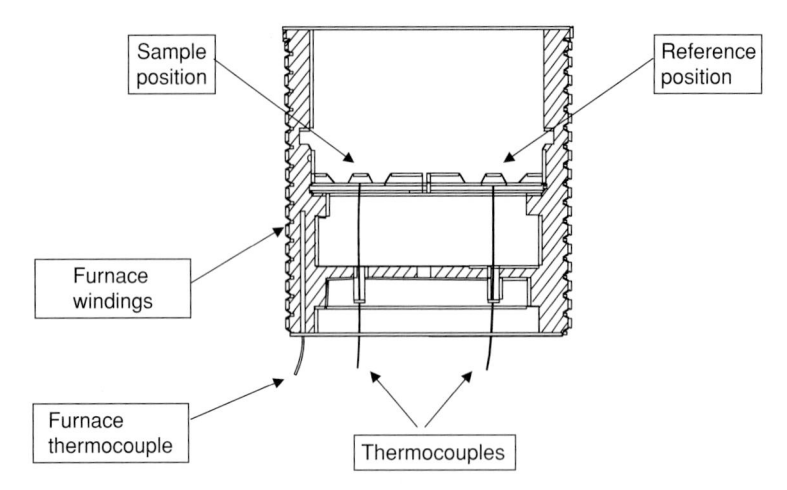

Figure 1.26 Diagram of a heat flux DSC. In this system, both the sample and reference experience the same heat flux, but as energy demands differ, the heating or cooling effect will differ resulting in a difference in temperature between sample and reference. This difference in temperature is converted to an energy equivalent by the analyser giving the familiar DSC signal in mW. (This figure was produced by Anantech in Holland.)

high-temperature DTA analysers normally offer a calibrated DTA (heat flow) signal giving a measurement derived from the heat flux to significantly higher temperatures.

1.7.3 Differential thermal analysis DTA

This design principle is similar to heat flux DSC, except that the Δt signal remains as a microvolt signal and is not converted to a heat flow equivalent. This was the original instrument approach used before quantitative energy measurements were established using DSC. Often instruments capable of heating to around 1500°C or higher use this principle and are referred to as DTA analysers. The furnace design is usually quite different to that of lower temperature systems, though modern equipment may offer a choice of heat flow or microvolt signals.

1.7.4 Differential photocalorimetry DPC

Reactions not only occur as a function of temperature but may also be initiated by irradiation, specifically ultraviolet (UV) light for materials that are photosensitive. UV systems have therefore been attached to DSC analysers to provide DPC systems. This can be done fairly crudely by shining a light on a material using fibre optics, but formal accessories are available from a number of manufacturers. Isothermal control of these systems is important, and the light source should not adversely affect the calorimetry. Applications are found in curable materials in the composites and manufacturing industries, dentistry and dental materials together with films, coatings and printing inks. The wavelength chosen and the intensity of the light are significant factors, which together with the temperature of reaction and length of exposure can be used to define a method. The effects of temperature, light intensity, and wavelength can be investigated on different materials and additives, and the kinetics of reaction investigated using isothermal kinetics models.

The effect of infrared light can also be investigated in a similar manner [20]. These authors also highlight the ability to measure a simultaneous non-contact TMA signal from a sample placed in the DSC cell.

1.7.5 High-pressure cells

Many manufacturers provide pressure cells as an accessory to the DSC. These normally work to fairly modest pressures of around 50–100 bar and are designed with a view to the suppression of volatiles in OIT tests involving oils and greases. A high-pressure cell capable of working to much higher pressures has been constructed by Höhne and co-workers and can be operated at pressures from 0.1 to 500 MPa over a temperature range from 20 to 300°C [21] and is potentially capable of being added to some commercial analysers.

Appendix: standard DSC methods

ISO Standards
11357-1:1997   Plastics – Differential Scanning Calorimetry (DSC)

Part 1: General Principles
11357-2:1999   Plastics – Differential Scanning Calorimetry (DSC)

Part 2: Determination of Glass Transition Temperature
11357-3:1999   Plastics – Differential Scanning Calorimetry (DSC)

Part 3: Determination of Temperature and Enthalpy of Melting and Crystallisation
11357-4:2005   Plastics – Differential Scanning Calorimetry (DSC)

Part 4: Determination of Specific Heat Capacity
11357-5:1999   Plastics – Differential Scanning Calorimetry (DSC)

Part 5: Determination of Characteristic Reaction Curve Temperatures and Times Enthalpy of Reaction and Degree of Conversion
11357-6:2002   Plastics – Differential Scanning Calorimetry (DSC)

Part 6: Determination of Oxidative Induction Time
11357-7:2002   Plastics – Differential Scanning Calorimetry (DSC)

Part7: Determination of Crystallisation kinetics

References

1. Pjpers TFJ, Mathot VBF, Goderis B, *et al.* High-speed calorimetry for the study of the kinetics of (de)vitrification, crystallization, and melting of macromolecules. *Macromolecules* 2002;35:3601–3613.

2. Gray AP. *Thermochim Acta* 1970;1:563.
3. Giuseppe DG, Della Gatta G, Richardson MJ, Sarge SM, Stølen S. Standards, calibration, and guidelines in microcalorimetry: Part 2. Standards for DSC. *Pure Appl Chem* 2006;78(7):1455–1476.
4. Lab of the Government Chemist (LGC), Queens Road, Teddington, Middlesex. TW11 0LY (www.lgc.co.uk).
5. Ehrenfest P. Phase changes in the ordinary and extended sense classified according to the corresponding singularities of the thermodynamic potential. *Proc Acad Sci Amsterdam* 1933;36:153–157.
6. Donovan JW. Phase transitions of the starch-water system. *Biopolymers* 1979;18(2):263–275.
7. Bair HE. Glass transition measurements by DSC. In: Seyler RJ (ed.), *Assignment of the Glass Transition, ASTM STP 1249*. Philadelphia: American Society for Testing and Materials, 1994; 50–74.
8. Wunderlich B. Reversible crystallization and the rigid-amorphous phase in semicrystalline macromolecules. *Prog Polym Sci* 2003;28(3):383–450.
9. Moynihan CT, Macedo PB, Montrose CJ, *et al*. Thermodynamic and transport properties of liquids near the glass transition temperature. Structural relaxation in vitreous materials. *Ann New York Acad Sci* 1976;279:15–35.
10. Gordon M, Taylor JS. Ideal copolymers and the second-order transitions of synthetic rubbers. I. Noncrystalline copolymers. *J Appl Chem* 1952;2:493–500.
11. Couchman PR. Compositional variation of glass transition temperatures. 2. Application of the thermodynamic theory to compatible polymer blends. *Macromolecules* 1978;11(6):1156–1161.
12. Ellis TS, Karasz FE, ten Brinke G. The influence of thermal properties on the glass transition temperature in styrene/divinylbenzene network-diluent systems. *J Appl Polym Sci* 1983;28(1):23–32.
13. Weyer S, Merzlyakov M, Schick C. Application of an extended Tool–Narayanaswamy–Moynihan model. Part 1. Description of vitrification and complex heat capacity measured by temperature-modulated DSC. *Thermochim Acta* 2001;377(1–2):85–96.
14. Reading M, Luget A, Wilson R. Modulated differential scanning calorimetry. *Thermochim Acta* 1994;238(1–2):295–307.
15. Reading M, Hourston DJ. *Modulated Temperature Differential Scanning Calorimetry: Theoretical and Practical Applications in Polymer Characterisation*. Berlin: Springer, 2006.
16. Schawe JEK, Hoehne GWH. The analysis of temperature modulated DSC measurements by means of the linear response theory. *Thermochim Acta* 1996;287(2):213–223.
17. Schawe JEK. Modulated temperature DSC measurements: the influence of the experimental conditions. *Thermochim Acta* 1996;271:127–140.
18. Cassel B. A stepwise specific heat technique for dynamic DSC. *Am Lab* 2000;32(1):23–26.
19. Merzlyakov M, Schick C. Step response analysis in DSC – a fast way to generate heat capacity spectra. *Thermochim Acta* 2001;380(1):5–12.
20. Degamber B, Winter D, Tetlow J, Teagle M, Fernando GF. 'Simultaneous thermal (DSC) spectral (FTIR) and physical (TMA)'. *J Meas Sci Technol* 2004;15(9):L5–L10.
21. Ledru J, Imrie CT, Hutchinson JM, Höhne GWH. High pressure differential scanning calorimetry: aspects of calibration. *Thermochim Acta* 2006;446(1–2):66–72.

Chapter 2
Fast Scanning DSC

Paul Gabbott

Contents

2.1	Introduction	52
2.2	Proof of performance	52
	2.2.1 Effect of high scan rates on standards	52
	2.2.2 Definition of HyperDSC™	54
	2.2.3 The initial transient	54
	2.2.4 Fast cooling rates	54
2.3	Benefits of fast scanning rates	57
	2.3.1 Sensitivity	57
	2.3.2 Measurement of sample properties without unwanted annealing effects	57
	2.3.3 Separate overlapping events based on different kinetics	59
	2.3.4 Speed of analysis	59
2.4	Application to polymers	61
	2.4.1 Melting and crystallisation processes	61
	2.4.2 Comparative studies	64
	2.4.3 Forensic studies	65
	2.4.4 Effect of heating rate on the sensitivity of the glass transition	67
	2.4.5 Effect of heating rate on the temperature of the glass transition	68
	2.4.6 Effect of heating rate on T_g of annealed materials (and enthalpic relaxation phenomena)	72
2.5	Application to pharmaceuticals	76
	2.5.1 Purity of polymorphic form	76
	2.5.2 Identifying polymorphs	78
	2.5.3 Determination of amorphous content of materials	79
	2.5.4 Measurements of solubility	81
2.6	Application to water-based solutions and the effect of moisture	82
	2.6.1 Measurement of T_g in frozen solutions and suspensions	82
	2.6.2 Material affected by moisture	83
2.7	Practical aspects of scanning at fast rates	83
	2.7.1 Purge gas	83
	2.7.2 Sample pans	84
	2.7.3 Sample size	85

2.7.4	Scan rate	85
2.7.5	Instrumental settings	85
2.7.6	Cleanliness	85
2.7.7	Getting started	86
References		86

2.1 Introduction

One of the most surprising recent developments in DSC is the discovery that it is possible to scan at hundreds of degrees centigrade per minute and still obtain excellent data. For some applications, this leads to much more accurate and meaningful data than information obtained at much slower traditional heating rates [1].

Since the introduction of DSC in the early 1960s the scan rate used for most DSC measurements has been 10°C/min. This rate has provided good data for many applications, though sometimes slower rates have been employed to give improved resolution of events, e.g. polymorphism, or time for a reaction to occur. Occasionally, faster rates of 20°C/min may have been used, but seldom anything faster than this; the fear being that thermal lags may cause inaccuracy of temperature measurement or that thermal gradients across a sample would make the data meaningless. However, this has proved not to be the case.

Pijpers and Mathot [2] first reported the use of fast scan rates whilst based at the DSM laboratories in Holland and this has given rise to a range of applications using scan rates of up to 500°C/min (at the time of writing the maximum available for commercially available analysers). Particularly significant are applications in the fields of polymers and pharmaceuticals, which will be described in this chapter. This approach was called high-performance DSC (HPer DSC) and was later commercialised by PerkinElmer and trademarked with HyperDSC™ to describe the technique as used with power compensation DSC, which is well suited to making these measurements. The data and applications shown in this chapter have been developed using power compensation DSC.

2.2 Proof of performance

2.2.1 Effect of high scan rates on standards

Some DSC analysers, particularly power compensation systems, have been capable of heating at high rates for many years; the DSC7 introduced by PerkinElmer in 1985 could be heated at 500°C/min, though these rates were normally employed when data were not being collected, since no one expected valid data to be collected at these rates. Part of the reason how such high rates are achievable is the small mass of the furnaces in a power compensation DSC. They are just less than a gram in weight, and being of very low mass can be readily heated at high scan rates. But the first issue that springs to mind is whether the sample itself can also be heated at high rates or whether thermal lags will be increased so as to make the data meaningless. For example, how much is the onset of melting of a standard such as indium affected by increasing scan rate, and what happens to the peak width and the heat of fusion

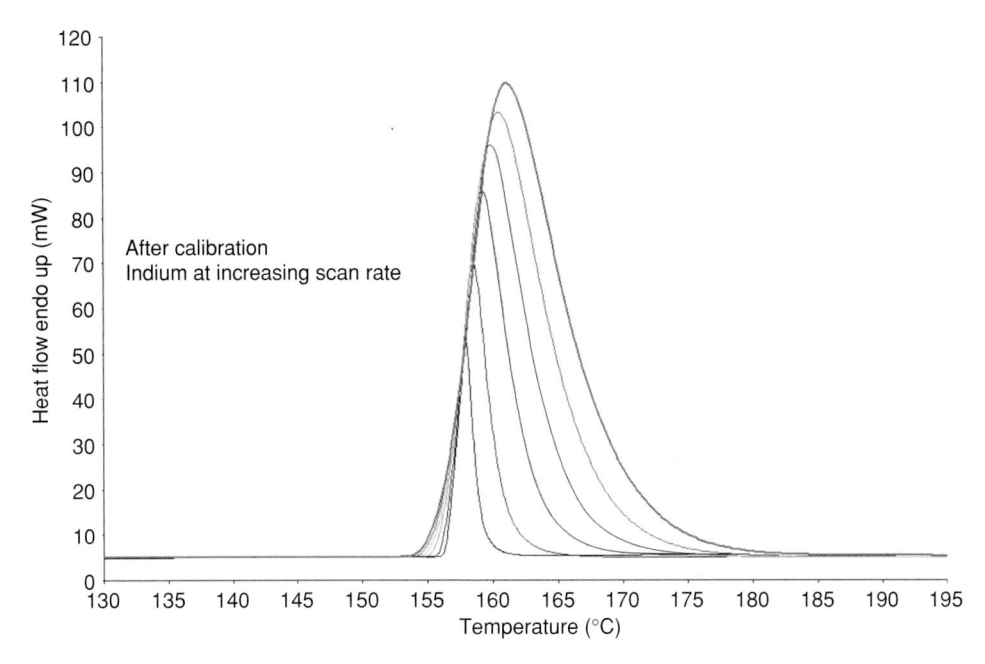

Figure 2.1 Indium after calibration, heated at rates of up to 500°C/min.

measurement? These questions can be readily checked by running an indium standard at increasing rate. To the very great surprise of those of us who have used DSC for many years, the fact is that the effects on indium and other standards are relatively small; for example, if a system is heated at 500°C/min the onset of melting is shifted by only 6–8° when compared to the value obtained when heating at the slow rate of 10°C/min, which is easily corrected for by normal calibration procedures. Following calibration, results are accurate both for temperature and energy; see Figure 2.1 and Table 2.1.

This indicates that measurement is not only practical but also quantitative, and there is no mathematical treatment of the data involved to enhance the appearance at high rates; it

Table 2.1 Onset and heat of fusion data from indium scanned at rates of up to 500°C/min after calibration showing that measurement of both temperature and energy is quantitative

Heating rate (°C/min)	Onset temperature (°C)	Heat flow (J/g)
20	156.60	28.81
100	156.74	28.35
200	156.67	28.43
300	156.74	28.58
400	156.69	28.32
500	156.75	28.57

is literally the direct heat flow measurement of the analyser. Power compensation circuitry responds at twice mains frequency, typically 100–120 Hz, and so the analyser can not only heat at high rates, but is also fully capable of making sensible and accurate measurements whilst heating at these rates so as to correctly characterise the events that occur. This leads to a definition of HyperDSC™ which refers to the ability to make meaningful measurements whilst heating at high rates.

2.2.2 Definition of HyperDSC™

It means making meaningful heat flow measurements when heating or cooling a sample with fast linear controlled scan rates, typically up to 500°C/min.

2.2.3 The initial transient

There are other practical issues to be aware of when using fast scan rates, and the duration of the initial transient is one of them. This is the time it takes for the analyser to control the scan at the required rate and shows as a period of instability at the start of a trace before flat baseline appears. The duration of the transient will depend to some extent on the size of sample, the type of pan and instrument set-up, and significantly upon the purge gas chosen. It must be fairly short if practical measurements are to be made since, for example, if the transient was of 1 min and the scan rate 500°C/min then the run would need to begin at least 500°C below the transition to be measured, which would make the technique of little value. With power compensation DSC the transient is typically of 6- to 8-s duration when using helium as the purge gas, independent of the scan rate (see Figure 2.2) which is an acceptably short period for sensible measurements to be made. Figure 2.3 shows the T_g of polyisobutylene at −70°C measured at 400°C/min. This would be impossible to measure if the instrument had a long transient, but is easily measured when using a DSC equipped with a liquid nitrogen system.

2.2.4 Fast cooling rates

Fast controlled cooling rates are also of great significance for DSC measurements, particularly with respect to crystallinity or glassy morphology within a material. Figure 2.4 shows an example of cooling rates that can be achieved using a power compensation DSC. The effect of different cooling rates on morphology is easily demonstrated with PET, which shows increasing amounts of amorphous material if cooled at increasingly fast rates from the melt; see the comparison in Figure 2.5. If cooled quickly PET remains completely amorphous, and when reheated slowly it shows a large glass transition from the amorphous material, followed by crystallisation which will then occur with slow heating rates. This is termed cold crystallisation when it occurs in this way. If cooled slowly from the melt the PET will be partially crystalline to start with, so will not show such a significant T_g nor a cold crystallisation event. Fast heating of these materials is reviewed in Section 2.4.1.

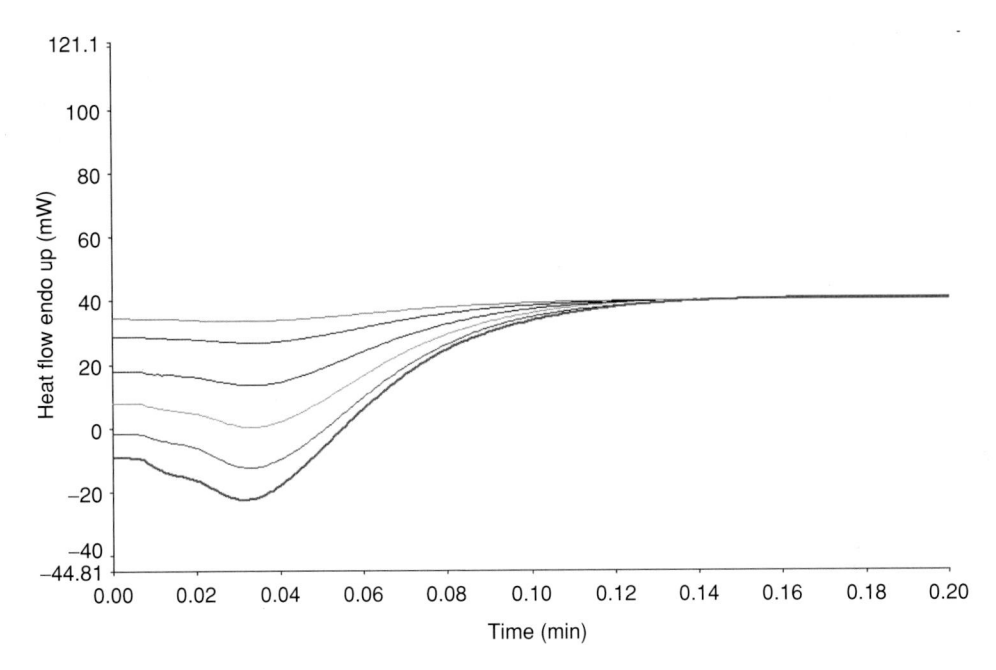

Figure 2.2 Transient duration is less than 8 s at rates of up to 500°C/min. At 500°C/min this means that the transient takes only about 60°C, which allows meaningful sub-ambient measurement.

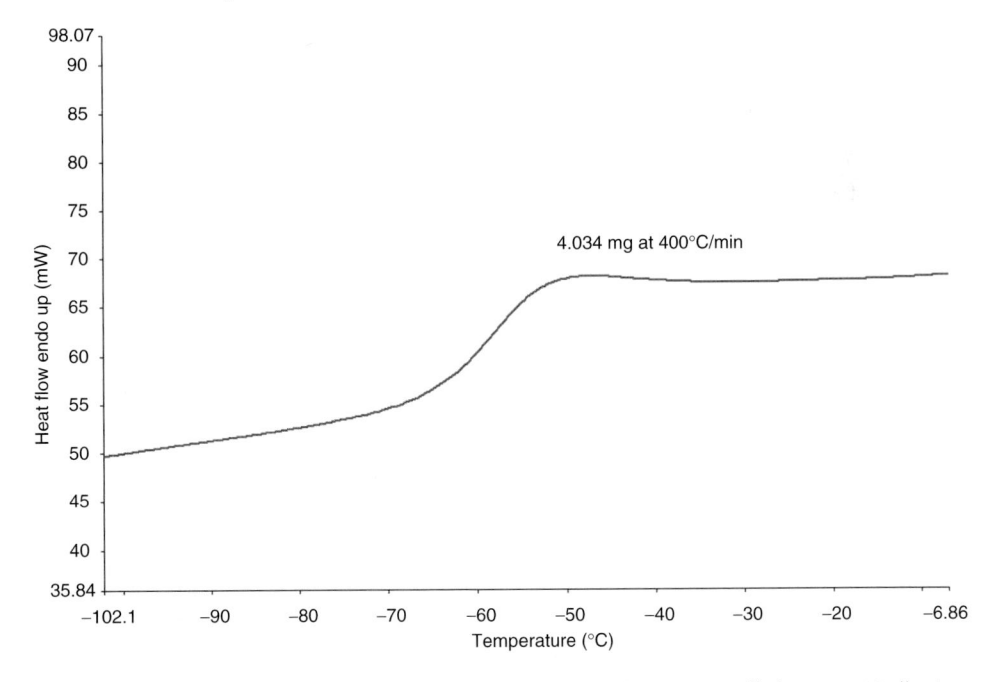

Figure 2.3 T_g of polyisobutylene scanned at 400°C/min. The analyser is controlled at −100°C allowing this T_g to be observed beginning around −70°C. Short initial transient times are essential for this type of measurement.

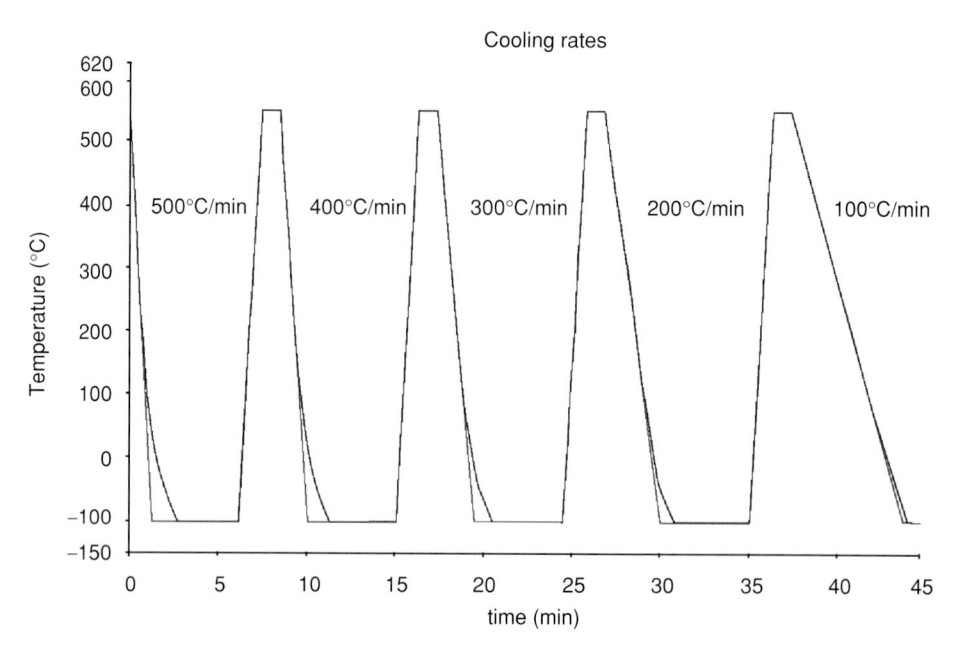

Figure 2.4 Cooling rates achieved with power compensation DSC equipped with liquid nitrogen cooling. The figure shows program temperature with the actual scan rate, which shows deviation at the lower temperatures. Rates of 200°C/min are achievable through the ambient region.

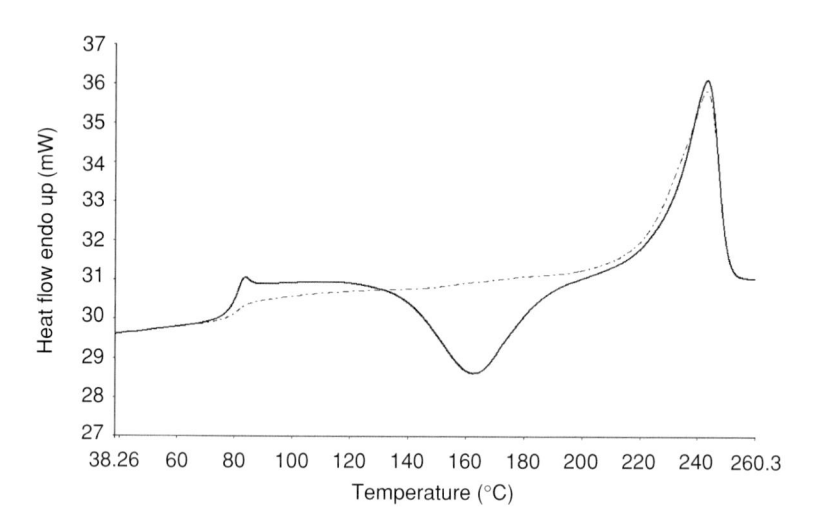

Figure 2.5 PET heated slowly after rapid cooling (solid line) and heated slowly after slow cooling (dashed line). After rapid cooling from the melt the PET remains fully amorphous, so shows a large T_g followed by cold crystallisation on heating. If cooled slowly the PET will partially crystallise, so showing a smaller T_g on heating and no further crystallisation.

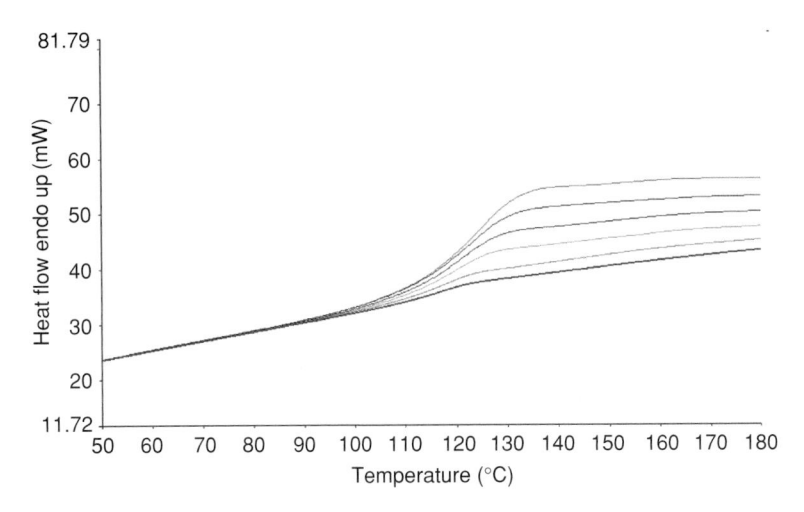

Figure 2.6 Increasing scan rate increases the size of the glass transition of PMMA. The increased sensitivity makes it possible to measure transitions that are difficult to see at low scan rates.

2.3 Benefits of fast scanning rates

2.3.1 Sensitivity

One of the most significant benefits from fast scanning is the resultant increase in sensitivity. The faster the scan rate, the bigger the size of the peak or step that is measured, which is evident from the melting of indium as shown in Figure 2.1 above and also the increase in the size of the T_g measurement in Figure 2.6.

The increased sensitivity has many benefits, allowing small sample masses to be used, small transitions to be easily observed, and potentially increased accuracy for measurement of specific heat of materials. The reason for the increased sensitivity is that energy flows more quickly at the higher scan rates. The amount of energy involved remains the same but the time during which it flows is reduced as the scan rate is increased, so the y-axis response of the DSC records the energy flow increases with scan rate. See Section 1.4.3 of Chapter 1 for a full explanation.

2.3.2 Measurement of sample properties without unwanted annealing effects

Sometimes there is a requirement to observe the changes that occur in a sample when it is heated slowly, for example when measuring cure or recrystallisation processes. On other occasions, these changes are undesirable and prevent measurement of sample properties at higher temperatures. Yet at the slow scan rates normally employed changes can easily occur,

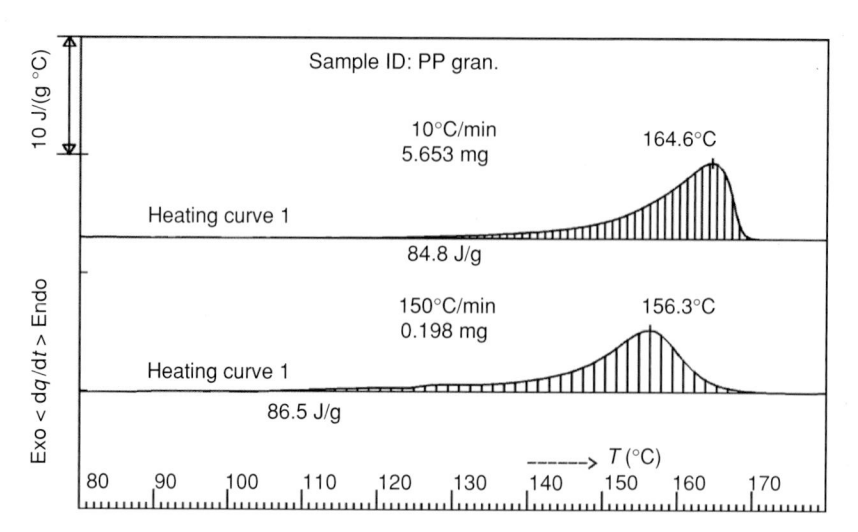

Figure 2.7 In the example above, polypropylene heated at 150°C/min melts at a lower temperature and with a broader profile than a sample heated at 10°C/min. Heating faster prevents any annealing processes which will displace the melting profile to higher temperatures. (Courtesy of Pijpers and Mathot.)

and with many polymer systems changes may occur without the analyst realising it. This can be seen very clearly with the melting of thermoplastic polymers such as polypropylene. The crystalline structure of polypropylene depends upon how it is cooled and the time given for crystals to develop. Upon fast cooling (typical of manufacturing processes) there is not enough time for large crystallites to form and a range of small less stable crystals will result. If then heated slowly, not only do the small less perfect crystals melt at lower temperatures, but larger more stable crystals develop from the molten material; the well-known process of annealing. However, this competing energetic process may lead to little if any net energy flow; in other words little if anything is observed in a DSC trace. The final melting profile is then that of the annealed material, which may be significantly different from that of the original un-annealed material placed in the pan. In this case, we have not measured the properties of the material put into the analyser, but those of the sample after it has changed during the slow heating rate. The actual material properties may be obtained by heating more quickly, though the scan rate required will vary from sample to sample depending upon the rate needed to prevent the sample changing before it is measured. An example is shown from the melt of polypropylene in Figure 2.7. The melting point and melting profile obtained at slow rates differ significantly from that obtained at 150°C/min since the material is annealed during the slow scan.

Many materials undergo changes in their crystalline structure when heated, including polymers, pharmaceuticals and foods. In some materials, particularly pharmaceutical materials, different crystal structures may exist (polymorphism), and on heating some materials can undergo changes from one form to another. Different changes may occur at different heating rates, but at faster rates such changes are less likely to happen due to the lack of time for crystal changes to occur. Thus, at high rates the original material is more likely to be characterised, just as for polypropylene referred to above.

If a material has been slowly cooled from the melt then the likelihood of changes occurring on slow heating is reduced. However, many materials are fast cooled, particularly in industrial processes, so that changes in structure under slow heating are likely to occur with many materials. This is particularly true of thermoplastic polymers where the crystal structure is likely to change during slow heating, the polymer having been annealed. Therefore, to prevent annealing and to measure the true properties of the material put into the DSC a fast heating rate is desirable, the actual scan rate being dependent on the kinetics of the processes involved.

2.3.3 Separate overlapping events based on different kinetics

Sometimes two events can occur over the same temperature range, often a large event obscuring a smaller event, or sometimes events are just poorly resolved. If the events are subject to different kinetics then altering the scan rate will allow separation on a temperature scale. This has been found to be effective in MTDSC where slow scan rates are employed but is also true where faster heating rates are used. One example is the loss of moisture from a damp material placed in a vented pan. In the case of a polyamide, Figure 2.8a, which normally contains some moisture, the glass transition of the amorphous material is completely obscured by the loss of the moisture. At faster heating rates not only is the T_g more obvious but the moisture loss, which is a slow process, is also displaced to much higher temperatures allowing the T_g to be clearly measured, Figure 2.8b.

This effect is also of advantage with materials which are prone to decomposition. When fast scan rates are used the decomposition process, which is normally slow to begin with, is displaced to higher temperatures allowing measurement of transitions that would previously be obscured by the decomposition process. This can permit clear melting profiles and measurement of heats of fusion materials from materials which would otherwise have decomposed.

If adopting this method, great care must be taken to prevent contamination of the analyser. Frequent checks and cleaning may be required, and for this reason small samples should be used together with high purge gas rates to remove volatiles.

2.3.4 Speed of analysis

Speed of analysis is not the main reason for using HyperDSC™, but there is always a benefit from increased speed of analysis, particularly when the improvement is very marked. In a situation where many samples need to be run for screening or other purposes the fast analysis times using HyperDSC™ are a significant benefit. Where 20–30 min are required at slow rates, 2–3 min (or less) may be all that is required at high rates. The curves presented in this chapter may look similar to those taken at slow rates yet many will have taken less than a minute to produce, compared to the tens of minutes for a typical scan, and this gives

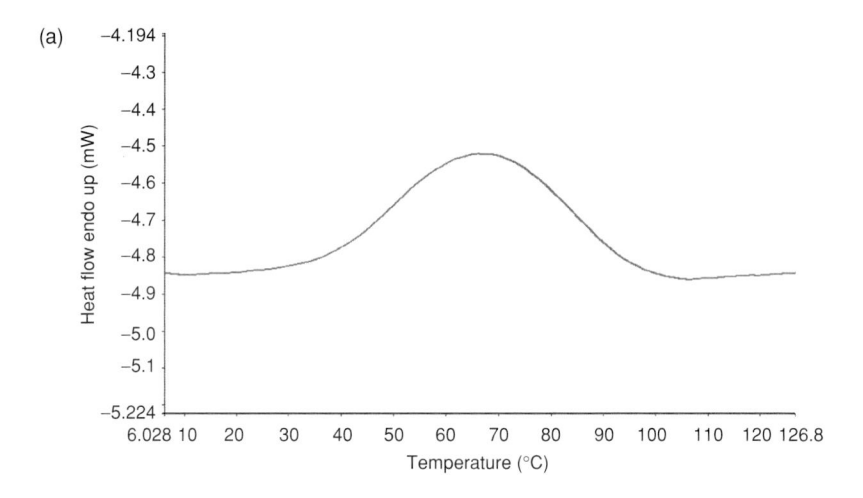

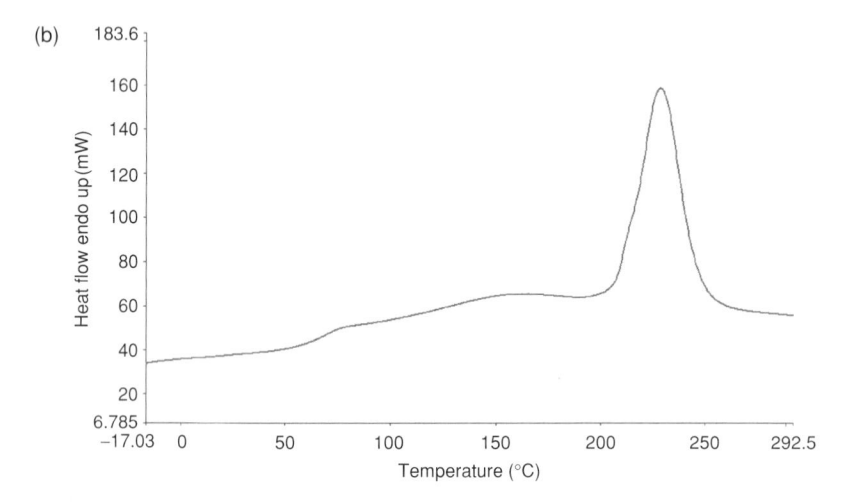

Figure 2.8 (a) Slow scan (10°C/min) of a polyamide sample. The T_g of this material is obscured by the moisture loss which shows as the broad endotherm. At faster rates the T_g can be observed; see part b. (b) Moisture loss from a polyamide is displaced to higher temperatures as the scan rate is increased to 500°C/min revealing the T_g of the polyamide, the step beginning around 50°C. Significantly, this is the true T_g of the material plasticised by the water content the loss of which is seen as the broad endotherm above T_g.

a real advantage to the technique. In industry, there is always a push to reduce times and improve throughput and this is achieved by fast scanning DSC. There may be questions concerning the effect of increased scan speed on the data measured and some of these concerns will be addressed in subsequent sections, particularly with regard to resolution, and onset temperatures of melting transitions and glass transitions. Note though that with regard to resolution one of the key aspects is the use of helium (or other high conductivity gas) as a sample purge, since it has a higher thermal conductivity and gives much faster heat transfer to the sample than a system running under nitrogen or air.

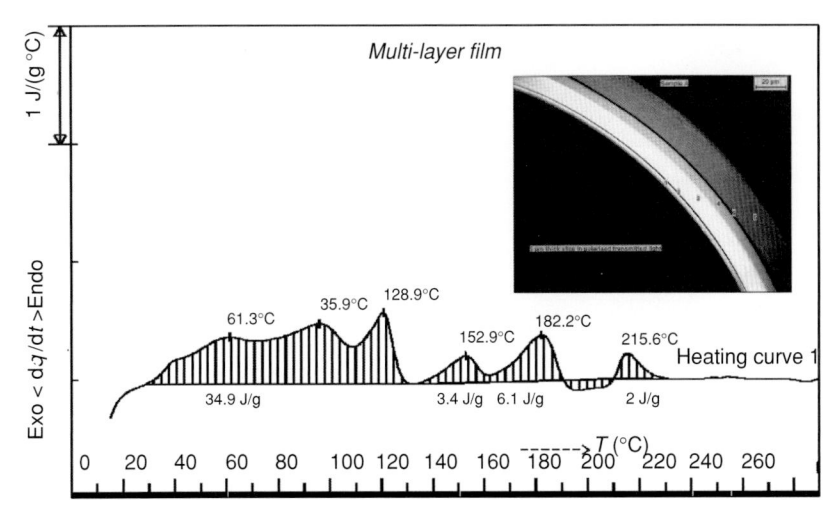

Figure 2.9 This shows the ability to distinguish a range of polymers of a multilayer film from their melting profile. In this case, the polymer is a thin film coated on the surface of a substrate from which it cannot easily be removed, heated at 150°C/min. At low rates sensitivity is insufficient to make these types of measurements. (Courtesy of Pijpers and Mathot.)

2.4 Application to polymers

2.4.1 Melting and crystallisation processes

Initial work with polymer melt performed by Pijpers and Mathot shows that the melting profile of polymers is not only measurable at fast rates, but also it is often different to that obtained on slow heating because of annealing processes [2]. Melting of any material usually gives a fairly large peak since a lot of energy is involved in the bond-breaking process, and there are concerns about how quickly this energy can be transmitted to the sample and the effect that this would have on the resolution obtained. Accordingly, low sample weights may be required if melting processes are to be observed at high rates. This may well be of significant advantage as in the case of films adhering to a substrate. Figure 2.9 shows a trace from a multilayer film with a large number of peaks all well resolved, indicating clearly the ability of HyperDSC™ both to heat fast and to retain good resolution.

Not only can the melting points of the constituent polymers be measured accurately, but the fact is that the increased sensitivity obtained at higher rates allows the melting process to be observed easily, whereas at low rates little would be seen.

Intuitively, there is an expectation that the faster a sample is heated the higher will be the observed melting point of the material due to an increase in thermal lags or thermal gradients across the sample during the melting process. However, this is not necessarily what happens. Annealing processes can allow growth of larger, more stable higher melting crystallites at the expense of less stable lower melting structures. This means that during slow heating the crystal structure of a material can change to give a higher melting point and narrower melting range than that of the material originally put into the DSC. This is

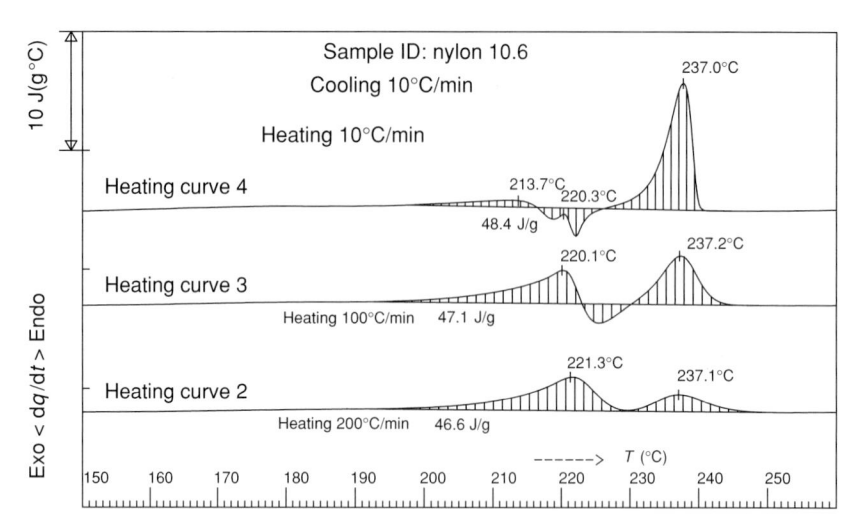

Figure 2.10 Nylon 10.6 heated at 10, 100, and 200°C/min. The faster the rate the less time the sample has to recrystallise and the trace is more representative of the original material put into the pan. (Courtesy of Pijpers and Mathot.)

the situation with the sample of polypropylene shown in Figure 2.7 where annealing during heating at 10°C/min has displaced the melting profile to higher temperatures compared to the scan at 150°C/min. The situation is even more evident if a polymer actually recrystallises on heating as is shown in Figure 2.10 where nylon 10.6 is analysed. In this case, the sample undergoes cold crystallisation during slow heating; a process that can be prevented by fast heating. Therefore, the lower trace in this figure more accurately represents the melting of the original material.

It is not possible to suggest any particular scan rate as being the best to use. The rate needed to prevent crystallisation from occurring and the true structure of the sample to be measured may vary from sample to sample. A rate should be sought where the profile does not change as a function of scan rate as this indicates that a sufficiently fast heating rate has been used. In the example in Figure 2.10 a rate of 200°C/min is fast enough to prevent an obvious exotherm from being observed, but that does not mean that all changes have been prevented, and it may be of value to examine faster scans to get a complete picture.

The rate of cooling is also important for examination of crystallisation processes and determination of material properties. The slower a material is cooled from the melt the more time for crystallisation to occur and the greater the extent of crystallisation that will result. Thus, an examination of increasing cooling rate and the need for fast cooling rates is important to both characterise and condition a material. For most thermoplastics it is important to heat, cool and reheat. So what cooling rate should be chosen? The answer is that a range of cooling rates may be useful as in the characterisation of PET shown in Figure 2.11. To obtain these data, the heating rate chosen (300°C/min) was fast enough to prevent changes occurring during heating. The differences in the melting profiles of PET are due to the different cooling rates employed which permit different levels of crystallinity to develop. Some of these profiles show distinctly binodal characteristics (see the curve

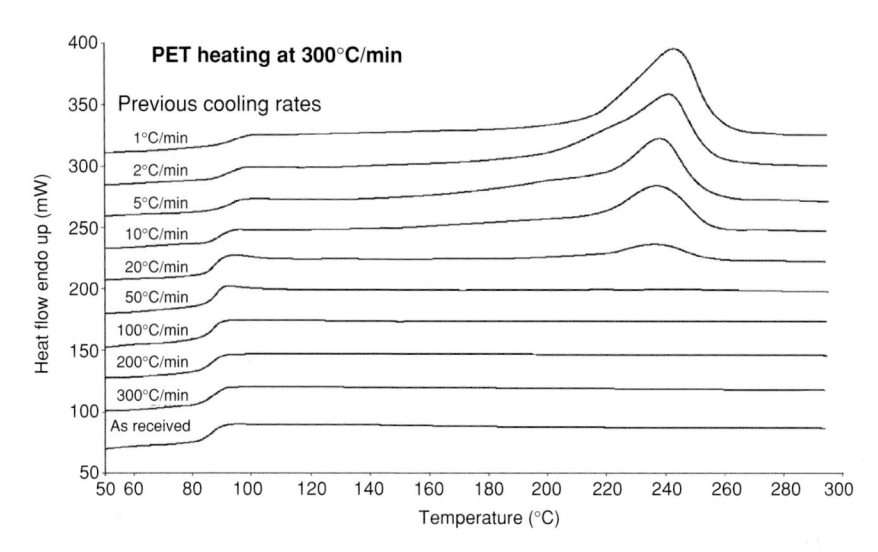

Figure 2.11 PET heated at 300°C/min after cooling at a range of different rates indicated. Increasingly slow cooling allows increased crystallinity, which can be measured with fast heating rates. Slow heating would allow further crystallisation so that all the traces would become similar.

following a 5°C/min cool) something not seen if slow heating rates are chosen because annealing alters the crystalline structure and the binodal characteristic disappears. In this way the material has been fully and accurately characterised. In Figure 2.12 the height of the glass transition and the heat of fusion of the melt are plotted as a function of cooling rate and this gives a view of the extent of crystallinity of the material. These measurements cannot be made at low rates since the material will recrystallise as shown in Figure 2.5. Though PET could be considered to be a well-known and well-characterised material, fast heating rates can further accurate information [3].

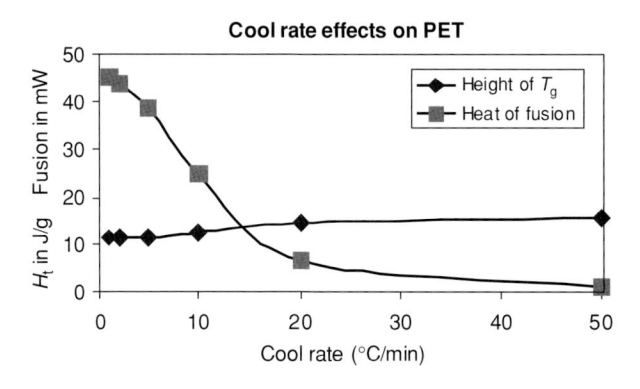

Figure 2.12 The height of the T_g and heat of fusion measured on heating PET at 300°C/min after cooling at different rates. The relative height of the T_g indicates that the sample of PET is still 60% amorphous even after cooling at 1°C/min.

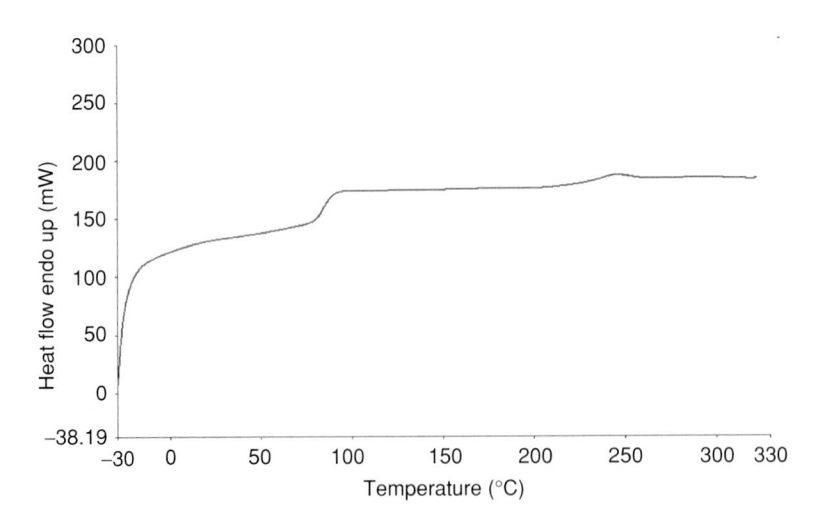

Figure 2.13 A small amount of crystallinity is measured in PET; see the small endotherm towards the end of the trace. At slow rates cold crystallisation will prevent such measurements.

Not only are the results of a fast or a slow heat of value but so are the trends that are observed as the material experiences different scan conditions. The cooling trace observed may also be of significant value; sometimes blends give greater resolution between crystallisation events during cooling than is found during corresponding melting events.

It should also be evident that measurements of the extent of crystallinity could be affected by slow scanning and the occurrence of annealing. Measurements of the extent of crystallinity are normally made from the heat of fusion of a material, and these may be better made at high rates before the crystalline structure changes. This may also apply to enthalpic methods which seek to give a value for the temperature dependence of crystallinity, by comparing the heat capacity of a material with known standards at a given temperature. These methods require that the analyser be fully quantitative for energy measurement at high rates, which modern power compensation systems have proven to be. Figure 2.13 shows an example of the measurement of crystallinity in PET. A small amount of crystalline structures can be observed even though the sample is expected to be fully amorphous after quenching in liquid nitrogen. To quench-cool the sample, a pan with molten PET was taken from a DSC furnace and dropped into liquid nitrogen, but this process was clearly not fast enough as the existence of a crystalline melt measured by the fast scan shows.

2.4.2 Comparative studies

One of the most significant uses of thermal equipment is in the comparison of samples, for example when comparing the effect of additives and fillers on a material, for troubleshooting when comparing good and bad samples, or when comparing a competitor's materials with

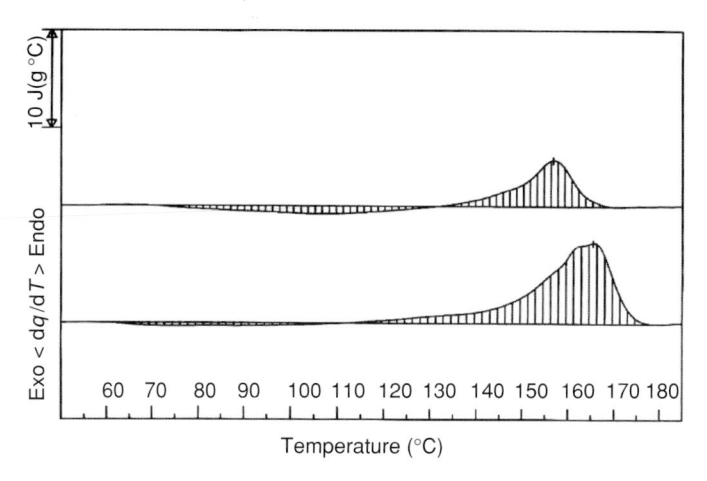

Figure 2.14 Biaxially oriented PP film compared with uniaxially oriented film both heated at 150°C/min. These show clear differences when scanned at high rates but are very similar when scanned at low rates. (Courtesy of Pijpers and Mathot.)

your own. Sometimes small annealing effects result from the way in which a material is used and the temperatures it is exposed to; for example, see Section 5.2.2. A lot of the resulting crystalline structures will be transitory and unstable in nature and the further annealing processes which take place under slow heating will cause them to disappear. Thus a lot of useful and important information can be lost on slow heating, as annealing processes allow crystalline structure to merge into a common more stable form.

Pijpers and Mathot compared the melting profile of biaxially oriented PP film with that of regular film and found them to be quite different, whereas at low rates the profiles are very similar; see Figure 2.14.

The same is true for many thermoplastic materials. A further example with nylon 66 is shown in Figure 2.15. A striking amount of detail is lost on slow heating. It is therefore important to use high rates when comparing different samples.

2.4.3 Forensic studies

As indicated in Figure 2.8 it is possible to make clear measurements from samples that ordinarily are too small to allow meaningful measurements. This has been the subject of a short study [4] where 50 μg of material has been found to give valuable information, Figure 2.16. In practice, the faster the scan rate, the smaller the sample that can be analysed. Melting transitions are normally large and so lower sample weights can be used than when measuring a small transition. However, even small transitions such as T_g can be identified from a few hundred micrograms when a fast scan rate is used; see Figure 2.17 where the T_g of PIB is identified from a 0.4-mg sample. It may seem academic to obtain transitions from

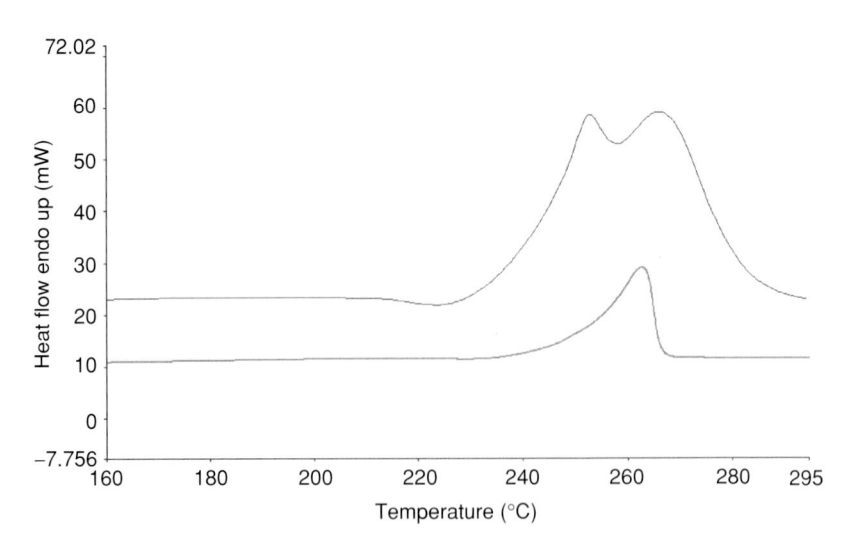

Figure 2.15 The upper trace shows 2.5-mg nylon 66 heated at 400°C/min and the lower trace 7.5 mg heated at 20°C/min. The faster scan shows more detail and in this case the binodal peak shape may be due to the α and β forms of nylon 66.

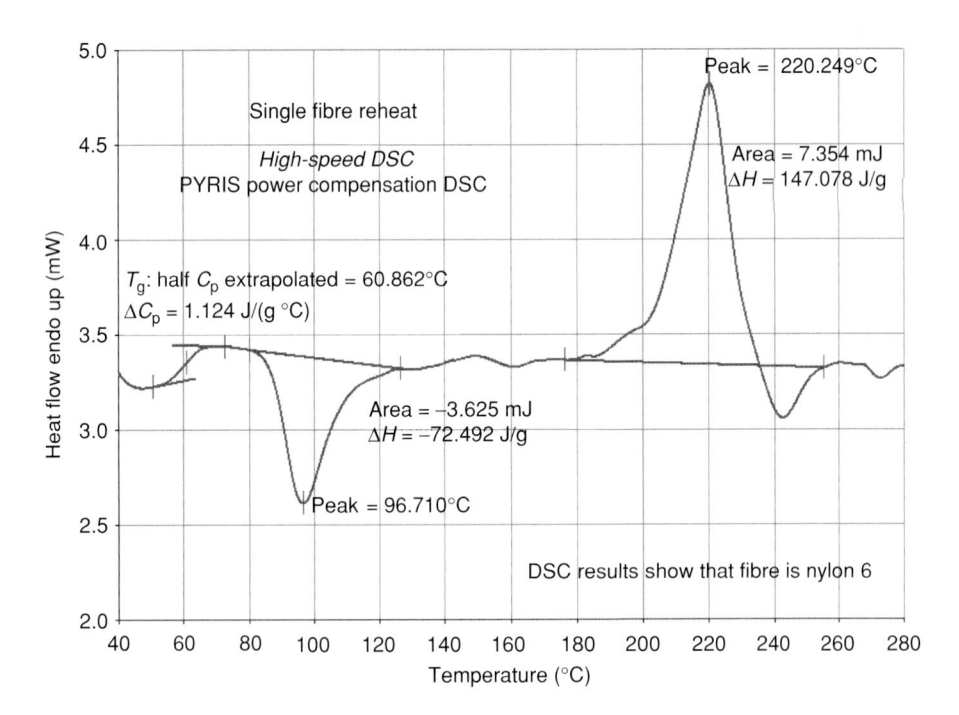

Figure 2.16 A single fibre of material weighing only 50 μg is identified from thermal transitions it undergoes.

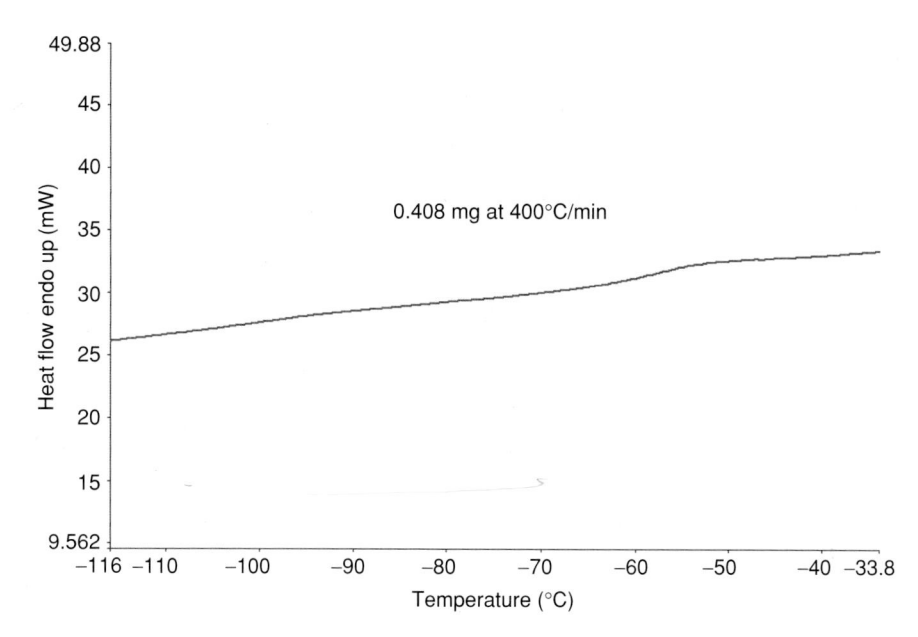

[Figure: Heat flow endo up (mW) vs Temperature (°C), showing a curve labeled "0.408 mg at 400°C/min"]

Figure 2.17 The T_g of this sample of PIB could not be found when using large weights at 10°C/min but is shown here with 400 μg scanning at 400°C/min.

such small weights, but in practice many samples are mixed and blended so that only small amounts of a given material may be present even within a large sample.

2.4.4 Effect of heating rate on the sensitivity of the glass transition

One of the main applications of HyperDSC™ has been in the measurement of the glass transition, where the increase in sensitivity has been found to be of very great value. Figure 2.6 shows how the T_g of PMMA is enhanced at higher rates, and this increase in sensitivity has meant vastly improved measurements in many materials. This is true for pharmaceuticals, foods and composites as well as thermoplastics; in other words any material where a T_g is to be measured. Composites are a particular area where scientists have often turned to DMA (see Chapter 4) in preference to DSC to make these measurements. This is because of the very small size of T_g in many composites. Modulated methods have afforded some improvement in sensitivity, but HyperDSC™ offers much greater sensitivity [5, 6]. Figure 2.18 shows an example from a composite material where the T_g could not previously be obtained by DSC. The measurements can be made rapidly, and give excellent reproducibility.

When analysing a glass transition there is always a question as to which of a number of parameters should be quoted as the temperature value of the T_g (see Section 1.5.4.) In practice, the onset value is very useful when high rates are used, since thermal gradients can extend the range of a transition if both high rates and large sample sizes are used. As always, the decision rests with the analyst who should quote the parameter chosen with the data.

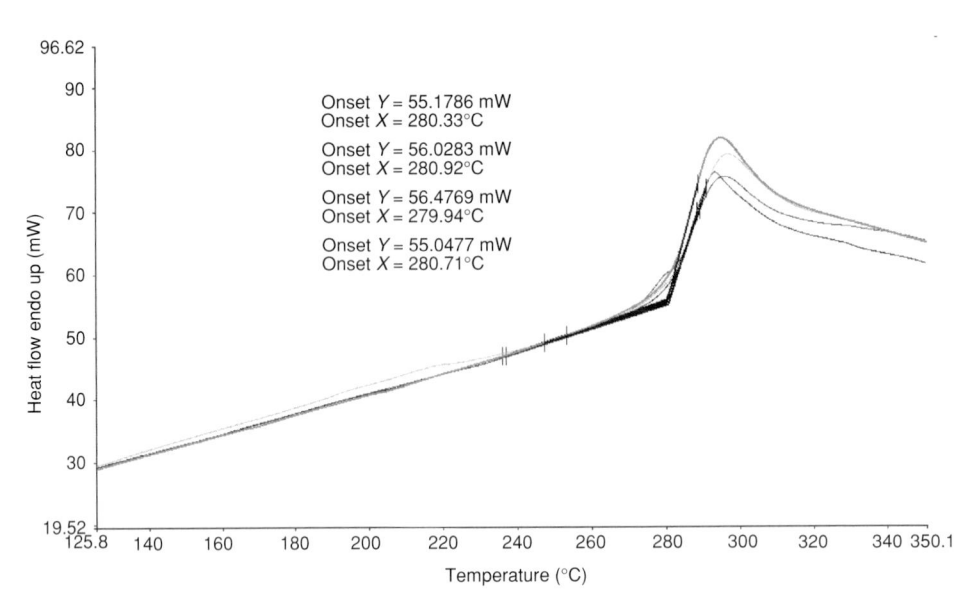

Figure 2.18 T_g of a composite material obtained at 400°C/min. Four samples are compared showing reproducibility of onset calculation to within 1°C.

2.4.5 Effect of heating rate on the temperature of the glass transition

Melt of a single crystal is a first-order transition which should be found to be independent of scan rate, provided equipment is suitably calibrated. Melting is a thermodynamically fixed event. The same cannot be said for the glass transition since T_g is found to vary with thermal conditioning; see Sections 1.5.6 and 10.2.1. In addition, the kinetics of the glass transition event are expected to be sensitive to the scan rate used resulting in a shift of T_g to higher temperatures as a function of increasing scan rate. An analogy is often drawn with the effect of frequency on the measurement of T_g using DMA where the temperature of the glass transition is increased as a function of increasing frequency; see Section 4.1.3. Yet despite these possible causes of variation in T_g, very little change is observed for many materials when traditional scan rates of 10–20°C/min are compared with fast scan rates of up to 500°C/min, provided thermal gradients are minimised or eliminated by the use of suitable sample sizes. This section considers the reason for these effects in more detail and presents evidence from a number of different materials that T_g changes little as a function of increasingly fast scan rate, provided any thermal lag is minimised. There is a potentially significant difference between the measurement of annealed materials where enthalpic relaxation phenomena can occur (see Section 1.5.6) and those which have been produced without annealing, typically by cooling rapidly through the T_g region, so these two categories of materials are considered separately. (The term annealing is applied to the thermal conditioning of materials generally and has been applied to thermal conditioning in the melt region in earlier sections of this chapter. In this context, it is applied to thermal conditioning in the glass transition region; see also Chapter 10 where the term is specifically applied to the thermal conditioning of glasses

PMMA effect of rate and weight

Onset temperature of PMMA at varying scan rates

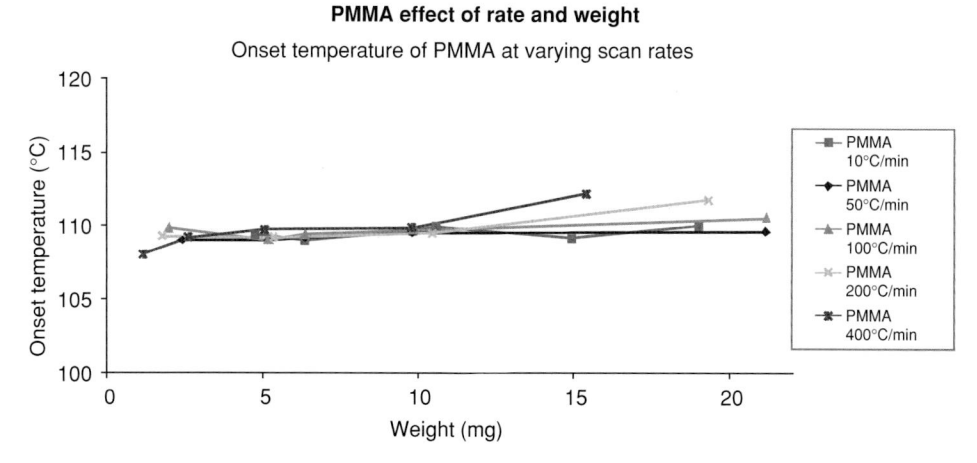

Figure 2.19 T_g of PMMA measured at increasing scan rate and sample weight. Measured onset temperatures remain within 1°C except for samples of high weight run at high rates of 200°C/min and above.

in the T_g region. Terminology such as ageing and relaxation is also used when polymers are conditioned in the T_g region but to be consistent the term annealing is used here.)

2.4.5.1 Un-annealed materials

In the author's experience, provided samples are not of excessive weight and are in good thermal contact with the DSC pan (to eliminate or minimise thermal gradients), then little shift in T_g has been observed with increasing scan rate between low scan rates of 10°C/min or 20°C/min and high scan rates of up to 500°C/min. This is a surprising observation, particularly in view of the frequency effect found when a T_g is measured by DMA. If very large sample sizes are used then some shift in the onset of T_g may be observed due to increased thermal lag, and some broadening of T_g would also be expected due to larger thermal gradients across the sample. If smaller sample sizes are used then the true influence of scan rate upon the kinetics of the T_g should be observed. Data obtained from a comparison is shown in Figures 2.19 and 2.20 where samples of PET and PMMA have been examined after cooling rapidly into an un-annealed state from above the T_g region [7]. In these examples, there is little if any change in the onset temperature as a function of scan rate, and any change that is observed is well within experimental error for such determinations, taking into account issues such as the reproducibility of onset calculations in themselves. This indicates that for many materials scanned at high rates, accurate values of T_g similar to those obtained at lower rates can be obtained. These are observations made with a range of polymers and pharmaceutical materials. A broad study may find materials where T_g changes more noticeably with scan rate, but for many materials in an un-annealed state the onset temperature of T_g does not in itself seem to be affected to any significant extent by increasing scan rate.

The effect of scan rate on T_g calculations other than onset temperature is also worth noting. If the data from which the calculations in Figures 2.19 and 2.20 are taken are examined, then

PET comparison of rate and weight

Onset temperature of PET at varying scan rates

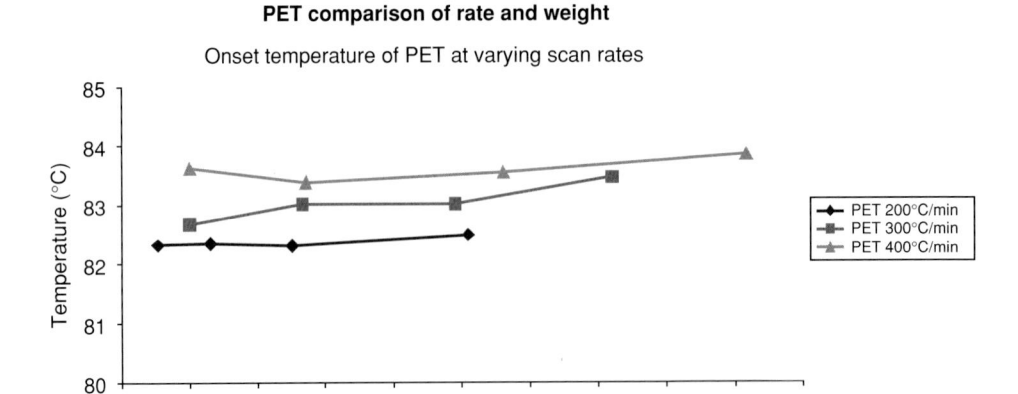

Figure 2.20 T_g of PET measured at increasing scan rate and sample weight. Measured onset temperatures remain within ±1°C except for samples of high weight run at high rates of 200°C/min and above.

it can be seen that not only is the onset reproducible but in fact the whole T_g range remains unchanged, provided lower weights are used, and for these examples this means weights of 10 mg or less, which in itself is a surprisingly large amount; see Figure 2.21.

This does indicate that a fast scan rate (heating rate) has very little effect on the measurement of the T_g except when very large samples are taken, or a poor experimental procedure results in larger thermal lags or gradients.

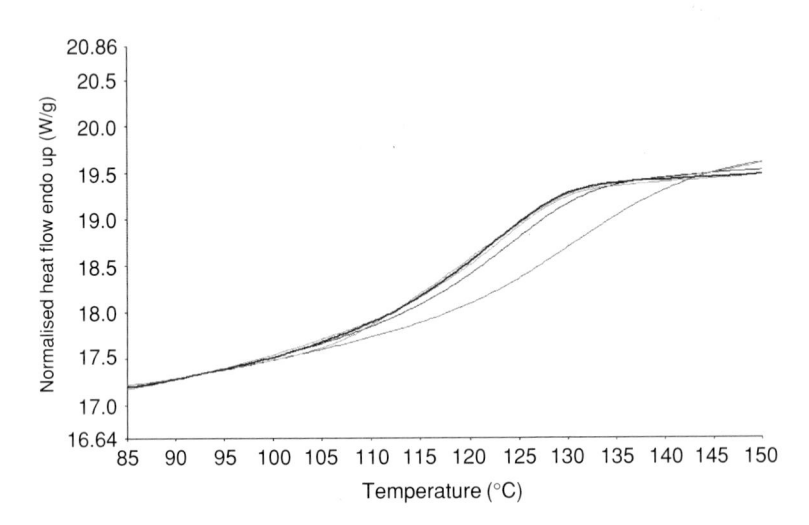

Figure 2.21 T_g of PMMA at 400°C/min. Sample weights from 0.1 to 23 mg were compared on a normalised y-axis. At this high rate, a slight shift is seen in the sample of weight 10 mg but a significant shift in onset temperature and range of transition is observed only for the higher weight of 23 mg. Lower weights all overlay each other.

Effect of scan rate on T_g of PS after fast cool

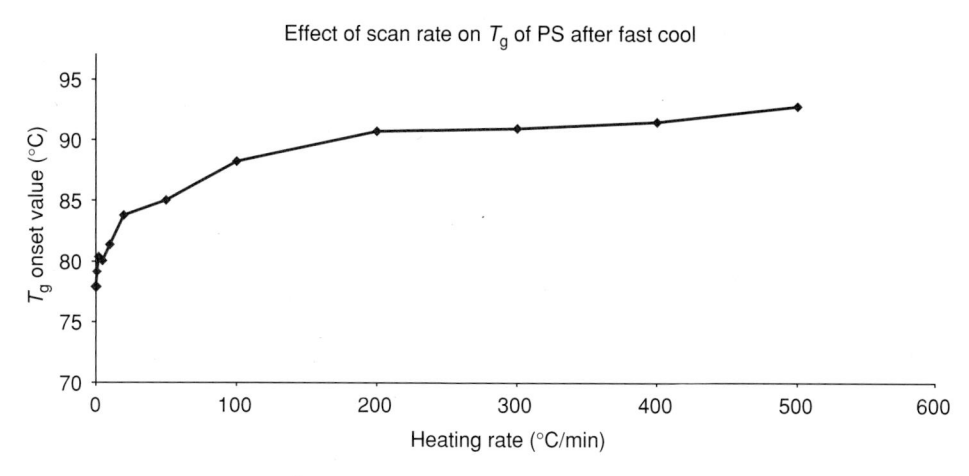

Figure 2.22 T_g of polystyrene after fast cool. At slow heating rates the material is being annealed and able to relax into a lower enthalpic state with a consequently lower value of T_g.

These observations are not expected, since increasing scan rate has long been held to have an effect on the measurement of T_g. Further light is shed on this issue if rates slower than the typical 10–20°C/min are also examined. In the case of polystyrene, Figure 2.22, a significant reduction in the value of T_g is observed at very low scan rates. The extent of this reduction in T_g for the polystyrene sample shown is over 12°C as scan rates are reduced to 0.1°C/min. The measured T_g onset at higher rates varies little, but at lower rates it is seen to reduce as a function of scan rate. A similar observation has been made with TMA techniques and reported in Section 10.5.3. Un-annealed samples not only show a decreased value of T_g but also a reduction in the coefficient of expansion as a sample is heated slowly into the T_g region. The explanation is that the sample is being annealed by slower scan rates into a lower enthalpic state, with a consequently lower T_g (and volume as measured by TMA). In effect, heating rate is shown to affect the value of T_g not because the measurement is shifted by interaction with the kinetics of the T_g transition itself but because the material is being annealed during slow scan rates into a lower enthalpic state which has a lower T_g associated with it. The lower value of T_g is an accurate measurement of the annealed material at the point it is made. This being the case, increasingly fast scan rates may not be expected to have a significant effect upon the measured value of T_g.

A further observation is that there are no enthalpic relaxation phenomena apparent in these data, nor have they been observed in our laboratories when un-annealed samples are heated through T_g at any given scan rate, fast or slow. This is a very significant point. If a material is heated so fast that it cannot undergo the glass transition at the correct equilibrium temperature, it is heated into a region where it is in a lower energy state than its equilibrium state requires. When it then goes through T_g it must also absorb energy to regain equilibrium. This is the source of the enthalpic relaxation peak (Section 1.5.6). The fact that no enthalpic relaxation peak is observed is further significant evidence that the sample has not been scanned at too fast a rate for the kinetics of the T_g and that the true T_g has been measured, even at the very high scanning rates used by HyperDSC™.

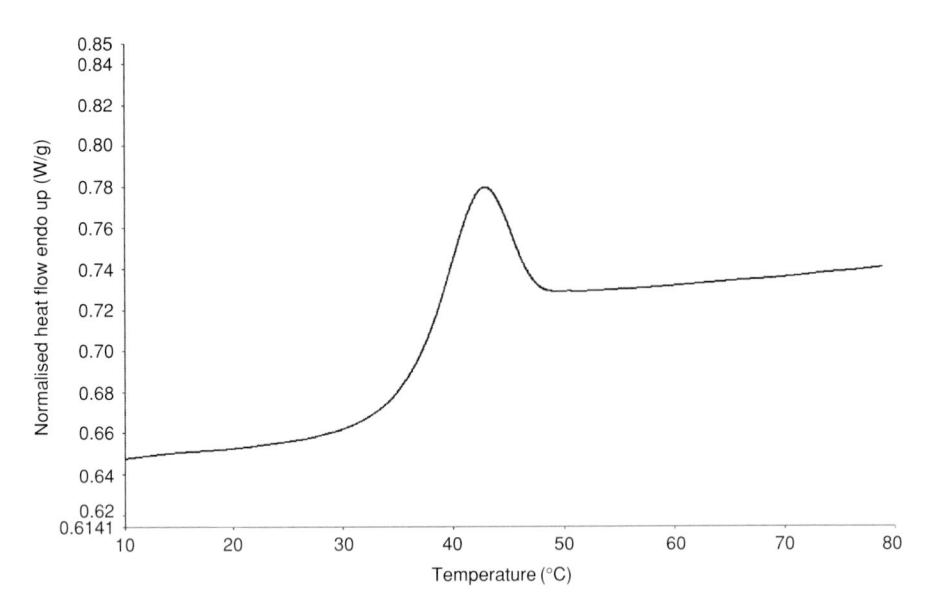

Figure 2.23 The T_g of indomethacin heated at 10°C/min after cooling from the melt at 1°C/min. This sample did not crystallise on cooling so low cooling rates could be investigated whilst retaining the amorphous structure. The enthalpic relaxation peak is clearly seen on top of the T_g.

2.4.6 Effect of heating rate on T_g of annealed materials (and enthalpic relaxation phenomena)

When a material is annealed in the region of T_g or just below, it is allowed to relax into a lower enthalpic state (Section 1.5.6). When materials enter a low enthalpic state the kinetics of the T_g also appear to be affected since even at modest DSC heating rates of 10 or 20°C the true T_g of the material (fictive temperature) is exceeded before T_g occurs and is measured by the DSC. The result is that the sample is heated into a non-equilibrium state of low energy so that when the material goes through T_g it also absorbs energy to regain equilibrium. This results in a step with a peak on top, Figure 2.23. This example is from indomethacin, a pharmaceutical material previously cooled slowly from the melt into a glass.

As scan rates are increased there is the natural assumption that the T_g of a material will be further exceeded before the glass transition actually occurs, due to the relatively slower kinetics, resulting in an increasing value for the area of the enthalpic relaxation peak. An initial investigation (Figure 2.24) shows what happens when this material is heated at increasingly fast scan rate.

Essentially, there is no significant change in the peak area; although there is a slight shift in onset values probably due to a thermal lag effect broadly similar peak areas are found. If the sample is cooled quickly then no enthalpic relaxation is observed at low or high rates; see Figure 2.25.

The effect of scan rate on the size of the enthalpic relaxation of indomethacin and the onset value of T_g have been measured over a wide range of scan rates [8], Figure 2.26. The area of the relaxation peak for indomethacin after cooling at 0.1°C/min was measured at heating

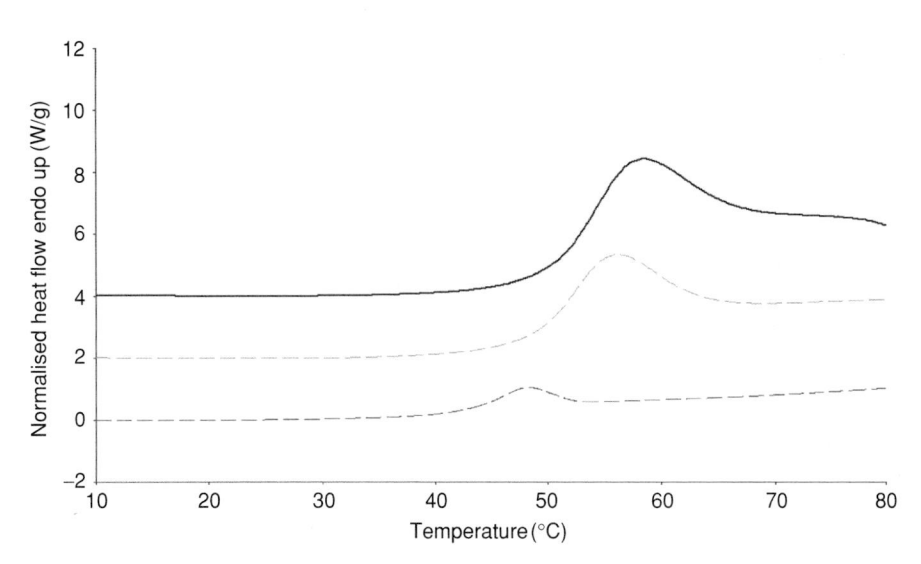

Figure 2.24 The T_g of indomethacin at scan rates of 50–300°C/min after cooling from the melt at 1°C/min. The enthalpic relaxation is apparent in all traces and is of a similar energy, possibly increasing slightly.

rates of 0.2–10°C/min. At rates above 2°C/min the area of the relaxation appeared not to change significantly. The enthalpic relaxation peak can to some extent affect or possibly mask T_g and this should be borne in mind when reviewing data, yet onset values are also found to level off. This seems to indicate that though the kinetics of the T_g of an annealed

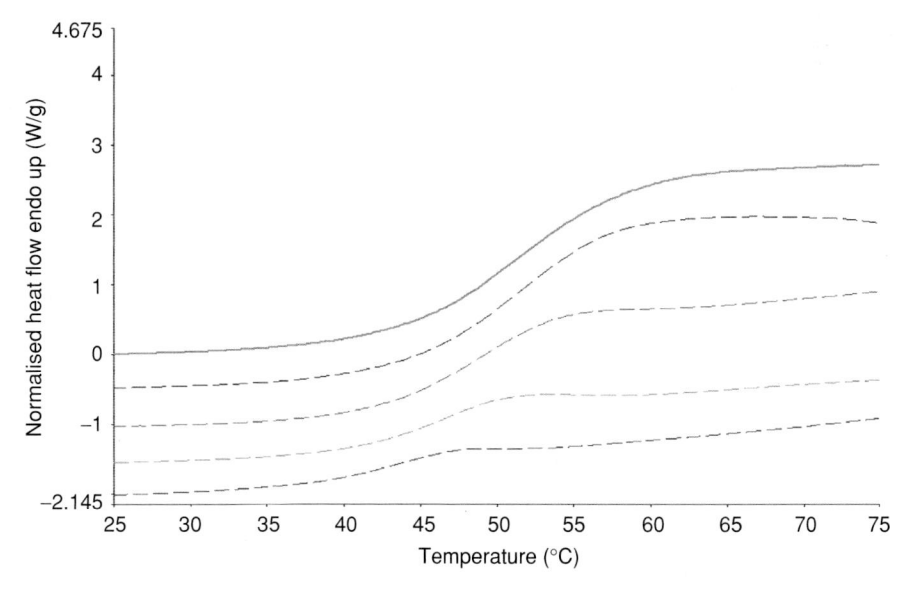

Figure 2.25 The T_g of indomethacin at scan rates of 50–400°C/min after rapid cooling from the melt. Little evidence of enthalpic relaxation is apparent in any of these traces.

Indometacin after slow cool

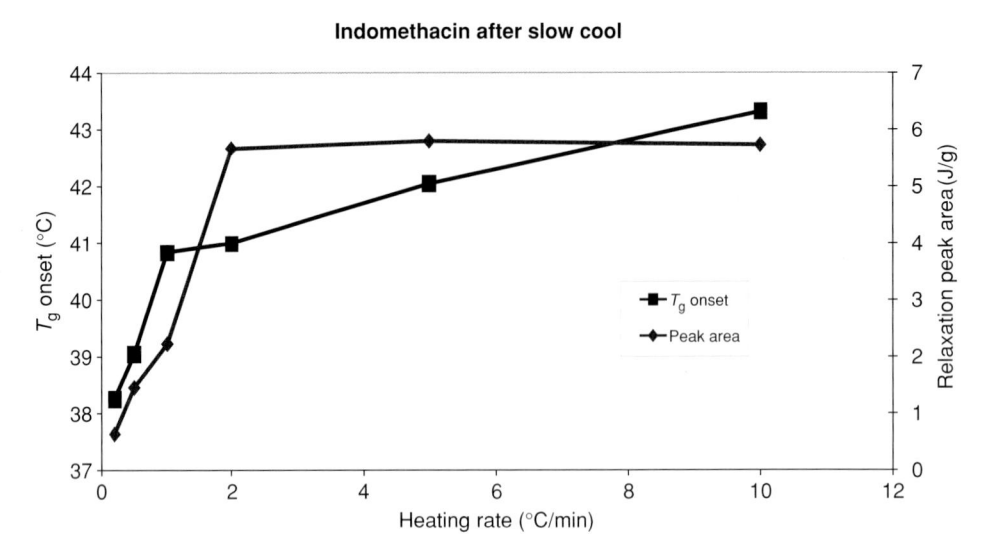

Figure 2.26 T_g of indomethacin after slow cool shown as a function of heating rate. The slow cool allows relaxation into a lower enthalpic state. On subsequent heating, even at modest rates the fictive temperature is exceeded resulting in a relaxation peak on top of the T_g. The area of this is shown (diamonds) together with onset temperature of T_g (squares). Both increase with heating rate but reach a fairly stable value. Onset temperatures are only a guide since the enthalpic peak swamps these to some extent; peak area is therefore more reliable.

material can be easily exceeded by slow scan rates, the overheating effect does not continue in a linear fashion, far from it there seems to be a given point past which there is little change.

Similar data have also been obtained from samples of polystyrene, Figure 2.27. A trace from a scan at low rate is shown in Figure 2.28.

The presence of an enthalpic relaxation peak does indicate that the fictive temperature of the material has been exceeded before the glass transition has occurred due to the speed of the scanning rate, resulting in an inaccurate value of T_g, yet this occurs at relatively slow rates and is an effect which is not significantly changed when using increasingly high scan rates.

2.4.6.1 Implications for the measurement of T_g

The enthalpic relaxation peak can cause difficulties in the measurement of the height of T_g because it is difficult to assess the correct position of the beginning and end of the transition. This has implications when trying to measure amorphous content of materials via the height of the T_g. For example, when trying to produce a fully amorphous material in order to measure the size of the T_g, it is essential to use the fastest cooling rates available to produce not only an amorphous material, but also a material in an un-annealed state with regard to T_g. If the T_g is obscured by relaxation phenomena, height and other calculations become more difficult and errors increase.

Effect of heat rate on T_g of PS after slow cool

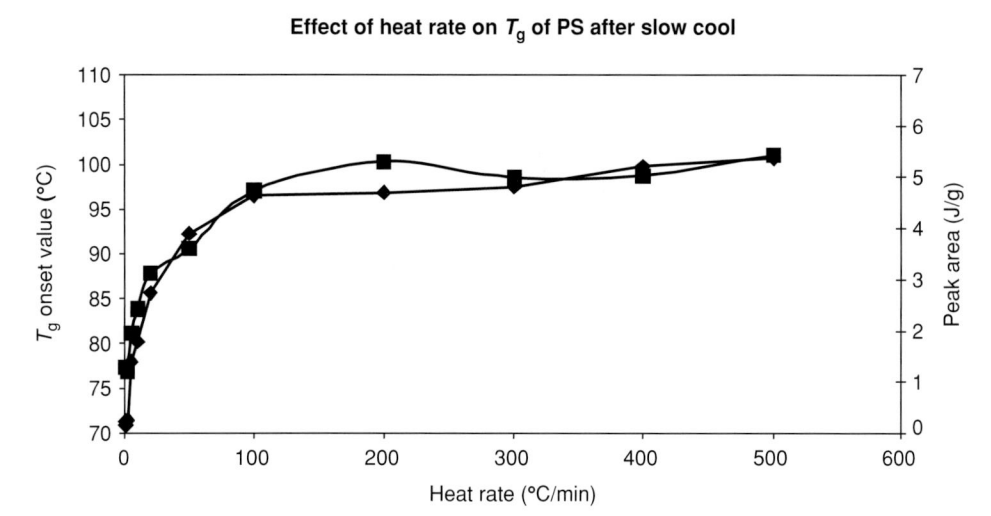

Figure 2.27 Effect of heating rate on the T_g of polystyrene after slow cool into an annealed state. The T_g onset is denoted by squares, the area by diamonds. Both the onset of T_g and the area of the enthalpic relaxation peak increase with scan rate to begin with but reach a nearly constant value at higher rates.

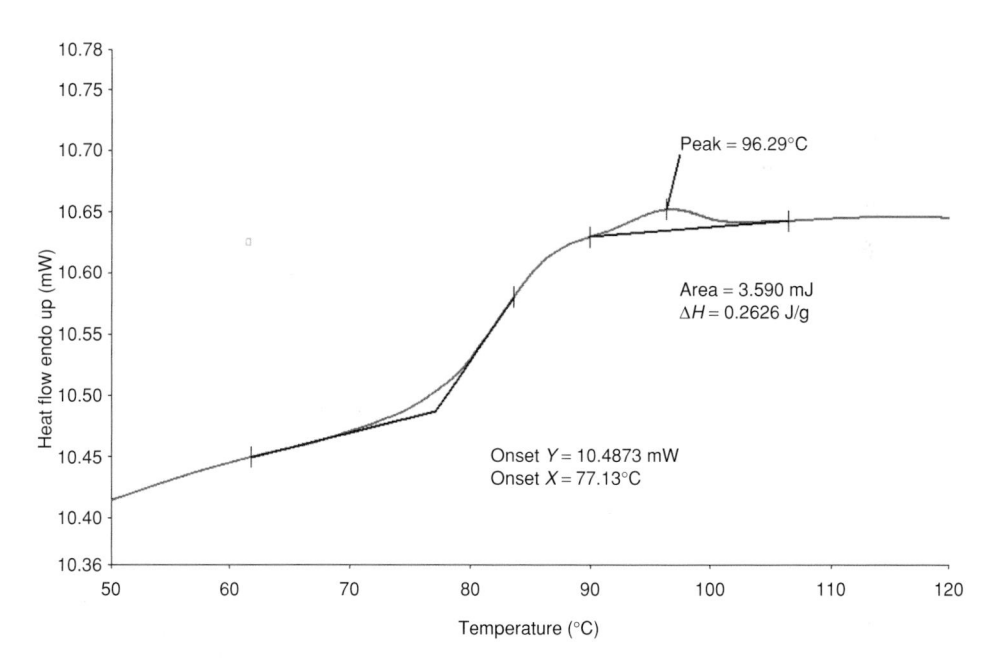

Figure 2.28 The T_g of polystyrene heating at 2°C/min after cooling at 0.1°C/min. The onset value of T_g is quite low and the size of the enthalpic relaxation is quite small indicating only a small shift from the true T_g value.

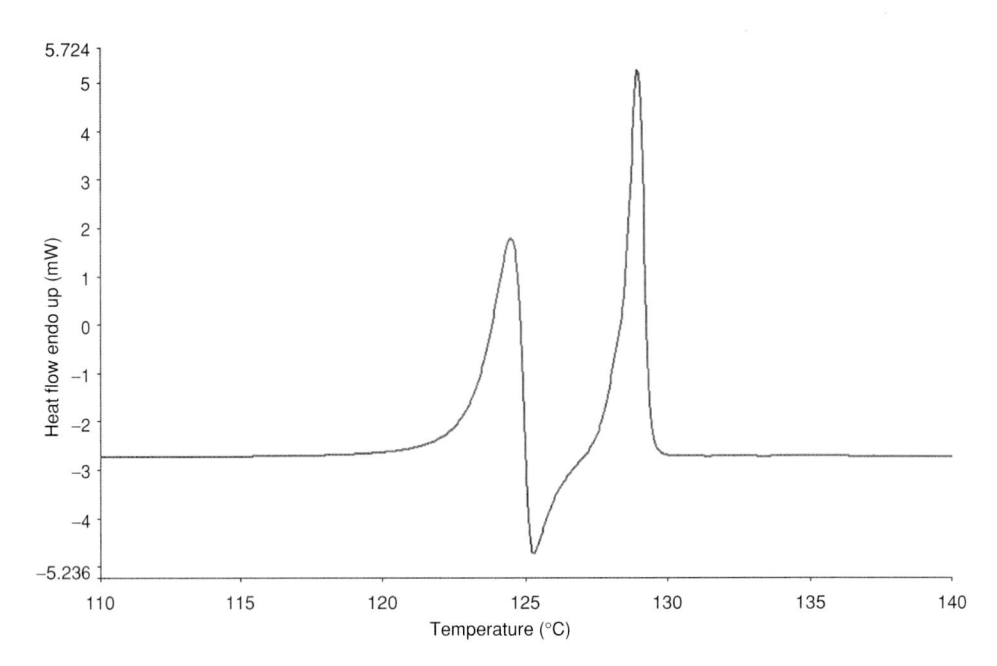

Figure 2.29 Chlorpropamide heated at 5°C/min showing melting, recrystallisation and then further melting. This is a classic example of polymorphism in a pharmaceutical material as observed by DSC.

2.5 Application to pharmaceuticals

2.5.1 Purity of polymorphic form

Many pharmaceutical materials exist in a range of different crystal forms and the classic type of DSC trace from such a material is shown in Figure 2.29. In this example, a sample of commercially available chlorpropamide has been heated at a slow rate and shows conversion from one form to another. The original crystal form melts, beginning around 120°C, then recrystallises, as shown by the exotherm around 125°C, and then the new crystal form melts around 127°C. In fact, the complexity of the second melt suggests that more than one form is melting in this region.

One of the uses of DSC is to confirm which crystal form of a drug is present. Since the melting point of a particular crystal form is specific to that crystal form, different polymorphs can be identified from their differing melting points. In some cases, melting points may be very close together so that distinguishing them requires good resolution, and one peak may appear as a shoulder on another, but this information is essentially available from observation of the melting profile. However, in a situation where a recrystallisation event has occurred, as with chlorpropamide (Figure 2.29), it is not possible to be sure whether some of the higher melting forms were present initially or not. Clearly some of the initial form is recrystallising, but it is not possible to tell from the slow scan whether all of the higher melting form resulted from the recrystallisation.

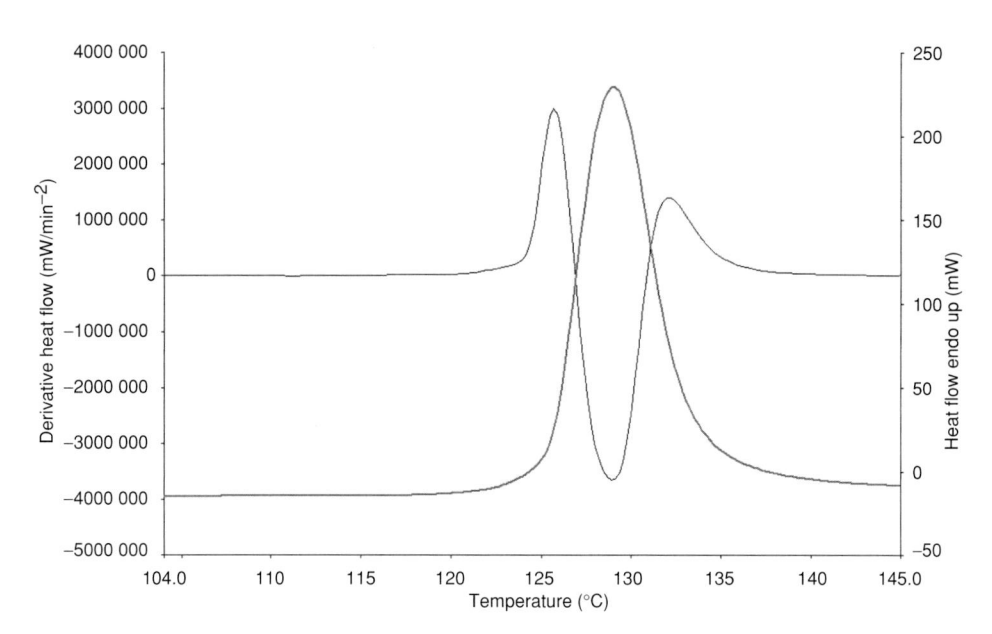

Figure 2.30 Chloropropamide heated at 300°C/min. The heat flow trace and first derivative are shown. This shows one single melt. The fast scanning rate has prevented recrystallisation and allowed confirmation of the purity of the original crystal form.

With increasing scan rate however, it may be possible to scan sufficiently fast such that recrystallisation is prevented and the analysis of the original material can be observed [9–11]. Figure 2.30 shows chloropropamide heated at 300°C/min.

At this scan rate, the material is unable to recrystallise and only the melting of crystal forms originally present will result. If only one form is found in such a situation it can be concluded that only one form was present to begin with so the material was pure with regard to crystal form. The second derivative trace can help when viewing a melting profile (Section 1.2.5) and in this case helps to show that there is only one transition.

It is sensible to consider the effect of increased thermal gradients at increased scanning rate, since melting peaks not only get bigger with increasing scan rate, they also get broader, so the question arises as to whether the lack of two or more peaks is just the lack of resolution. The effectiveness of this approach relies on the fact that there is a sufficient difference in the melting point between the forms in order to distinguish them. Decreasing the sample weight in order to improve resolution is an important aspect as it is in all polymorphic studies. Another aspect to look at is the trends that occur as the scan rate is increased, and often the higher melting forms can be clearly distinguished to the point that they are no longer formed. If pure samples of different forms are available it may also be possible to prepare a mixture to see how the different forms can be distinguished. If a particular crystal form is present to begin with, then in general it will become more obvious as the scan rate is increased since the peak size will increase. The same is true with any transition. Figure 2.31 shows an example with carbamazepine. A sample of a commercially available material was run at increasing scan rate and it can be seen that the lower temperature transitions become

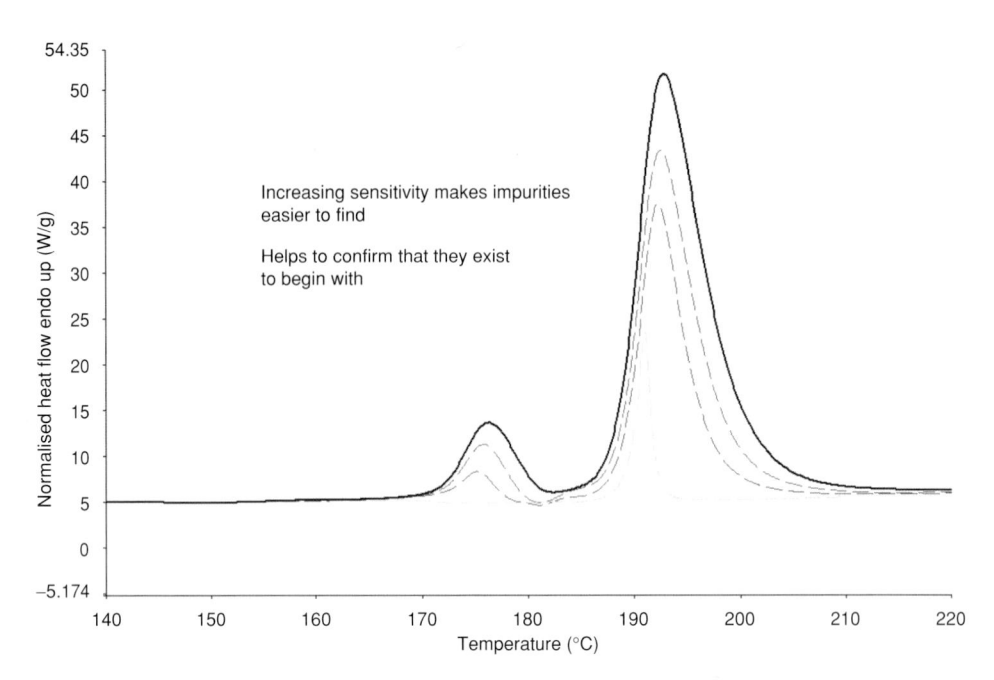

Figure 2.31 Commercially available carbamazepine. Increasing scan rate shows increasing size of major transitions.

much more obvious as the scan rate increases. In general, if more than one form is present it will be distinguishable at high rates, though note that rates fast enough to prevent a particular recrystallisation may not always be achievable.

2.5.2 Identifying polymorphs

Different crystal structures can be identified by their melting points, so in principle by scanning a sample in a DSC the different crystal forms present can be distinguished. Whereas in the above application the aim is to prevent recrystallisation, when recrystallisation does occur it may be of interest to characterise the forms produced, and so identify a number of different polymorphs. By scanning at a range of different rates, different forms may be produced depending upon the kinetics of the crystallisation reaction. Therefore, it can be informative to scan a sample from slow to high rates to see what transitions can be found.

In addition, cooling a sample at different rates can also be useful. Not all samples recrystallise, but those that do may form different crystal forms depending upon the cooling rate. These can be examined upon reheating, so the analyst has a range of experimental permutations to choose from for the initial heating, cooling from the melt, and reheating. Figure 2.32 shows chlorpropamide on reheating after cooling at different rates; note how the ratio of melting peaks changes as a function of cooling rate.

Where a material does not recrystallise on cooling or where the cooling rate is sufficient to prevent any recrystallisation, it may cool into a glass, and the T_g can be measured on heating.

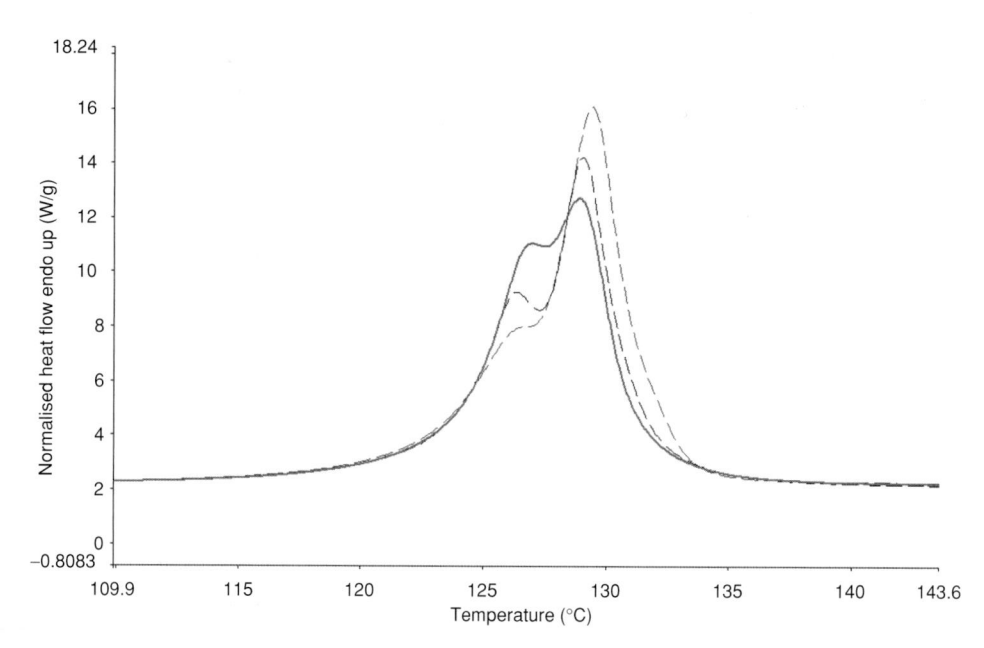

Figure 2.32 Chlorpropamide reheated at 50°C/min after cooling at 10°C/min (solid line), 20°C/min and 50°C/min. Rate of cooling can influence the polymorph formed.

Subsequent crystallisation may occur with these materials and melting measurements can then be made.

On some occasions, melting and recrystallisation can occur almost simultaneously so that at a slow heating rate neither the melting nor recrystallisation processes may be distinguished, or even noticed. As with the carbamazepine sample in Figure 2.31 small, apparently unnoticed, overlapping transitions can be separated or enhanced into much larger, clearly measurable, transitions at higher rates. Therefore, it is worth checking materials at high scan rates to see what additional information may be available. The resulting information is more likely to tie in with other techniques, such as X-ray crystallography, because at high rates the sample does not have time to change and true values can be measured.

2.5.3 Determination of amorphous content of materials

Measurements of the glass transition have proven to be greatly enhanced at high scan rates due to the increase in sensitivity, and it is possible to study this transition in a wide range of materials where the T_g could not previously be measured; in particular those with only a small percentage of amorphous content where the T_g is very small, possibly as low as 1% [12–14]. The need to analyse for low amorphous content arises for a variety of reasons: for example, development of materials and the effect of manufacturing processes such as milling, which is used to produce the desired particulate size. There is concern that these processes could increase the non-crystalline content, so there is need to measure it. There are already a number of well-defined methods for doing this, for example vapour sorption

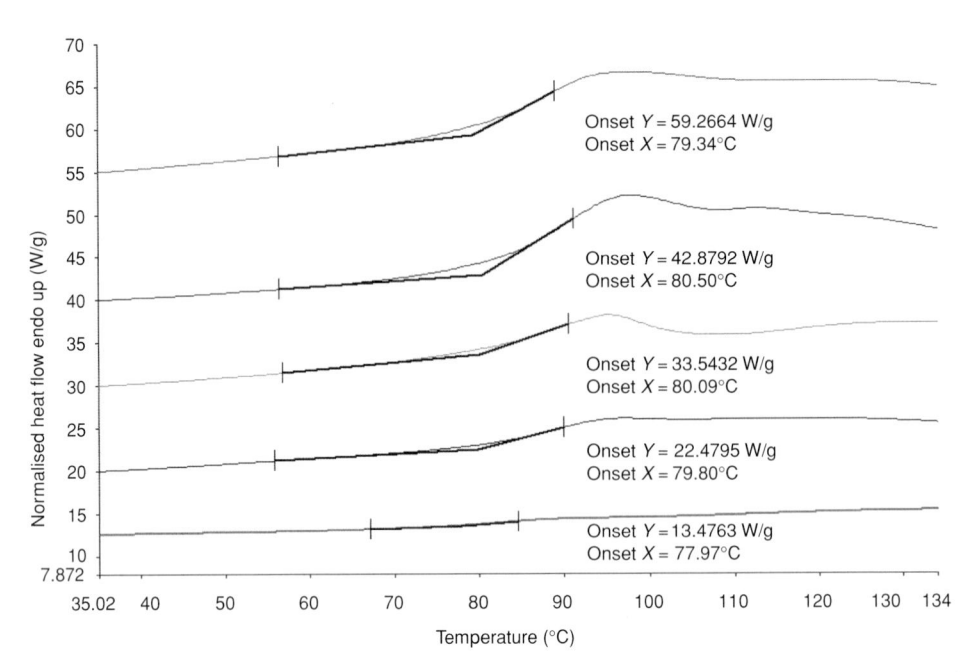

Figure 2.33 T_g of spray-dried lactose measured at increasing scan rate from 100°C/min (bottom curve) to 500°C/min (top two curves). The faster the scan rate the larger the T_g.

or solution calorimetry, but these can be very time consuming and not all methods will work well for a given material so there is need for a further rapid method.

A study has been performed with lactose [15] which shows that low amounts of amorphous content can be quickly detected by HyperDSC™. Fast scanning rates greatly increase sensitivity, Figure 2.33, and by measuring the height of the glass transition a linear relationship has been shown between height of T_g and per cent amorphous content, Figure 2.34.

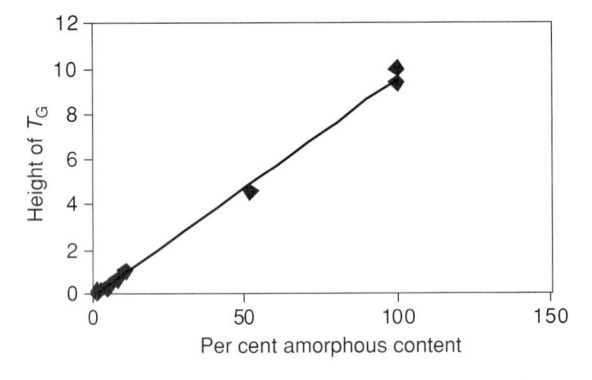

Figure 2.34 Height of T_g as a function of amorphous content of spray-dried lactose mixed with crystalline α-monohydrate. This shows a linear relationship between height of T_g and amorphous content.

Thus, if the relative height of T_g can be determined this can be related to the amorphous content.

Materials which are 100% crystalline should have no glass transition, so if a sample thought to be 100% crystalline is scanned at high rates by DSC and a T_g is found, this immediately means that there is some amorphous material present which needs further investigation. A measurement like this need only take 2 or 3 min to prepare and complete. To obtain a quantitative value, the height of the T_g needs to be calculated and compared with that of a known standard. A value for a 100% amorphous material can frequently be obtained by cooling the material rapidly from the melt into its amorphous state and then reheating it. This is best done in the analyser since taking a sample out, even to drop into liquid nitrogen, is sufficiently uncontrolled such that crystallinity can occur; see Figure 2.13. The sample needs to be cooled as rapidly as possible through the T_g region and into the glassy state both to avoid crystallinity and enthalpic relaxations on top of the T_g (see Section 2.4.5) which cause difficulties for accurate measurement. This approach requires that the material does not decompose on melting, nor that any crystallisation occurs. This can be checked by subsequent melting.

2.5.3.1   A practical approach for measuring small amounts of amorphous content

1. It is always helpful to know as much about a sample as possible, and in particular the T_g region in question. If this is not known, then cool a sample from the melt rapidly into the glass and reheat it as described above. The sample mass should not be so large that it will seep out of the pan when melted.
2. Sample mass generally will need to be maximised in order to make the best measurements since glass transitions are usually very small. Use a large pan and compress to give good thermal contact. Heat as fast as possible through the T_g region, but do not melt, as large sample masses may cause contamination issues when melted. Make sure that the pan is vented to allow volatiles to escape and to prevent pan distortion under pressure. Make sure that the sample has fully equilibrated in the purge gas atmosphere before the start of the run, this may take a few minutes depending upon the flow rate.
3. To calculate the height of the T_g use a consistent method. One approach would be to normalise the data (this divides by the sample weight) and slope the curve before T_g so that the baseline approaching T_g appears flat. Any variation in the position of the start of the calculation has no effect. Select the height from the baseline to the maximum of the step and compare the height of T_g from the as-received sample with the height measurement from a sample of known content, typically a sample previously cooled from the melt into a glass.

2.5.4   Measurements of solubility

One study looking at the measurement of undissolved drug in a matrix showed that high scan rates are essential for accurate measurement [16]. This study reports that low heating rates allow further solution of the drug in the matrix, whereas fast rates permit measurement of the

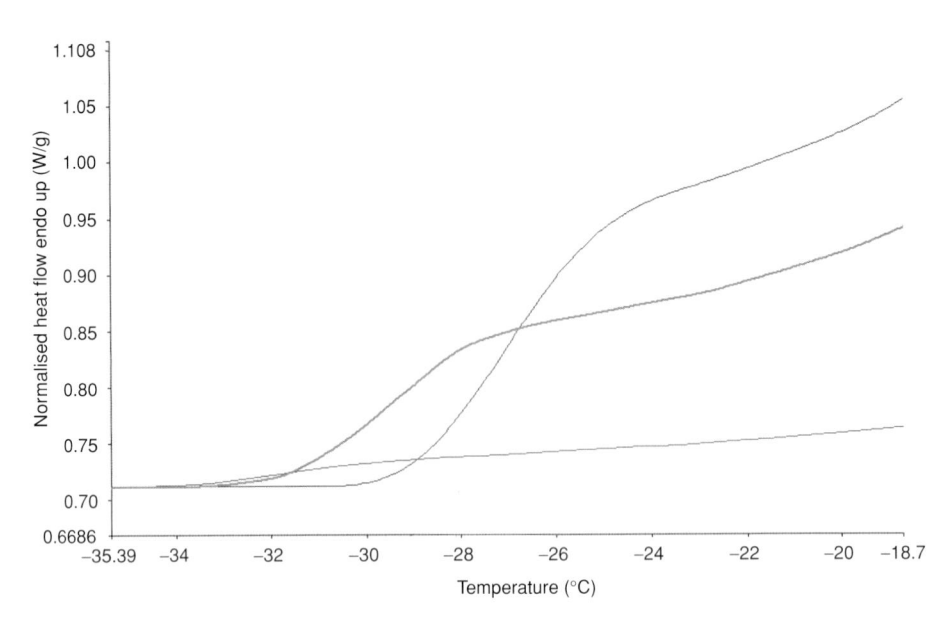

Figure 2.35 The glass transition from a 10-mL sample of a 5% sucrose solution scanned at 10°C/min (lower trace), 50°C/min and 100°C/min (largest step). This shows the large increase in sensitivity possible, though water-based materials exhibit larger thermal lag effects which significantly shift the measured T_g as a function of scan rate.

melting profile of the drug without further solution occurring. The amount of undissolved material can be obtained from the enthalpy of melting; see Chapter 8 for further details. The same approach can be used with foods where, for example, the amount of undissolved sugar needs to be measured.

2.6 Application to water-based solutions and the effect of moisture

2.6.1 Measurement of T_g in frozen solutions and suspensions

As with polymers and pharmaceuticals, one of the most useful measurements made using fast scanning is that of the glass transition, and again with this is found to be enhanced at high rates; see Figure 2.35. In this case a shift in the position of the T_g is also observed, probably due to thermal lag associated with the large thermal mass of water/ice. Whist at very low mass the transition may not be any more scan-rate dependent than the polymers described in Section 2.4.5, for most practical purposes the thermal mass of water/ice materials means that the onset temperature will be influenced to some extent by the mass of material present.

With many life science materials where a T_g is to be measured, high rates are very helpful. However, where cell tissue may be destroyed by freezing there are limitations to the technique, since low temperatures may not be used at the start of a run. In many cases, this means that the duration of the initial transient is likely to overrun the transition of interest so lower scan rates may be needed. If a system can be frozen without harming the sample, the duration of the melting peak of water should be taken into account. This can mask events close to ambient because of its size and the time taken for it to complete, so this may limit the usefulness of the technique and mean that lower rates need to be chosen. However, even at rates of 50 or 100°C/min more useful information may be gained than scanning at very slow rates.

2.6.2 Material affected by moisture

Keeping materials of any sort absolutely dry is very difficult given that most materials absorb moisture from the atmosphere, and this is particularly true of amorphous materials where water molecules can more easily penetrate and surround the host molecules than in a crystalline material. Furthermore freeze-dried or spray-dried materials are not completely dry when produced, and contain variable amounts of moisture depending upon the production conditions. This directly affects the stability of these materials, and it is important to measure the T_g, plasticised by the moisture content, to determine the actual value as it would be found in storage.

Using slow scanning rates a DSC may give some information, but in an unsealed pan a material is likely to dry out before T_g is measured, and moisture loss from the material prior to T_g will not only change the position of T_g but can mask it altogether. In a sealed pan there is no guarantee that moisture will remain in the sample, since it could come out of the sample, though remain in the pan. One approach is to measure the moisture content of the sample, and then to measure the T_g of the dried material. Using the Gordon–Taylor equation the T_g of the plasticised material can then be calculated (see Section 1.5.8).

With fast scanning the T_g of the non-dried material (complete with its moisture content) can easily be measured since the material does not have time to dry out. Furthermore, the peak from moisture loss is displaced to higher temperature so the T_g may be more clearly seen, as in the case of heparin, Figure 2.36. High scan rates therefore provide a simple and rapid method for the determination of stability of glassy materials affected by moisture.

2.7 Practical aspects of scanning at fast rates

2.7.1 Purge gas

One of the main issues involved in fast scanning DSC is getting good thermal transfer between the analyser and the sample and in practice this can be achieved by using helium as a purge gas. Helium has much higher thermal conductivity than nitrogen, air or oxygen and results in much better performance. It is not worth using nitrogen unless unavoidable. At higher temperatures, instrumental limitations may not permit the use of helium and gas mixtures such as helium/neon have been used to give both high performance and a broad temperature range. The rate of flow of purge gas used is normally very low, e.g. 30–40 mL/min, so the costs involved are small. After a sample is loaded into the furnace, sufficient time should be given for all air to purge from the system before the run begins; noting that vented pans should also be given time to fully purge. This usually takes a few

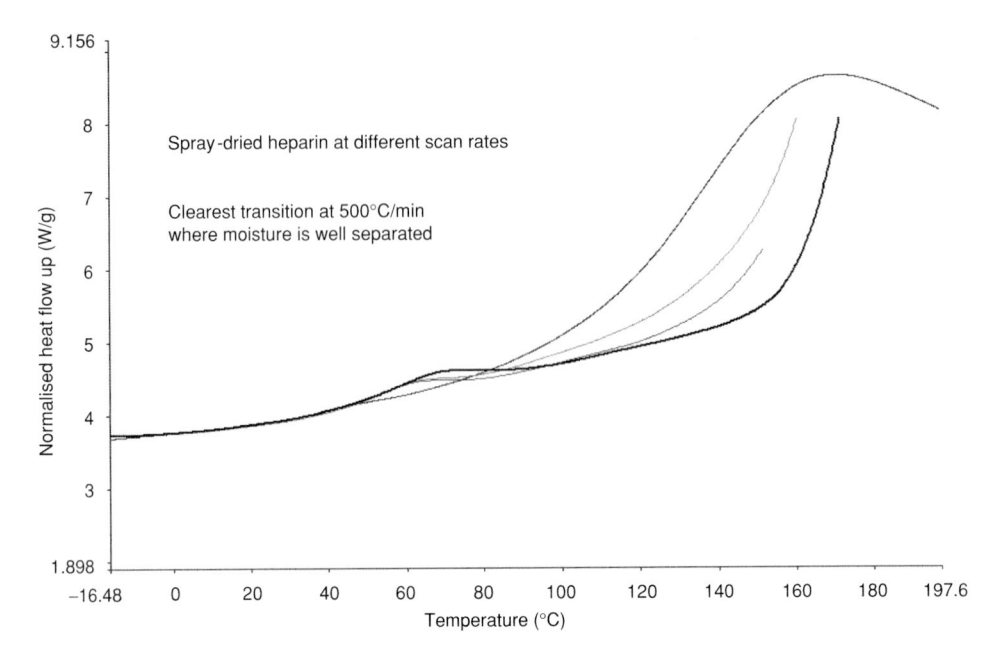

Figure 2.36 T_g of heparin determined by HyperDSC™. As the scan rate is increased moisture loss is delayed to higher temperatures revealing the glass transition, plasticised by the moisture.

minutes, and can be gauged by the stability of the heat flow signal. A delay of a few minutes can be inserted into a standard method, and a typical SOP should suggest 3–4 min.

2.7.2 Sample pans

Sample pans should have a flat base with the sample well pressed down to give good thermal contact. Pan selection is likely to be dependent upon the type of transition being measured. When looking for a glass transition resolution issues are generally irrelevant, so the size and thickness of pan have little influence on the result. Large, thicker aluminium pans are usually best because they hold large amounts of sample and in general retain a flat base; the flatness of the base is far more important than thickness of the pan. If melting transitions are to be measured then the traditional flat type of pan is best, or if using dish-shaped autosampler pans make sure of good internal contact. Very thin aluminium pans are available for HyperDSC™ work where resolution issues are important and where thickness of pan is suspected of having an effect, but for most work this is unlikely to be the case. In effect, the same pans are used at high scan rates as for low scan rates, flatten or discard pans that do not have a flat base, and use good sample loading techniques.

Beware of contamination of pans. Contamination will have more effect at high rates because of increased sensitivity. Pre-heat pans to 300°C to volatilise off potentially contaminating oils before looking for very small transitions.

2.7.3 Sample size

The best size of sample to use is often difficult to predict since there are so many potentially influencing factors. However, the main issue is to decide what type of transition is involved. Minimise sample size when melting a material; 1–2 mg is probably a good starting point, and avoid decomposition and resulting contamination issues. The faster the scan rate the smaller the sample needed. Maximise sample size when looking for a small T_g. The actual mass will vary greatly with density of sample. It may not be possible to optimise conditions such that satisfactory measurement of a T_g and a melting profile can be obtained from a single run. Do not try to melt large sample sizes which will potentially collapse and so give poor data, apart from any contamination issues.

2.7.4 Scan rate

As shown with most applications in this chapter, a wide range of scan rates can be very useful, from very slow for reactions which need more time to complete (not excluding isothermal work either) to very fast. The biggest question is where to start. Generally, it is best to avoid extremes and for much of the work shown here a scan rate of 300°C/min was chosen for initial investigations. For small hard-to-find glass transitions 500°C/min may be immediately chosen, though for melting profiles, particularly of polymers, this may prove too rapid for optimal results.

Remember to calibrate appropriately for the scan rates selected.

2.7.5 Instrumental settings

In general, work with the widest dynamic range available since transients and melting peaks can be very large, and set data point collection rates to be as fast as possible. For most work in this chapter, data were collected at a rate of 20 data points per second.

2.7.6 Cleanliness

As remarked with regard to sample pans, small contamination effects which are not noticeable at low rates can become evident at fast rates; hence the need for a clean and careful technique, ensuring that the working environment is also clean. Small amounts of volatilisation or sublimation can cause contamination around the furnace area which may be noticeable on subsequent runs, so bear in mind the need to check and clean furnaces between runs, particularly when melting materials. If there are any anomalies in the instrumental baseline it may be useful to use a baseline subtraction technique, particularly where very small glass transition measurements are being made. Subtraction should only be used if the analyser is clean and the baseline is totally reproducible, but baseline subtraction is normally worth using since it only takes a minute or two to perform at high rates.

2.7.7 Getting started

Start with simple samples that have known transitions and will not easily decompose and so cause contamination issues. Look at the effects of increasing scan rate and varying sample weight and begin to get a feel for the practice of the technique. Do not start with the most difficult or challenging sample that is available.

References

1. Gabbott PV, Goth S, Li L. HyperDSC™ – a breakthrough method for materials characterisation. *PerkinElmer Applications Note*.
2. Pijpers TFJ, Mathot VBF, Goderis B, Scherrenberg RL, van der Vegte EW. High-speed calorimetry for the study of the kinetics of (de)vitrification, crystallization, and melting of macromolecules. *Macromolecules* 2002;35:3601–3613.
3. Minakov AA, Mordvintsev DA, Schick C. Melting and reorganization of poly(ethylene terephthalate) on fast heating. *Polymer* 2004;45:3755–3763.
4. Sichina WJ. High sensitivity characterization of transparency films using high speed DSC. *PerkinElmer Application Note*.
5. Saklatvala RD, Saunders MH, Fitzpatrick S, Buckton G. A comparison of high speed differential scanning calorimetry (HyperDSC™) and modulated differential scanning calorimetry to detect the glass transition of polyvinylpyrrolidone: the effect of water content and detection sensitivity in powder mixtures (a model formulation). *J Drug Deliv Sci Technol* 2005;15(4):257–260.
6. Bilyeu B, Brostow W, Keselman M, Menard K. Characterization of epoxy curing using high heating rate DSC. In *ANTEC 2003*.
7. Gabbott P. Examining the effect of weight and rate with HyperDSC™. *TAC 2005*.
8. Gabbott P. Does enthalpic relaxation affect Tg measurement? *TAC 2006*.
9. Gabbott P. Confirming polymorphic purity with HyperDSC™. *PerkinElmer Application Note* 2004.
10. McGregor C, Saunders MH, Buckton G, Saklatvala R. The use of high-speed differential scanning calorimetry (Hyper-DSC) to study the thermal properties of carbamazepine polymorphs. *Thermochim Acta* 2004;417:231–237.
11. Giannellini V, Bambagiotti-Alberti M, Bartolucci G, et al. Solid-state study of mepivacaine hydrochloride. *J Pharm Biomed Anal* 2005;39:444–454.
12. Saunders M, Podluii K, Shergill S, Blatchford C, Buckton G, Royall P. The potential of high speed DSC (HyperDSC™) for the detection and quantification of small amounts of amorphous content in predominantly crystalline samples. *Int J Pharm* 2004;274:35–40.
13. Hurtta M, Pitkänen I. Quantification of low levels of amorphous content in maltitol. *Thermochim Acta* 2004;419:19–29.
14. Lappalainen M, Pitkanen I, Harjunen P. Quantification of low levels of amorphous content in sucrose by HyperDSC™. *Int J Pharm* 2006;307:150–155.
15. Gabbott P, Clarke P, Mann T, Royall P, Shergill S. A high-sensitivity, high-speed DSC technique: measurement of amorphous lactose. Am Lab 2003, reprint from PerkinElmer.
16. Gramaglia D, Conway BR, Kett VL, Malcolm RK, Batchelor HK. High speed DSC (HyperDSC™) as a tool to measure the solubility of a drug within a solid or semi-solid matrix. *Int J Pharm* 2005;301:1–5.

Chapter 3
Thermogravimetric Analysis

Rod Bottom

Contents

3.1	Introduction	88
3.2	Design and measuring principle	89
	3.2.1 Buoyancy correction	90
3.3	Sample preparation	92
3.4	Performing measurements	93
	3.4.1 Influence of heating rate	93
	3.4.2 Influence of crucible	94
	3.4.3 Influence of furnace atmosphere	95
	3.4.4 Influence of residual oxygen in inert atmosphere	95
	3.4.5 Influence of reduced pressure	96
	3.4.6 Influence of humidity control	97
	3.4.7 Special points in connection with automatic sample changers	97
	3.4.8 Inhomogeneous samples and samples with very small changes in mass	98
3.5	Interpreting TGA curves	98
	3.5.1 Chemical reactions	99
	3.5.2 Gravimetric effects on melting	101
	3.5.3 Other gravimetric effects	101
	3.5.4 Identifying artefacts	103
	3.5.5 Final comments on the interpretation of TGA curves	104
3.6	Quantitative evaluation of TGA data	104
	3.6.1 Horizontal or tangential step evaluation	104
	3.6.2 Determination of content	106
	3.6.3 The empirical content	107
	3.6.4 Reaction conversion, α	110
3.7	Stoichiometric considerations	111
3.8	Typical application: rubber analysis	111
3.9	Analysis overview	112
3.10	Calibration and adjustment	112
	3.10.1 Standard TGA methods	113

3.11 Evolved gas analysis   114
    3.11.1 Brief introduction to mass spectrometry   115
    3.11.2 Brief introduction to Fourier transform infrared spectrometry   115
    3.11.3 Examples   117
Reference   118

3.1 Introduction

Thermogravimetric analysis (TGA) is an experimental technique in which the weight or, strictly speaking, the mass of a sample is measured as a function of sample temperature or time. The sample is typically heated at a constant heating rate (so-called dynamic measurement) or held at a constant temperature (isothermal measurement), but may also be subjected to non-linear temperature programs such as those used in sample controlled TGA (so-called SCTA) experiments. The choice of temperature program will depend upon the type of information required about the sample. Additionally, the atmosphere used in the TGA experiment plays an important role and can be reactive, oxidising or inert. Changes in the atmosphere during a measurement may also be made.

The results of a TGA measurement are usually displayed as a TGA curve in which mass or per cent mass is plotted against temperature and/or time. An alternative and complementary presentation is to use the first derivative of the TGA curve with respect to temperature or time. This shows the rate at which the mass changes and is known as the differential thermogravimetric or DTG curve.

Mass changes occur when the sample looses material in one of several different ways or reacts with the surrounding atmosphere. This produces steps in the TGA curve or peaks in the DTG curve. Different effects can cause a sample to lose, or even gain, mass and so produce steps in the TGA curve. These include the following:

- Evaporation of volatile constituents; drying; desorption and adsorption of gases, moisture and other volatile substances; loss of water of crystallisation. See Figure 3.1.
- Oxidation of metals in air or oxygen.
- Oxidative decomposition of organic substances in air or oxygen.
- Thermal decomposition in an inert atmosphere with the formation of gaseous products. With organic compounds, this process is known as pyrolysis or carbonisation.
- Heterogeneous chemical reactions in which a starting material is taken up from the atmosphere, for example reduction reactions with a purge gas containing hydrogen. Furthermore, reactions in which a product is evolved, for example decarboxylation or condensation reactions.
- Ferromagnetic materials. The magnetic properties of some materials change with temperature (Curie transition). If the sample is measured in an inhomogeneous magnetic field, the change in magnetic attraction at the transition generates a TGA signal. The magnetic field is produced by placing a permanent magnet in close proximity to the furnace close to the sample.
- Uptake or loss of water in a humidity controlled experiment.

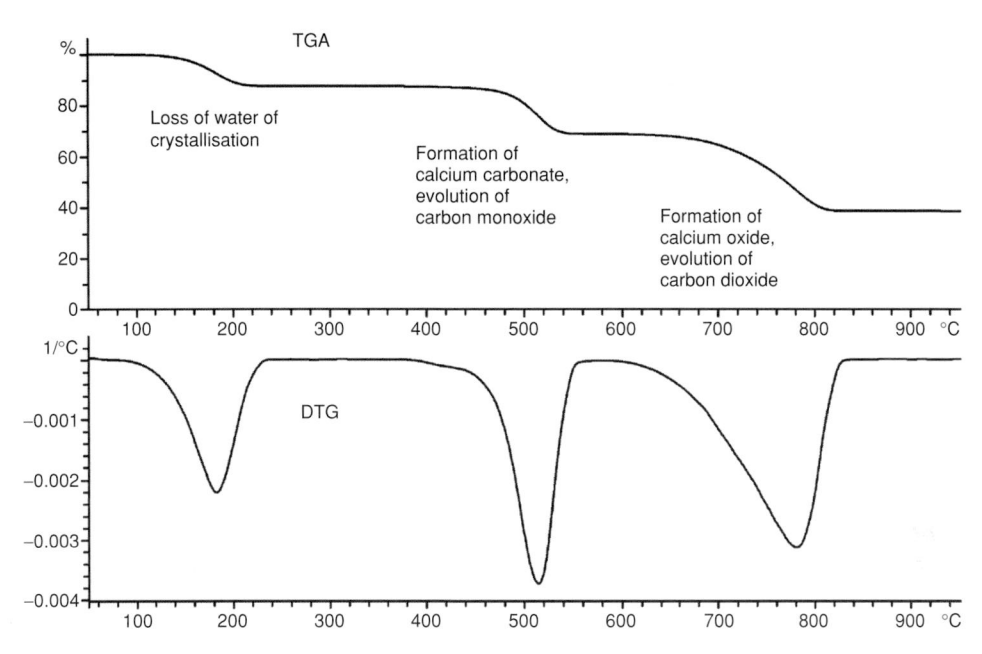

Figure 3.1 Stepwise decomposition of calcium oxalate monohydrate: sample mass 19 mg, heating rate 30 K/min, nitrogen. The TGA curve has been normalised (divided by the sample weight) and therefore begins at 100%. The temperature range of the three mass losses is particularly clear in the normalised first derivative or DTG curve.

3.2 Design and measuring principle

Three different designs of thermobalances are shown schematically in Figure 3.2. Nowadays, mainly compensation balances are used. With this type of balance the position of the sample in the furnace remains exactly the same even when the mass changes. At the same time, however, one can distinguish between simple moving coil measuring systems and more sophisticated weighing cells (e.g. parallel guided). In the horizontal arrangement, simple moving coil systems have the disadvantage that samples that move horizontally during

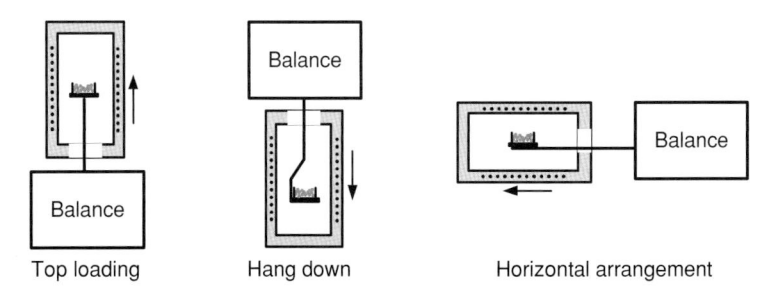

Figure 3.2 Thermobalance designs showing the top loading, hang down and horizontal arrangements.

heating (e.g. during melting) generate an apparent change in mass. Use of a parallel-guided system overcomes this problem.

Constructional measures have to be incorporated between the balance and the furnace to protect the balance against the effects of heat radiation and the ingress of corrosive decomposition products. In most cases the balance housing is purged with a protective gas.

In some thermobalances, an 'external' furnace is used whereby the furnace is not in contact with the atmosphere used in the experiment and this can be useful if measurements are made in pure hydrogen atmospheres.

Depending on the resolution they provide, balances are classed as semimicro- (10 μg), micro- (1 μg) or ultramicro- (0.1 μg) balances. Besides resolution, the (continuously measurable) maximum capacity of the balance is also an important factor. This is particularly the case when measuring inhomogeneous materials where a few milligrams are often hardly representative and a larger sample mass is desirable.

3.2.1 Buoyancy correction

Due to the change in density of a gas as the temperature changes, buoyancy corrections must be made in TGA measurements. Without corrections every sample will appear to show a mass increase during a heating experiment. TGA measurements are usually corrected for the effect of buoyancy by performing a blank measurement. A blank experiment uses the same temperature program and crucible as the experiment but without a sample. The resulting blank curve (also called a baseline) is then subtracted from the sample measurement curve. In some instruments, a 'standard' baseline is automatically subtracted from all measurements.

Buoyancy correction is essential for tests such as ash content where the residue at the end of the test needs to be determined accurately, and where very small weight losses are expected.

The temperature dependence of density at constant pressure is given by the equation

$$\rho = \rho_0 \frac{T_0}{T} \qquad (3.1)$$

where ρ_0 is the density of the gas at the reference temperature, T_0, of 25°C (298 K) and T is the temperature in kelvin.

From Table 3.1 it follows that a body with a volume of 1 mL experiences a buoyancy force in dry air of 1.184 mg at 25°C or 0.269 mg at 1000°C. This means that the body appears to become 0.915 mg heavier, i.e. to weigh more, when it is heated from 25 to 1000°C.

3.2.1.1 Simultaneous TGA/DTA

Modern thermobalances are often equipped so that they can record the DTA (differential thermal analysis; see Section 1.3) signal at the same time as the actual thermogravimetric measurement.

In addition to showing the energetic nature of weight loss events, the DTA signal can also show thermal effects that are not accompanied by a change in mass, e.g. melting,

Table 3.1 Density of several gases at 25, 500 and 1000°C at a standard pressure of 101.3 kPa

Gas	Density (mg/mL) at 25°C	Density (mg/mL) at 500°C	Density (mg/mL) at 1000°C
Dry air	1.184	0.457	0.269
Nitrogen	1.146	0.441	0.268
Oxygen	1.308	0.504	0.306
Argon	1.634	0.630	0.383
Helium	0.164	0.063	0.038
Carbon dioxide	1.811	0.698	0.424

crystallisation or a glass transition. The evaluation is usually restricted to the determination of onset and peak temperatures. Processes involving a loss of mass usually give rise to endothermic DTA effects because of the work of expansion. There are exceptions however. If combustible gases are formed at sufficiently high temperature (spontaneous ignition, catalytic effects on the surface of the platinum crucible) and sufficient oxygen is available, then the enthalpy of combustion is greater and the net effect is exothermic; see Figure 3.3.

If the DTA signal is calibrated and adjusted using the melting peaks of reference substances, the DTA curve (in °C) can be converted to a DSC-type curve (in mW). This then allows enthalpy changes to be quantitatively determined.

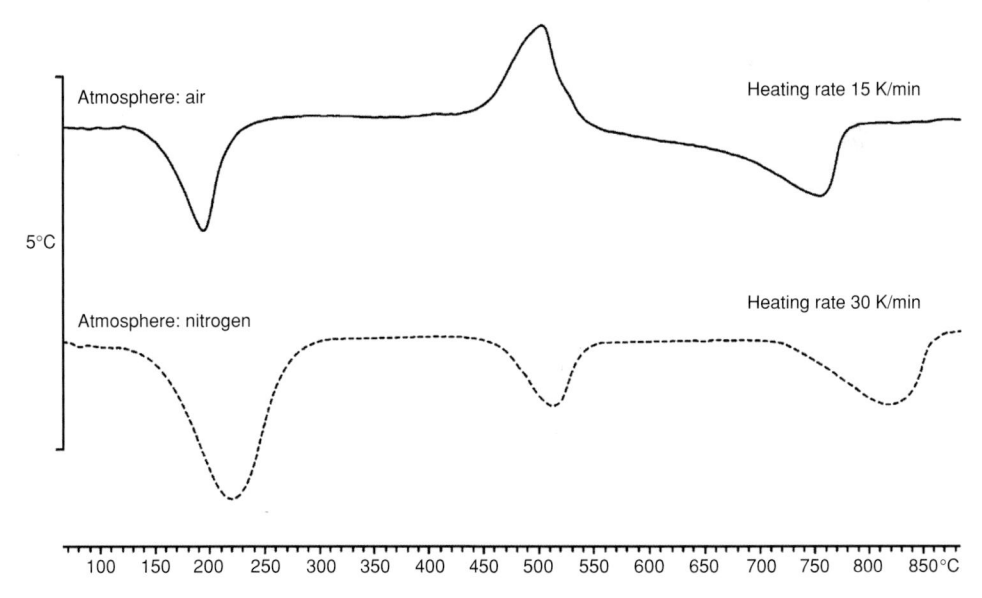

Figure 3.3 DTA curves of calcium oxalate monohydrate measured in air (upper curve) and in nitrogen (lower curve). The second effect at 500°C corresponds to the elimination of carbon monoxide, a process that is endothermic in an inert atmosphere. If oxygen is present, the CO immediately burns to CO_2, which produces an exothermic effect (above).

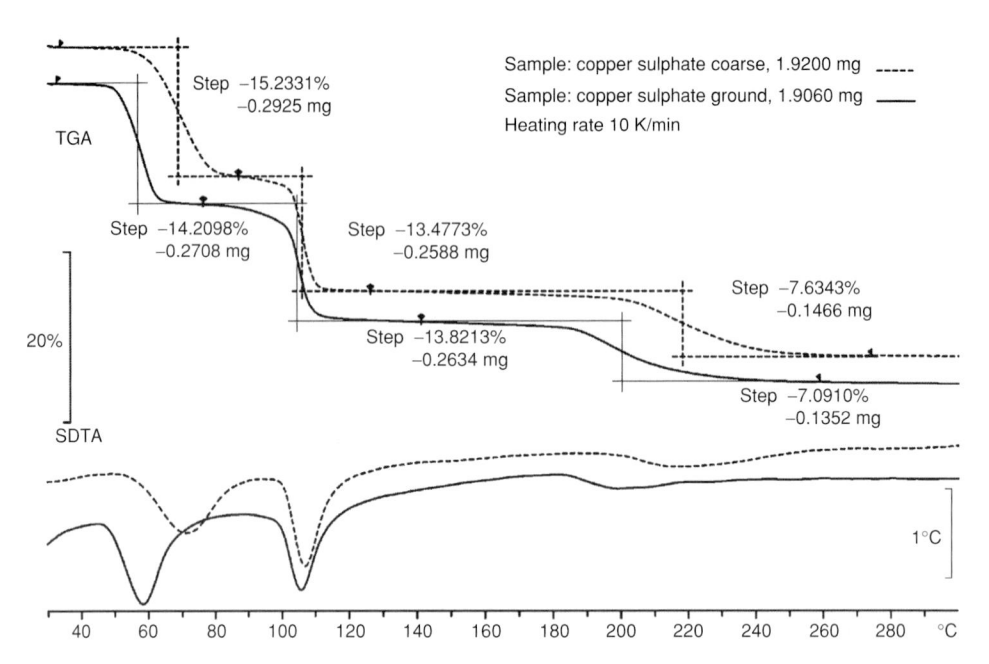

Figure 3.4 TGA and DTA curves showing the elimination of water of crystallisation from copper sulphate pentahydrate. The dotted curves were measured with rather coarse crystals, the solid curves with finely ground crystals. The first and third steps in the curve of the finely ground crystals are shifted more than 10 K to lower temperature. The heating rate was 10 K/min. Curves are offset for comparison.

3.3 Sample preparation

A number of factors should be considered when preparing samples for TGA experiments:

- The sample should be representative of the material being analysed.
- The mass of the sample should be adequate for the precision required for the test.
- The sample should be changed as little as possible by the sample preparation process.
- The sample should not be contaminated by the sample preparation process.

The morphology of the sample influences the diffusion rate of reaction products and in turn the course of the reaction. At the same time, it also affects heat transfer within the sample.

The mass used in the experiment also influences the rate of weight loss due the same diffusion and heat transfer processes.

Therefore, it is important in quality control measurements to use a consistent and reproducible sample preparation technique. In designing quality control methods, the robustness of the method to external factors such as sample preparation must be investigated. Using consistent sample masses is very important in producing comparable TGA data.

In Figure 3.4, the decomposition of finely ground copper sulphate pentahydrate is compared with the decomposition of more coarse crystals. The reaction is faster in the powdered sample because the gaseous reaction products reach the surface of the individual powder grains more quickly. This example illustrates the importance of sample preparation in thermogravimetric measurements.

3.4 Performing measurements

Thermogravimetric measurements are influenced by various factors, such as the following:

- Method parameters – heating rate, atmosphere (air, nitrogen, argon; pressure, humidity).
- Sample preparation – sample size, homogeneity and morphology of the sample: coarse crystals, fine powder.
- Choice of crucible
- Instrumental effects such as buoyancy and gas flow. These effects can be reduced or eliminated by performing blank curve subtraction.
- Changes in the physical properties of the sample during the measurement. For example, a change in emissivity (which affects the heat transfer within the sample and from the furnace to the sample) or the volume (which leads to a change in buoyancy).
- This sample may 'spit' or move and artefacts caused by such events can be minimised by grinding the sample or covering with a platinum mesh.

3.4.1 Influence of heating rate

The systematic deviation between the true sample temperature and the measured temperature, which is heating rate dependent, can be determined and corrected through temperature calibration and adjustment. This is usually done using pure metals that have good thermal conductivity properties. Real samples (e.g. polymers) can of course exhibit quite different thermal conductivity behaviour. For this reason, the measured sample temperature will still be expected to show a slight dependence on the heating rate, even if the instrument has been properly adjusted. The effect is very small with the onset temperature, but is more pronounced with the peak temperature.

If the sample undergoes chemical reactions, the temperature region in which the reaction occurs is very much dependent on the heating rate. In general, higher heating rates cause reactions to shift to higher temperatures. The choice of the heating rate is particularly important if secondary reactions occur with starting temperatures that differ only slightly from each other. If unsuitable heating rates are used, the reactions may overlap and remain undetected. It is, however, often possible to separate different reactions by choosing favourable heating rates (in general lower, sometimes higher).

3.4.1.1 Sample controlled TGA

A quite different approach for separating overlapping reactions makes use of rate of change in sample weight to automatically control the heating rate: the faster the change in mass, the slower the heating rate. Pioneers in this field include Jean Roquerol and Ole Toft Sorensen [1].

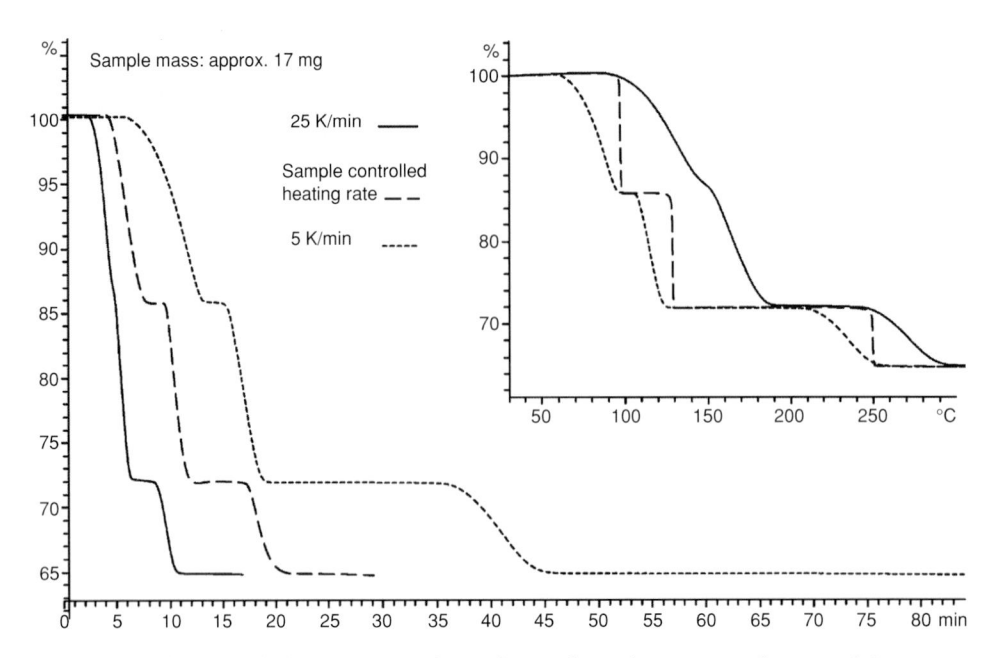

Figure 3.5 Influence of the heating rate on the resolution of partial reactions. In the inserted diagram on the right, the dotted and solid TGA curves of copper sulphate pentahydrate were measured conventionally at 5 and 25 K/min, whereas the dashed curve was recorded using the sample controlled heating rate. In this presentation of mass against temperature, the steps in the curve appear to be nearly vertical because, at low heating rates, the reaction takes place almost isothermally. In contrast, in the mass against time presentation (main diagram), the shapes of the three curves at first sight appear similar. On closer inspection, the better separation obtained using sample controlled heating rates – especially in the first two steps – becomes apparent.

Nowadays, software is available that can control the heating rate in this way. This technique is generally known as sample controlled or constrained rate thermal analysis (SCTA); see Figure 3.5. Care must be taken when using these techniques not to produce artificial steps in the data.

3.4.2 Influence of crucible

During the TGA measurement, crucibles must of course be 'open' to the atmosphere. It can be important, however, to seal the sample hermetically first, before the actual measurement is performed in order to prevent it from coming into contact with air. The lid of the crucible is then pierced immediately before the start of the measurement (e.g. in the sample changer).

Reactions in the gas phase proceed more rapidly in completely open crucibles than in a so-called self-generated atmosphere. In a sealed crucible with a very small hole in the lid, or in a crucible with a lid without a hole placed loosely over the sample, the weight loss is shifted to a higher temperature.

The material of which the crucible is made must not influence the reaction of the sample. In general, alumina (aluminium oxide) crucibles are used for TGA measurements. These have the advantage that they can be heated to over 1600°C. Sapphire crucibles are even more resistant and are especially suitable for the measurement of metals with high melting points, such as iron, which partially dissolve and penetrate ordinary alumina crucibles at high temperatures.

Platinum crucibles have the advantage of good thermal conductivity, which improves DTA performance. Finally, platinum is not always inert. It has a catalytic effect and can, for example, promote combustion reactions. This is also discussed in Section 10.3.2.

3.4.2.1 Note on metal samples in platinum crucibles

Damage to platinum crucibles due to alloy formation with metal samples can be prevented by covering the bottom of the crucible with a very thin layer of α-aluminium oxide powder before inserting the metal sample.

3.4.3 Influence of furnace atmosphere

Clearly, the mass of a sample in a closed system remains constant and cannot be a function of temperature or time. Thermogravimetric measurements are only possible if the sample is free to exchange material with its immediate surroundings. An important requirement is therefore that the gas atmosphere surrounding the sample can be changed to suit the experimental requirements.

First, a protective gas is required to protect the balance against any corrosive gases that may be evolved. Typically, dry inert gases such as nitrogen or argon at flow rates of 30 mL/min are used but users should always refer to the manufacturer's instructions. Besides the protective gas, a purge gas and/or reactive gas can be led into the furnace chamber via separate gas lines. The purge gas removes the gaseous reaction products from the furnace chamber. If helium is used as purge gas, heat transfer from the wall of the furnace to the sample improves, especially at temperatures below about 700°C. Reactive gases can be delivered to the sample in order to observe the interaction of the reactive gas with the sample. Examples of reactive gases are air or oxygen (oxidation) or hydrogen (catalysis, reduction) diluted with argon (usually 4% hydrogen with 96% argon) to prevent the possibility of an explosion. Typically, flow rates of 30 mL/min are used for reactive and purge gases.

If high concentrations of hydrogen are being used then special care must be taken to remove the risk of explosions. Instruments where the furnace does not come into direct contact with the atmosphere can be useful for these measurements.

3.4.4 Influence of residual oxygen in inert atmosphere

Very often the question arises as to the amount of residual oxygen in the system. This can very easily be determined by measuring the combustion rate of activated carbon at 700°C in the thermobalance.

Table 3.2 Sources of residual oxygen and precautions

Sources of residual oxygen	Precautions
Oxygen content of the purge gas Leaks in the gas supply tubing and other fittings and connections	Use an inert gas containing less than 10 ppm oxygen. Oxygen can diffuse through plastic tubing! Either use very short sections of plastic tubing (less than 50 cm) or metal tubing with a minimum number of connections and joints. Carefully check all joints for leaks.
Outgassing of constructional components (oxygen adsorbed on parts of the measuring cell) and dead volume	Switch on the flow of protective gas to the microbalance several hours before measurement. Purge the vacuum connection as well. Only open the furnace briefly to insert the sample. Active replacement of air through slight evacuation to about 1 kPa. Then flood with the desired purge gas (if necessary twice).
Ingress of atmospheric oxygen due to leaks	A possible cause of leakage is the furnace seal which could be damaged or dirty.
Ingress of atmospheric oxygen due to back diffusion at the purge gas outlet	Attach a long narrow tube to the outlet (this functions as a diffusion baffle).

For example, given a purge gas flow rate of 100 mL/min a 5 µg/min weight loss corresponds to 90 ppm concentration of oxygen. A routine check might require the weight loss to be less than, say, 10 µg/min; see Table 3.2.

3.4.5 Influence of reduced pressure

Mass losses through vaporisation or evaporation often occur at the same time as a decomposition reaction and are therefore difficult to distinguish from one another. The separation of the effects can often be improved by reducing the pressure in the measuring cell. A typical example is shown in Figure 3.6. The measurement at normal (atmospheric) pressure shows that mass is lost from about 320°C onwards and the process possibly occurs in two steps. When the measurement is performed under reduced pressure at 1.5 kPa (15 mbar), the separation of the two steps is greatly improved.

Even at reduced pressure, a purge gas must still be used to protect the microbalance against the condensation of possibly corrosive decomposition products.

In vacuum operation, the vacuum pump is normally in continuous operation because the reaction products, air from possible leakages, and the purge gas have to be removed in order to achieve a constant vacuum. To obtain realistic values for the pressure in the furnace chamber, the pressure meter should be installed close to the furnace chamber and not in the vacuum line leading to the vacuum pump. The working pressure is typically in the range 0.1–10 kPa.

Thermogravimetric Analysis

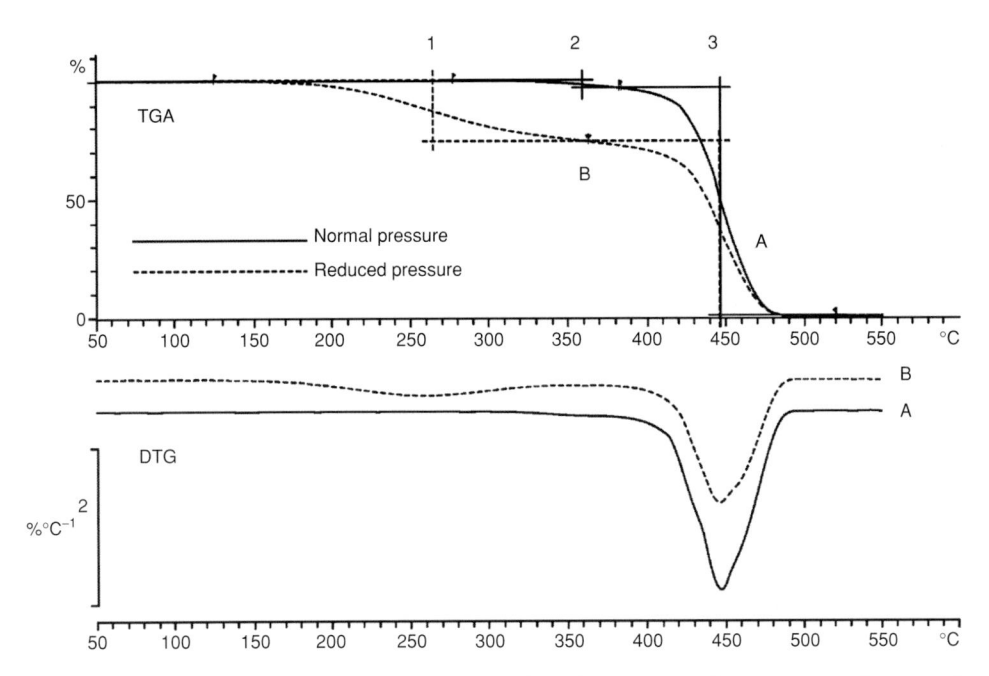

Figure 3.6 Effect of pressure on the decomposition of an elastomer: sample A is measured at normal (atmospheric) pressure, and B under reduced pressure, 1.5 kPa (15 mbar). Under reduced pressure, the vaporisation of the volatile components (additives, 1) is clearly separated from the decomposition of the elastomer (2,3). At normal pressure, the step height evaluated (on an expanded scale) from the stable region of the baseline at 270°C to next DTG maximum at 380°C does not correspond to the true additive content.

Use of reduced pressure may require recalibration of the temperature scale if high temperature accuracy is required.

3.4.6 Influence of humidity control

The interaction of water with many materials can have a significant impact on the material properties, for example plasticisation of the glass transition temperature and stability. TGA measurements under controlled humidity are useful in studying the adsorption and desorption of water and for distinguishing bound and unbound water. Dynamic vapour sorption (DVS) is a similar technique and has its own class of dedicated instrumentation. A full description of this is outside the scope of this book.

3.4.7 Special points in connection with automatic sample changers

Increasingly, automatic sample changers are being used. This poses particular problems since the samples on the turntable waiting for measurement must be protected against loss of volatile components such as moisture, and against the uptake of water or oxygen from the surrounding air. Ideally, the crucible should only be opened just before it is actually measured.

This can be done in a number of ways:

1. The sample is protected against direct contact with the surrounding atmosphere by placing an *aluminium lid* on top of the crucible. The lid is then removed and held by the sample robot's gripper during the measurement. In this case, alumina or other high-temperature crucibles can be used.
2. An aluminium crucible is sealed with a *perforable aluminium lid* and then automatically pierced, or opened, immediately before insertion into the measuring cell. In this case, the maximum temperature of the experiment is limited to 640°C to prevent melting of the crucible.
3. The carousel may be purged with a protective gas.

3.4.8 Inhomogeneous samples and samples with very small changes in mass

If the volatile content of the material is very low, or if the material is inhomogeneous, then clearly a large sample must be used.

An idea of the sample mass required to detect small changes in mass can be obtained by considering the following imaginary experiment.

You want to determine an ash residue of 1% with an accuracy of about 1%. If the reproducibility of the blank curve is about 10 μg, then an ash residue of about 1 mg is required to obtain 1% accuracy. It follows that the quantity of sample required is 100 mg.

3.5 Interpreting TGA curves

In addition to the TGA curve, a number of other curves can be used for interpretation purposes:

- The first derivative (DTG curve, rate of change of mass).
- The DTA curve (shows exothermic or endothermic events similar to DSC); DTA curves can be interpreted in much the same way as DSC curves.
- EGA, evolved gas analysis, online FTIR or MS measurements of evolved gases.

The visual inspection of the sample after the measurement – if possible using a reflected light microscope – can yield qualitative information about the residue (ash like, glassy, white or coloured powder, soot particles).

The effects described in the following sections show the typical shape of TGA curves. The curves are blank curve corrected.

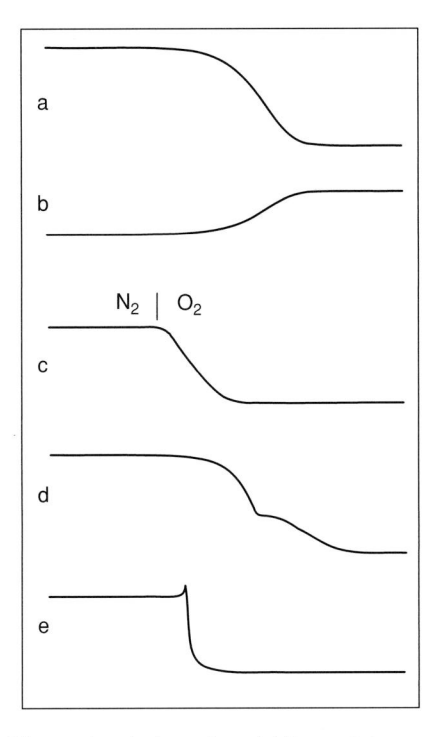

Figure 3.7 TGA curves of different chemical reactions. (a) Thermal decomposition with the formation of volatile reaction products. (b) Corrosion, oxidation of metals (formation of non-volatile oxides). (c) Combustion of carbon black on switching from N_2 to O_2. (d) Multi-step decomposition. (e) Explosive decomposition with recoil effect.

3.5.1 Chemical reactions

- The width of the mass loss step of a chemical reaction can be about 100 K (from 1 to 99% conversion). Usually, the step develops quite slowly from the initially horizontal part of the TGA curve. The point of inflection is at about 60% conversion. The radius of curvature at the end of the reaction is significantly smaller at the beginning (Figure 3.7a).
- If the reaction occurs stoichiometrically, the molar mass of the eliminated molecule can be calculated. For example, the molar mass of aspartame (Figure 3.9) is 312 g/mol; the step of 10.4% at 180°C therefore corresponds to 32 g/mol. The possibilities are O_2, S or CH_3OH. However, since aspartame is not a peroxide and does not contain sulphur, methanol is the most likely decomposition product.
- Oxidation reactions, for example the rusting of metals, lead to an increase in mass (Figures 3.7b and 3.8) and may also affect Curie point calibration substances.
- Diffusion-controlled reactions occur at an almost constant rate (i.e. the slope of the TGA curve is almost constant, Figure 3.7c). Here the transport of reactants or products is the limiting factor.
- The decomposition process often occurs in several steps (Figure 3.7d).

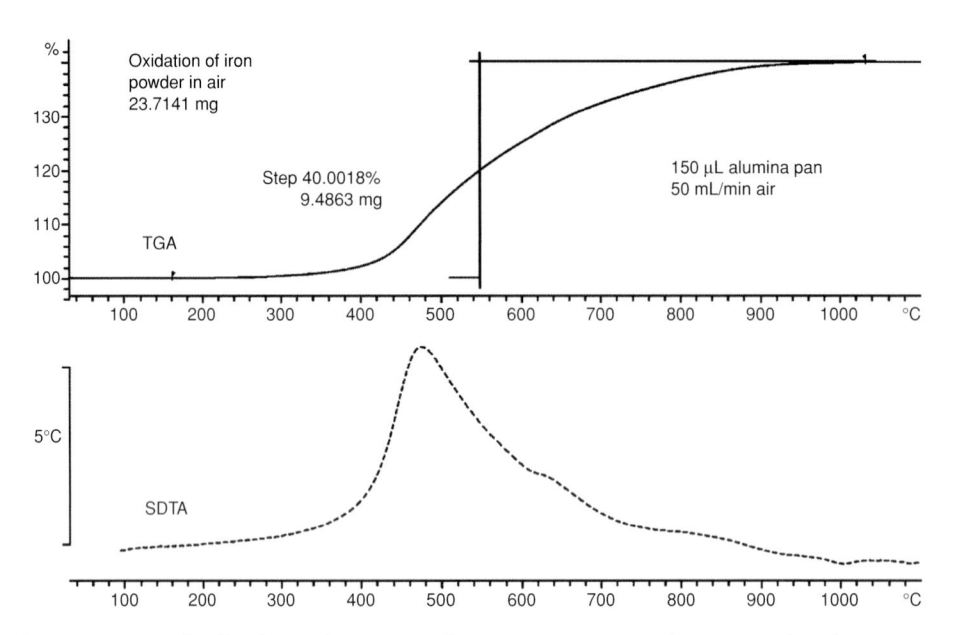

Figure 3.8 Example of a chemical reaction with an increase in mass. The iron powder takes up 40% oxygen in air and forms Fe_3O_4 and Fe_2O_3. The SDTA curve below confirms that the reaction is strongly exothermic; heating rate 20 K/min, Al_2O_3 crucible 150 µL. This reaction has an exceptionally wide temperature range of 600 K for 1–99% conversion.

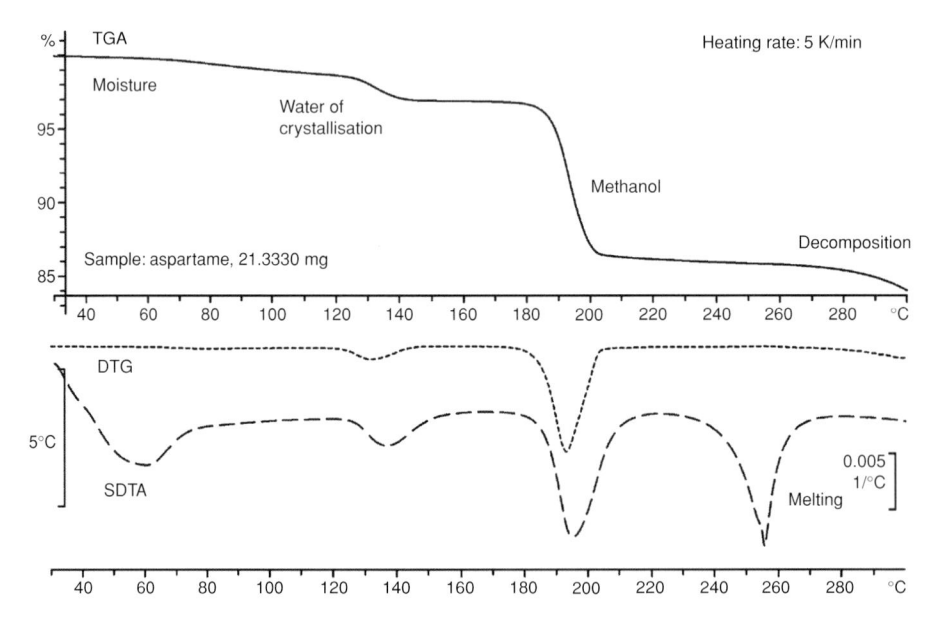

Figure 3.9 The decomposition of the artificial sweetener aspartame appears to be relatively complex. First, the water of crystallisation is lost at about 130°C. This is followed by the elimination of methanol at 180°C and the formation of a piperazine ring. This reaction takes place in an extremely narrow temperature range of just 20 K for 1–99% conversion. The SDTA curve shows that the piperazine derivative melts at 250°C.

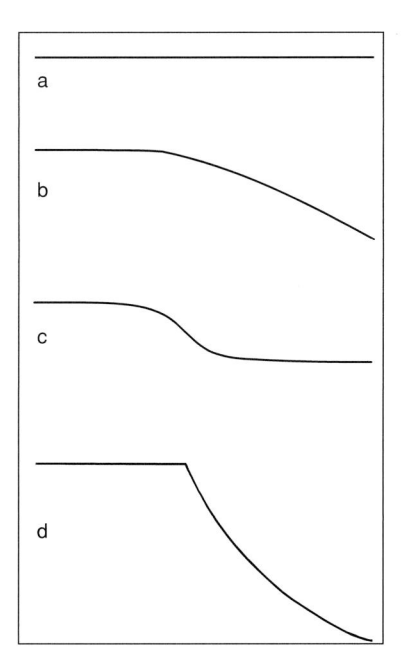

Figure 3.10 Thermogravimetric effects on melting. (a) Sample with low vapour pressure (no TGA effect). (b) Volatile melt (the liquid sample evaporates). (c) Moisture escapes on melting. (d) Sample melts and decomposes.

- Explosive materials sometimes decompose so rapidly that the recoil disturbs the TGA signal (Figure 3.7e). This problem can be overcome by using smaller sample quantities or diluting the sample with an inert material.

3.5.2 Gravimetric effects on melting

In general, one does not expect to observe a gravimetric effect on melting. The change in buoyancy due to the small change of sample density on melting is usually less than 1 µg. Even so, the point at which the sample melts can often be seen in the TGA curve. This is due to effects such as increasing vapour pressure or more rapid decomposition in the liquid phase (Figures 3.10 and 3.11). In some cases, however, the melting process can immediately be followed by a loss of trapped solvent. Simultaneous TGA/DTA instruments are useful in identifying such events.

3.5.3 Other gravimetric effects

These include drying steps, which typically occur at the beginning of the temperature program and have a width of about 100 K (Figure 3.12a). The desorption of other substances such as solvent residues or monomers occurs in much the same way. Organic compounds

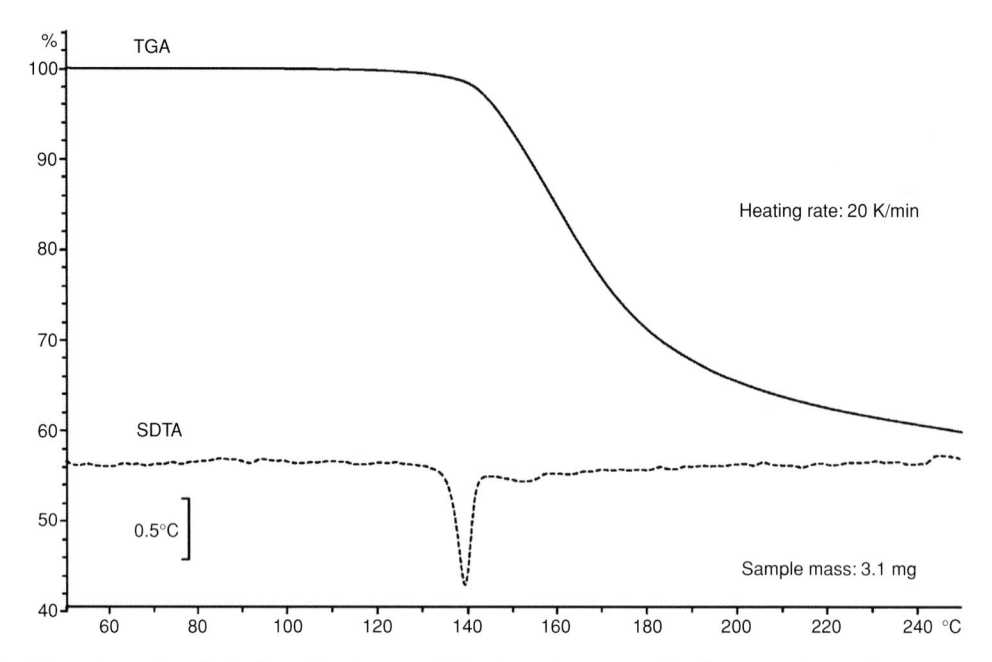

Figure 3.11 Acetylsalicylic acid melts at 140°C begins to decompose. The decomposition product acetic acid would be expected to yield a mass loss of 33.3%. The measured loss of mass is clearly greater. This is due to the simultaneous vaporisation of other substances.

consisting of relatively small molecules show a tendency to sublime, that is, they pass directly from the solid phase into the gas phase.

Liquids evaporate in a completely open crucible over a wide temperature range below their boiling point. If the crucible lid has a small hole in it, a so-called self-generated atmosphere is created in which the vapour molecules remain in equilibrium with the liquid phase right

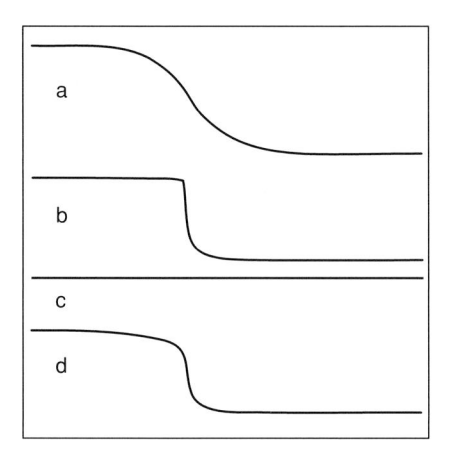

Figure 3.12 Other thermogravimetric effects. (a) Drying, desorption, sublimation. (b) Boiling in a crucible with a small hole in the lid. (c) Ferromagnetic Curie transition without a magnet: no TGA effect. (d) The same sample with a permanent magnet below the furnace.

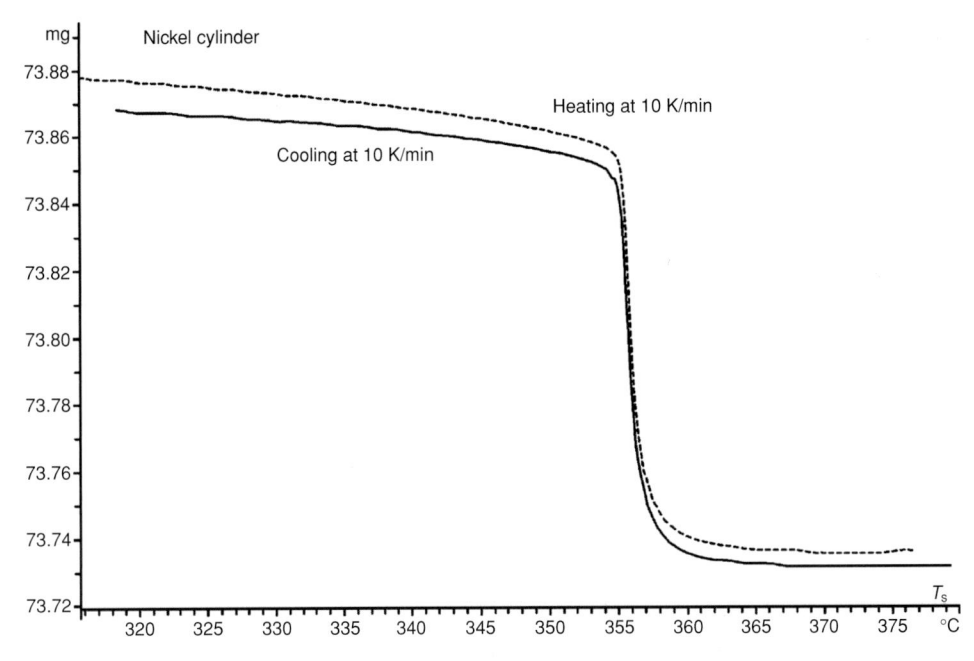

Figure 3.13 A permanent magnet placed below the furnace of the thermobalance attracts a ferromagnetic material and causes an apparent increase in mass. When the Curie temperature of the sample is exceeded, the force is no longer exerted and there is a sudden loss in apparent mass. This effect is reversible and occurs again on cooling. The abscissa shows the sample temperature, T_s.

up until the boiling point is reached. The liquid then quickly evaporates and produces a sharp TGA step whose onset temperature corresponds to the boiling point (Figure 3.12b).

Ferromagnetic materials become paramagnetic above their Curie temperature. In an inhomogeneous magnetic field this transition causes a gravimetric signal. The magnetic field is produced by placing a permanent magnet at a cool place close to the furnace. Its magnetic field exerts an attractive force on the ferromagnetic sample and a different mass is recorded. When the Curie temperature is exceeded, the force is no longer exerted and there is a sudden change in apparent mass (Figures 3.12d and 3.13).

3.5.4 Identifying artefacts

Artefacts are effects observed on the measurement curve that are not directly caused by the sample, that is, effects that have nothing to do with the sample properties you want to measure.

The main types of TGA artefacts are as follows:

- Buoyancy effects caused by the density of the surrounding gas decreasing on heating. Typically this results in an apparent mass gain of 50–200 µg. Since buoyancy effects are reproducible, the curves can be corrected by performing automatic blank curve subtraction. This also applies to buoyancy effects due to gas switching, a technique often used in TGA.

- Fluctuations of the purge gas flow rate can also affect the measurement curve. The flow rate should therefore be kept as constant as possible and not be changed during a measurement.
- Sudden loss of mass arising from ejection of part of the sample. This effect often occurs when a sample decomposes with the formation of gas. It can be prevented by covering the sample with coarse grain Al_2O_3 or using a crucible lid with a hole.
- Apparent mass gain caused by samples that foam and make contact with the furnace wall. This problem can be overcome by using smaller samples.
- Artefacts can be identified if repeat measurements are made, and blank experiments are performed.

3.5.5 Final comments on the interpretation of TGA curves

If the interpretation of TGA effect is uncertain, other measurement techniques can be used to gain additional information. These include the following:

- Analysis of evolved gaseous compounds (EGA).
- DSC measurements.
- Observation of the sample under the hot-stage microscope.

3.6 Quantitative evaluation of TGA data

Most TGA evaluations involve the determination of step heights of TGA effects. Since several effects occur one after another and often partially overlap, the question is how do we choose the limits for evaluating the step. Here, the first derivative or DTG curve is very helpful. In the DTG curve, the mass loss steps appear as more or less distinct peaks. The evaluation software provides methods for evaluating the steps in the TGA curves. These yield the absolute and percentage change in sample mass (the step height) of the different effects.

3.6.1 Horizontal or tangential step evaluation

Normally, the steps are interpreted as a change in mass before and after the effect, so that the corresponding 'baseline' is horizontal. The sum of all the steps plus the residue at the end of the last step is equal to the original sample mass or 100% (assuming blank curve correction was properly performed and that no oxidation of the sample has occurred).

The steps are defined manually or, in some cases, automatically. Depending on the sample the steps are easily identified (Figure 3.14). When this is not true then the DTG curve is often used to help define the two evaluation limits required for each TGA step (Figure 3.15).

The residue is generally only of interest after the last step. Sometimes relatively sharp decomposition steps of other components occur during broad vaporisation or decomposition processes. In such cases, the step evaluation with a 'tangential' baseline is used (Figures 3.16 and 3.17).

From the step heights, you can determine the content or check the stoichiometry of reactions using pure starting materials.

Thermogravimetric Analysis

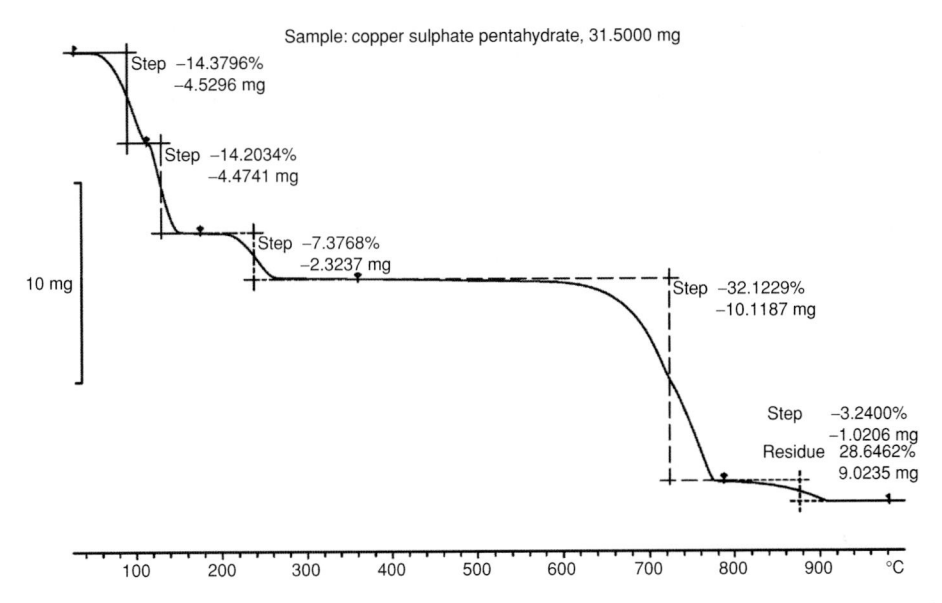

Figure 3.14 Separation in a TGA curve of copper sulphate pentahydrate at 10 K/min. The individual steps correspond to the consecutive elimination of the following molecules: 2 H_2O, 2 H_2O, 1 H_2O, SO_3, 0.5 O_2. The residue is Cu_2O.

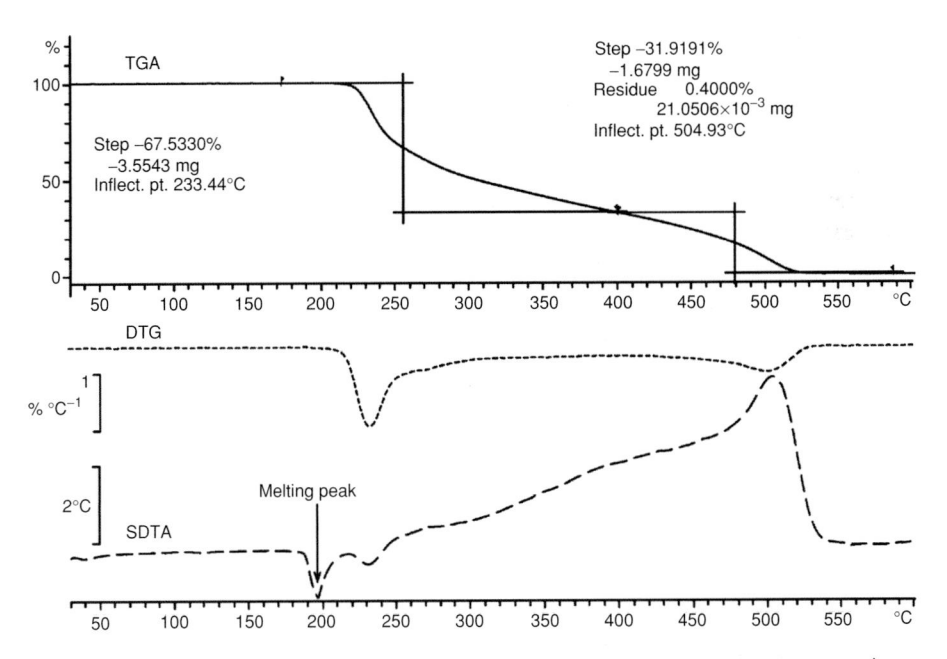

Figure 3.15 The decomposition of sucrose at 10 K/min in air occurs in two consecutive steps, the second directly following the first. In the evaluation, a value of 405°C was used for the end of the first step and the beginning of the second step. The simultaneously measured SDTA curve shows the endothermic melting peak followed by the endothermic caramelisation, at which stage the exothermic decomposition/combustion occurs. An almost white residue of 0.4% remains at the end of the measurement. This corresponds to the ash content.

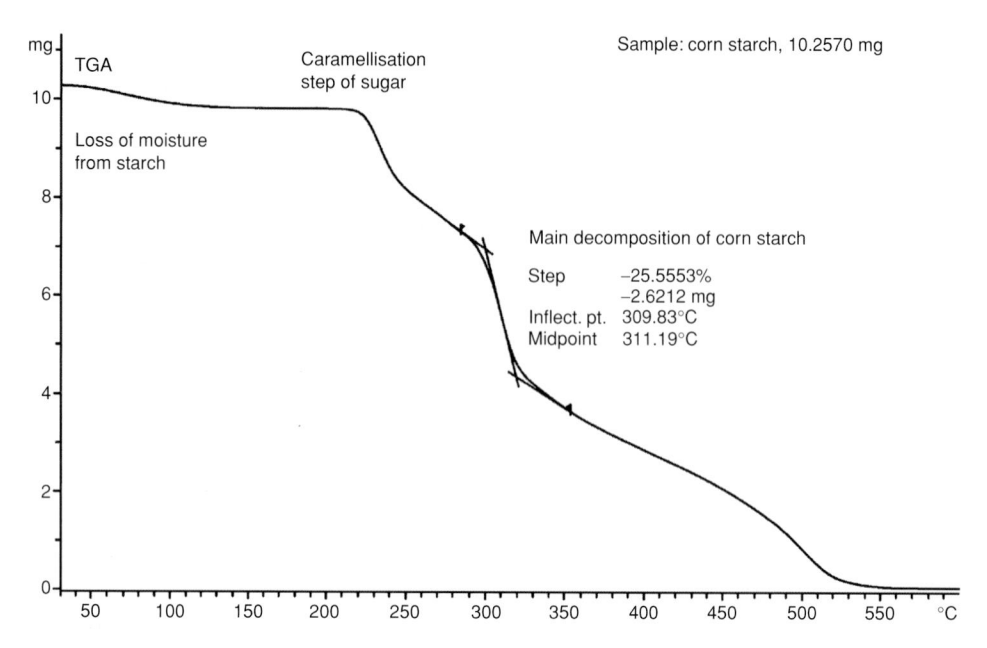

```
mg                                  Sample: corn starch, 10.2570 mg
      TGA           Caramellisation
                    step of sugar
10
     Loss of moisture
     from starch
 8
                              Main decomposition of corn starch
                              Step         −25.5553%
 6                                          −2.6212 mg
                              Inflect. pt.  309.83°C
                              Midpoint     311.19°C

 4

 2

 0
     50   100  150  200  250  300  350  400  450  500  550   °C
```

Figure 3.16 TGA curve of a 1:1 mixture of maize-corn starch and crystalline sugar: heating rate 10 K/min, purge gas air at 50 mL/min. The main decomposition step of corn starch occurs at roughly 310°C (midpoint). In a mixture with crystalline sugar, the 'starch' step occurs in a region in which the mass of the sugar continuously decreases. The step obtained using a 'horizontal' evaluation would be too large. This error can be avoided using a 'tangential' baseline evaluation. The measured step of 25.5% refers to the initial total mass of the mixture at the beginning of the measurement (see Section 3.6.3).

3.6.2 Determination of content

Content determination is easiest when the constituent you want to determine disappears completely during the measurement (e.g. moisture or other volatile substances, residue-free decomposition reactions as well as certain depolymerisation reactions, or the combustion of carbon black). In this case, the percentage content, G, can be calculated from the mass loss, Δm, and the initial sample mass, m_0, as follows:

$$G[\%] = \frac{\Delta m}{m_0} \times 100\%. \tag{3.2}$$

See Figure 3.18.

In the case of stoichiometric reactions where there is only partial mass loss (e.g. dehydration, decarboxylation), G can be calculated using the equation

$$G(\%) = \frac{\Delta m}{m_0} \times \frac{M}{n \times M_{gas}} \times 100\% \tag{3.3}$$

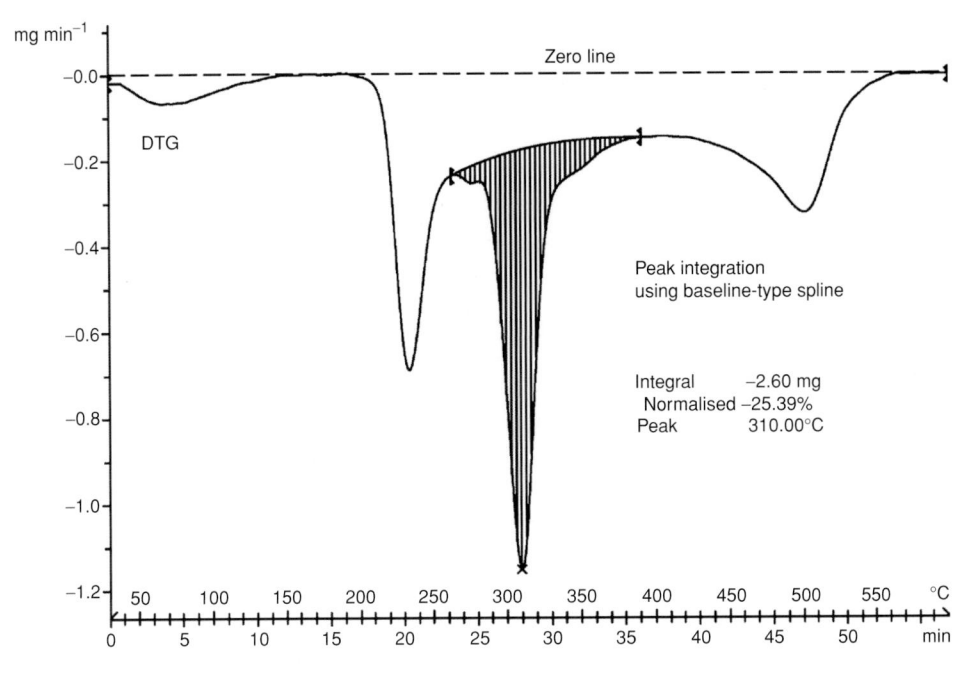

Figure 3.17 Instead of using a step evaluation of the TGA curve, a mass loss can be quantified by integrating the DTG peak that corresponds to the TGA step. This is illustrated using the example described in Figure 3.16. This method is useful for evaluating a relatively sharp decomposition step on the background of a broad step. It has the advantage that different baseline types can be used for the evaluation. In this example, the 'spline' baseline of the DTG curve simulates the decomposition of the sugar.

where M_{gas} and M are the molar masses of the eliminated gases and original constituent of interest in the sample material, and n is the number of moles of gas eliminated for each mole of sample material.

Example. Determination of the calcium carbonate content of a sample of limestone:

$$CaCO_3 \rightarrow CaO + CO_2 \quad (3.4)$$
$$100 \text{g/mol} \quad 56 \text{g/mol} \quad 44 \text{g/mol}$$

One CO_2 molecule is eliminated for every $CaCO_3$ molecule. This means that n is 1, M is 100 g/mol, and M_{gas} is 44 g/mol. The sample mass, m_0, is 10.1360 mg, the measured mass loss, Δm, is 4.2439 mg. Entering these numbers in the above equation gives a $CaCO_3$ content of 95.16%; see Figure 3.19.

3.6.3 The empirical content

Non-stoichiometric or complex reactions require a reference sample of known purity (content) whose TGA curve is measured beforehand under the same conditions. The empirical

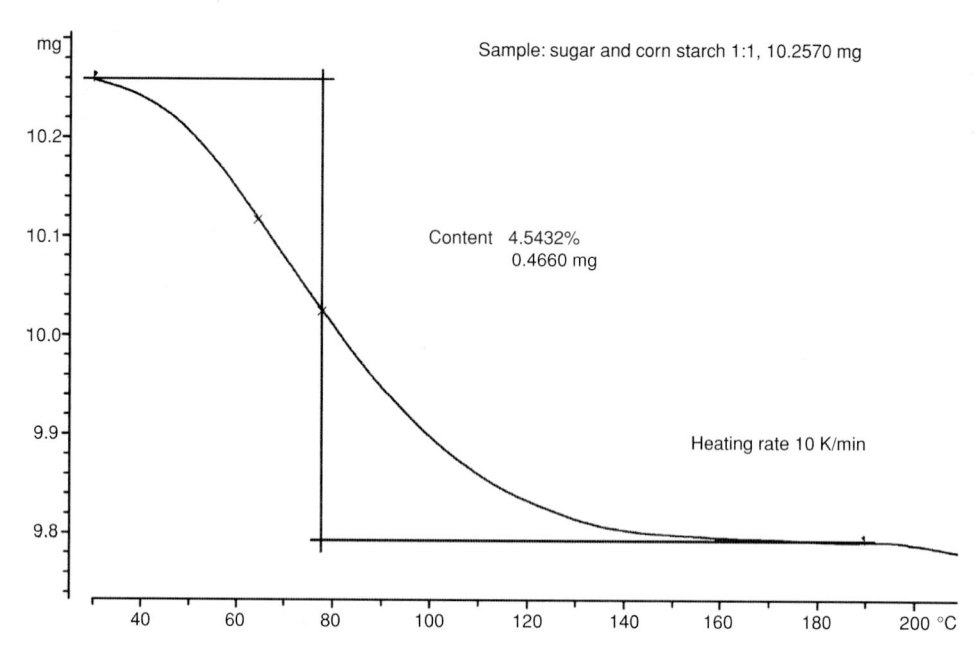

Figure 3.18 The moisture content of the starch–sugar mixture used in Figure 3.16. Sample mass is 10.2570 mg, mass loss is 0.4660 mg, and the content 4.54%.

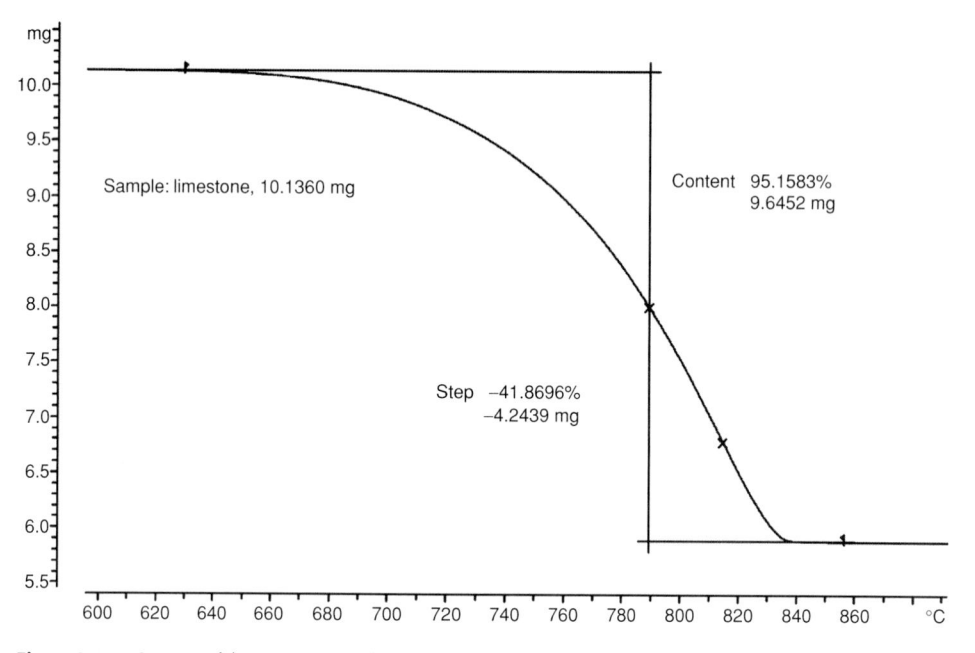

Figure 3.19 Section of the TGA curve of limestone used for stoichiometric content determination; heating rate 20 K/min. The CO_2 elimination resulted in a mass loss of 4.2439 mg, which corresponds to an original $CaCO_3$ content of 95.16%. The residue is mainly of CaO.

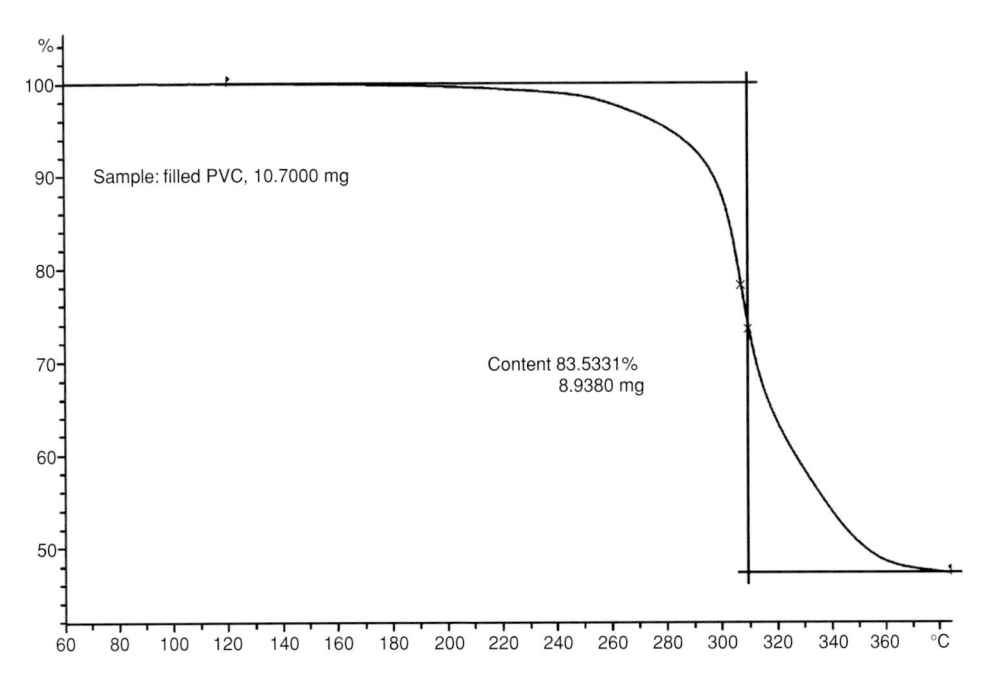

%
100-
90- Sample: filled PVC, 10.7000 mg
80-
70-   Content 83.5331%
                                   8.9380 mg
60-
50-
   60 80 100 120 140 160 180 200 220 240 260 280 300 320 340 360   °C

Figure 3.20 Pyrolysis reaction of PVC containing a filling material in nitrogen. A mass loss of 52.79% occurs. Divided by the reference step of the pure PVC (63.2%), this gives a PVC content of 83.5%. PVC eliminates HCl in the temperature range shown in Figure 3.20. Stoichiometrically this would result in mass loss of 58.3%. In fact, a mass loss of 63.2% was determined in the range 120–385°C. The value of R_{emp} is then 63.2%/100% or 0.632. (The content of unfilled PVC reference sample was assumed to be 100%.)

'reference step' R_{emp} is calculated as follows:

$$R_{emp} = \frac{\Delta m_r}{m\, G_0} \qquad (3.5)$$

Here Δm is the measured mass loss step in %, m the sample mass used, and G_0 is the purity (or content) of the reference sample in %.

The reference step is needed for the evaluation of the empirical content determination:

$$G\,(\%) = \frac{\Delta m}{m_0\, R_{emp}} \times 100\% \qquad (3.6)$$

PVC eliminates HCl in the temperature range shown in Figure 3.20. Stoichiometrically this would result in mass loss of 58.3%. In fact, a mass loss of 63.2% was determined in the

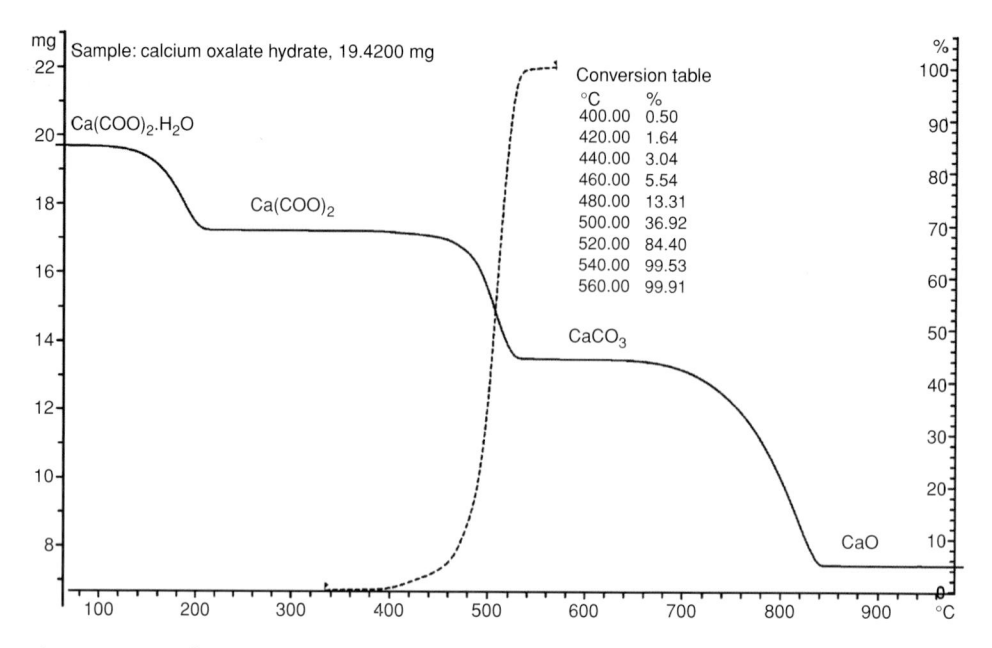

Figure 3.21 At a heating rate of 20 K/min, calcium oxalate eliminates CO and forms calcium carbonate between 350 and 600°C. The dotted curve displays the reaction conversion.

range 120–385°C. The value of R_{emp} is then 63.2%/100% or 0.632. (The content of unfilled PVC reference sample was assumed to be 100%.)

3.6.4 Reaction conversion, α

On the assumption that a mass loss step can be attributed to a single chemical reaction, the reaction conversion, α, can be calculated as a function of temperature.

If the mass loss of a sample is calculated in relation to the total mass loss step, one obtains the reaction conversion, α, relating to the temperature T:

$$\alpha(T) = \frac{\Delta m_T}{\Delta m_{tot}} \quad (3.7)$$

Here Δm_T is the mass loss at temperature T and Δm_{tot} is the step height.

The conversion calculated in this way can be calculated and displayed over the entire mass loss step. It is in fact a normalised presentation of this TGA step, which begins at 0 and ends at 1 (or 100%); see Figure 3.21.

From such conversion curves it is possible to obtain information about the reaction kinetics of the system observed using kinetic models or also with model-free kinetics (similar to with DSC).

3.7 Stoichiometric considerations

With a pure compound of known composition and molar mass, M, it is possible to calculate the molar mass M_{gas} of the eliminated gaseous elimination product of a step using the equation below, where n is the number of gas molecules for each original molecule:

$$n M_{gas} = \frac{\Delta m}{m_0} M \qquad (3.8)$$

Example. The second mass loss step of calcium oxalate monohydrate is 19.2%, and M is 146 g/mol. This means that $n M_{gas} = 0.192$ times 146 g/mol, which equals 28.03 g/mol. Since with this oxalate it is not possible to form a gas with a mol mass of 14 or 56 g/mol, n must be 1 and the eliminated compound is therefore carbon monoxide (see Figures 3.1, 3.3 and 3.21).

3.8 Typical application: rubber analysis

The aim of rubber analysis is to quantitatively determine the main constituents of technical elastomer systems. Basically, it is possible to distinguish between the following main components:

- Volatile substances such as moisture, plasticisers and other additives.
- The actual elastomers involved.
- Carbon black.
- Inorganic fillers and ash.

Using an optimum choice of measuring conditions (atmosphere, heating rate) it is, in fact, possible to determine all the constituents. Under nitrogen, the volatile constituents (plasticisers) vaporise up until about 300°C. Elastomers and other organic compounds undergo pyrolysis up to about 600°C. At 600°C the gas is switched from nitrogen to air or oxygen leading to the combustion of the carbon black. The inorganic fillers (typically zinc oxide, calcium carbonate or calcium oxide) and the elastomer ash are left behind as a residue. The results of such a rubber analysis are shown in Figure 3.22.

Besides the quantitative information obtained from the mass loss steps, qualitative results can also be obtained from the DTG curve. For example, the measured peak temperatures are characteristic of certain types of elastomers; some examples are summarised in Table 12.3.

Other methods can also be used including lowering the temperature before switching to the oxidising atmosphere.

Please also refer to Chapter 6 for additional discussion of the analysis of rubber compounds.

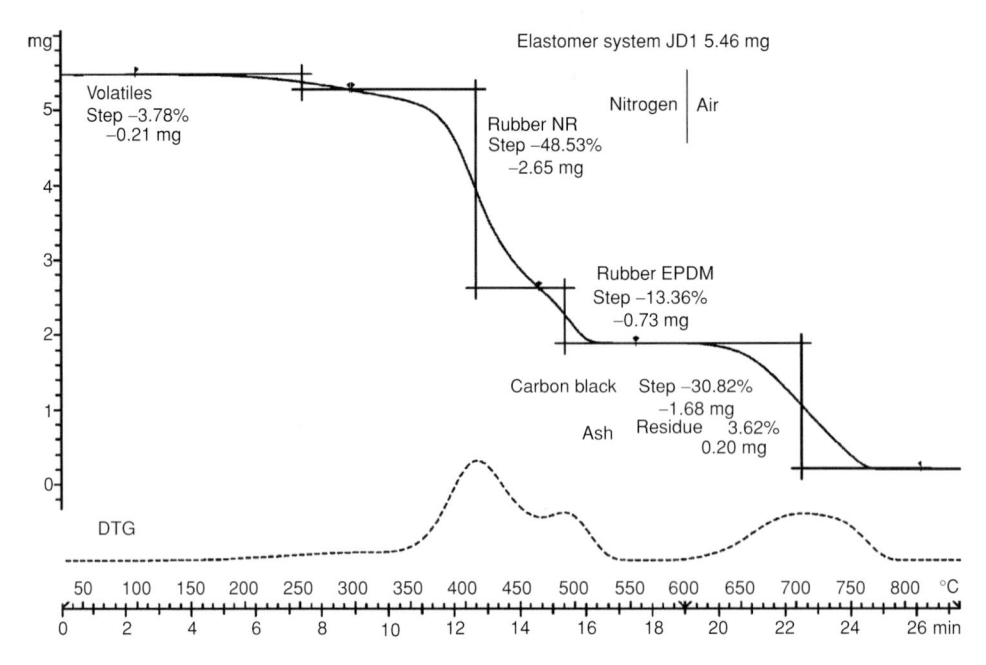

Figure 3.22 Compositional analysis of an elastomer system by TGA showing typical results; heating rate 30 K/min.

3.9 Analysis overview

Effect, property of interest	Evaluation used
Composition	Step evaluation, residue
Thermal stability, decomposition	Step evaluation
Stoichiometry of reactions	Content
Kinetics of reactions	Conversion, kinetics
Adsorption/desorption processes	Step evaluation
Vaporisation behaviour	Step evaluation, DTA, integration
Influence of reactive gases	Step evaluation
Moisture	Step evaluation
Oxidation stability	Onset

3.10 Calibration and adjustment

A thermobalance measures the sample mass as a function of sample temperature. It follows that the user must make sure that the mass and temperature values measured are in fact correct. This is done through frequent checking using a suitable reference sample (determining the deviation of a measured value with respect to the known true value). If the deviation is

Table 3.3 DTG peak temperatures under pyrolysis conditions, measured at 10 K/min

Elastomer	DTG peak temperature, measured at 10 K/min
NR, natural rubber	373°C
SBR, styrene butadiene	445°C
BR, polybutadiene	460°C
EPDM, ethylene-propylene terpolymer	461°C

too large, the instrument must be adjusted (the instrument parameters are changed in such a way as to eliminate the deviation). Following any adjustment, the check procedure must be repeated to ensure that the deviation is now within the allowed limits.

The balance is checked and adjusted using reference masses. Some TGAs are automatically adjusted with built-in reference masses when switched on and during longer periods of inactivity.

The temperature may be calibrated using a method based on Curie point transitions of certain ferromagnetic materials (e.g. Ni, Fe, and a range of alloys). This is done by placing a permanent magnet in close proximity to the reference sample (outside the furnace). The sample experiences a magnetic force of attraction and a higher mass is recorded. The sample is then heated in the normal way. At the so-called Curie temperature the sample loses its ferromagnetic properties and is no longer attracted by the magnet, resulting in a sudden loss in apparent mass. This transition occurs over a relatively narrow temperature range and produces a step in the TGA curve. The temperature is recorded at the point of inflection or the end set of the step and compared with the Curie temperature of the reference material. The temperature accuracy of such Curie points is limited to a few degree centigrades due to physical reasons (hysteresis, heating/cooling rate dependence, magnetic field dependence) and is not as accurate as the use of melting point standards.

The temperature can be calibrated more accurately using the DTA signal and melting point standards. Very pure samples of In, Zn, Al, Au and Pd are often used for this purpose. The onset temperatures of the melting peak are used for calibration and adjustment. A number of typical DTA melting peaks are shown in Figure 3.23 (lower curves). Consideration should also be made of potential reaction of the calibration materials with the purge gas; especially if oxidising or reactive purge gases are used.

3.10.1 Standard TGA methods

A number of international standard test methods exist for TGA and these are listed below:

E1131-03 Standard Test Method for Compositional Analysis by Thermogravimetry
E1582-04 Standard Practice for Calibration of Temperature Scale for Thermogravimetry
E1641-04 Standard Test Method for Decomposition Kinetics by Thermogravimetry

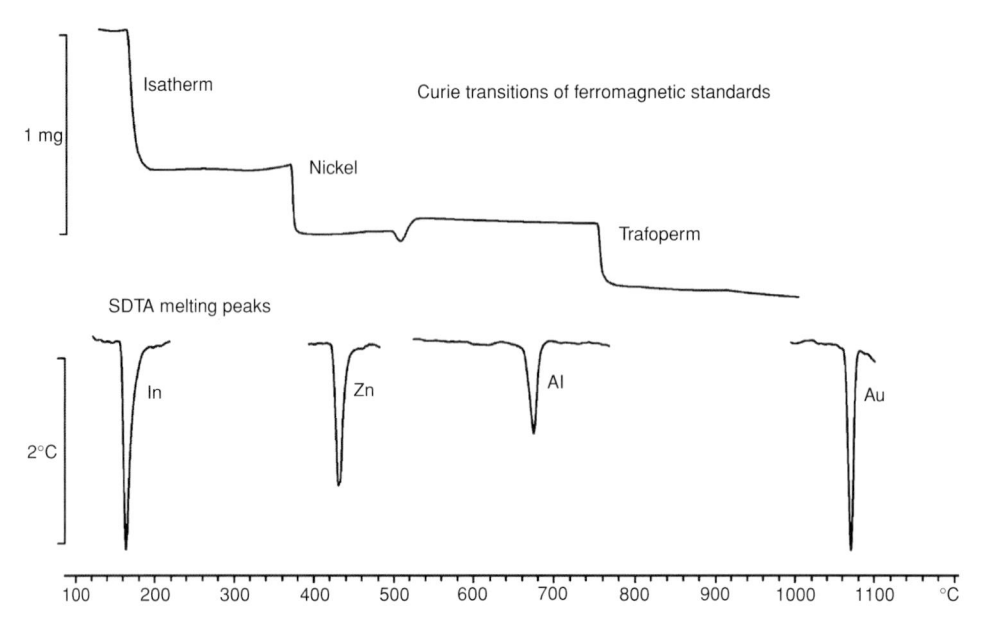

Figure 3.23 Temperature calibration by measuring ferromagnetic materials in a magnetic field (above) and using the DTA signal of pure metal melting point standards (below).

E1868-04 Standard Test Method for Loss-on-Drying by Thermogravimetry
E2008-04 Standard Test Method for Volatility Rate by Thermogravimetry
E2403-04 Standard Test Method for Sulfated Ash of Organic Materials by Thermogravimetry
ISO 9924-1/2 Rubber and Rubber Products – Determination of the Composition of Vulcanizates and Uncured Compounds by Thermogravimetry
ISO 11358 Plastics: Thermogravimetry (TG) of Polymers – General Principles
ISO 21870:2005 Rubber Compounding Ingredients: Carbon Black – Determination of High-Temperature Loss on Heating by Thermogravimetry
DIN 51006

3.11 Evolved gas analysis

Thermogravimetric analysis (TGA) is a technique that provides quantitative information on the change in mass of a sample as a function of time as it is heated, cooled, or held at constant temperature. TGA alone, however, is not an identification technique for unknown samples.

The combination of TGA with a mass spectrometer (MS) or a Fourier transform infrared spectrometer (FTIR) allows the nature of the gaseous products formed in the TGA to be investigated online. When several compounds are evolved, the MS or FTIR can track their evolution profiles. Mass spectra and infrared spectra are substance specific. The spectra can be used to characterise the substance or class of substance through spectral interpretation and comparison with database reference spectra. Decomposition pathways can thereby be elucidated.

	Features	Benefits
TGA-MS	High sensitivity coupled with very fast measurement.	Extremely small amounts of substances can be detected. Ideal for the online characterisation of all types of volatile compounds.
TGA-FTIR	High chemical specificity and fast measurement.	Characterises substances by identifying their functional groups, i.e. the class of substance. Ideal for the online measurement of substances that exhibit medium to strong infrared absorption.

3.11.1 Brief introduction to mass spectrometry

Mass spectrometry characterises substances by identifying and measuring the intensity of molecular fragment ions of different mass-to-charge ratio (m/z).

The gas molecules entering the mass spectrometer from the thermobalance are first ionised in the ion source. The positive molecular ion and fragment ions formed are then separated according to their m/z value by a combination of magnetic and electrostatic fields. A mass spectrum is recorded by scanning the field strength so that ions of increasing m/z ratio arrive at the detector.

In a TGA-MS system, the mass spectrometer is usually set to monitor individual m/z values that are characteristic of specific structural features.

3.11.2 Brief introduction to Fourier transform infrared spectrometry

Infrared spectroscopy measures the light absorbed by different types of vibrations in molecules.

Infrared radiation from the light source is divided into two beams by the beam splitter. One beam is reflected onto a moving mirror and the other onto a stationary mirror. Both beams are then recombined and pass through the sample to the detector. Fourier transformation of the resulting interferogram yields an infrared transmission spectrum.

In TGA-FTIR, the absorption bands of each spectrum are usually simultaneously integrated over the entire spectral range or over characteristic spectral regions. The intensity is presented as a function of time as so-called Gram–Schmidt curves or chemigrams.

3.11.2.1 Coupling the TGA to a gas analyser

The TGA is coupled to the MS using a fused silica capillary tube heated to 200°C to prevent condensation. Often just a small fraction of the evolved gas is taken into the mass

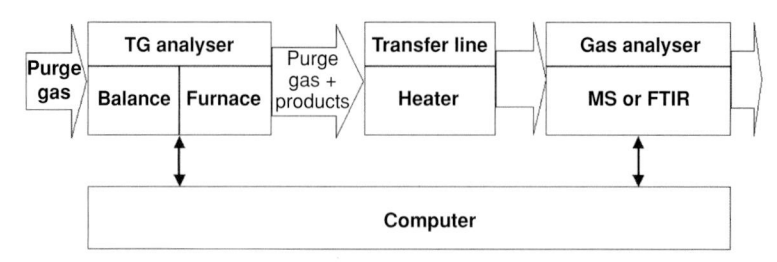

Figure 3.24 Coupling the thermobalance to a gas analyser.

spectrometer, but this can be increased by using larger diameter capillaries to increase sensitivity. The purge gas is often helium or argon (Figure 3.24).

In contrast, the TGA-FTIR combination uses the total volume of purge gas and gaseous decomposition products from the TGA. The gases are transferred through a heated glass-coated steel transfer capillary line into a heated gas cell installed in the FTIR spectrometer. Nitrogen, which does not exhibit IR absorption, is used as purge gas.

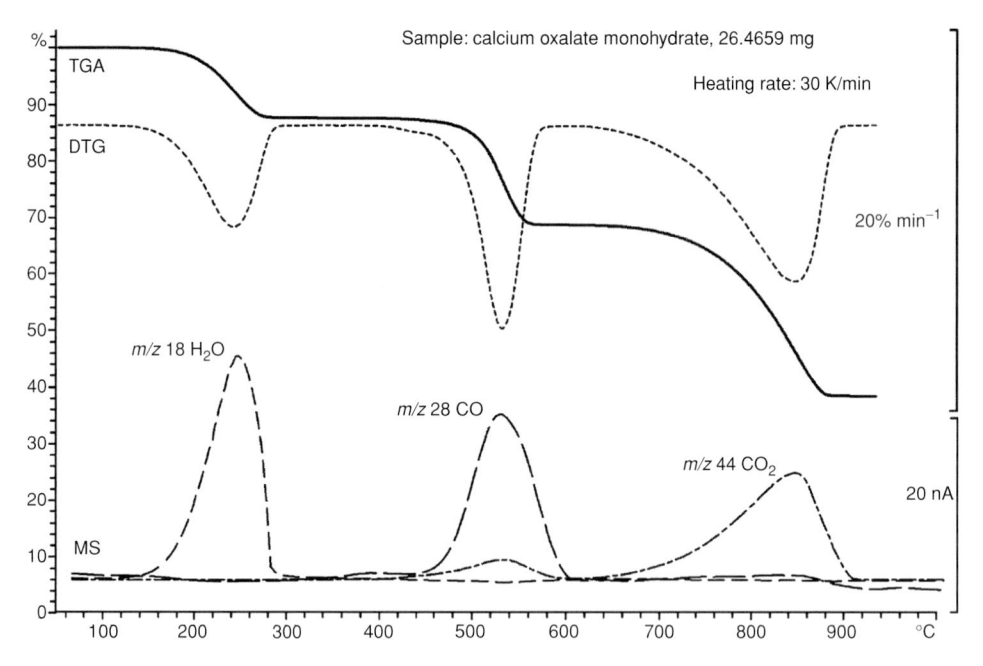

Figure 3.25 TGA–DTG–MS curves of the thermal decomposition of calcium oxalate monohydrate measured at 30 K/min in a 70-μL alumina pan. Purge gas argon, 50 mL/min. The diagram shows that calcium oxalate monohydrate decomposes in three distinct steps. The MS fragment ion curves for water (m/z 18), CO (m/z 28) and CO_2 (m/z 44) display peaks that correspond closely to the individual steps in the TGA curve. The first mass loss step relates to the elimination and vaporisation of water of crystallisation (1); the second step to the decomposition of anhydrous calcium oxalate with formation of CO (2); and the third step to the decomposition of calcium carbonate to calcium oxide and CO_2 (3). The m/z 44 ion curve shows that CO_2 is also formed in the second step at 550°C (besides CO). This is a result of the disproportion reaction of CO to CO_2 and carbon.

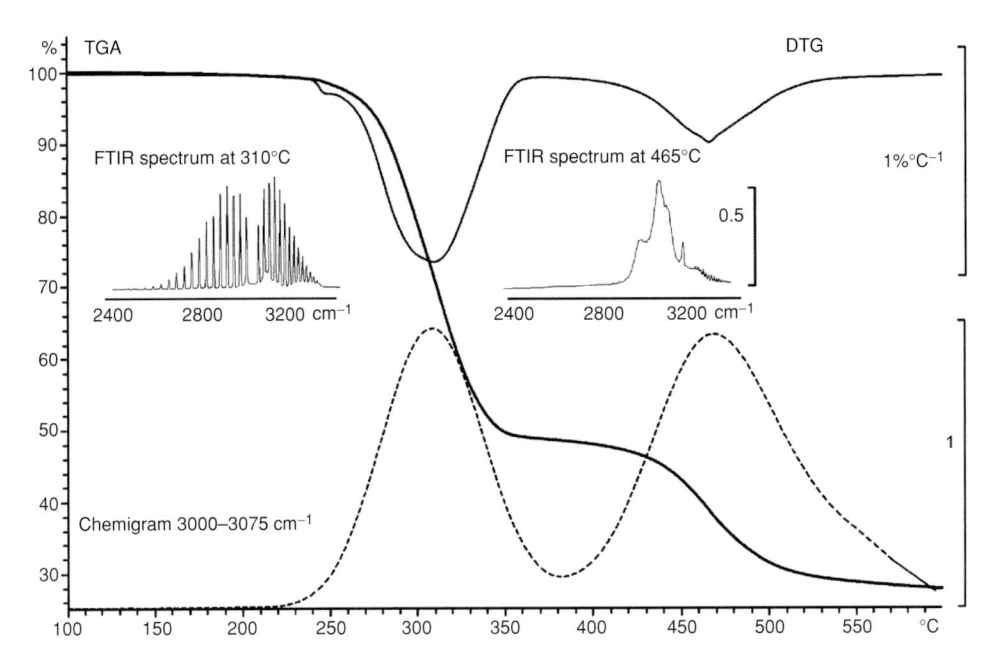

Figure 3.26 Thermal degradation of polyvinylchloride. This example describes the thermal degradation of PVC. The TGA mass loss curve exhibits two clear steps. The IR spectrum (above left) measured at the maximum of the peak at 310°C corresponds to HCl formed through the reaction $(CH_2-CHCl)_n \rightarrow (CH=CH)_n + n \cdot HCl$. The spectrum measured at the maximum of the second peak at 465°C (above right) is due to benzene formed through the cyclisation of $(CH-CH)_n$. The curve in the lower part of the diagram shows a chemigram in the wavenumber range 3090–3075 cm^{-1}. Absorption bands in this region are characteristic of molecules with aromatic rings (C–H stretching vibrations).

These so-called hyphenated techniques are used in research and development, in quality control, and to investigate material failure. Typical applications are as follows:

- Thermal degradation processes (oxidation, pyrolysis)
- Vaporisation and sublimation
- Detection of additives in a matrix
- Characterisation of starting materials and end products
- Investigation of chemical reactions (catalysis, syntheses, polymerisation)
- Outgassing and adsorption/desorption behaviour
- Confirmation of evolved products in the pharmaceutical industry.

3.11.3 Examples

Decomposition of calcium oxalate monohydrate (TGA-MS) (Figure 3.25)
Pyrolysis of PVC (TGA-FTIR) (Figure 3.26)
Detection of solvents in pharmaceutical substances (TGA-MS) (Figure 3.27)

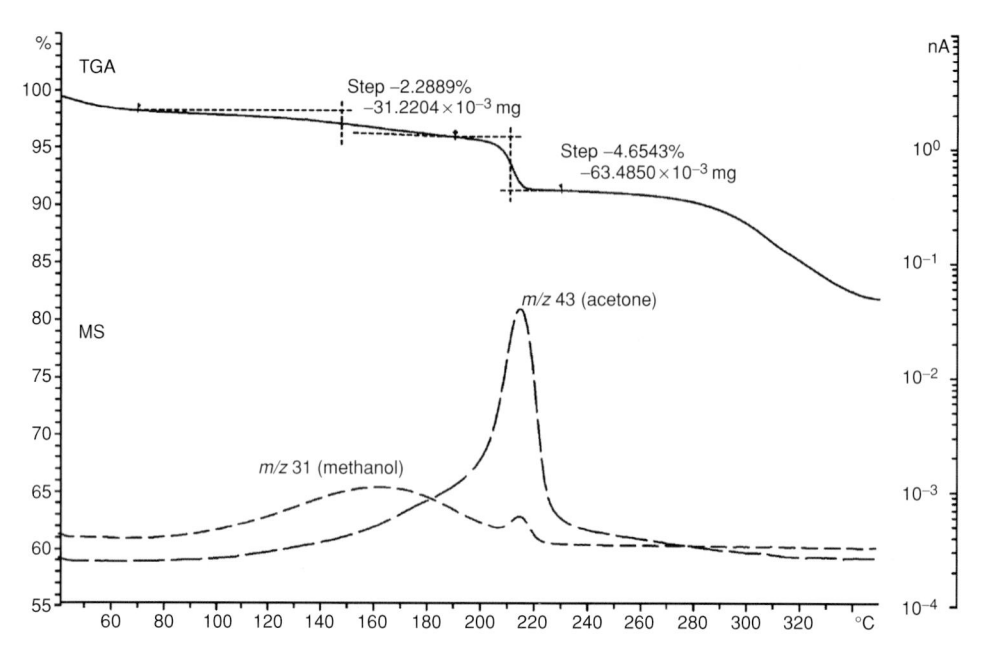

Figure 3.27 Thermogravimetric analysis of a pharmaceutical substance. Different solvents are often used both in the synthesis of a pharmaceutical substance and afterwards for purification/recrystallisation. The presence of residual amounts of solvents can influence the properties of the substance and must therefore be kept as low as possible. The TGA curve exhibits several mass loss steps. In the final step above 250°C, the substance begins to decompose. The two steps in the range 70–240°C indicate that moisture or solvents are lost through heating. The simultaneously recorded MS ion curves confirm that the mass loss steps correspond to methanol (m/z 31) and acetone (m/z 43, the main fragment ion of acetone). The methanol is released over a wide temperature range. In comparison, the acetone is eliminated in a much narrower temperature range at a higher temperature. This indicates that the acetone is more firmly bound, possibly as a solvate.

Reference

1. Toft Sørensen O, Rouquerol J (eds). *Sample Controlled Thermal Analysis – Origin, Goals, Multiple Forms, Applications and Future.* Norwell, MA: Kluwer Academic Publishers, 2004; 224 p.

Chapter 4
Principles and Applications of Mechanical Thermal Analysis

John Duncan

Contents

4.1	Thermal analysis using mechanical property measurement	120
	4.1.1 Introduction	120
	4.1.2 Viscoelastic behaviour	121
	4.1.3 The glass transition, T_g	123
	4.1.4 Sub-T_g relaxations	124
4.2	Theoretical considerations	125
	4.2.1 Principles of DMA	125
	4.2.2 Moduli and damping factor	127
	4.2.3 Dynamic mechanical parameters	127
4.3	Practical considerations	128
	4.3.1 Usage of DMA instruments	128
	4.3.2 Choosing the best geometry	129
	4.3.3 Considerations for each mode of geometry	133
	4.3.4 Static force control	134
	4.3.5 Consideration of applied strain and strain field	134
	4.3.6 Other important factors	135
	4.3.7 The first experiment – what to do?	136
	4.3.8 Thermal scanning experiments	137
	4.3.9 Isothermal experiments	137
	4.3.10 Strain scanning experiments	138
	4.3.11 Frequency scanning experiments	138
	4.3.12 Step-isotherm experiments	138
	4.3.13 Creep – recovery tests	138
	4.3.14 Determination of the glass transition temperature, T_g	139
4.4	Instrument details and calibration	140
	4.4.1 Instrument drives and transducers	140
	4.4.2 Force and displacement calibration	141
	4.4.3 Temperature calibration	141
	4.4.4 Effect of heating rate	142
	4.4.5 Modulus determination	142

4.5	Example data	143
	4.5.1 Amorphous polymers	143
	4.5.2 Semi-crystalline polymers	145
	4.5.3 Example of α and β activation energy calculations using PMMA	147
	4.5.4 Glass transition, T_g, measurements	150
	4.5.5 Measurements on powder samples	152
	4.5.6 Effect of moisture on samples	155
4.6	Thermomechanical analysis	156
	4.6.1 Introduction	156
	4.6.2 Calibration procedures	157
	4.6.3 TMA usage	157
Appendix: sample geometry constants		162
References		163

4.1 Thermal analysis using mechanical property measurement

4.1.1 Introduction

The aim of this chapter is to present mechanical property measurement as a thermal analysis technique. The subject of mechanical property measurement is extensive, but a majority of the testing that occurs is only conventionally carried out at a fixed temperature. Here the main focus will be on how mechanical properties change during temperature ramp tests. There are many good reasons for performing such tests. A rather tragic example of the variation of material performance with temperature is the Challenger space shuttle disaster. Here the rubber O-rings used to seal the booster rocket's individual sections had lost their normal rubbery behaviour, since the cold temperatures on the launch pad had caused the material to approach its glass transition (see Section 4.1.3). In this stiffened condition, the seal was impaired and this was one of the contributing factors to the tragic accident. Many materials applications in the aerospace industry have critical temperature limits and materials for biomedical applications should ideally be tested at 37°C and in an appropriate body fluid.

There are two main types of mechanical thermal analysis instruments, namely thermomechanical analysis and dynamic mechanical analysis, TMA and DMA respectively. The first is a simple technique that has been available for many years and simply records change of sample length as a function of changing temperature. Despite this simplicity it enables the measurement of phase transitions, glass transition temperature and coefficient of thermal expansion. It has the advantage of being simple to use and perhaps more importantly the interpretation of results is quite straightforward.

The second method is DMA. What is DMA? DMA is a technique for measuring the modulus and damping factor of a sample [1]. The modulus is a measure of how stiff or flimsy a sample is and the amount of damping a material can provide is related to the energy it can absorb; see Glossary for definitions. DMA is commonly used on a variety of materials, for example thermoplastics, thermosets, composites and biomaterials. The samples may be presented in a variety of forms including bars, strips, discs, fibres and films. Even powders can be tested when suitable containment is arranged.

How does it work? DMA applies a force and measures the displacement response to this force. This results in a stiffness measurement that can be converted into a modulus value if the sample dimensions and deformation geometry are known [2]. When the temperature is changed these measured properties change quite markedly and this yields important information about the materials' molecular structure [1].

The aim of this guide is to cover as many practically important aspects of TMA and DMA as possible. Many theoretical texts exist, describing detailed deformational mechanisms of specific polymer systems and the molecular relaxations that are responsible for the observed behaviour. Such studies are generally made over a long period of time and with considerable experience on the part of the research investigator. Whilst DMA is an invaluable tool for this work, its routine application is generally far more trivial. I should estimate that over 50% of measurements carried out are to determine the glass transition, or an equally identifiable feature that affects the use of the material under test. It is likely that a good deal less than 20% of all measurements are devoted to a detailed structural investigation of the material at the molecular level.

The practical choice of sample geometry, techniques for specifying the glass transition temperature and dealing with errors from geometry and heating rate will all be presented. Where possible this will be shown on representative data from real samples.

A mention of materials that are suitable candidates for dynamic mechanical analysis would be useful. The simple answer is any material where there is a need to know its modulus or damping factor under periodic and small strain loading conditions. Of course, the class of materials that would be most usefully described by these parameters is the one having viscoelastic behaviour. Materials having long molecules, such as synthetic and natural polymers, are immediate candidates, but are by no means the only ones. Any material that forms a glass will have viscoelastic behaviour. Typically these are referred to as 'amorphous', in that they have no regular crystalline structure. There are also large numbers of such materials that are classified as semi-crystalline (and by implication, semi-amorphous). Such structures usually contain one phase within the other. This will be further discussed under the subject of glass transition (see Section 4.1.3). Generally inorganics, such as metals and ceramics, will predominantly exhibit elastic behaviour, but under special circumstances such as elevated temperature, unusual processing routes, e.g. fast quenching, this may not be the case. However, standard DMA equipment may not cover a suitable temperature or force range for these materials.

4.1.2 Viscoelastic behaviour

Many mechanical property measurement tests do not deal with samples that change as a function of time. For example, the application of force to a steel sample at room temperature causes an instantaneous extension that does not change with time. If the load is increased the extension immediately increases to a new value, proportional to the load increase. This is Hooke's law and describes the behaviour of materials within their linear elastic range.

The opposite end of the spectrum to elastic behaviour is viscous behaviour. This is readily evident in liquids where they flow in response to an applied force. Frequently, the action of gravity on the fluid mass is sufficient to cause significant flow (witness the phenomenon of spilt milk!). If we consider a simple fluid such as water, it follows Newtonian behaviour. The ratio of force to the rate of the applied strain defines a viscosity, which is a constant over a wide range of force and strain rates.

One of the major differences between elastic and viscous behaviour is that in the former case a sample will always return to its original size upon removal of the load; i.e. the deformation is fully recoverable. Clearly this does not occur in the case of a liquid. The flow in a Newtonian liquid is permanent and the fluid will retain its new shape, even if the load is removed.

Some materials behave between the elastic and viscous regime [3]. Such materials are described as viscoelastic and many glass forming or amorphous materials fall into this category. For example, the continuous increase in length with time under the influence of a constant applied force defines a phenomenon called 'creep' [4]. This occurs since the materials' molecules do not have a simple instantaneous response to the applied force. Frequently these materials can be modelled as having three response mechanisms. One is elastic (just like a spring), the other is viscous (just like a dashpot, or hydraulic damper such as a shock absorber) and the other is a combination of these elements in parallel (e.g. a Voigt element) [3]. A material can be modelled as a summation of many of these elements, in order to arrive at a mathematical expression for the observed relaxation time of the material.

The spring part represents force or energy that is put into the sample that will be fully and instantaneously recoverable upon removal of the load. This can be identified as elastic behaviour. The dashpot part represents the force that is put into the sample that results in permanent deformation and is lost. Neither the sample deformation nor this energy can be recovered. This can be identified as viscous behaviour.

The parallel spring and dashpot element can be identified with viscoelastic behaviour. Here the deformation is more complex. This element successfully describes the phenomenon of creep and recovery observed in many materials. As the force is applied little happens at first. The spring cannot respond as it is limited by the dashpot, which only has a slow response to any applied force. As time increases, the dashpot starts to extend and therefore the sample extends as well. This process continues until the extension of the spring element gradually accounts for a greater proportion of the applied force. During this process the sample continues to lengthen, but at a slowing rate. This is creep. Eventually, the spring accounts for the total applied force. At this point there is no driving force to extend the dashpot, so the creep process stops. In practice this may take many years, or even hundreds of years of creep, since the sample is required to reach equilibrium behaviour for this to occur.

It is interesting if we now consider the removal of the force from the parallel spring and dashpot element. The spring is now fully extended and as the load is removed the stretched spring will start to press back on the dashpot. We now see the process occurring in reverse. Again the dashpot is slow to respond, but as it starts to move the sample will gradually return to the unloaded position. The spring and dashpot element will eventually reach its original position. This is creep recovery [4].

In practice, the question of whether a sample fully recovers will depend on the model that describes how the molecular structure responds to force. If there is a large contribution from a purely dashpot element, then much of the deformation will be permanent. This is the situation in linear polymers with no cross-linking, LDPE for example. In the absence of a purely dashpot element, much of the deformation will be recoverable. This is the situation in highly cross-linked polymers, epoxy resins for example.

Therefore, a mathematical model to fit the time-dependent response to an applied force can be derived for a material. If the chemical structure is known, from FTIR or NMR studies for example, then it may be possible to relate real chemical elements, or specific moieties to the elements of the mathematical model. Since DMA is routed in periodic loading it affords excellent opportunities for studying time-dependent behaviour. This will be described in Section 4.2.1.

4.1.3 The glass transition, T_g

First, it is useful to know which materials exhibit a glass transition. Those that form glasses have common characteristics, namely those that hinder the formation of crystalline phases, usually due to having high molecular weight molecules, whose unwieldy nature precludes their easy alignment to form crystalline phases. Most polymers can exist in an amorphous condition (amorphous – having no crystalline structure).

When a solid undergoes the transition between its glassy and rubbery state, the key parameter that changes is the molecular relaxation time. The amount of molecular motion that can occur at any instant is determined by the mobility of the structure, which in turn depends upon the energy available to move the molecules.

Each molecular deformation process that occurs within a material will have a characteristic relaxation time, and depending upon the temperature and frequency of test the sample will respond in a relaxed or unrelaxed manner. Put more simply, the molecular deformation process either occurs or it does not. The glass transition temperature T_g is the region where the transition from glassy (unrelaxed) to rubbery (relaxed) deformation occurs. In a dynamic mechanical modulus determination carried out at a certain frequency (see Section 4.2.2), if the test temperature is above the T_g for the frequency being used, then the measured polymer modulus will be low and commensurate with that of a rubber. Rubbers are characterised as easily extensible materials due to their fully mobile molecular bonds; i.e. their relaxation time is very short compared to the applied loading timescale. These flexible bonds and an availability of 'free volume' allow co-operative motion of the material's main backbone. This means that the structure is able to respond to the applied stress and the molecules immediately (well very fast anyway) take up a new equilibrium position, resulting in little resistance to the applied force, which is the reason that the rubbery modulus is low, typically 1–10 MPa. Compare this to a sample tested below its glass transition. Now the relaxation time will be very long and it will be impossible for the molecules to take up new equilibrium positions. Consequently, a different molecular deformation mechanism occurs. Instead of free rotation around the main backbone of the polymer, occurring with little resistance, there is a regime of bond stretching and opening and displacement against Van de Waals forces. The resultant modulus is much higher (typically 1–5 GPa) since only a small deformation results for the level of force, meaning that the material is much stiffer than when it is above T_g. This partly explains why so many glassy polymers have the same modulus, since their glassy deformational mechanisms are very similar, at least for the low strains used in DMA.

Many factors must be considered when attempting to define the glass transition process. The importance of relaxation time has been mentioned. Free volume also plays an important part in polymer behaviour and in the T_g process. The concept of free volume is the existence of so much space between neighbouring molecular chains that facilitates their cooperative motion. Flory and Fox [5] have postulated a definition of T_g based upon the temperature where this quantity becomes constant. Below T_g, glassy behaviour is observed since the free volume has a constant value that is too small to allow the conformational arrangements associated with rubbery behaviour [3]. Above T_g the free volume increases with increasing temperature and facilitates the large-scale conformational rearrangements that define rubber elasticity. In this argument, the attainment of a critical free volume is the definition of the glass transition temperature.

The glass transition can also be considered on an energetic basis. Glassy behaviour results when the thermal energy available is insufficient to overcome the potential barriers for segmental motion to occur [3]. In this condition we have a 'frozen' liquid, where we are in the regime of bond opening and stretching as discussed above. As the temperature increases the vibrational amplitude increases and when it becomes comparable to the energy barriers, we are in the T_g region. Here marked frequency dependence is observable in parameters such as the storage and loss modulus (E' and E''), tan δ. Low frequencies yield values expected from a rubbery material, whereas high frequencies yield values typical of glassy material. This is due to the response of the molecular motion of the material having a specific relaxation time, which is sympathetic to the applied frequency. When the material's relaxation time is fast, it will respond as a rubber and when it is very slow, compared to the timescale of the experiment, it will exhibit glassy behaviour. Once through the T_g region, all relaxation times become short, with respect to the applied loading frequency and therefore all frequencies produce a rubbery response. This explains why we see frequency dispersion of the complex modulus properties as a result of molecular relaxations.

This explains two key features of the glass transition, or T_g. It must be regarded as a range of behaviour rather than a single point and secondly it is sensitive to the measurement technique employed to evaluate it. Therefore there is no such thing as a single T_g value for any material, unlike a melting point, which can be uniquely defined. Therefore it is always important to state all measurement conditions, such as method of determination, heating rate and frequency of test, as these will all affect the result. Finally, the parameter (E'', tan δ) chosen to represent the T_g can also change significantly from material to material. This will be discussed in Section 4.5.4.

4.1.4   Sub-T_g relaxations

The collective motion of regions of the material observed at the T_g is not the only deformational mechanism that can occur in amorphous materials. With long molecule materials, such as polymers, sugars and proteins, limited sections of the polymer backbone, side groups, or parts of side groups, can also move. These are referred to as secondary relaxations (β, γ, etc. processes). The normal convention for naming these is to assign the highest temperature process, which is always the glass transition as the α process, the first one below T_g as the β process, the next lowest as the γ, and so on. These sub-T_g processes are due to limited motion, for example the rotation of a pendant side group, or some other process, such as deformation at an interface. The latter is a macroscopic structural feature, whilst the former is a molecular or microscopic structural feature.

4.2 Theoretical considerations

4.2.1 Principles of DMA

The usefulness of dynamic mechanical analysis over other forms of mechanical testing is the well-defined periodic loading regime that is used. A periodic force or displacement, usually sinusoidal, is applied to the sample and the resultant displacement or force is measured. This measurement includes the amplitude of the signals and also the phase difference between them [6].

Note that only instruments producing linear displacement will be considered here. Rotational instruments are required to produce torsional shear and these are mainly sold as rheometers and normally used for the analysis of liquids. Solid samples can be tested on such instruments with suitable bar clamps, but it may be difficult to obtain measurements through the glass transition, due to stiffness range considerations.

The essential measurement from DMA is complex stiffness. The stiffness is always returned as a value in N/m (newtons per metre). It is the range of stiffness that can be covered in a single experiment that determines a DMA's usefulness. This range should be a minimum of four decades. Most commercial DMAs will cope with stiff samples very well and this makes the upper force limit unimportant. It is the lower stiffness limit that frequently causes problems. A good instrument will measure a stiffness of at least 500 N/m. Most instruments measure this quantity with high precision, as their force and displacement calibrations are traceable to secondary standards. The conversion of this stiffness to an accurate modulus is frequently a major source of errors, usually due to an inappropriate choice of sample geometry. The measured stiffness is as much a function of sample geometry as modulus. Therefore, the correct choice of sample size and geometry is of paramount importance. This is explained in Section 4.3.2.

Figure 4.1 shows the response to a perfectly elastic material. The phase lag, δ, between the force and displacement is zero. This is not the case for viscoelastic materials where a

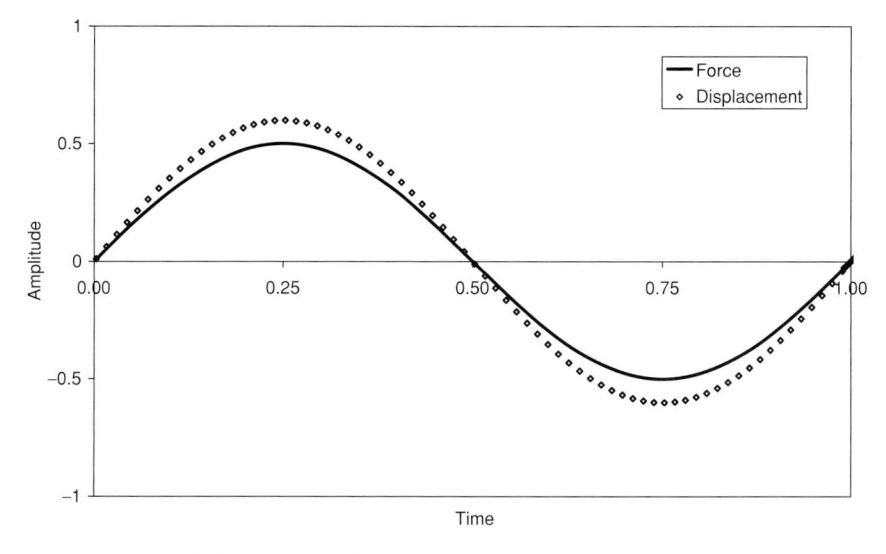

Figure 4.1 Force and displacement signals for elastic material ($\delta = 0°$).

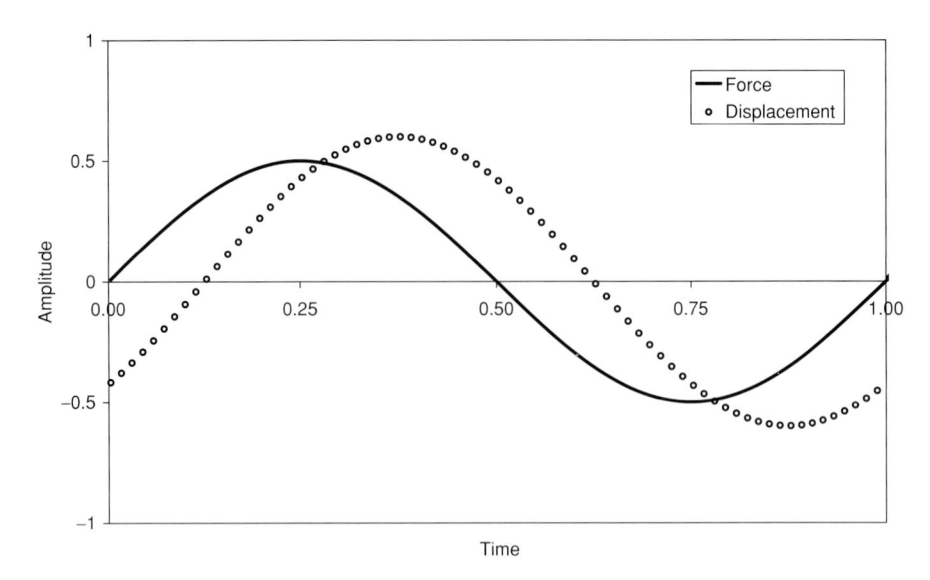

Figure 4.2 Force and displacement signals for viscous material ($\delta = 90°$).

delay occurs in the displacement response to an applied force. Figure 4.2 shows the response for a perfectly viscous material, such as liquid that flows freely. The phase lag, δ, is now 90°; i.e. the displacement amplitude is exactly one quarter cycle behind the applied force amplitude. In DMA instruments, the force and displacement are resolved into in- and out-of-phase components (see Figure 4.3), thus defining the storage or real modulus and the

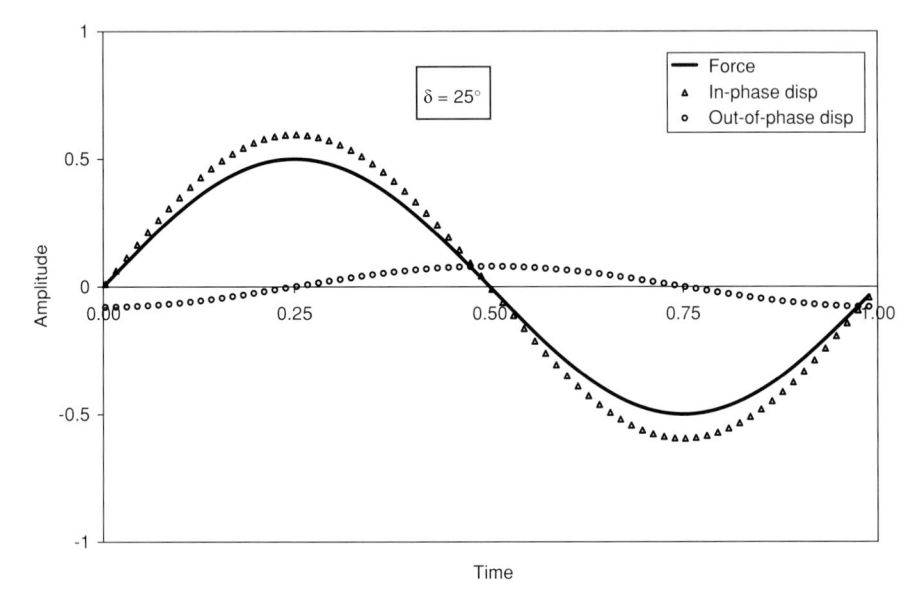

Figure 4.3 Resolution of measured signals into in- and out-of-phase components.

loss or imaginary modulus respectively. The proportion of deformation that is in-phase represents energy elastically stored and recoverable (hence storage or real modulus), whilst the proportion of 90° out-of-phase deformation is due to viscous flow, or other dissipative energy processes (hence imaginary or loss modulus). The damping factor, tan δ, is defined as the ratio of loss to storage modulus and by definition the ratio of energy lost to energy stored. The mathematical relationships between moduli used in DMA are shown in Section 4.2.2.

4.2.2 Moduli and damping factor

$$E^* \text{ or } G^* = |F/y| \, 1/k$$

where F is the max force (N), y is the maximum displacement (m), k is the geometry constant (m^{-3}) (see Glossary). E^* is the complex Young's modulus, from tension, compression or bending geometries. G^* is the complex shear modulus from simple shear geometry [2]. Appendix p162 lists the constants for each mode of geometry (tension, shear, bending, etc.).

The storage component is calculated thus (E has been used, but replace with G for shear measurements)

$$E' = \cos \delta \, |F/y| \, 1/k$$

and the loss component is calculated thus (E has been used, but replace with G for shear measurements)

$$E'' = \sin \delta \, |F/y| \, 1/k$$

The damping factor $\tan \delta = E''/E'$, or for shear G''/G'; see Figure 4.4.

Sometimes the tan δ is distinguished as $\tan \delta_E$ or $\tan \delta_G$. Generally they have a similar magnitude, so comparison is usually straightforward. An approximate relationship between these is given below [7]:

$$\tan \delta_G \approx \tan \delta_E + (\nu' \tan \delta_\nu)/(1 + \nu')$$

Normally $\tan \delta_\nu$ will be considerably smaller than $\tan \delta_E$ and the approximation that

$$\tan \delta_G \approx \tan \delta_E$$

introduces only a small error.

4.2.3 Dynamic mechanical parameters

Users of DMA will be familiar with the typical outputs of such instruments, namely M' (storage or real modulus, E' or G'), M'' (loss or imaginary modulus, E'' or G'') and tan δ, which is the ratio M''/M'. One of the reasons that DMA is such a powerful technique for exploring the properties of polymeric materials and others that show time-dependent behaviour is that all of the above parameters are influenced greatly by the materials' relaxation

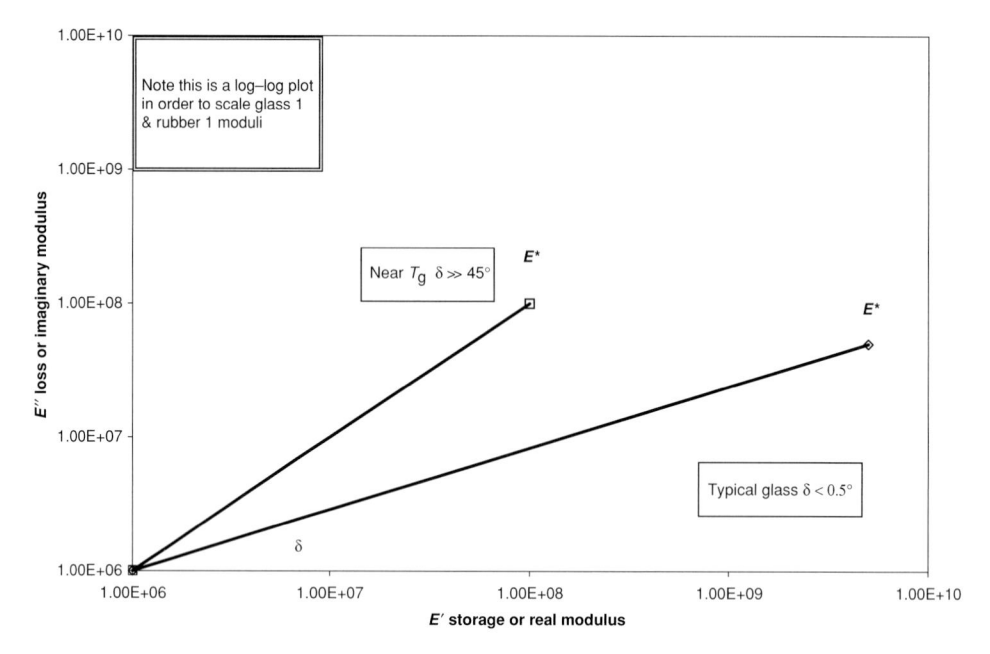

Figure 4.4 Complex modulus Argand diagram. (see Glossary.)

time. The glass transition process was explained in Section 4.1.3 and the frequency M', M'' and tan δ is readily apparent in Figure 4.12.

Therefore the parameters M', M'' and tan δ give a view of a material's relaxational behaviour, which in turn reveals its molecular structure. Subtle differences between grades of the same material and different structures arising from a variant polymerisation will all be visible if the molecular structure has been altered.

4.3 Practical considerations

4.3.1 Usage of DMA instruments

Molecular structure characterisation is the main reason that DMA was developed. The ability to explore the molecular structure via a simple thermal scanning test, requiring a relatively small amount of sample (0.5–2 g), gave the polymer chemists a powerful characterisation tool. It can be argued that solid state NMR, dielectric studies and electron spin resonance give more detailed information, which they invariably do, but generally the apparatus is more expensive, harder to use, experiments take longer and specimens frequently require special preparation. This is why DMA has emerged as the dominant tool for the structural evaluation of polymers.

Now there are four main areas of DMA use, which are molecular structure characterisation, general material analysis, food and biomedical testing and the derivation of engineering data. The first category is explained above. Clearly, this explains the use of DMA in polymer synthesis labs where it is important to check what has been made. It is hard to estimate how much DMA use this accounts for, but I estimate this to be below 40% of all instruments, split approximately equally between industry and academia.

The vast expansion of the use of DMA has occurred in the field of polymeric material analysis. First, DMA is one of the best techniques for assessing the amorphous content of a material. It is important to know how much amorphous material is present in a number of situations. Since DMA is sensitive to molecular structure it is frequently used to check one sample against another that is meant to be the same. Also, processing can have a large effect on final properties. For thermoplastics DMA is sensitive to the level of crystallinity, physical age state and polymerisation. For thermosets, the state of cure can be readily determined and as DMA is a mechanical test useful information can be obtained on interfacial properties for composite materials. To a certain extent this category overlaps with the first, except that fewer measurements are made and a complete molecular profile of the material under test is not obtained. This general usage probably accounts for 40–50% of all DMA.

An important and growing sector of DMA application is within the food and bioscience sector. Many samples are measured for the reasons given above, namely the determination of T_g, checking similarity of samples. Generally, the techniques are similar to those required for the analysis of polymers. However, the temperature range is usually less (-50–$200°C$) and the water content frequently plays a pivotal part in the sample's properties. It is for this reason that controlled humidity testing is being incorporated with TMA and DMA apparatus [8]. I would not like to put a figure on how much these sectors contribute to overall DMA usage, but it is over 10%.

A more specialised area of use of DMA is the derivation of engineering data. DMAs are capable of generation of modulus and damping factor (tan δ) data over a wide range of frequency and temperature. In such applications great care has to be taken that meaningful data are generated. Instruments such as tensile testing machines have undergone considerable development to ensure that accurate moduli are generated. For example, large sample test pieces with favourable geometry for the mode of deformation are used; clip-on extensometers measure the strain very accurately and this avoids any strain that may occur within the clamping mechanism. This is hard to achieve in DMA as one of the main design considerations is to keep the sample small to facilitate changing the temperature rapidly and lightness for high-frequency measurements. Therefore much thought has to be given to sample geometry and the sample size for accurate modulus determinations. Generally, this tends to be a more specialised area of DMA use. It is mainly the preserve of producers of acoustic damping material (including the military), critical component designers and software designers who need accurate values for software packages such as those used for mould filling and finite element analysis. Other mechanical testing equipment is normally used to source static modulus or viscosity data, but DMA is usually the only source when high frequency or damping factor data are required. In my opinion such use accounts for less than 10% of all measurements.

4.3.2 Choosing the best geometry

Choosing a sample to suit the stiffness range of your DMA is the single most important factor in obtaining good results. You will need to know the stiffness range of the dynamic mechanical analyser being used together with your sample's highest and lowest expected value of modulus, or an approximation if the material has not been measured before. From this information the ideal sample size can be calculated.

Different geometric arrangements have inherently different stiffnesses. If they are ranked from the highest to the lowest, the following results:

$$\text{Simple shear} > \text{tension} > \text{clamped bending} > \text{three-point bending}$$

All commercially available DMAs support these geometries.

The stiffness of the different geometries is quite intuitive. Consider deforming a 150-mm steel rule between your fingers. It bends with only a small applied force but requires considerably more force to achieve the same displacement in tension and finally it is very hard to shear (from one surface to the other) as it is very stiff. This is exactly how the DMA sees any sample presented to it. Therefore a wide range of moduli can be covered by the appropriate choice of geometry. *Clamped bending* is a good general choice as it covers the working range of most DMAs for convenient sized sample bars and it is the easiest to use. Three-point (or simply supported) bending is a good choice for high-modulus materials that exhibit a high rubbery modulus, for example fibre reinforced composite samples. Tension is suitable for film samples, where the natural high stiffness of this form compensates for the low sample thickness, in order to properly exploit the stiffness range of the DMA instrument being used. *Simple shear* is a good choice for rubbers and gels, as it is a stiff geometry and suits their low moduli. Rubbers and gels can also be tested in compression and whilst samples are easy to mount in this mode of geometry, modulus accuracy is generally poor (see below).

Most dynamic mechanical analysers display the valid range of modulus for a particular sample geometry. Ensure that the results obtained are well within these limits or adjust the sample size accordingly to better fit the range of the DMA being used. The lower stiffness limit of the instrument will dictate when the measurement has to stop.

Table 4.1 gives an indication of the choice of sample geometry and dimensions as appropriate to the modulus range being measured. The best choice indications are the experiments that will generate the best data with the least adjustment of instrument parameters. These data will suit most DMAs having a 10–20 N load range. Instruments with a larger load capacity may be able to accommodate larger samples, but thermal equilibrium will be poor, unless very slow heating rates are used. Therefore the indicated sizes are preferred. Aim to keep the dynamic strain amplitude between 0.1 and 0.2% (see Section 4.3.5 for a detailed explanation on choice of strain). Figure 4.5 shows a schematic of available sample geometries.

The combination of modulus and sample size must be chosen to suit. Tension mode, see Figure 4.5c, is the best mode for thin films (<20 μm) having a modulus between 10^9 and 10^6 Pa. The stiff nature of tension geometry compensates for the low sample stiffness and best suits the stiffness range of the DMA. If there is a need to measure the rubbery modulus of films, then thicker films (>100 μm) will enable the use of longer free lengths, which will yield more accurate moduli. A dynamic displacement amplitude of 10 μm will result in a 0.1% peak strain for a 10 mm long sample.

Bar samples are easily tested using single cantilever mode, see Figure 4.5a. A sample 2 mm thick, 5–10 mm wide and 50 mm long is easy to mould and handle. Such a bar can produce two samples for single cantilever geometry, having a free length of 10 mm, or one sample for dual cantilever mode; see Figure 4.5b. For a sample of free length 10 mm,

Table 4.1 Typical geometry, sample dimensions and heating rates for samples of given modulus

Best choice	Sample modulus (Pa)	Preferred geometry (for indicated sample size)	Sample thickness (mm)	Free length (mm)	Ideal heating/ cooling rate (°C/min)
	10^{10}–10^{6}	Tension	<0.02	2	5
X	10^{10}–10^{5}	Tension	0.02–1	2–10	5
X	10^{10}–10^{6}	Single cantilever	1–2	5–10	3
X	10^{10}–10^{6}	Single cantilever	2–4	10–15	2
	10^{10}–10^{6}	Single cantilever	>4	15–20	1
X*	10^{10}–10^{6}	Dual cantilever	2–4	10–15	2
X	10^{12}–10^{8}	Three-point bending	1–3	10–20	3
	10^{11}–10^{7}	Three-point bending	>4	15–20	2
X	10^{7}–10^{2}	Simple shear	0.5–2	5–10 (dia)	£2
	10^{7}–10^{2}	Compression (good for irregularly shaped samples and any others that are difficult to mount)	0.5–10 (height or thickness)	5–10 (dia)	£2

Width: Generally, sample width is uncritical and 5 mm is recommended (a wider sample may not be held uniformly in the clamps). A smaller value should be used for stiff samples in tension (1–2 mm).
* For highly orientated samples that are likely to retract above T_g.

thickness 2 mm and width 5 mm, both single and dual cantilever modes with a dynamic displacement amplitude of 25 μm will result in a 0.15% peak strain for a 10 mm long sample. Single cantilever bending is preferred over dual, since the latter can result in large residual stress build-up. This is discussed in Section 4.3.3. Note also that a dual cantilever sample is effectively two single cantilever samples (see Appendix p162). Heating or cooling rates of 1 or 2°C/min will produce reasonably accurate transition temperatures. Rates of 5–10°C/min will be in error (see Section 4.4.4), but provide rapid results for comparative testing. Samples with thickness greater than 2 mm are not optimum. Ideally the sample should be machined to reduce the thickness to 2 mm, unless the sample is inhomogeneous, where reduction of thickness would alter its properties. The reason for keeping the thickness to a minimum is to avoid excessive thermal lag errors (see Section 4.4.4). Another advantage of using clamped bending modes, such as single or dual cantilever, is that no static force control is necessary (see Section 4.3.4). This enormously simplifies the experiment.

Three-point (or simply supported) bending (see Figure 4.5e) is a good choice for high-modulus materials that exhibit only a small change in modulus throughout the test, for example measurements in the sub-T_g region only. A typical analysis to measure T_g of an amorphous polymer would give a poor result in this mode as the sample would collapse under its own weight after passing through the glass transition. This mode of deformation will yield the most accurate modulus values, especially for high-stiffness samples, as the clamping errors associated with this geometry are the lowest (see Section 4.4.5). This geometry also applies a small strain amplitude for a given dynamic displacement amplitude and is therefore suitable for brittle samples, such as inorganic glass and ceramics. For a sample with a free length of 15 mm, 5 wide and 2 mm thick, an 80 μm dynamic displacement amplitude results in a 0.1% dynamic strain.

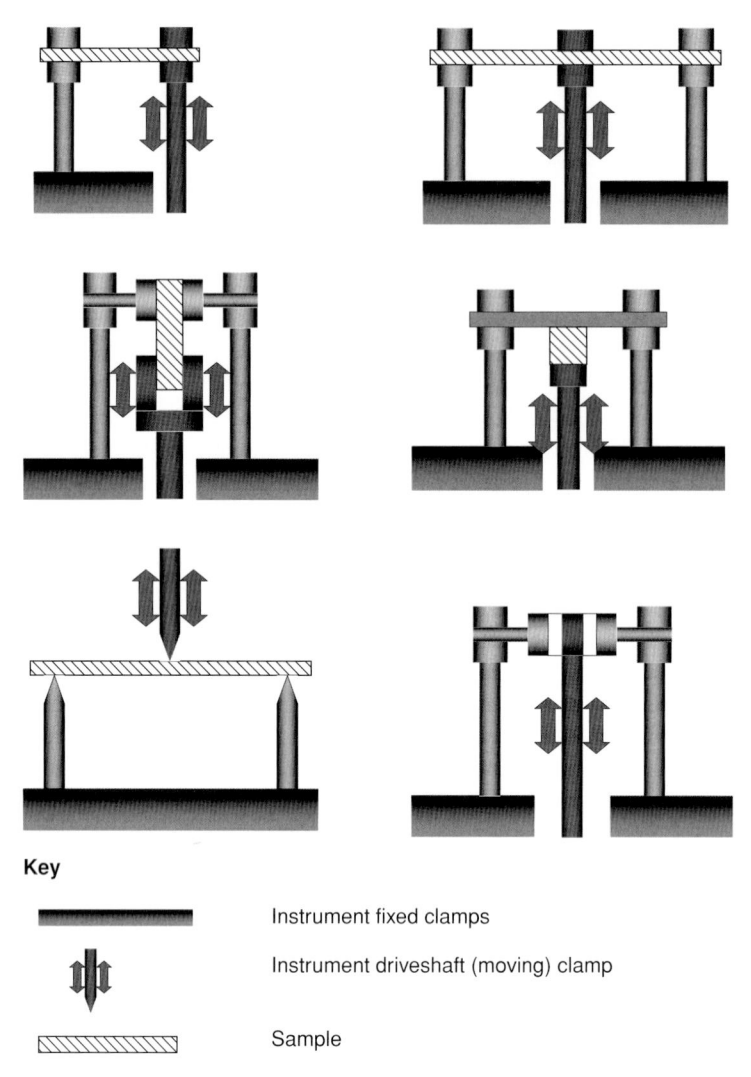

Figure 4.5 Schematic of available sample geometries. (a) Top left – single cantilever bending; (b) top right – dual cantilever bending; (c) middle left – tension; (d) middle right – compression; (e) bottom left – three-point bending; (f) bottom right – shear. Note that for these definitions the sample length, *l*, is always taken as the distance between the fixed clamp and the driveshaft clamp.

Simple shear, see Figure 4.5f, is a good choice for rubbers and gels, as it is a stiff geometry. Accurate modulus values for samples with moduli about 10^3–10^7 Pa can be obtained. For stiffer samples, results are generally better if the sample is bonded to the shear holder. This avoids the sample slipping. Cyanoacrylate adhesives (superglue®) form a rapid setting and rigid glue layer that is suitable for room temperature testing of rubbers (the T_g of superglue is approximately 140°C). As with clamped bending, there is no need for imposition of a static preload (see Section 4.3.4). For a sample with a thickness of 1 mm and a diameter of

10 mm, a 10 μm dynamic displacement amplitude results in a 1% dynamic shear strain. Generally this mode of geometry will be used for lower modulus, where a higher strain amplitude may be beneficial (see Section 4.3.5 for a detailed explanation on choice of strain).

Finally there is compression mode; see Figure 4.5d. Generally, it is better to use the other modes discussed above, but compression mode does find application in giving an easy determination of the transition temperature of irregularly shaped, small samples and materials such as powders. The reason for this is that this mode invariably suffers the most errors. In a compression test the sample should freely grow in area as it is squashed, just as a tension sample freely contracts in its width and thickness (due to Poisson's ratio). In practice, this condition is rarely satisfied and a geometric error results, leading to inaccurate modulus values. Secondly, only relatively low modulus materials (rubbers) can be accommodated by the instrument's stiffness range (just as with shear mode). However, this does lead to one favourable application area for compression, namely the testing of foams. Here we have low-modulus materials that can be tested with varying degrees of 'crush', which affects the measured modulus. Therefore foams are a special class of materials ideally suited to testing in compression mode.

4.3.3 Considerations for each mode of geometry

Single and dual cantilever clamped bending modes will give the best results for measurements through the glass transition for samples having a thickness of at least 2 mm. Good data can be obtained with minimum preparation or precautions. Single cantilever is always preferred, since with dual cantilever mode as the sample is held firmly at either end it is impossible for there to be any movement to relax thermal stresses. This causes a build-up of either tension or compression along the sample length, which may yield a false modulus result. Similarly, DMA instruments whose driveshaft has a degree of lateral compliance are preferred as this automatically compensates for the change in sample length due to thermal expansion or contraction when using single cantilever mode. The only time when dual cantilever mode has an advantage is when a sample has significant molecular orientation. This would cause a large amount of retraction (or extension) as it passes through the glass transition, which in turn would cause excessive movement of the driveshaft. The fixed clamps at either end of the dual cantilever geometry prevent any movement of the sample or driveshaft and preserve the orientation.

Shear mode is also an easy mode to use, with minimal sample precautions required. Only samples having a modulus 10^7 Pa or less can be tested in this mode, since higher modulus materials would be too stiff in this geometry.

Tension mode generally gives good results, but it is complicated by the need to superimpose a static force on the dynamic, in order to establish correct measurement conditions. The same is true for compression mode. This topic is discussed fully in Section 4.3.4.

Three-point bending is the best choice of geometry for accurate modulus determination. In fact, testing an accurately machined steel sample (ordinary steel, not stainless) is certainly the best way of verifying a DMAs' modulus measuring accuracy. Despite this, a poor choice of sample dimensions or displacement amplitude can yield results which are wrong. First, it is best that no sample is over 5 mm in width. It is very hard to clamp or uniformly load wide samples, unless they are very flat. Keeping the width to about 5 mm avoids the problem and only has a small effect on sample stiffness. Second, the strain developed in three-point (or simply supported) bending is four times lower than that for an equivalent length in clamped bending. Also, the fact that stiff samples are more likely to be tested means that the length will probably be longer and these factors reduce the strain for a given dynamic displacement amplitude. Aim for a strain of at least 0.1%. For a sample length 15 mm, 5 mm wide and 2 mm thick, an 80 μm dynamic displacement amplitude results in a 0.1% dynamic strain. If very small displacements (<10 μm) are used, there is a distinct possibility that the sample simply 'settles' on the clamps and we are not testing it in bending mode at all, but a complex bending/torsional mode that yields an incorrect value of modulus. Note that friction can occur in three-point bending, so the measured tan δ value is sometimes higher. The effect on E'' is generally small and as tan δ increases around T_g the frictional contribution is insignificant.

4.3.4 Static force control

In tension, compression and three-point bending modes, it is necessary to superimpose a static force larger than the applied dynamic force, to ensure that the sample remains under a net tensile or compressive force. Failure to satisfy this condition will either cause the sample to buckle (in tension) or to lose contact with the sample clamps in compression or three-point bending. In either case an erroneous modulus results. All instruments provide various methods of dealing with this problem, ranging from manual control to a number of automatic loading regimes. However, successful measurements are difficult to obtain on materials that creep excessively. This will be especially true close to the glass transition for amorphous polymers and close to the melting point for semi-crystalline polymers. For this reason, materials such as low-density polyethylene (LDPE) may be difficult to measure in these modes. If creep is excessive, little can be done and it is much simpler to obtain results in clamped bending mode for example.

4.3.5 Consideration of applied strain and strain field

Dynamic mechanical analysis is an investigative tool. It is a probing technique to determine the material's structure. Normally the loads and strains applied are small (0.1–1%) and at the lower levels they do not influence the materials' structure at all; this is desirable. The behaviour of most materials studied by DMA can be described as viscoelastic. Normally we strive to keep the deformation within the linear viscoelastic region. Here the modulus has a constant value, independent of the applied strain. As the strain is increased the modulus will fall if we move into the non-linear behaviour region. Also, a large third harmonic distortion can be observed in the strain signal [9]. Strictly speaking, tan δ is no longer defined under these conditions, since the harmonic distortion means that we effectively have more than one frequency. However, measurements are made in this region and tan δ values are quoted. As long as the strain remains below ≈0.2%, most materials will be within the linear viscoelastic region (except carbon-filled rubbers; see below). Typical force ranges and samples sizes for DMA usually ensure that these conditions can be met. One range of materials that do exhibit strong non-linear behaviour at smaller strains is carbon-filled rubber. With this class of materials the modulus can change by a factor of 10 when the strain changes from 0.01 to 0.1%. This is due to an interaction of the carbon black with the rubber [10]. It is normal practice to condition or 'scrag' such samples by exposing them to a high strain, often at a higher frequency before measurements start. At high strain amplitudes the temporary structure attained between the carbon filler particles and the rubber breaks down, thereby removing some of the strain dependence seen in the modulus. After several days the temporary structure it regained. This conditioning of the rubber leads to more reproducible measurements.

It is desirable to check whether a material is within its linear viscoelastic range before commencing a series of measurements. This can be done by performing a strain scan on the glassy material for about 0.01–5% (or the maximum possible strain). If a constant modulus is obtained over the whole range, then we are within the linear range. If the modulus starts to drop at a certain level then this is the onset of non-linear behaviour, and unless there is a specific reason to study this effect strains should be kept below this level. Most rubbers will be within the linear viscoelastic range up to at least 10% strain. The important exception is carbon-filled rubbers, as mentioned above. If your DMA has the potential to give the magnitude of the third harmonic component (see Glossary) to the displacement, then this should also follow the strain-dependent modulus and increase as non-linear behaviour is encountered, as described above. For normal testing the magnitude of the third harmonic should be below 2%.

There are some theoretical reasons why certain modes of geometry are preferable, in addition to the practical ones discussed so far. The first consideration is that some modes have affine deformation (each unit of structure experiences the same strain), whilst others are inhomogeneous (a gradient of strain exists throughout the sample, with a neutral axis in the central plane). Tension, compression and shear cause affine deformation, whilst bending (all types) and torsion result in inhomogeneous strain. This usually has very little practical significance, especially if the sample strain is small ($<0.1\%$). However, for highly filled (carbon black) rubber materials that exhibit significant strain dependence, a difference may be observed in changing from one mode of deformation to the other as the strain magnitude will be different.

If dealing with highly strain-dependent materials the strain equivalence must be taken into account. A strain of 0.1% in simple shear, for example, would be equivalent to a strain $= 0.1/\sqrt{3}\%$ in tension.

4.3.6 Other important factors

Generally, experiments work better if the sample is heated from the lowest to the highest temperature. This is due to the fact that the sample will shrink as it cools and becomes loose in the clamps, worsening the modulus accuracy. Therefore the clamps are preferably tightened at the lowest temperature and always below the glass transition temperature, if this is possible. Most DMA manufacturers provide a torque driver to ensure the correct clamping pressure on glassy samples. This ensures that the highest value of the glassy modulus is obtained and that reproducibility is improved, especially with different operators. If the sample becomes excessively loose, strong spring washers can be used under the sample clamps in order to apply a constant pressure. This can be quite successful, although such measures are unnecessary for most samples.

If sample testing is to be carried out only in the rubbery region then clamps should be tightened only finger tight. Use of the torque driver will cause excessive deformation of the sample. Also, since these materials have a lower modulus, it is unnecessary to clamp them so tightly. Another technique for testing rubbers is to prepare them with metal ends that can be securely clamped or alternatively glue them directly to the instrument clamps. The latter method is frequently used for shear samples.

4.3.7 The first experiment – what to do?

Some typical experiments are detailed below, but first we must check the basics:

- What is the goal of this measurement?
- Do you know what the material is?
- What geometry and sample size should you use for an ideal sample stiffness?
- What temperature range should be used?

These are the most basic questions. Starting with the first, what data are required? If the value of modulus and tan δ are needed at room temperature, then this is a very quick and simple measurement. For samples >1 mm thick, three-point bending will almost certainly be the best geometry to use. It is easy to satisfy a low applied strain level to avoid any non-linear behaviour (typically keep dynamic strain below 0.1% – see Section 4.3.10) and strain levels can be varied, if required, by changing applied amplitude and also sample length (remember that the strain depends on the square of the sample length and thickness). For the second question, if the material is known, then the expected modulus value and possibly damping factor (tan δ) will be known. Therefore question 3 can be answered: based upon the modulus, choose what stiffness you would like the sample to have and calculate the sample size to achieve this. Most software accommodates this, or advises on the valid modulus range for the geometry entered. Question 4 is easy to answer as room temperature data have been requested.

Now let us assume some different answers. First, say, we do not know what the material is. A guess can usually be made at the modulus, since we may well have an idea of the likely material. Is it a glassy polymer for example, in which case the modulus will be close to 5 GPa (at room temperature)? On the other hand, is it a metal? Here the likely modulus will be around 100 GPa. Alternatively, it may be a rubber and therefore its modulus will lie between 1 and 10 MPa (typical values for industrial rubbers). If you still do not know, measure it in three-point bending with a long free length for stiff samples or single cantilever bending, medium free length for flimsy samples (e.g. rubbers). This should give you a reasonably accurate figure.

Second scenario: now say the purpose of the measurement is to find the T_g and T_m for the material. If the material is known we can answer all of the questions. If not, we can see how to arrive at a suitable sample size from the procedure above. So we can go straight to question 4. Is the sample crystalline, glassy or rubbery? A metal would be crystalline and polymer could be all three depending on the temperature! It should be easy to guess what state the material is in, just from the feel of it. If it does not appear to be stiff, then it is probably a rubber and therefore testing should commence from at least $-100°C$. To be on the safe side testing should continue until about $300°C$, as melting points are frequently in excess of $200°C$. Some DMAs allow experiment termination when the stiffness falls to a preset limit. This can automatically stop the instrument once melting has occurred and avoids decomposition of the sample. Once T_g and T_m have been established a sensible range can be set for comparative measurements on other samples. Suggest at least $50°C$ below T_g for the start temperature and about $10-20°C$ above the melting point. More information can be gathered if two frequencies are used (1 and 10 Hz for example) as this allows easy distinction between T_g and T_m via the frequency dependence of data.

The above is only an illustration of likely scenarios. However, it is often useful to perform a wide-range thermal scan ($-100-300°C$, or the melting point, if known) at two frequencies (1 and 10 Hz for example). This allows easy distinction between T_g and T_m via the frequency dependence of data. Generally, the experimental methods are best considered with real data and such examples are given in Section 4.5. However, a brief description of experiment types is given below.

4.3.8 Thermal scanning experiments

Thermal scanning experiments are the most common ones performed with DMA and, based on my personal experience with users, I estimate that a simple, single frequency, single strain amplitude test accounts for some 80% of all DMA measurements made. Such an experiment will yield the key identifying parameters for any material. Figure 4.11 shows the tan δ curve for LDPE and LLDPE which exhibit the features shown by semi-crystalline polymers. Starting from high temperature and working downwards, we start with the melting point, a crystalline relaxation process α around $90°C$, a β relaxation at $-20°C$ due to chain branching and finally the γ relaxation at $-120°C$ (see Section 4.5.2). Transitions are shown as peaks in the tan δ trace.

4.3.9 Isothermal experiments

These experiments are mostly performed where the sample changes as a result of a stimulus. Examples of this may include the drying of a sample, perhaps as studied in a controlled humidity environment (see Section 4.5.6), the post-cure of a thermoset resin or the decomposition of a sample at high temperature. There is usually little restriction with regard to data collection for these experiments, as they are made over relatively long time periods. Therefore, they are frequently performed at multiple frequencies and this can be used to evaluate half-lives or relaxation times for the process under investigation (see Section 4.5.3).

4.3.10 Strain scanning experiments

These have largely been explained under Section 4.3.5 and have most application for materials whose modulus is observed to be highly strain dependent, carbon-filled rubbers for example.

4.3.11 Frequency scanning experiments

These experiments have a specific utility in the generation of materials data, for example for sound damping applications. In such cases, it is important to know the modulus and damping factor at certain frequencies and temperatures. Considerations must be given to ideal sample size (see Section 4.3.2) and any errors that may spoil the measurement accuracy (see Section 4.4.5). Note that many DMAs cannot measure above the resonant frequency. Some DMAs will have higher resonant frequencies than others; the mass of the driveshaft is the main parameter that will determine this. The lighter the driveshaft the higher the resonant frequency. Some can measure above the resonant frequency anyway, thereby extending the useful range of measurement. However, it is usually best to avoid measurement at or near this value as results can be significantly in error.

4.3.12 Step-isotherm experiments

These are a combination of the isothermal and frequency scanning experiments. They are commonly used for detailed investigation of the glass transition process (see Section 4.5.4) using time–temperature superposition (TTS) analysis (see Apendix 2). As such this is the most complicated experiment to perform, especially if using tension mode with thin films, for example. It is only recommended for experienced users of DMA and these experiments frequently last for 24 h. There is also much consideration needed for the sample and whether or not it is suitable for TTS analysis. Only samples that remain in the same state throughout the measurement and have a clearly identifiable single relaxation are suitable. This immediately excludes a large group of materials, in that semi-crystalline polymers rarely satisfy this condition. There are data-quality checks that can be made before TTS analysis (see Appendix p450), such as the wicket plot, and these are discussed in Section 4.5.4. It is highly recommendable to make a simple multi-frequency temperature scan before performing such long tests as this can show problems that occur with the sample and this avoids much lost time if the desired experiment needs modification.

4.3.13 Creep – recovery tests

DMA can be used to make creep, creep-recovery and sometimes stress-relaxation experiments. In some respects these tests defeat the purpose of a DMA, in that generally the dynamic data (E', tan δ) are more useful and give greater insight into the materials' mechanical behaviour than these constant load tests. However, they can be useful where a component is used at a specific temperature and load, e.g. a fixing lug, and it may be useful to test the material under the exact loading conditions to be used in the application.

4.3.14 Determination of the glass transition temperature, T_g

The glass transition has been explained and defined in earlier Sections 4.1.3 and 4.2.3, respectively. The determination of the glass transition temperature is certainly the main use of dynamic mechanical analysis. It is sensitive to the process by virtue of the large differences that exist between the glassy and rubbery state, a 1000-fold change in storage modulus, E', for amorphous materials, for example. Compare this to the change in heat capacity, as measured by differential scanning calorimetry (DSC), which is typically only one order of magnitude.

A simple experiment to determine T_g could use a single frequency and a relatively fast heating rate (5°C/min). This is a good procedure for routine T_g measurement. Note that the fast heating rate will cause a measurement error (see Section 4.4.4). Better temperature accuracy will be obtained using a slower heating rate (1 or 2°C/min), but this is only necessary when an absolute value is required. For general comparative work, using the faster rate and obtaining a value with the associated lag error is acceptable. A nominal correction can be made, based on the procedures detailed in Section 4.4.4.

A more complex investigation of the glass transition process would involve the use of multiple frequencies during the heating ramp. Any loss process is seen to be frequency dependent, as discussed in Section 4.2.3. The temperature difference between one loss peak (E'' or $\tan \delta$) and one at another frequency is a measure of the activation energy of the process. If a simple Arrhenius plot is made of these data (the T_g process is not well represented by an Arrhenius fit; see later), an activation energy can be calculated. A polymer having a T_g at 100°C at 1 Hz and showing a 7°C shift per decade of frequency will have an activation energy of about 400 kJ/mol. This is the expected value for the glass transition process, in an amorphous polymer.

With preliminary or complex measurements it is useful to carry out experiments at a minimum of two frequencies, say, 1 and 10 Hz. The results should be tested for frequency dependence. The magnitude of any observed shift between data will confirm the nature of the relaxation. Such frequency dependence is most evident in the position of loss peaks, either the E'' or $\tan \delta$ peaks. A shift of over 15°C shift per decade of frequency means that the activation energy would be lower than expected for a glass transition process and therefore the effect observed is either a secondary relaxation (β process), due to limited motion, e.g. the rotation of a pendant side group, or some other process. Shifts in the range of 7°C per decade of frequency are likely to be glass transitions. This is usually self-evident in a singe polymer, e.g. PMMA or PS, but if a polymer blend is under investigation a small peak may indicate a secondary relaxation (β process), or it could be the T_g of a minor component. Checking the peak shift as a function of frequency should enable the correct assignment of such a process. If no shift is observed and different frequency data are coincident, then the process being observed is that of a first-order reaction, e.g. a phase change or some physical process. First-order reactions include the melting of crystalline material present, or a solid state phase transition. Frequently, the drying of water or solvent from a sample causes a frequency-independent loss process, which is a simple, physical effect. Chemical reactions, such as decomposition, will usually be seen as a frequency-independent loss process too.

However, care must be exercised in assuming that all chemical reactions behave this way, since the cure of an epoxy resin, for example, will affect the glass transition of the resin as the cross-linking reaction progresses and this causes a loss peak that will be frequency dependent.

4.4 Instrument details and calibration

Let us consider standards relating to DMA. ISO 6721 is a multi-part standard covering principles of measurement, geometry definition and measurement of T_g for composites (ISO CD 6721 part 11).

There are several ASTM specifications that relate to DMA. The most useful are ATSM D4065-95 'Standard Practice for Determining and Reporting Dynamic Mechanical Properties of Plastics' and D4092-96 'Standard Terminology Relating to Dynamic Mechanical Measurements on Plastics'. ASTM E1867 and E2254 deal with the calibration of temperature and storage modulus respectively. Whilst not concerning the topic of calibration, ASTM E1640-04 'Standard Test Method for Assignment of the Glass Transition Temperature by Dynamic Mechanical Analysis' is also highly relevant.

The calibration of the main variables, force, displacement and temperature will be covered below. First though some discussion on the components of TMAs and DMAs would be useful.

4.4.1 Instrument drives and transducers

Both the TMA and DMA apply force and measure the displacement that results from this stimulus. Most TMAs and DMAs currently in production use linear variable displacement transducers (LVDTs) to measure displacement. There is one exception, which is the TA Instruments 2980, or later variant Q800, which uses an optical displacement transducer instead.

Older TMAs are not fitted with motors, but rely upon a top-loading scalepan for the application of force. In TMA the force is constant for all classic experiments, therefore the manual arrangement is quite satisfactory for this work. In DMAs and more modern TMAs a voice coil type motor is employed. This comprises a light coil and former, connected via fine wires, often made from Litz wire. This assembly is either supported by springs or the drive arrangement is supported by an air bearing. Conceptually it is similar to the design of a loudspeaker.

In the case of DMAs the lighter the moving mass of the drive arrangement is, the lower the inertia correction will be. The inertia correction is given by $m\omega^2$, where m is the drive mass in kg and ω is the angular frequency ($2\pi f$). This factor becomes particularly significant with measurements of low stiffness materials at high frequencies.

Whilst all DMAs utilise a voice coil type motor, some are fitted with a force transducer to measure the force transmitted through the sample. This requires a different design of instrument. It is frequently and incorrectly thought that the omission of a force transducer leads to poor modulus accuracy. Provided that the voice coil motor is correctly calibrated and its own contribution to the total stiffness measurement is removed, it does not make any difference, since the force measured by the transducer and the one taken from the motor calibration are both accurate to within better than ±1%, which is adequate. Far larger errors occur from the poor choice of sample size and geometry, which leads to making measurements at the extreme ranges of the instrument. The instruments fitted with a force transducer may sometimes do better at measuring very low stiffness samples, as the contribution from the motor does not have to be subtracted. However, the force transducer is also a source of instrument compliance, which will lead to errors at high stiffnesses. Corrections can be made for compliance and in one particular DMA there is a unique design that automatically removes this contribution from the measurement.

4.4.2 Force and displacement calibration

The top-loading scalepan TMAs require no calibration. Instruments fitted with the voice coil type motors will require calibration. Most instruments achieve this by applying one or more known masses (which can be accurately measured on a balance calibrated by traceable standards) and nulling the displacement that occurs. This way the force required to rebalance the applied mass is obtained independently from the displacement measurement. The instrument measuring system will adjust the voltage or current applied to the coil, and when the displacement has been successfully nulled this value is equivalent to the force due to the applied mass. The value of force to current should be linear over the range of operation of the instrument.

Displacement calibration is usually made by inserting known thickness samples, such as slip gauges, into a suitable fixture and noting the resultant displacement. A calibration can be created from these measurements. The displacement applied should be traceable to primary or secondary standards. LVDTs are usually linear over their range of operation and typically cover $\pm 1-$ to ± 5 mm with resolution to ≈ 0.1 μm or better, which is usually sufficient for most experiments.

4.4.3 Temperature calibration

Temperature calibration is arguably one of the most contentious and important issues when dealing with thermal analysis instrumentation. DMAs in particular have very good sensitivity for the measurement of glass transition temperatures and the tan δ peak, for example, would provide an excellent parameter to calibrate the instrument. Unfortunately, any other technique that would produce a tan δ value would give a different result (see Section 4.1.3). Therefore users frequently resort to a melting point measurement, which can be verified by other instruments, for example DSC; see Section 1.4. Unfortunately large samples are required for DMA, which can be difficult to obtain or expensive and the large weakness of these tests is that they do not accurately simulate typical samples measured. Many samples will be poor thermal conductors, whereas the melting point standards are usually metals that have high thermal conductivities. Since the DMA measurement will stop when the metal melts, this occurs even if a large gradient exists across the sample.

Frequently, 'standard' polymers are used to check the accuracy of a DMA. Such materials include polymethylmethacrylate (PMMA), polyetherimide (PEI) and polyethersulphone (PES). The tan δ or E'' loss peaks can be used to compare the temperature measured with other users. Unfortunately these temperatures cannot be determined on any instrument other than a DMA, so they cannot be used as calibrations, but they are useful checks on reproducibility. Details of how to quote the T_g value can be found in ISO6721 [6] part 11 – in preparation.

4.4.4 Effect of heating rate

For samples approximately 2 mm thick, most DMAs return thermal lag errors of about 1°C for each 1°C/min of heating rate. This was established in a round robin survey carried out by the NPL [11]. This is easy to check. Three T_g determinations should be made at heating rates of say 1°, 3° and 5°C/min. A new sample should be used each time. The measured T_g temperature is recorded against heating rate and a plot extrapolating back to 0°C/min should be constructed. The value at 0°C/min is the true T_g and the slope of the line is the thermal lag error as a function of heating rate.

Therefore a 1 or 2°C/min test will provide more accurate temperature information, but will of course take longer. If tests are only made for comparative purposes the temperature error is largely insignificant. Thicker samples are not preferred as they will cause larger thermal lag errors.

4.4.5 Modulus determination

The main purpose of DMA is to make measurements as a function of temperature. This invariably results in a compromise between optimum modulus determination and the best use of the DMA stiffness range, as discussed in Section 4.3.2. Three-point bending and tension geometry are the best modes for accurate modulus determination.

Modulus errors occur for all clamped geometries. They are greatest for clamped bending samples having a short length and will be least significant for tension measurements on thin films. This is due to the relative stiffness of the sample. The error is due to movement of the sample within the clamps, which is not factored in the geometry constant. This is a consequence of the clamps' stiffness being inadequate for the stiffness of the sample under test and consequently the clamping arrangement deforms in the test. Both the measured E' and E'' values will be in error by a similar amount (this is true where the phase angle δ is small), therefore the tan δ value will be accurate (see below).

Three-point (or simply supported) bending is free of such clamping errors and therefore the measured moduli are correct. In fact, testing an accurately machined steel sample (ordinary steel, not stainless) is certainly the best way of verifying a DMA's modulus measuring accuracy.

The clamping error is difficult to correct since this error depends upon the sample stiffness, its modulus and its aspect ratio. By the time the sample reaches the rubbery condition this error is insignificant, due to the low sample stiffness.

Since the error is a constant value which adds to the length, it can be mitigated by ensuring that stiff samples are run with long free lengths, at least 10 mm. Fortunately, such samples tend to show the least drop in modulus at T_g, especially fibre reinforced composites, and therefore measurement of the minimum modulus value is not an issue with such samples. Alternatively thinner samples can be produced, which therefore have a lower sample stiffness.

Remember that doubling the length decreases the sample stiffness by 8 times, as does halving the sample thickness. Doing both reduces the stiffness by 16 times.

This error can be quantified in a series of experiments that are quick to perform. The sample should be measured over at least three lengths and a plot of $1/E^{1/3}$ versus l is constructed. The slope of this line $(1/E^{1/3})$ then yields the true modulus whilst the intercept with the x-axis yields the effective sample length deforming within the clamps. This length can be quite significant, >1 mm for stiff, thick samples. This is a very significant error on, say, a 5 mm length, considering the cubic dependence of the geometry constant on length. The modulus value obtained from the slope should agree very well with a determination made with three-point (or simply supported) bending mode.

4.5 Example data

4.5.1 Amorphous polymers

Figures 4.6 and 4.7 show the storage modulus (E'), loss modulus (E'') and tan δ plots for a general grade of poly(styrene) (PS) and poly(carbonate) (PC). All data are measured at an applied frequency of 1 Hz & 10 Hz and a strain amplitude of ≈0.5%. Both materials show distinct T_g processes, but the PC sample additionally exhibits a β process below T_g. No such β process relaxation is evident for PS. This shows that a potential high-frequency energy dispersion mechanism exists in PC, whereas it does not exist in PS. This explains the high toughness possessed by PC, since high-frequency impact effects can be dissipated by the material. No melting point is observed in either material, since they are amorphous. A gradual softening occurs, the final viscosity depending upon samples' molecular weight.

Figure 4.8 compares the data between a general grade and a high impact grade (HIPS) of poly(styrene). HIPS has butadiene added in the form of small spheres. The low temperature

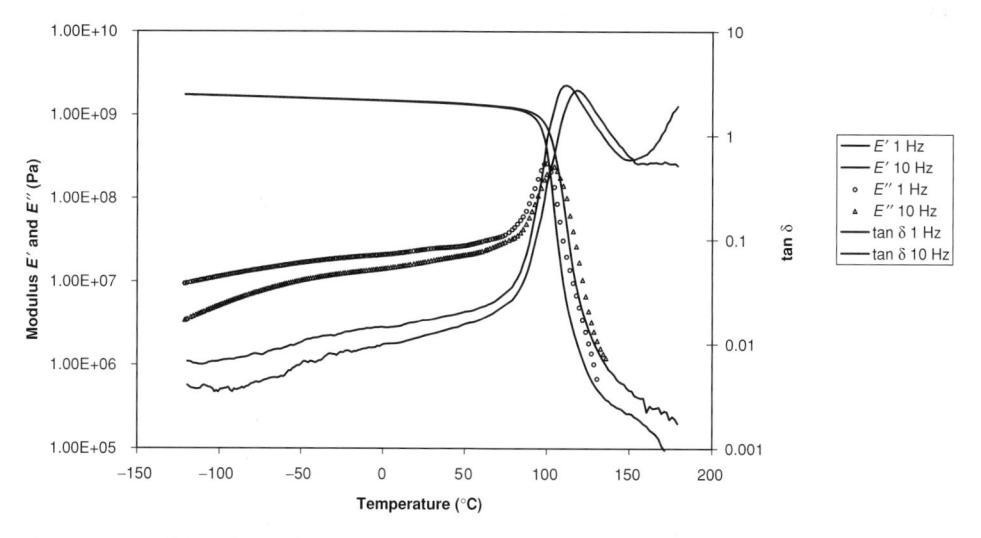

Figure 4.6 Moduli and tan δ data for poly(styrene). Storage modulus E' shows a decrease above ~100°C whereas both loss modulus and tan δ show a peak.

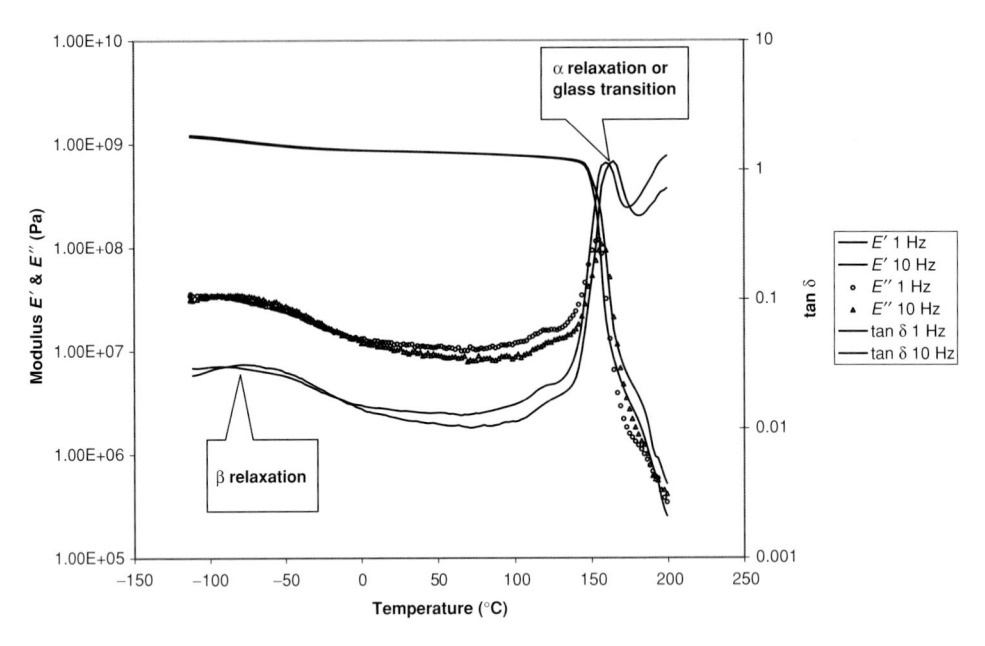

Figure 4.7 Moduli and tan δ data for poly(carbonate). Note the β relaxation which occurs below zero, shown by the increase in tan δ and loss modulus.

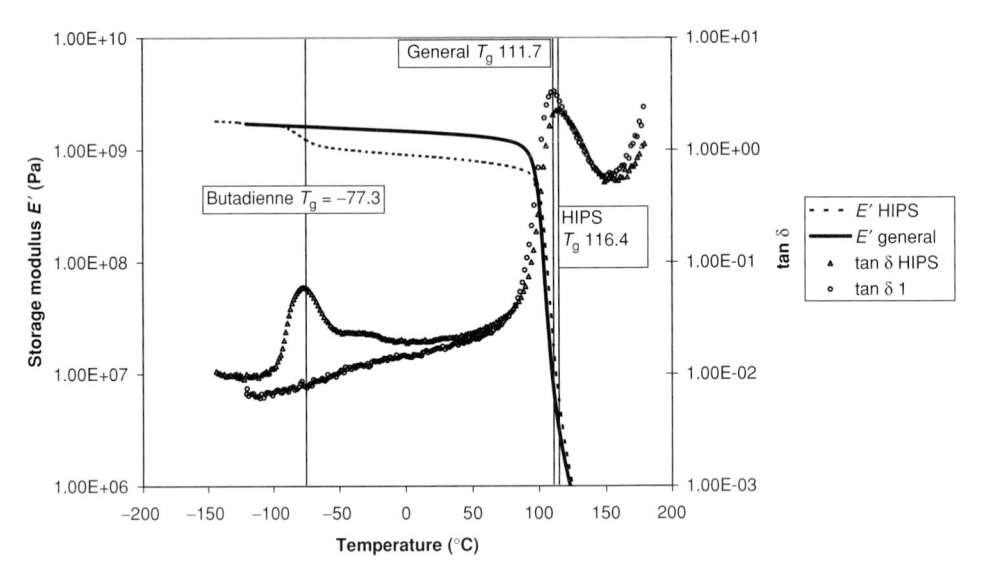

Figure 4.8 Moduli and tan δ data for general grade and high impact poly(styrene). The low temperature T_g of the butadiene component of the HIPS sample, seen as a peak in tan delta below −50°C, provides a mechanism for energy dissipation.

T_g of the butadiene ($-77°$C) and the T_g of PS are readily evident in the tan δ plots for the two materials. The fact that we see two distinct T_g's immediately tells us that the blend is immiscible. This is generally the case with most polymers, where only limited solubility of one polymer within the other occurs. In the few polymers that are miscible (poly(styrene) and poly(phenylene oxide) for example) only a single T_g is observed, whose temperature lies in between the two component T_g's, in proportion to the composition.

Glass transition temperatures are frequency dependent and since the T_g of the butadiene is low (as measured at 1 Hz – data shown here), its T_g will be greater at higher frequencies. In a typical impact situation the frequency of loading is high. This means that there is large energy dissipation capability, due to the T_g of the butadiene, which confers toughness. Compare this to the data for PC which has its own built-in energy absorption mechanism in the form of a β process. The form of the butadiene as small spheres also adds to impact resistance by acting as a crack stopping mechanism; any cracks finding it difficult to propagate through the butadiene phase.

4.5.2 Semi-crystalline polymers

The two main thermal events for semi-crystalline materials will be the T_g and the melting point, T_m. The latter will be the highest temperature event, will have the greatest change of modulus, and will be independent of frequency, since this is a first-order transition. It is also independent of the measurement technique, unlike the glass transition, T_g. Therefore T_m values obtained from various instruments, e.g. DMA, DSC and TMA, will be directly comparable. This can be a useful guide to the temperature accuracy of the instrument being used.

Figures 4.9–4.11 show the E' E'' and tan δ plots for all of the polyolefine family: LDPE, HDPE, LLDPE and PP.

Figure 4.11 shows the tan δ curve for LDPE and exhibits the features shown by semi-crystalline polymers. Starting from high temperature and working downwards, we start with the melting point, a crystalline relaxation process α around 90°C, a β relaxation at $-20°$C due to chain branching and finally the γ relaxation at $-120°$C. Both the γ and the β process are derived from the amorphous phase and they diminish as the crystalline content increases, as would be expected.

The T_g value for poly(ethylene)s is controversial [12, 13]. Some people take the β relaxation around $-20°$C as T_g, whilst others choose the γ relaxation at $-120°$C. The γ value is very low for a T_g and in the author's opinion is not the T_g. It is more likely to be the $-20°$C value, but this depends heavily on the extent of branching and can disappear when little branching is present, as seen with HDPE. The T_g of PP occurs around 10°C (tested at 1 Hz and measured by tan δ peak position). Since PP has a very similar structure to PE, it seems consistent to assign the $-20°$C value as the T_g of PE.

Depending on the polymer and the level of crystallinity, the T_g process may be obvious or undetectable. Polymers that are approximately 50% crystalline, PET for example, show a T_g process proportional to the amorphous content [14].

What is the structure of a semi-crystalline polymer? Essentially all semi-crystalline polymers behave as a two-phase system: one being the amorphous fraction and the other being the crystalline fraction. If the sample were fully crystalline, then no relaxations due to the amorphous phase would be observed [14]. The only thermal features would be a gradual

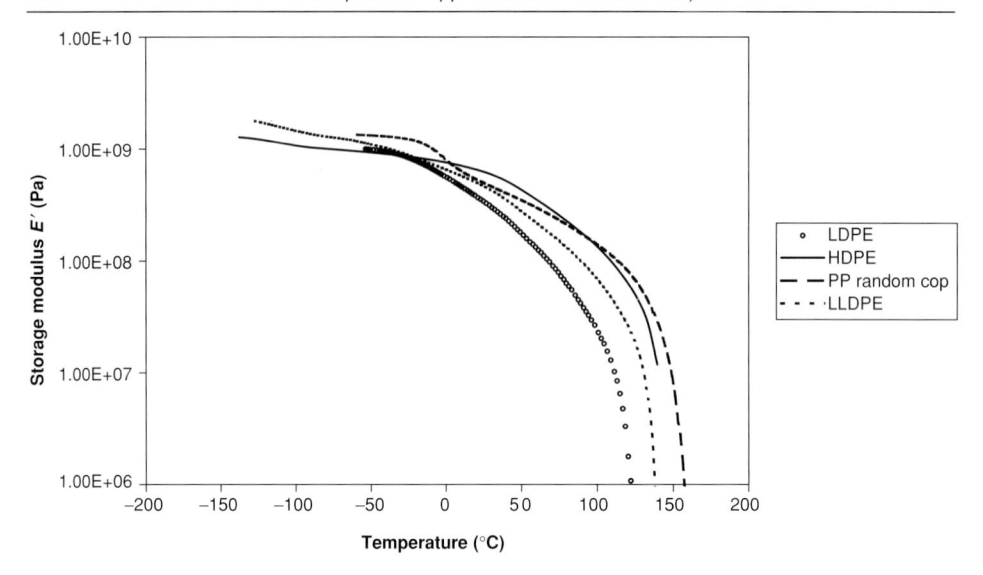

Figure 4.9 Storage modulus (E') for various polyolefins showing a β relaxation at $-20°C$ and the large decrease in modulus as the sample is heated into the melt region.

softening of the sample close to the melting point, T_m. This is not a sharp transition as would be observed for a small inorganic molecule, since there is a range of melting points, due to the range of molecular weights present in a sample. Nevertheless, it is significantly sharper than the glass transition which is observed in amorphous polymers.

A fully amorphous sample will exhibit large changes of modulus at the glass transition and frequently will be seen to crystallise at a temperature above the T_g, but below T_m. This will cause a relaxation above the glass transition temperature. When both amorphous and

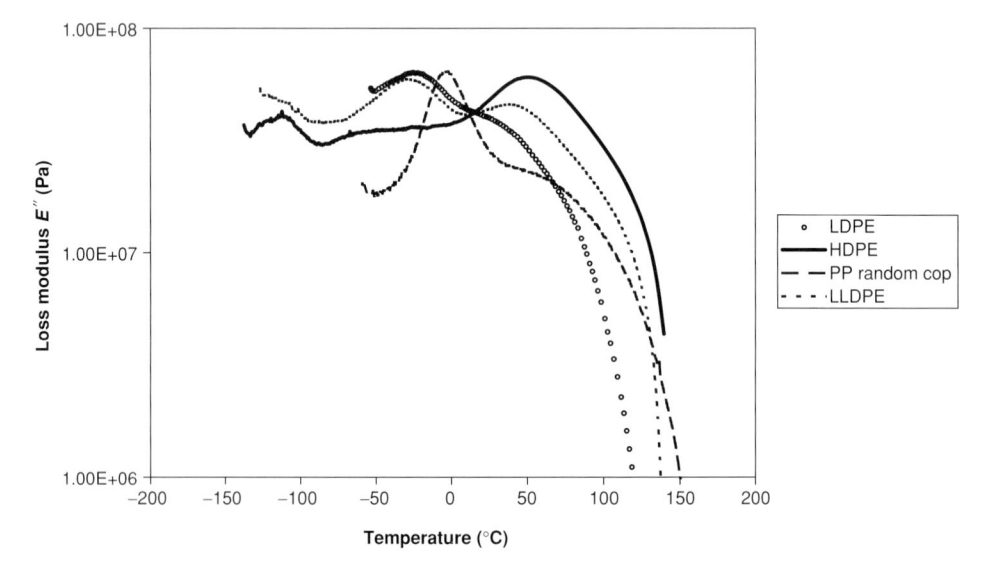

Figure 4.10 Loss modulus (E'') for various polyolefins. The loss modulus shows a peak at $-20°C$ due to the β relaxations and then decreases as the sample is heated into the melt region.

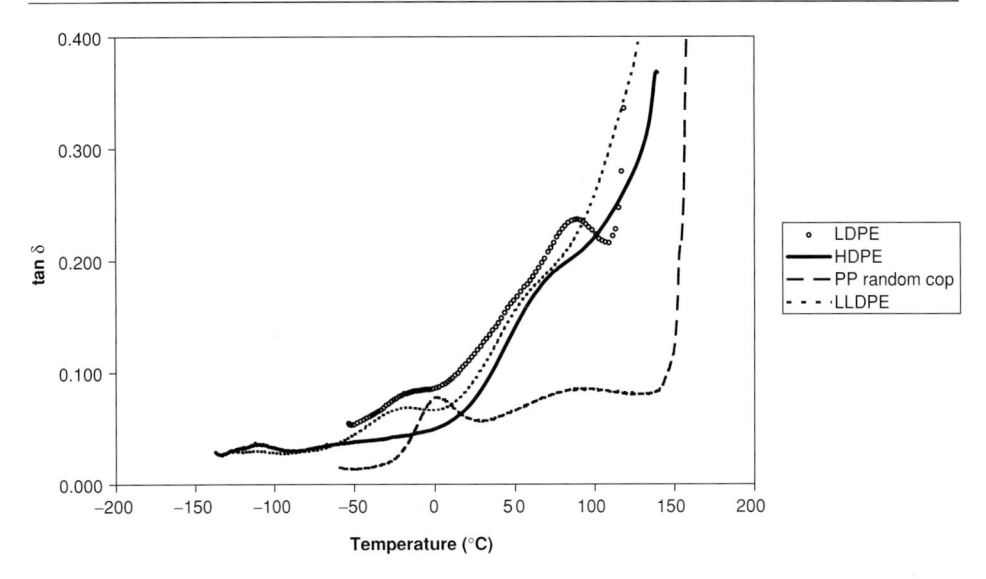

Figure 4.11 Tan δ for various polyolefins. Transitions are shown as peaks in tan δ which increases as the sample is heated into the melt region.

crystalline phases are present a mixture of the above is observed. First the magnitude of the glass transition process is smaller, in terms of the change in E' (or G' for shear) and the magnitude of the loss modulus and tan δ peak. The tan δ peak height is proportional to the difference between the difference $E'_{glassy} - E'_{rubbery}$. The glassy modulus will not be noticeably higher unless the level of crystallinity is very high; however, the rubbery modulus may be one or two orders of magnitude greater. The reason for this is that the crystalline regions act as cross-links within the rubbery matrix of amorphous material. This is analogous to the level of chemical cross-linking in thermosets, where a high degree of cross-linking leads to a high rubbery modulus plateau.

4.5.3 Example of α and β activation energy calculations using PMMA

The calculation of a specific relaxation's activation energy is the main reason why DMA was developed. When the molecular structure is known, theoretical calculations can be made for certain bonds rotating and/or stretching. These theoretical results can then be compared to the measured activation energies. If agreement is obtained, then we can be confident that this is the mechanism that operates. Where differences arise, either the deformational mechanism may be more complex than the modelling or, alternatively, the material may not have the structure postulated. Therefore this is an extremely powerful tool for polymer synthesists.

Figure 4.12 shows variation of tan δ with temperature for multiple frequencies, where the α (T_g) and the β process can be readily observed. Figure 4.13 is a zoom-in of the β process. The values of tan δ for the α process were modelled using an exponential curve fit. Figure 4.14 shows the residual tan δ data value after subtracting the modelling for glass transition, leaving just the tan δ contribution for the β process. Figure 4.15 is an activation energy plot, using the peak temperature values for each frequency for both relaxational processes.

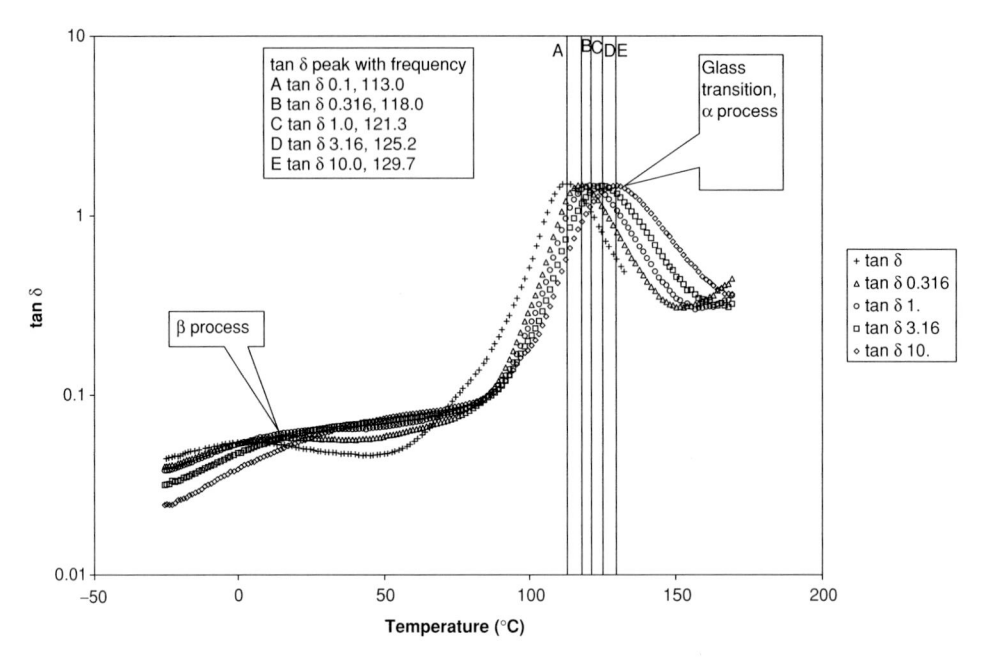

Figure 4.12 Variation of tan δ with temperature for multiple frequencies (PMMA).

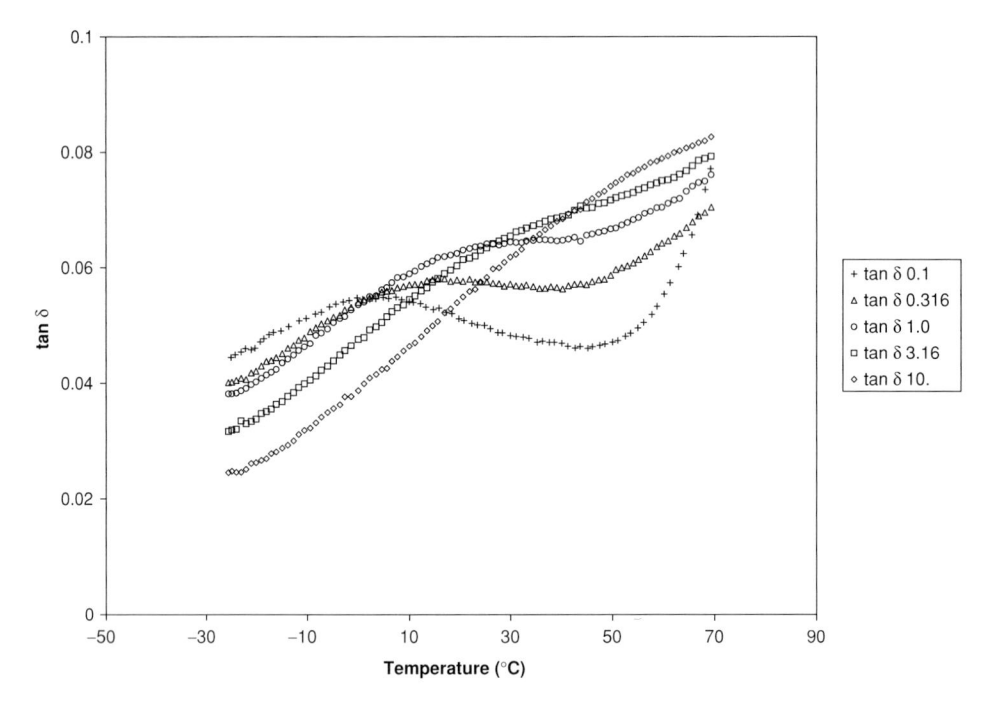

Figure 4.13 Variation of tan δ between −30 and 70°C (β relaxation region of PMMA).

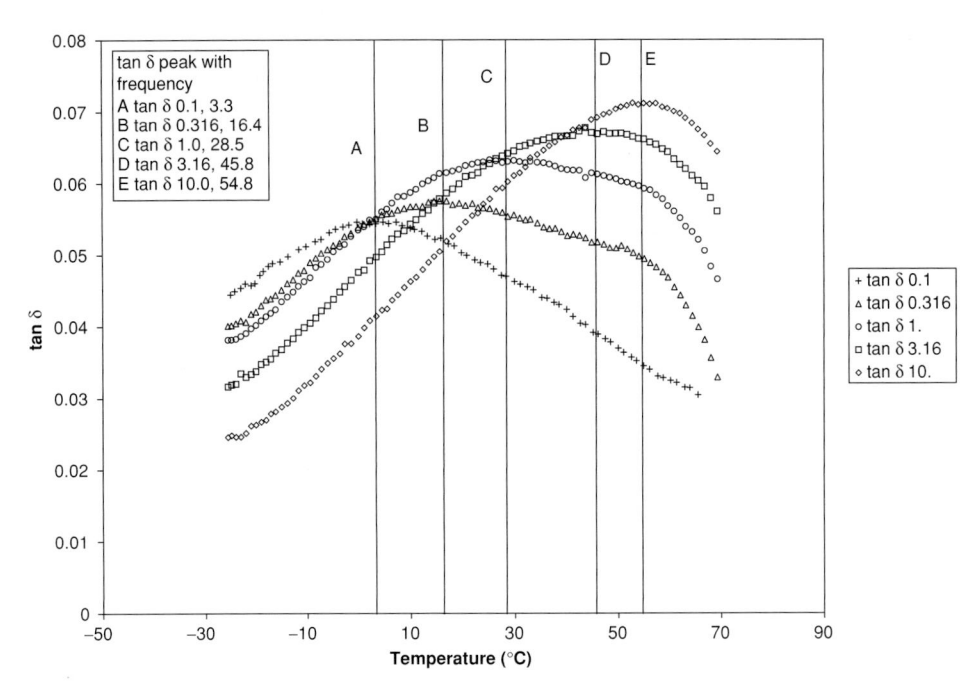

Principles and Applications of Mechanical Thermal Analysis   149

Figure 4.14 The β relaxation in PMMA, after subtraction of the α contribution.

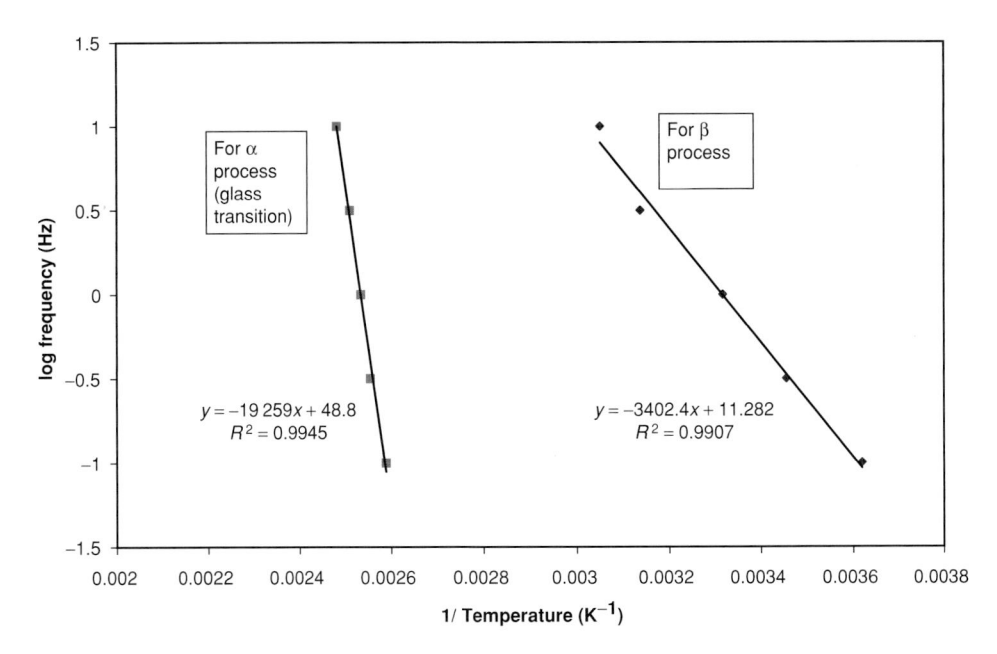

Figure 4.15 Activation energy plot for α and β processes in PMMA.

4.5.4 Glass transition, T_g, measurements

Figure 4.16 shows data for a filled composite sample and details some of the ways T_g is defined. ISO 6721 part 11 states that the inflection point of the modulus curve (E') should be quoted as the glass transition temperature. This is easiest to measure as the derivative of the linear storage modulus (E'). The temperature measured should be close to that of the loss modulus (E'') peak. Indeed for a single Debye relaxation process they would be identical and for materials showing a well-defined T_g (with no overlapping relaxation processes) these two values are very similar (see Figure 4.19). Engineers frequently use the onset method to define the glass transition. The modulus is just beginning to fall at the onset temperature and therefore it would be undesirable to use the material beyond this temperature. It can be defined as the onset (see construction in Figure 4.16) of modulus plotted on a linear or log scale. The onset temperature from a linear plot will give the lowest value of T_g at any given frequency. The remaining parameter that is frequently used to specify the glass transition temperature is the tan δ peak. This is preferred by chemists, but not of much use to engineers, since the modulus will be almost at its rubbery value by this point, unless the material is being used as a rubber of course. The tan δ peak will give the highest value for T_g at any given frequency.

Figures 4.17 and 4.18 show the same parameters for PMMA, which has a significantly merged β relaxation process with the α, or T_g process. Note how the onset could be picked as either 67 or 84°C, depending on whether the α or β region is chosen for the tangent construction. Generally, the onset can vary significantly and is therefore not necessarily the most reproducible parameter to use for T_g measurement. For PMMA there is a large difference between the loss modulus and the derivative of the storage modulus (E') peaks.

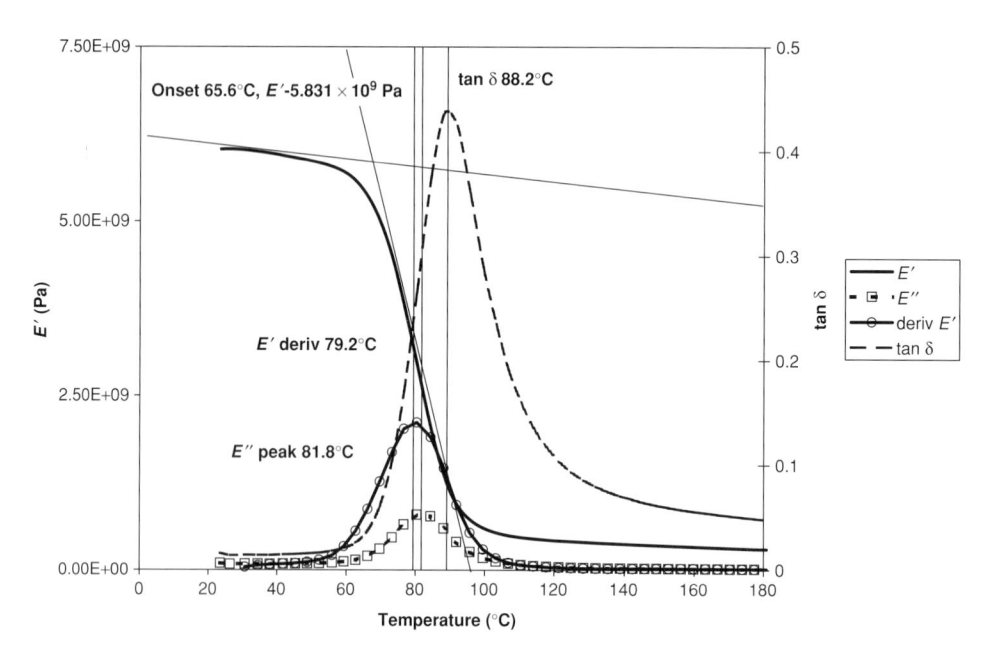

Figure 4.16 Determination of the glass transition temperature. Different methods of calculation include onset of storage modulus decrease, inflection point (derivative maximum), loss modulus maximum, or tan δ maximum.

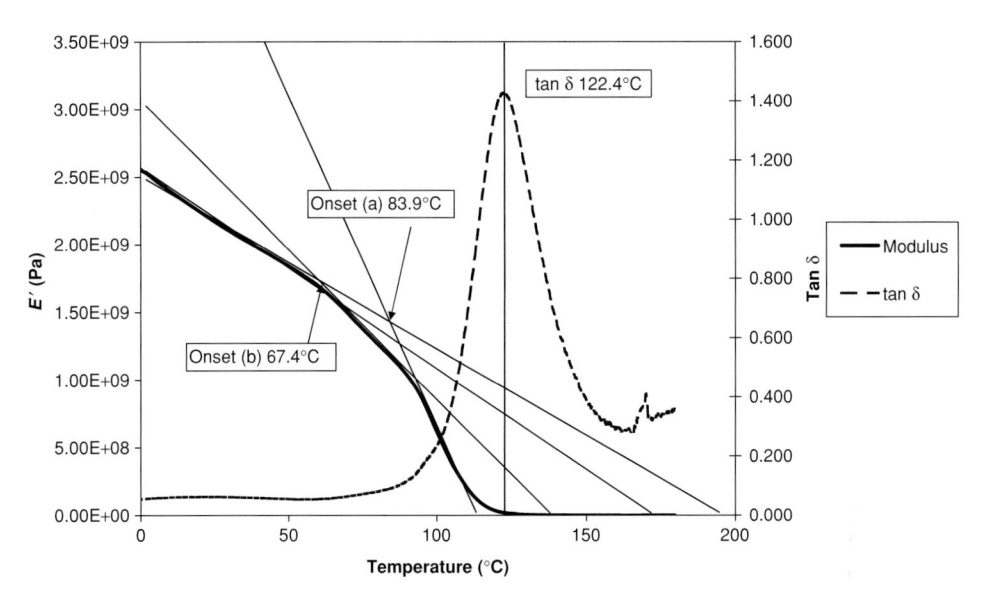

Figure 4.17 T_g determination for a sample with merged α and β processes; E' and tan δ.

This is due to the strongly overlapping relaxations. The presentation of both parameters is useful where materials show such differences and may explain why certain results do not reproduce well and indeed whether or not a certain parameter is a good choice for material specification.

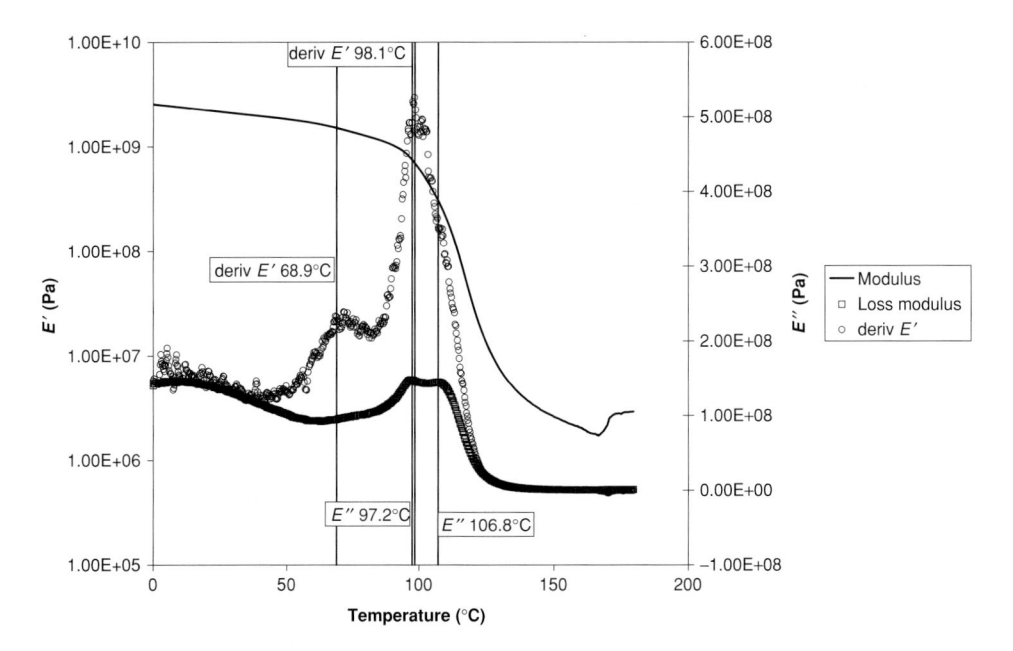

Figure 4.18 T_g determination for a sample with merged α and β processes; E', E'' and E' derivative.

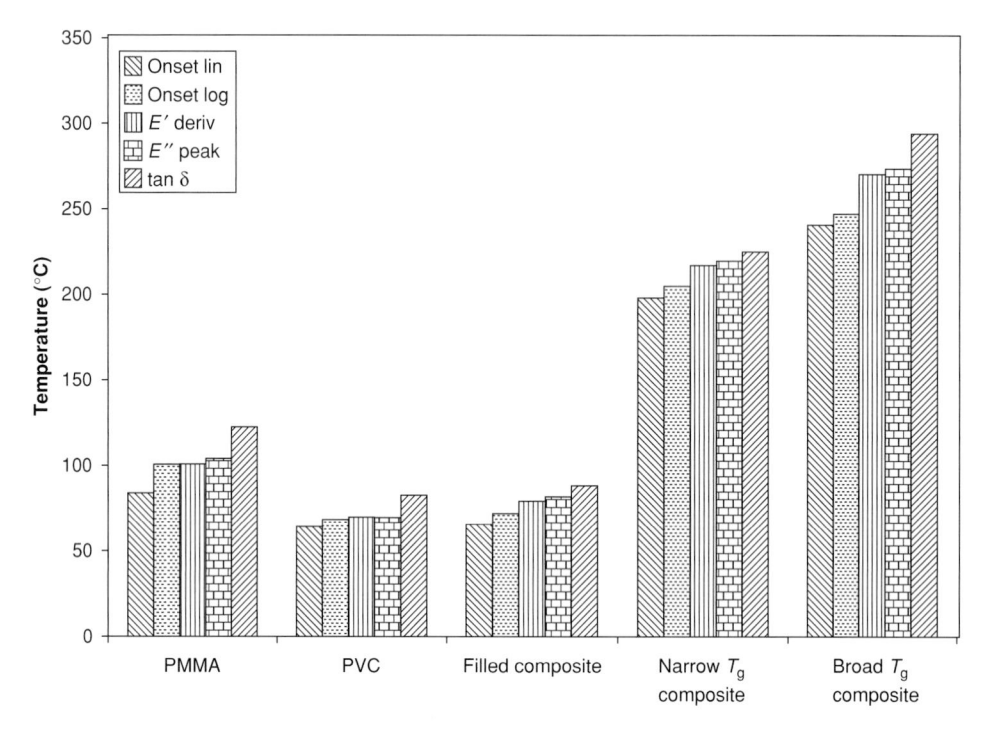

Figure 4.19 Comparison of T_g determination methods for various materials.

Figure 4.19 shows how these techniques of T_g determination can produce different results. The range between these different methods can be over 25°C, depending on the presence of overlapping relaxations, whether it is filled or not and if any further chemical or physical effects occur as we pass through T_g, e.g. further cure in the case of a composite. Generally the tan δ peak yields the highest reproducibility, and also gives the highest value for a given frequency. The loss modulus (E'') peak in particular can be variable, or broad, depending on a polymer's age state, or if some chemistry is occurring at T_g. The onset methods also suffer from poor reproducibility if testing a material where the slope before or after T_g varies significantly from sample to sample. The difference could be real, but may also reflect small and insignificant changes in physical age state for example. They may also be attributable to poor clamping or sample preparation.

4.5.5 Measurements on powder samples

Powder samples are not usually measured by DMA due to difficulty in containing them. However, a recent innovation by Triton Technology Ltd [15] enables such samples to be analysed in a simple manner. The Materials Pocket for powders and films consists of a foldable metal sheet (see Figure 4.20). Alternatively measurements are possible in compression, but this geometry never has the sensitivity of normal bending modes.

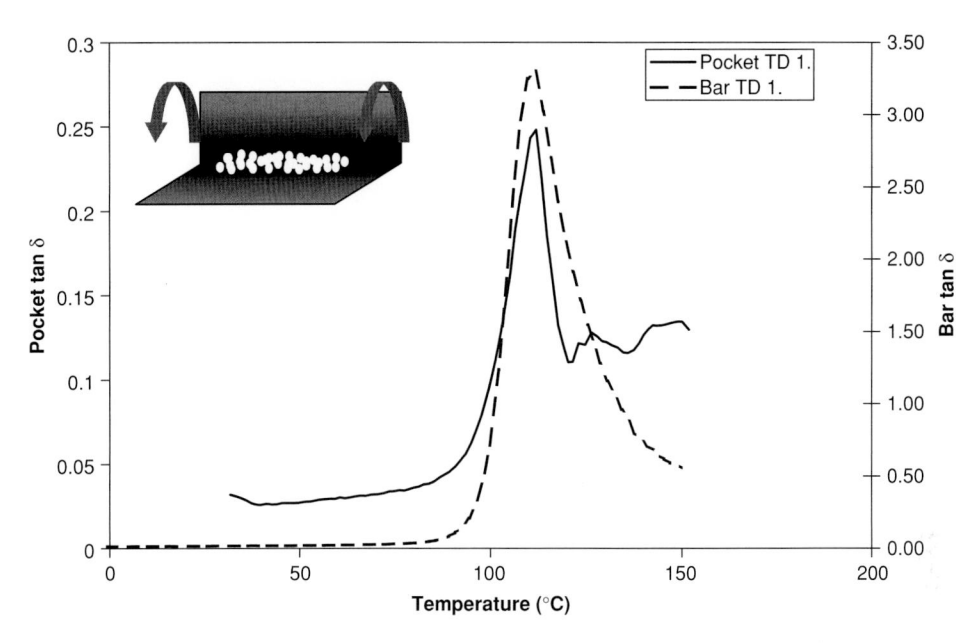

Figure 4.20 Comparison of tan δ results from poly(styrene) bar sample and Material Pocket.

Frequently, T_g detection is a requirement in pharmaceutical and food studies, but it is difficult to obtain solid bar samples. The greater sensitivity of DMA over DSC offers excellent T_g determination, especially where only small amounts of amorphous material are present. To use the Material Pocket, 10–50 mg of sample is poured into the V-shaped trough and the pocket is tightly closed with a pair of pliers or a small vice. The absolute values of the storage modulus (E') and loss modulus (E'') will not be quantitative, due to the stiffness of the metal Material Pocket. However, during a T_g process the change of each of these parameters is quantitative.

Figure 4.20 shows a comparison of tan δ data for a poly(styrene) powder analysed in a Material Pocket with a poly(styrene) bar sample ($4 \times 5 \times 10$ mm^3). Whilst the magnitudes differ by a factor of 10, the temperature information and the glass transition detection are almost identical. The lower tan δ is due to the effective dilution of the signal by the stiffness of the Material Pocket.

Figures 4.21 and 4.22 illustrate a DMA method for quantitatively determining the amorphic content of lactose using the Material Pocket. Lactose is a very important pharmaceutical excipient used in tablet and inhalation products. It is prone to forming amorphous regions on processing, however, and can be problematical to characterise the exact amount of amorphic material in a sample. Figure 4.21 shows a typical DMA response from both a fully crystalline and a fully amorphous sample of lactose. The amorphous sample shows various regions of interest in tan δ corresponding to loss of water, the glass transition, recrystallisation of the amorphous material and finally melting. The crystalline material has no peak corresponding to the initial loss of water as it is less hygroscopic. It obviously has no glass transition due to the absence of any amorphic material. The higher temperature peaks correspond to the loss of hydration water, followed by melting. Whilst many of the features of the above tan

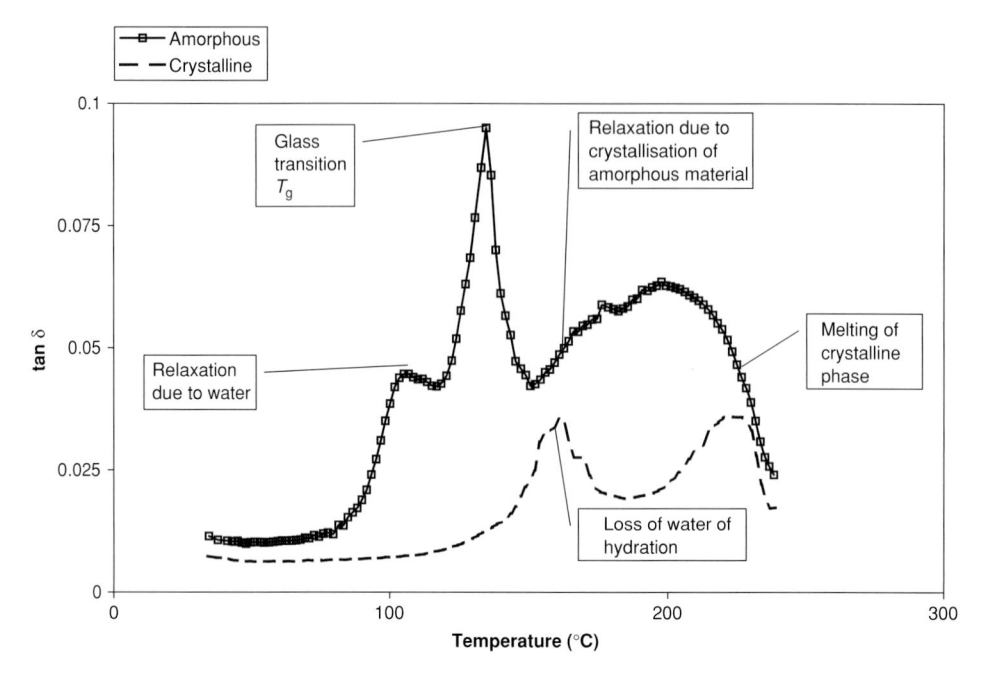

Figure 4.21 Amorphous and crystalline lactose measured in Material Pocket.

δ plot are attributable to the amorphous content, the peak observed for the T_g is the best measure of amorphicity. Figure 4.22 shows the variation of tan δ magnitude as a function of amorphous content. The calibration curve shows a good correlation of fit. It was calculated that this technique has a theoretical limit of detection of 2.8% w/w for amorphous

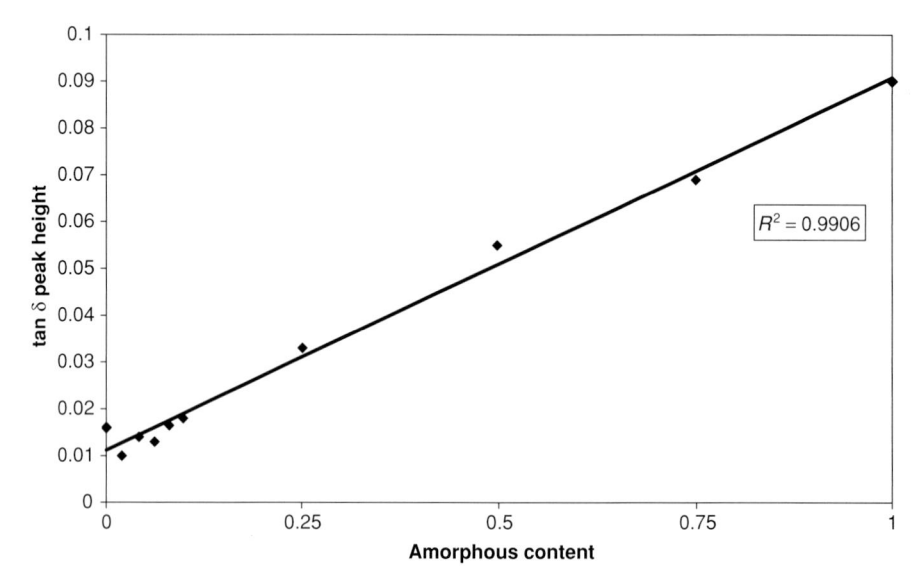

Figure 4.22 DMA relaxation strength for fraction of amorphous phase versus amorphous content for lactose.

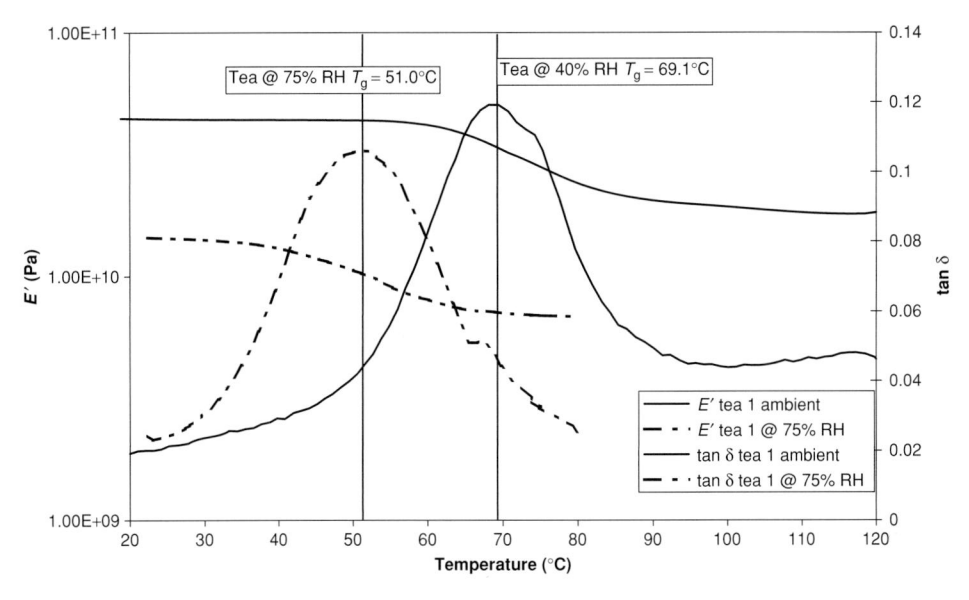

Figure 4.23 Tea 1 thermal scans at ambient and 75% RH.

lactose. This is comparative to other techniques that have been used to quantify amorphous lactose.

A more comprehensive description of both the technique and the data can be found in the publication by Royall et al [15].

4.5.6 Effect of moisture on samples

One variable that is increasingly studied within thermal analysis is the effect of humidity. Frequently, this can have a profound effect upon composite materials, foodstuffs, pharmaceuticals, etc. and can be measured in a number of ways. Specifically gravimetric vapour sorption (GVS) is the term given to the technique when used to measure weight changes in a controlled humidity environment, and this has wide application in the field of foods and pharmaceuticals. Analogous measurements can also be made with calorimetric equipment using static or flowing environments. A number of companies now supply automated humidity controllers for use in DMA experiments. The data from a combination of this accessory and DMA will be presented here.

Two tea samples in caked form were tested in the Triton Material Pocket[15]. These samples were kindly supplied by Unilever Foods for evaluation and the results are reproduced here with their permission. Figures 4.23 and 4.24 show the effect of moisture as a thermal scan is made. The first scan in each figure was carried out under ambient humidity conditions (i.e. a normal heated oven) and the second was performed with a controlled 75% RH environment during heating. The plasticising effect of the moisture is clear, with the glass transitions for teas 1 and 2 being reduced by 18°C and 37°C respectively. Evaluations using an ordinary air oven, where the humidity decreases as the temperature is raised, would have yielded an incorrect value for T_g.

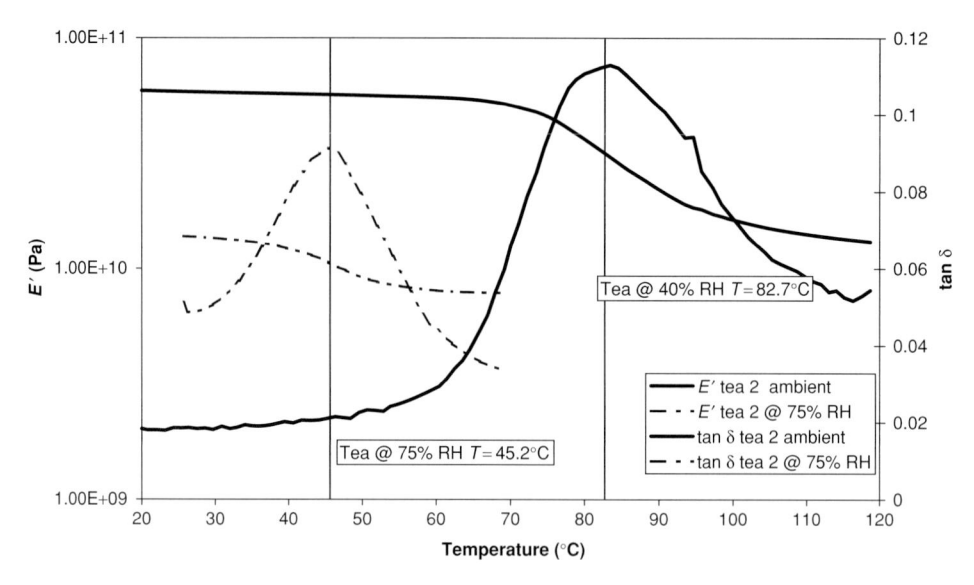

Figure 4.24 Tea 2 thermal scans at ambient and 75% RH.

Various experiments are possible as the humidity sensor is situated inside the environmental chamber and measurements are valid over the operational temperature range from 10 to 85°C and humidity range 10–80% RH. Normally, it is best to make measurements as a function of time at constant temperature and humidity. This gives the sample ample time to equilibrate. If thin films or samples that have high diffusion rates are being tested, then either the humidity or temperature can be scanned during an experiment, but care should be taken to ensure that the sample is truly in equilibrium. Such experiments enable an accurate picture to be built up of the material's true properties. This has serious implications when considering the storage life of such products.

4.6 Thermomechanical analysis

4.6.1 Introduction

Thermomechanical analysis (TMA) predates the use of dynamic mechanical analysis techniques. TMA is used for T_g determination, but is significantly less sensitive than DMA and cannot be used for studying the weaker β relaxations, as seen in many polymers. In many respects TMA is the simplest form of thermal analysis equipment. A small sample is mounted in the instrument, which is surrounded by a furnace and the variation of sample length is recorded as a function of time or temperature.

One important application area is the derivation of the coefficient of thermal expansion (CTE). This is usually performed using tension or compression geometries and measuring the expansion; although modern TMAs generally have the same range of deformation geometries as used by DMA, described in Section 4.3.2.

Another variant of TMAs is the dynamic TMA. These function with half the facilities of a DMA, in that they apply a small constant dynamic load throughout a thermal scan, usually made from low to high temperature. When the sample is glassy, the force is insufficient to cause significant deformation, but as the modulus falls through the glass transition, then the resultant displacement becomes larger and can be seen as an envelope of the TMA signal. Note that the dynamic TMA mode does not measure the phase difference between the force and displacement signals. It just reports the displacement amplitude resulting from the dynamic force.

4.6.2 Calibration procedures

Calibration of TMAs is relatively straightforward. The displacement (height) signal can be calibrated by inserting accurately measured height samples and noting the instrument output. The offset and gain can then be set to match the true sample size. The temperature is set by the same principle as that used in DSC, namely by melting point standards. Force can usually be calibrated by placing a known weight on the system and noting the force needed to offset it.

For accurate quantitative measurements the effect of the baseline may also need to be taken into account. Normally this only need be carried out when CTE measurements are being made. If the softening point only is required then baseline subtraction is unnecessary. Results of CTE measurements may also be compared to a standard, in a similar manner to specific heat measurements, but provided care is taken with calibration and that scan rates used are fairly low to avoid thermal lag, results obtained should be accurate.

If a film is being measured in extension then the baseline may be measured with a known sample in place. The normal procedure here is to use a well-characterised material with a known CTE [16]. A thermal scan can be performed with quartz or aluminium, for example, using the same sample length and heating rate as will be used in the actual determinations. The known sample expansion can be subtracted from the measured result and this difference is due to the instrument's baseline, which in turn can be subtracted from future experiments on unknown material. Note that this baseline usually depends on sample length, geometry, heating/cooling rate, thermal conductivity and heat capacity. If any of these parameters change, then the baseline should be remeasured. With respect to the sample's thermal properties it is best that a material closely resembling the sample under test is used for the calibration.

4.6.3 TMA usage

4.6.3.1 Experiment with zero load

Figure 4.25 shows a classic TMA trace for a polymeric sample being heated through its glass transition. Here the polymer is the amorphous material, poly(carbonate). A 10 mm long bar sample, with a cross-section of approximately 4×4 mm^2 was mounted in the tension clamps and heated at a rate of 2°C/min. The ordinate axis shows the sample displacement, with temperature as the abscissa. The indicated temperature in the figure shows the onset calculation, which is the intersection of the glassy sample thermal expansion, from 30 to 130°C and the rubbery sample expansion from 150 to 170°C. The intersection can be used as a measure of the glass transition. These data were obtained by mounting a sample in tension clamps and applying zero load. Such an experiment allows free expansion of the sample, which permits the determination of the CTE.

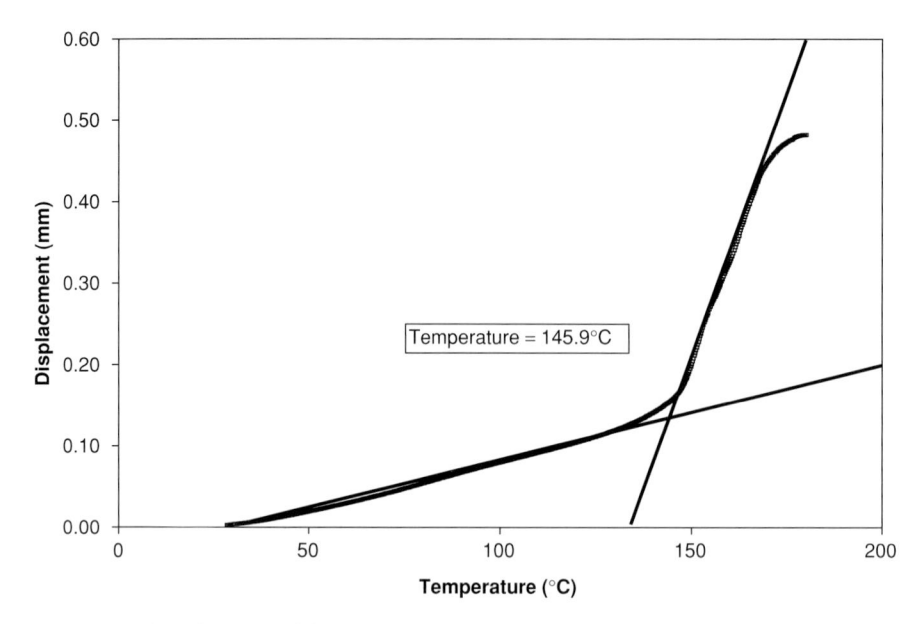

Figure 4.25 T_g from the onset of thermal expansion plot. The sample in the case is under zero force. A film or fibre may need to be under very slight tension (a few mN) to hold the clamps apart.

4.6.3.2 Experiment with constant load

Figure 4.26 shows thermal expansion data for the same polymer (polycarbonate), tested as a 10 mm long bar sample mounted in the tension clamps, but this time with a compressive load of 1 N. The glassy expansion occurs as before, but when the sample softens at the T_g it is progressively squashed under the compressive load. The onset temperature for the T_g is the same, 145.9°C. Figure 4.27 shows the two curves together.

4.6.3.3 Coefficient of thermal expansion determination

Figure 4.28 shows the coefficients of thermal expansion (CTEs) calculated from the slopes of the displacement versus temperature plots, from the above experiments. The CTE is the slope divided by the sample length, i.e. the amount of expansion (mm/mm/°C). Both glassy CTEs are comparable, but only the data from the no-load experiment have been used to provide the rubbery CTE data. The data under load represent how easily the sample deforms and have nothing to do with the thermal expansion of the sample.

4.6.3.4 Experiment with constant and dynamic load

Figure 4.29 shows data from the poly(carbonate) sample run with a compressive load. In addition to the static load, a constant dynamic load at a frequency of 1 Hz was applied throughout the temperature scan. At the start, the sample is very rigid and the dynamic displacement is less than 1 μm. After the static T_g determined above, the dynamic displacement is observed to increase over 100-fold. This shows the increased sensitivity that is available from dynamic equipment.

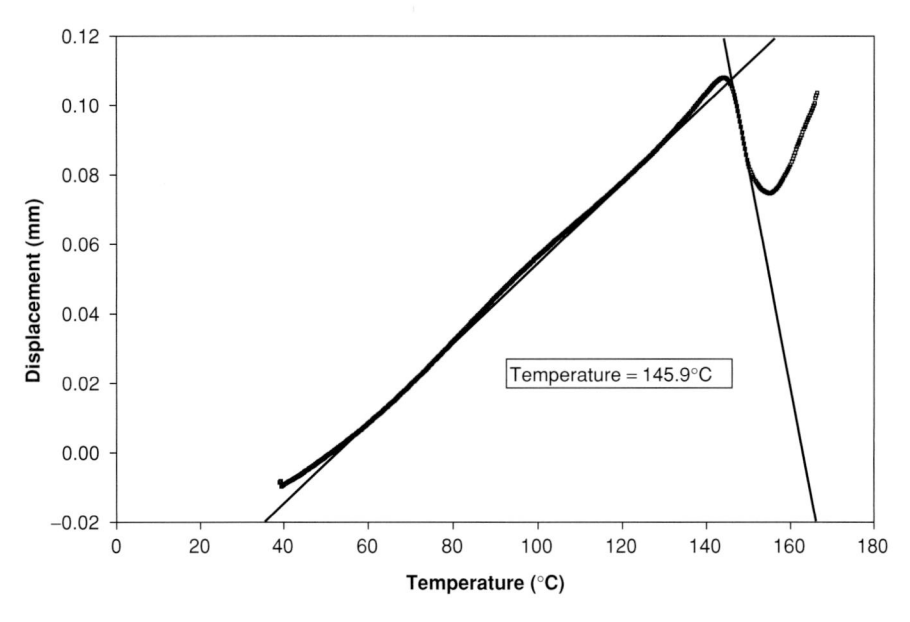

Figure 4.26 T_g from the onset of thermal expansion plot with the sample under compressive load. At T_g the samples softens and is compressed.

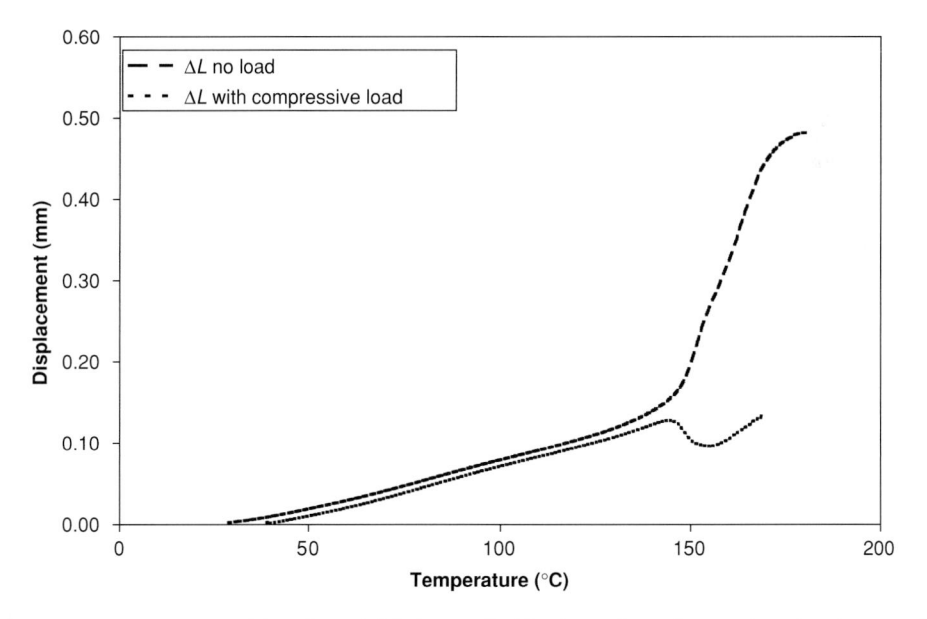

Figure 4.27 Comparison of the effect of differing applied load to a thermal expansion plot. A range of curves could be produced depending upon the force applied to the sample.

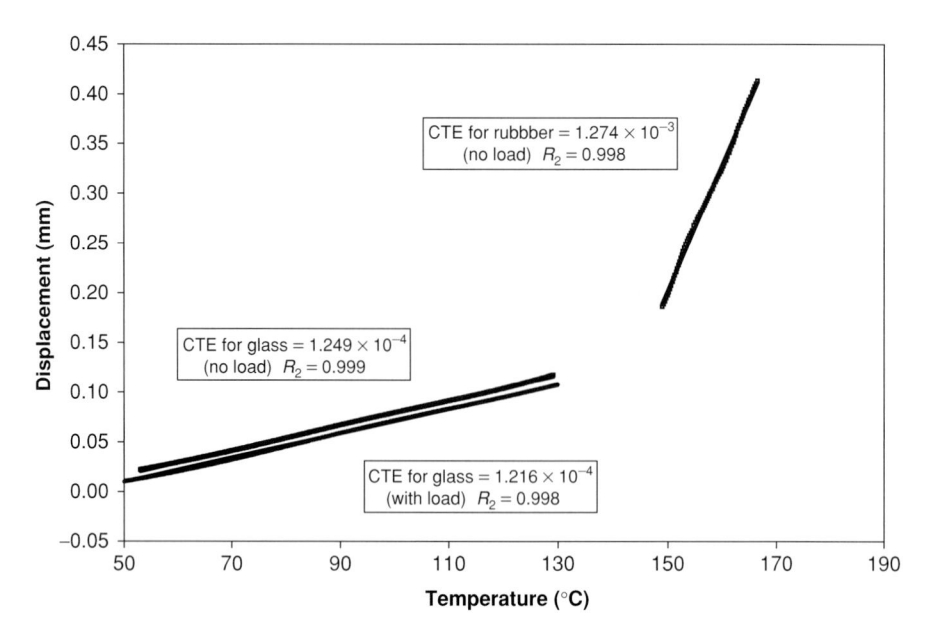

160 *Principles and Applications of Thermal Analysis*

Figure 4.28 Thermal expansion coefficients for glassy and rubbery states.

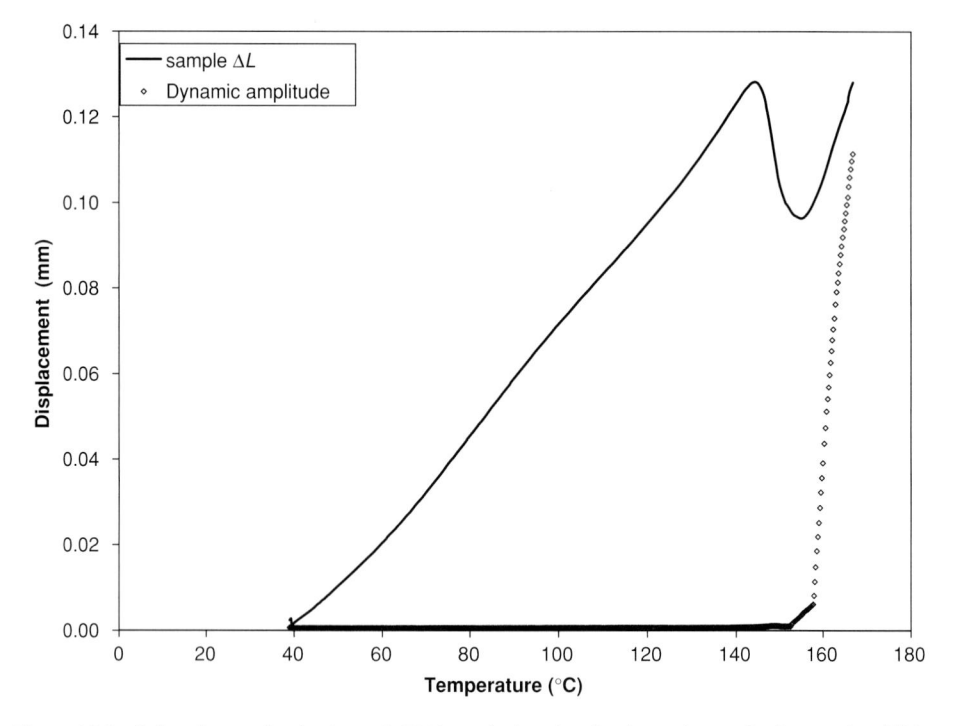

Figure 4.29 Polycarbonate bar in dynamic TMA mode showing the dynamic amplitude trace in addition to the thermal expansion trace observed under compressive load.

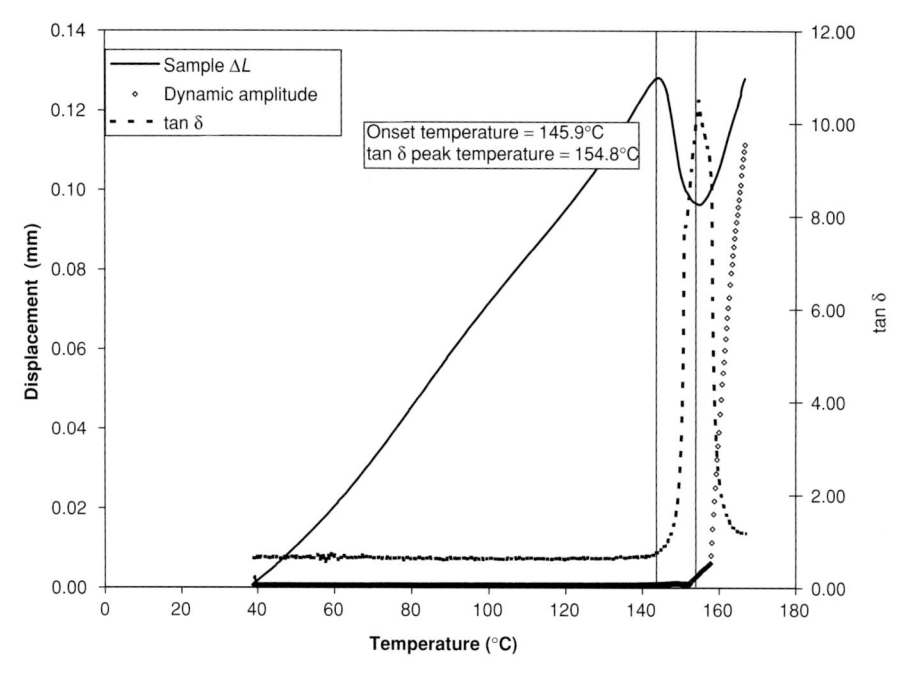

Figure 4.30 Dynamic TMA mode also showing tan δ plot.

One advantage of performing the dynamic load TMA experiment on a DMA is that we can also view the tan δ values. Figure 4.30 shows these data. The immediate difference apparent from the two traces is the difference between the T_g as defined by the onset from the displacement measurement and the T_g as defined by tan δ. The onset T_g is essentially obtained from a static or low-frequency test. This difference is due to the frequency dependence of the T_g. This is fully discussed in Sections 1.3 and 5.3.

4.6.3.5 Baseline measurement example

Figure 4.31 shows baseline correction data obtained using the method described in Section 4.6.2. An aluminium bar sample was used, with a 10 mm length and with a cross-section of approximately 3.5×4 mm^2. This was mounted in the tension clamps and heated at a rate of 2°C/min. A value of 25×10^{-6} mm/mm/°C [16] was used for the CTE of the aluminium sample, which is calculated over the entire ramp by multiplying the temperature difference by this value and the sample length. This value is then subtracted from the observed expansion/contraction values and the resultant curve is the instrument baseline. This has then been fitted to a fourth-order polynomial (or a straight line fit can be used if it is linear) and this polynomial is used in future experiments, the difference between the total displacement and this value being the sample expansion/contraction. As the total data are less than the calculated curve for aluminium, we can conclude that the instrument baseline shows a shrinkage on heating. This is quite normal. Due to various parts of the instrument being at different temperatures, either expansion or contraction is commonly observed. However, sample expansion should always be positive on heating, unless the sample is highly anisotropic.

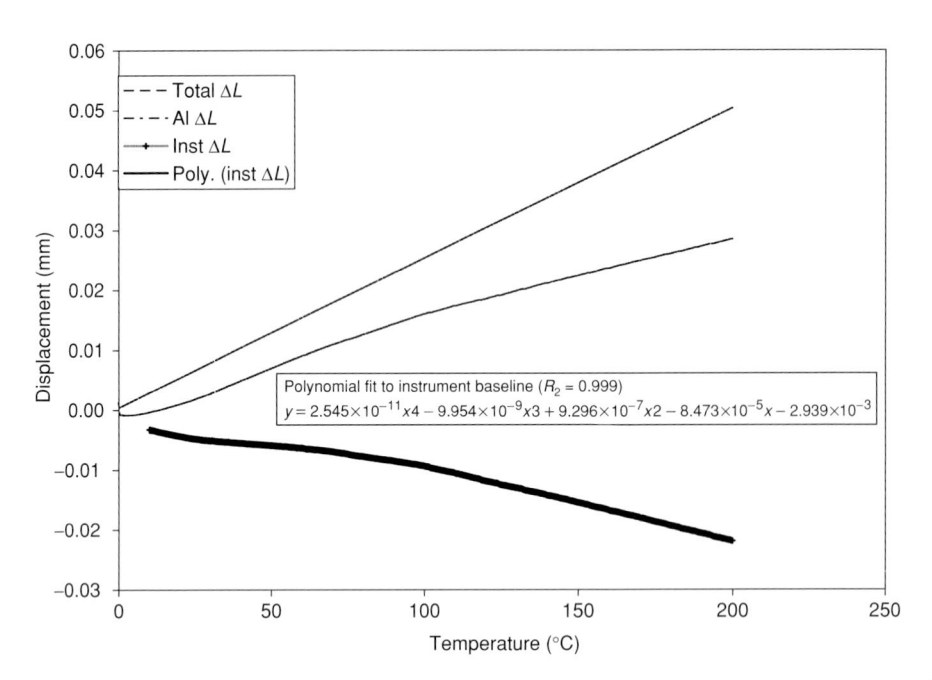

Figure 4.31 Instrument baseline determination in extension mode. The upper curve is the theoretical aluminium response, the middle curve the actual data. The subtracted curve shown in bold is the calculated instrumental baseline.

Appendix: sample geometry constants

The geometry constant (k) is calculated using the following basic equations:

Geometry	Geometry factor	Dimensions
Tension and compression	$k = A/l$	A is the cross-sectional area and l is the free length or thickness
Single cantilever	$k = w(t/l)^3$	w is the width, t is the thickness, and l is the free length
Dual cantilever	$k = 2w(t/l)^3$	w is the width, t is the thickness, and l is the free length
Three-point bending	$k = w(t/l)^3/2$	w is the width, t is the thickness, and l is the free length
Simple shear	$k = 2A/t$	A is the cross-sectional area and t is the thickness

Note that for these definitions the sample length, l, is always taken as the distance between the fixed clamp and the driveshaft clamp. Therefore samples in single and dual cantilever mode will both have the same length for a given clamp spacing. Similarly, the sample length in three-point bending is the half length of the sample supported between the two fixed clamps.

A small shear correction is normally applied to clamped and three-point bending data by dividing k by $1 + 2.9(t/l)^2$ for clamped bending and $1 + 2.9(t/2\,l)^2$ for three-point bending. This correction is based on an average Poisson ratio of 0.33 for glassy polymers and 0.5 for rubbers, namely 0.45.

References

1. McCrum NG, Read BE, Williams G. *Inelastic and Dielectric Effects in Polymeric Solids.* New York: John Wiley and Sons, 1967.
2. Read BE, Dean GD, Duncan JC. Determination of dynamic moduli and loss factors. In: *Physical Methods of Chemistry*, Vol. 7. New York: John Wiley and Sons, 1991.
3. Aklonis JA, MacKnight WJ. *Introduction to Polymer Viscoelasticity.* New York: Wiley-Interscience Publication, 1983.
4. Ferry JD. *Viscoelastic Properties of Polymers.* New York: Wiley, 1961.
5. Fox TG, Flory PJ. *J Phys Chem* 1951;55:221.
6. ISO 6721 Plastics. Determination of dynamic mechanical properties.
7. Read BE, Duncan JC. Measurement of dynamic properties of polymeric glasses for different modes of deformation. *Polym Test* 1981;2:135.
8. Duncan JC, Sauerbrunn S. Polymer glass transition by RH-DMA. In *NATAS Conference*, Williamsburg, USA, September, 2004.
9. Hyun K, Nam JG, Ahn KH, Lee SJ. *Rheol Acta* 2006;45:239–249.
10. Dean GD, Duncan JC, Johnson AF. Determination of non-linear dynamic properties of carbon filled rubbers. *Polym Test* 1984;4:225.
11. Mulligan D, Gnaniah S, Simms G. *Thermal Analysis Techniques for Composites and Adhesives*, 2nd edn, NPL Measurement Good Practice Guide No. 62, March, 2003; p. 18.
12. Boyd RH. Relaxation processes in crystalline polymers: experimental behaviour – a review. *Polymer* 1985;26(March):323.
13. Boyd RH. Relaxation processes in crystalline polymers: molecular interpretation – a review. *Polymer* 1985;26(August):1123.
14. Coburn JC, Boyd RH. Dielectric relaxation in poly(ethylene terephthalate). *Macromolecules* 1986;19:2238–2245.
15. Royall PG, Huang C-Y, Tang SJ, Duncan JC, Van-de-Velde G, Brown MB. The development of DMA for the detection of amorphous content in pharmaceutical powdered materials. *Int J Pharm* 2005;301:181–191.
16. Kaye GWC, Laby TH. *Tables of Physical and Chemical Constants.* Harlow: Longman, 1995; pp. 73–75.

Chapter 5
Applications of Thermal Analysis in Electrical Cable Manufacture

John A. Bevis

Contents

5.1	Introduction	165
5.2	Differential scanning calorimetry	165
	5.2.1 Oxidation studies (OIT test)	165
	5.2.2 Thermal history studies	166
	5.2.3 Cross-linking processes	168
	5.2.4 Investigation of unknowns	171
	5.2.5 Rapid scanning with DSC	172
5.3	Thermomechanical analysis	175
	5.3.1 Investigation of extrusion defects	176
	5.3.2 Cross-linking	177
	5.3.3 Material identification	177
	5.3.4 Extrusion studies	178
	5.3.5 Fire-retardant mineral insulations	179
5.4	Thermogravimetric analysis	180
	5.4.1 Practical comments	180
	5.4.2 Investigation of composition	181
	5.4.3 Carbon content	185
	5.4.4 Rapid scanning with TGA	186
5.5	Combined studies	188
5.6	Concluding remarks	189
	References	189

This chapter gives a series of practical examples of how thermal methods have been applied in an industrial laboratory when a series of troubleshooting or development tasks have been brought to the chemist. It reflects principles and approaches that can be applied in the cable industry and broadly over many industries.

5.1 Introduction

Cabling for power transmission and distribution must give reliable continuous service over many years. In addition to maintaining electrical integrity it may be required to operate at relatively elevated temperatures and to continue providing service under fire conditions. Many of the materials used in cable manufacture are polymeric, and some are cross-linked to permit service at elevated temperature and to provide required electrical and mechanical strength.

Thermal analysis offers the possibility of investigating ageing properties, cross-linking processes, service conditions and behaviour during the burning process, as well as being an invaluable tool for the analysis of materials to verify composition of particular formulations for special applications. In addition it is possible to make use of thermograms of unknown materials for identification purposes. The discussion below gives examples of these applications.

Although the emphasis here is on applications to electrical cabling, it is envisaged that they would be readily transferred to other industries engaged in the manufacture of polymeric products.

5.2 Differential scanning calorimetry

5.2.1 Oxidation studies (OIT test)

For polymeric products required to operate at temperatures appreciably higher than ambient, degradation processes, whether purely thermal or as a result of accelerated oxidation, are important and can significantly affect life expectancy if appropriate grades, adequately protected by suitable antidegradants, are not selected. Oven ageing trials, with periodic mechanical testing of test pieces, are good predictors of performance but are necessarily slow. The measurement of oxidation onset by differential scanning calorimetry (DSC), at higher temperatures than those used for oven ageing trials, is useful in some cases.

The basis of the measurement is that of holding a sample under isothermal conditions under an air or oxygen atmosphere, the sample being contained in a pan which is open to the purge gas, and to measure the time taken for the oxidation exotherm to appear in the DSC trace (the oxidation induction time or OIT test, see Section 1.5.9). The experiment is repeated at several temperatures so that data from different samples can be compared, either directly or by using the reciprocal of the OIT as a pseudo rate and plotting the logarithm of the pseudo rate against reciprocal absolute temperature in an Arrhenius plot.

This approach was adopted for a study of polypropylene used as the filler in the interstices between the cores of some types of three-phase power distribution cables. The cores consisted of a conductor with polymeric insulation and extruded screens which were in turn screened with copper tape. Polypropylene twine is used to pack the interstices as the cable is laid up and then the outer parts of the cable are applied later. Thus the polypropylene is in direct contact with the copper tape.

Copper is well known as a catalyst of the autocatalytic oxidation of polyolefins, catalysing hydroperoxide breakdown and chain scission; polypropylene is particularly susceptible to this degradation route and it was therefore vital that grades were selected which contained a suitable metal deactivator as well as adequate antioxidant.

OIT was measured at temperatures in the range 180–210°C for a range of samples, either in aluminium pans alone or with 3% of metallic copper also added to the pan. Sample size was kept as close as possible to 10 mg for the sake of consistency between tests. Samples were heated to the test temperature under nitrogen and the purge changed to oxygen at the start of the isothermal part of the measurement.

In addition some samples were reformed by pressing into sheets; small samples were then cut for ageing in air-circulated ovens at various temperatures, between 120 and 150°C, both in and out of contact with copper. These oven ageing trials were conducted in order to validate the extrapolation to higher temperatures so that future material selection based solely on DSC measurement of OIT could be validated. The time to failure in the oven ageing trials was assessed by a simple three-point bending test, since it was found that the polypropylene lost all mechanical integrity rather rapidly once the oxidation process was under way. The time to failure was reported as OIT at the lower temperatures and these results incorporated into the Arrhenius plots.

Interestingly, the extrapolation from the DSC measurement temperature range to that of the oven ageing trials proved successful, and useful Arrhenius plots were obtained. The initial trial permitted the selection of suitable grades of polypropylene, with adequate resistance to copper-catalysed oxidation, for the designated use with confidence. Subsequently, OIT studies were used as a screening process for other candidate grades for this application, thus saving considerably time and effort which might have been spent on oven ageing trials.

Other materials such as certain types of cable sheath also have to give specified levels of performance in oven ageing trials. It was decided to experiment with some fire-retardant cable sheathing in purely oxidative studies since it was known that the oven ageing trials for this material gave very convincing Arrhenius plots. The DSC measurements were carried out in the range 190–220°C, and very consistent replicates and good Arrhenius plots were obtained. Unfortunately, the two sets of Arrhenius data had different slopes, indicating that the reaction mechanisms for degradation at the higher temperature range differed from those at the lower oven ageing range, unlike the oxidation of polypropylene which had followed the same reaction mechanism over a very wide range of temperature.

This second trial demonstrated the necessity of making adequate measurements at oven ageing temperature ranges to test the validity of using data obtained at higher temperatures in the DSC for extrapolation purposes. It is clearly vital to ensure that the reaction mechanism remains the same over the whole range of interest.

5.2.2 Thermal history studies

When power cables are energised, some heating takes place as a result of resistive and other effects. Cables are therefore designed with these effects in mind and such that they can accommodate the rated load for the circuit without overheating and consequent accelerated

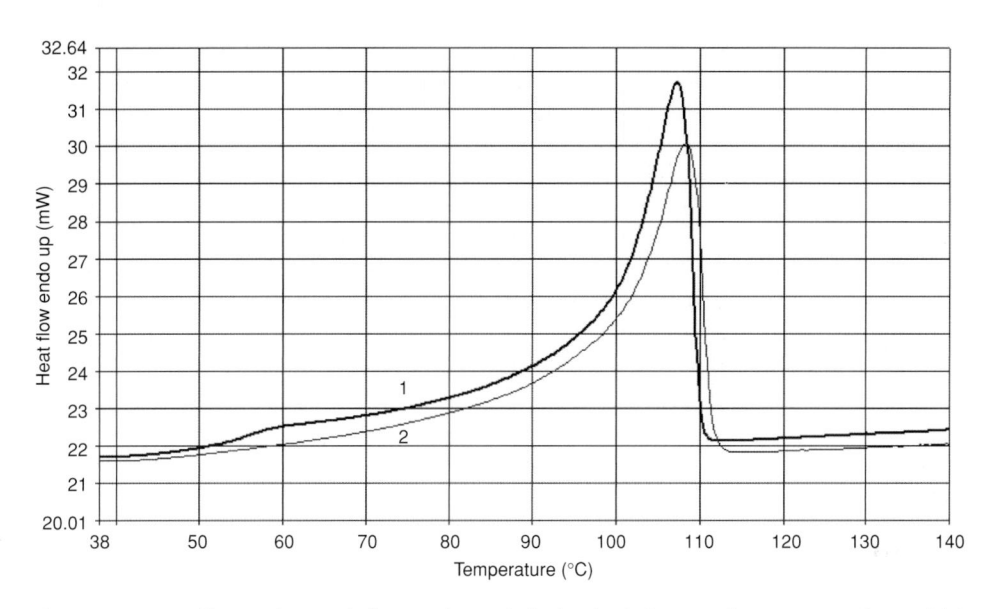

Figure 5.1 Crystalline melting endotherms of cross-linked polyethylene insulation as manufactured (1) and after recrystallisation at 140°C (2).

ageing. It is useful, therefore, to have some means of investigating the thermal history of cable insulation.

Much cable is insulated with cross-linked polyethylene and in this case the crystalline melting envelope gives an opportunity to investigate past heating effects. When polyethylene is heated at a temperature within the crystalline melting envelope, reorganisation occurs for that part of the structure which can melt at the temperature achieved. On cooling, this new structure is retained. As a result the shape of the crystalline melting envelope, obtained in a DSC experiment, has a pronounced irregularity at the final highest temperature of heating, when compared to a sample cooled from a temperature above the crystalline melting point. This work was based on a report by ERA Technology [1].

It thus becomes possible to study cross-linked polyethylene insulations with a view to investigating possible past overheating events in service and also, in the case of very large cables with thick insulation, to investigate temperature gradients across the thickness of the insulation by making measurements at intervals across the thickness. Data are required from the cable in the as-manufactured condition and in the fully recrystallised state. Figure 5.1 shows an example.

It will be noted that there is a small event between 50 and 60°C. This is a result of a thermal cycle in manufacturing to remove by-products of the cross-linking process.

Curves for insulation, which have been oven aged at temperatures that might be expected in normal light service and under conditions of overheating, are shown in Figure 5.2.

These measurements were carried out under nitrogen purging, at a heating rate of 10 K/min. Samples were approximately 7 mg in weight and held in crimped (unsealed) pans

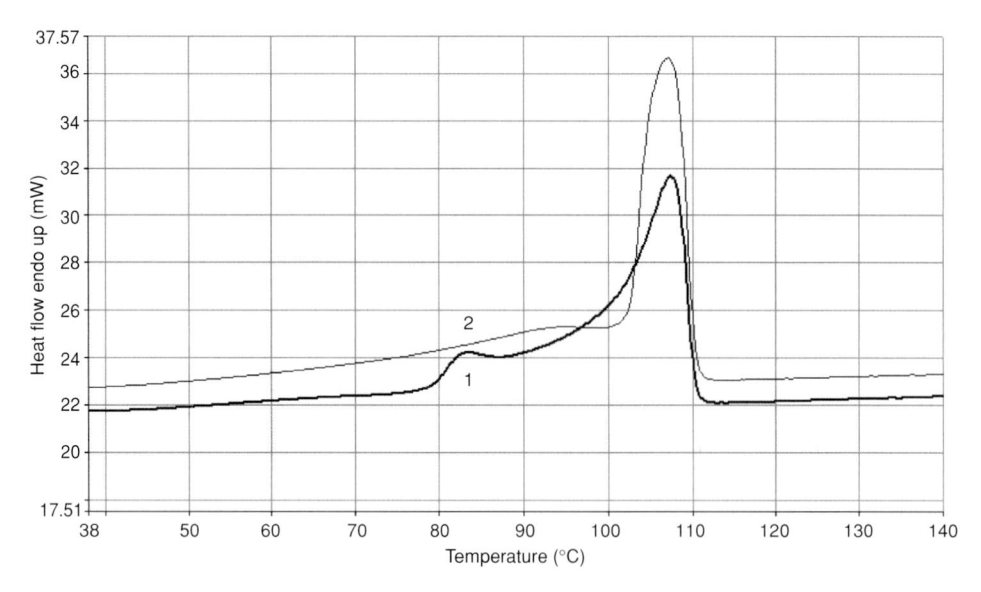

Figure 5.2 Crystalline melting endotherms for cross-linked polyethylene which had been oven aged at 78°C (1) and 102°C (2).

with unperforated lids. A previously measured baseline, obtained with empty pans in both reference and sample positions, was subtracted from the raw data.

It is notable how closely the onsets of the crystalline melting event match the measured oven ageing temperatures, and it is also worth noticing that the previous thermal history of the manufacturing cycle has, in all cases where a subsequent heating had taken place, been recrystallised out. Thus if two thermal histories were observed, it would be possible to conclude that the lower temperature event occurred after the higher. If no thermal history was observed at all, it would be reasonable to conclude that the insulation had fully recrystallised because the thermal history of the manufacturing process would have disappeared (see Figure 5.1). Since the normal service rating for cross-linked polyethylene-insulated cables is designed to give a maximum temperature of 85–90°C at the conductor, depending on design, full recrystallisation would be indicative of a period of being run under appreciable overload conditions, although the actual temperature achieved cannot be measured. In this case it has sometimes been possible to examine the outer cover of the cable, where these are manufactured from medium- or high-density polyethene which both have higher peak temperatures for their crystalline melting points than does the low-density polyethene usually employed for insulation purposes.

5.2.3 Cross-linking processes

The cross-linking of cable insulation is frequently accomplished by means of a process in which an organic peroxide is decomposed, forming radicals which abstract hydrogen atoms from the polymer molecules. The polymer radical sites then react together to form cross-links. It will be clear that the cross-linking must be adequate to ensure that the insulation

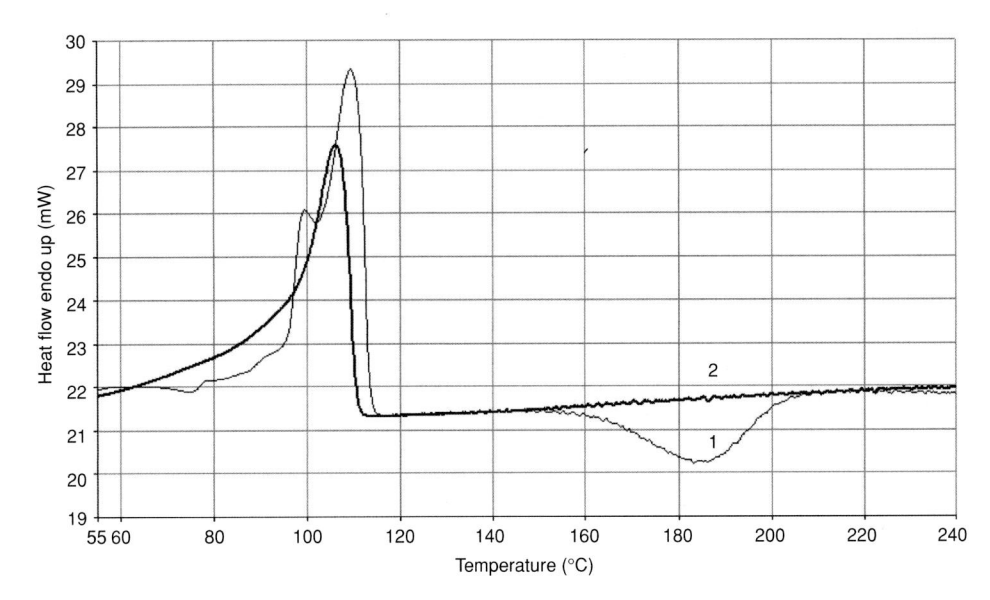

Figure 5.3 DSC curves for initial heating of polyethylene granule with peroxide incorporated (1) and subsequent reheating (2).

does not deform, despite the cable operating within the crystalline melting envelope. Thus control of the cross-linking process is necessary and DSC has been employed in several ways to assist in this control.

Figure 5.3 shows a DSC curve obtained for a precompounded commercially available polyethene containing a suitable peroxide. The sample (approximately 6 mg, weighed accurately) was held in a hermetically sealed aluminium pan, the instrument was purged with nitrogen and the heating rate was 10 K/min. After the first heating, the sample was allowed to cool and reheated. As for the thermal history work the data are baseline subtracted. The data also show the thermal history impressed on the polymer by the manufacturing process of the granules. A similar exotherm to that observed in Figure 5.3(1) was found from a scan of the pure peroxide, and for the purpose of this work, this exotherm was assumed to be due to the decomposition of the peroxide.

It has been found necessary to use hermetically sealed pans for this type of work. The decomposition products of the peroxide are appreciably volatile and if not contained, will volatilise, resulting in erroneously small peak integrations for the reaction.

These types of data have proved useful in several ways.

First, the peroxide type can be determined using peak onset and peak position data for the exothermic peroxide decomposition. This information is useful for the prediction of decomposition products to be removed from the cable in subsequent process and also in the continuous vulcanisation plant used to effect the heating under pressure and to carry out cross-linking in the production process. Build-up of these decomposition products in the plant may otherwise cause a number of safety hazards.

Second, integration of the peroxide decomposition peak can be used to make a quantitative measurement of peroxide present. It is necessary to obtain curves with different amounts

Table 5.1 Extent of reaction for peroxide decomposition under isothermal conditions

Time at 155°C/min	Extent of reaction
1	0.25
4	0.65
6	0.75
10	0.9
20	1.0

of peroxide and a suitable quantity of peroxide-free polyethene, ideally the base polymer, in order to produce a suitable calibration curve. This information can be used for material verification purposes but also for determination of residual peroxide where the cross-linking process is incomplete.

The investigation of incomplete reactions has also been employed to study dwell time requirements in the vulcanisation plant. Samples (10 mg) were heated isothermally in hermetically sealed pans at a temperature determined as being that of the coolest part of the cable during vulcanisation and for different short times of the order of minutes. After the short isothermal heating the DSC furnace was rapidly cooled, using liquid nitrogen, and then an experiment of the type shown in Figure 5.3 was run to measure the residual peroxide so that the extent of reaction for each time/temperature combination could be measured. Data for a series of experiments at 155°C are given in Table 5.1.

Peroxide content measurement proved very useful when investigating alternative peroxides, some of which would not cross-link the polyethylene used in the experimental mixtures. It was possible to demonstrate that the peroxide was present in the mixtures but that it was insufficiently reactive to achieve adequate cross-linking.

Finally, the onset of peroxide decomposition is of considerable interest during the manufacturing process. The temperature of the polyethylene in the extrusion plant and dies must remain below the decomposition onset of the peroxide or else cross-linking will begin to occur there. This process results in inhomogeneities in the extruded insulation, which present a risk of electrical failure in service and are therefore unacceptable.

DSC measurements were also found useful when investigating the extent of cure of large epoxy castings used in the insulation of high-voltage cable joints and terminations. These are manufactured with a two-stage cure cycle and it proved possible to monitor cure at the end of each stage during some investigative work. Both residual reaction exotherm and thermodynamic glass transition temperature were monitored since the glass transition temperature of the resin increased as cure progressed. Figure 5.4 gives the glass transitions of two examples.

Interestingly, it was found that the thermodynamic glass transition matched not only the change in coefficient of expansion which accompanies the glass transition found by thermomechanical analysis (TMA), but also the heat distortion temperature as measured on macroscopic test bars. This finding permitted the use of DSC data with confidence for cure checks, in preference to the more complex and time-consuming preparation of heat distortion test bars.

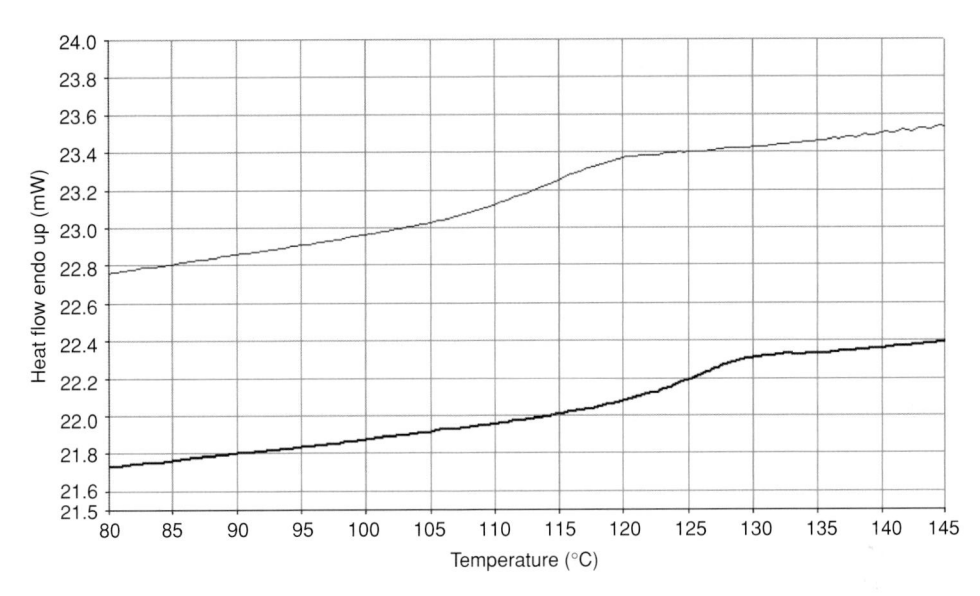

Figure 5.4 Thermodynamic glass transition of two cast epoxy samples showing the increase in T_g (lower curve) due to increased cure.

5.2.4 Investigation of unknowns

From time to time it happens that problems arise in most manufacturing processes and cable manufacture is no exception. Faults in cables are not acceptable, owing to the long service life and reliability requirements, and therefore these must be investigated so that the cause of the fault can be remedied. Various analytical strategies are employed in such investigations, often in combination, since this gives better confidence in interpretation. DSC is a valuable tool for this type of work, either alone or in conjunction with other techniques, particularly where the fault is caused by a contaminant.

One such example involved a product with a thin insulation for low-voltage applications; this product is tested online for electrical integrity at a voltage considerably in excess of the service voltage and failures are cut out and rejected. At the time of the incident, online faults were at an unacceptable level, leading to costs in locating faults and removing them. The extrusion filters used to screen out foreign bodies were examined and a number of small pieces of plastic sheeting, which were identified as polyethene, using Fourier transform infrared (FTIR) spectroscopy, were observed. Samples were therefore subjected to DSC scans, at 10 K/min, in closed aluminium pans and under nitrogen purging between 30 and 200°C. In addition, samples of candidate sources of this contamination were gathered together and were also analysed, by both DSC and FTIR spectroscopy. The DSC data are given in Figure 5.5.

The thermograms given in Figure 5.5 permitted the identification of the extrusion material, trace 1, and the similar trace just below it to be identified as the source of contaminant. This correlated well with FTIR data, which would have been insufficient in itself

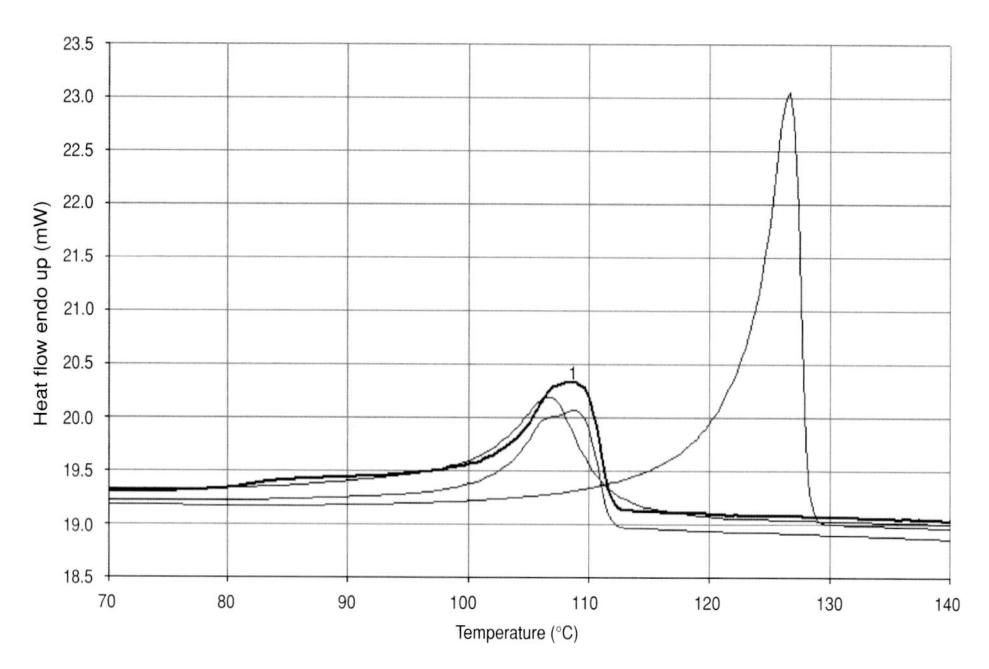

Figure 5.5 DSC data for crystalline melting point of polyethene extrusion contaminant (1) and candidate sources.

for identification because the actual analytical samples were themselves contaminated with matrix and so the FTIR spectra contained matrix peaks in addition to sample peaks.

Taking together the two data sets DSC and FTIR gave good confidence in the identification, whereas neither was fully conclusive alone. The practice of using more than one analytical techniques is very helpful when analysing unknowns because it permits the testing of the interpretation of one data set against that from other techniques until a consistent and supportable interpretation of all data is achieved.

5.2.5 Rapid scanning with DSC

High-speed calorimetry, as reported by Mathot and co-workers [2], has also proved extremely useful on a number of occasions for identification of very small inclusions. Figure 5.6 shows the melting point of a sample cut out of an extrusion filter.

Experimental details were as follows: The sample was contained in a crimped standard Perkin Elmer pan; the instrument used was a Perkin Elmer DSC7, with water cooling and nitrogen purging, which had been calibrated to run at the scan rate of 250 K/min.

The sample had been removed from the filter using dissection tools whilst observing with a light microscope. It had first been placed in a diamond anvil cell for FTIR microspectroscopy, and the infrared (IR)-absorption spectrum, which permitted identification of a polyamide, was obtained. The sample was then recovered from the diamond anvil cell, again with the aid of the microscope and transferred to a pre-weighed sample pan for the DSC experiment,

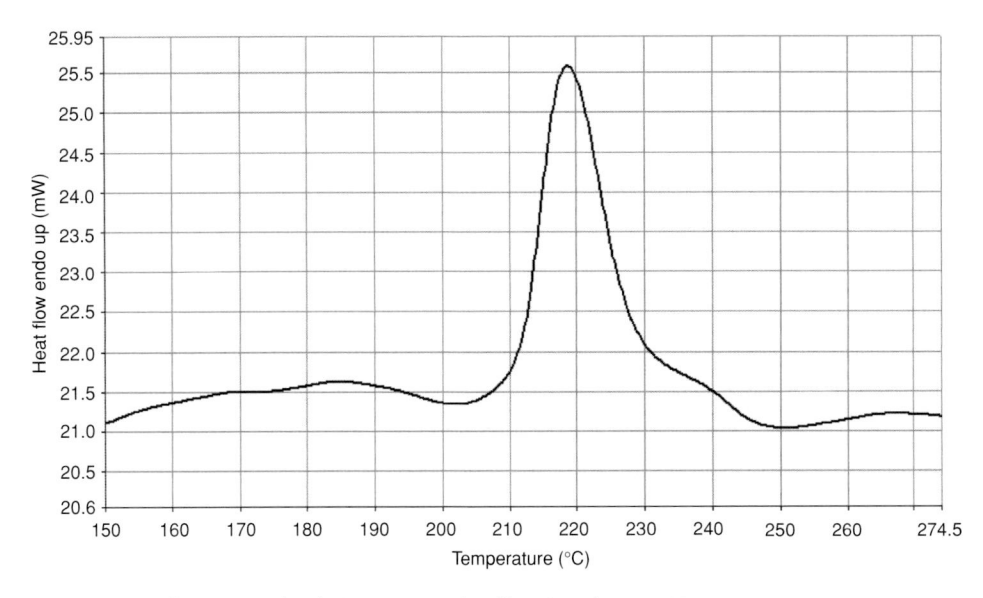

Figure 5.6 Melting point of inclusion on extrusion filter. Sample mass 89 µg.

which gives confidence in the IR spectroscopy and suggests, on the grounds of melting point, that this is a nylon-6 grade.

This analytical strategy has also been applied to very small polyethene inclusions of the same material as those referred to in Figure 5.5, with some success. In both cases the additional confidence of having consistent data and interpretations from two techniques proved invaluable in supporting the production engineering investigations which were under way to eliminate the fault.

A different opportunity to make use of higher scan rates arose when some very small defects in an extruded cable screen were presented for analysis. The screen material is difficult to analyse by DSC because it contains a high (>30%) concentration of carbon for electrical purposes and exhibits very weak DSC responses to any transitions. Data were acquired in a Perkin Elmer Pyris 1 DSC instrument, scanning at 50 K/min with helium purging on samples of about 0.35 mg. The initial data (Figure 5.7) gave no particular suggestion of any contaminant, and derivative curves were obtained for matrix and defect. It was concluded that the defect was pre-cross-linked gel because of the similarity of the trace from the defect to that of the pre-cross-linked gel.

Interestingly, a second fault of this type also appeared and the analysis was repeated. This fault was, however, shown to be due to a contaminant as the difference between the data for matrix and defect show in Figure 5.8. In both investigations the data obtained was of considerable importance to the resolution of the defect-producing mechanism.

These samples were also subjected to some TMA, which reinforced the interpretation of the DSC data and is described below. This experience, as well as demonstrating the value of at least attempting a rapid scan approach with difficult samples, also shows that investigation of derivative curves can be of great assistance in interpreting data, particularly from difficult samples.

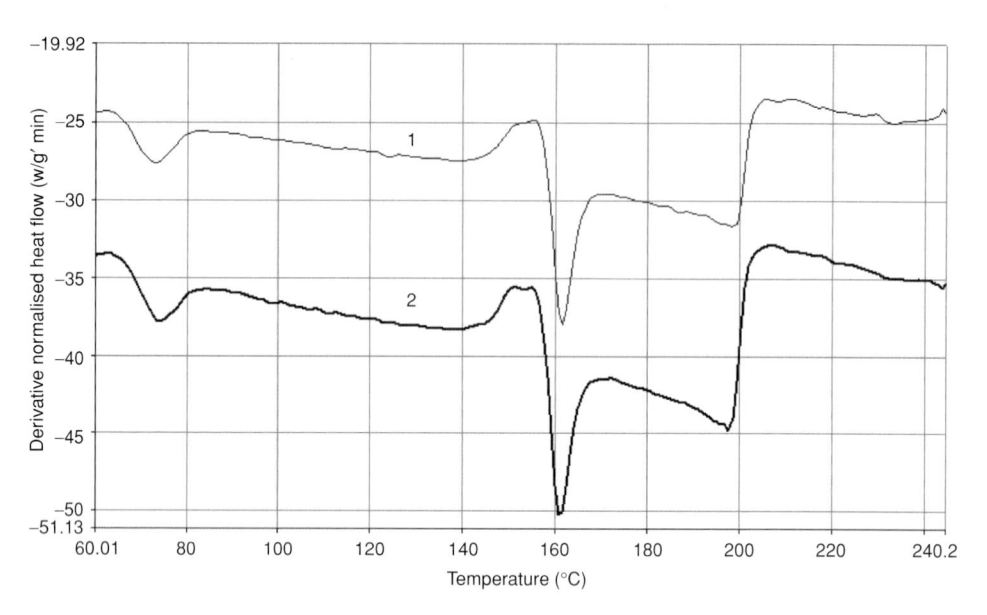

Figure 5.7 First derivative DSC curves for matrix (1) and defect (2) in extruded cable screen.

Measurement of the glass transition sometimes causes considerable difficulties if attempted by the conventional heating rate approach to DSC, particularly in the case of formulations carrying heavy loads of filler as either reinforcement or fire retardant. It was desired to determine if a highly filled cable-jacketing material was suitable for low-temperature

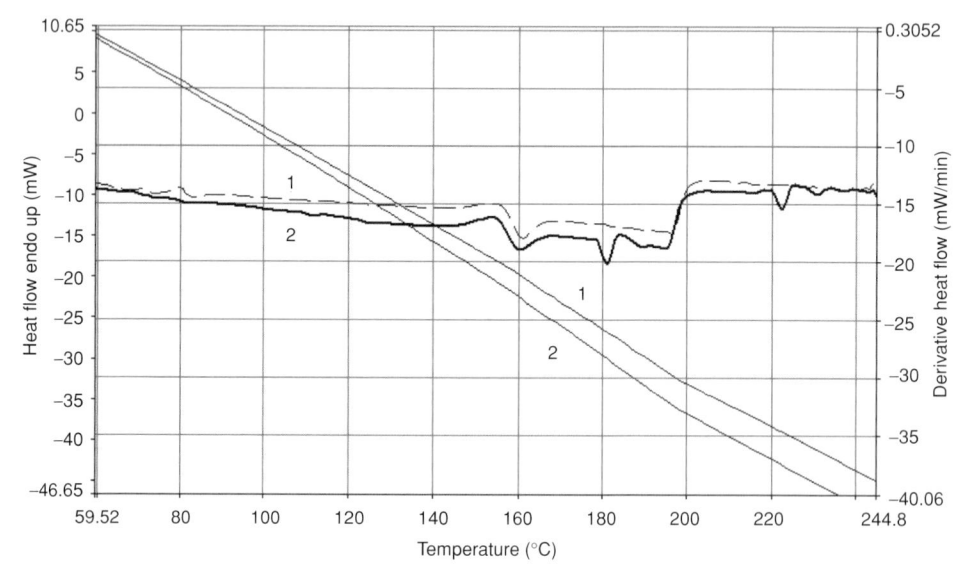

Figure 5.8 DSC data, with derivatives, for matrix (1) and defect position (2) for extruded cable screen with contaminant.

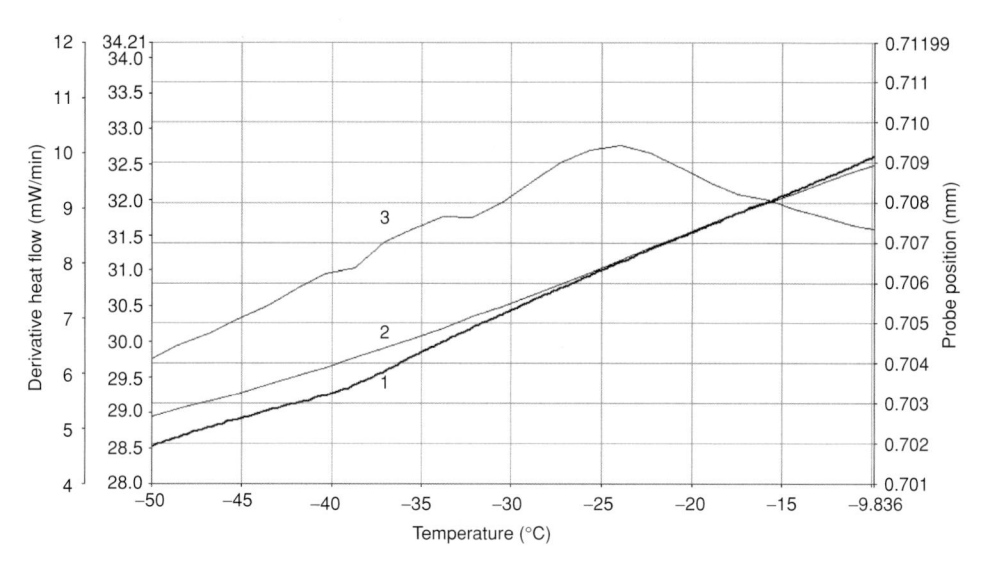

Figure 5.9 TMA signal (1) for cable sheath compared with DSC signal (2) and derivative (3).

service. In this case the Pyris 1 DSC, with liquid nitrogen cooling and helium purging, was used from −100 to 25°C at a heating rate of 100 K/min for 0.5-mg samples. An example of the data is given in Figure 5.9, together with an expansion curve obtained by TMA using a 3.7-mm-diameter probe and 10-mN load at 10 K/min. It is observed that all three curves presented show a transition between −40 and −30°C and that whilst the data from either experiment might be regarded with some caution, the combination of data is more persuasive of the presence of a transition in that region, permitting advice to be provided about minimum permissible temperatures of service once a suitable factor of safety had also been considered.

It is arguable that none of the applications of rapid-scanning DSC given above would have been accessible by a more conventional approach. In the case of the identifications described, the technique was used to corroborate FTIR spectroscopic analysis on very small samples, whilst in the case of the other applications TMA was used to corroborate the DSC investigation of difficult samples. These corroborations were deemed necessary as part of the learning cycle in use of a novel application but are also desirable when dealing with unknowns, particularly those contaminated with the matrix. At the time of writing, the strategy appears to present an opportunity to extend the DSC experiment to samples which would have been previously considered too small or otherwise intractable.

5.3 Thermomechanical analysis

TMA has already been considered in conjunction with glass transition measurements in applications such as degree of cure or heat-distortion temperatures of epoxy resin castings. The technique has proved useful for many other applications, some of which are described below.

5.3.1 Investigation of extrusion defects

The screen defects which were discussed in relation to Figures 5.7 and 5.8 were removed from a cable which had not been cross-linked. Thus the matrix would be expected to deform in a TMA experiment if heated under an appropriate load. Accordingly, prior to the DSC experiments described above, the samples were placed in DSC sample pans, and the lids rested on the samples, but not crimped, and the TMA load then applied. This allowed the TMA experiment to take place on the uncross-linked sample and then the same sample in the same pan to be used for the DSC experiments where composition investigation was of more interest. Data were obtained by heating at 10 K/min between ambient and 200°C and the TMA furnace purged with nitrogen throughout. A load of 110 mN was applied to the DSC pan lid for each measurement. The height was zeroed using an empty pan and lid.

It will be recalled that no contaminant could be demonstrated to be present, based on the DSC curves, and that the sample had not been subjected to any cross-linking process. The data in Figure 5.10 show that the matrix yielded freely at the melting point but that the sample containing the defect did not. Since the defect was concluded not to be a contaminant, based on the DSC evidence, it was possible to conclude that it was a pre-cross-linked gel and thus able to resist the deformation that the relatively light probe force could impart to a fully molten sample of uncross-linked matrix.

In the case of the second fault, in which DSC indicated a contaminant, no significant difference in TMA response was found between matrix and fault position, and this led to the conclusion that this fault was not a consequence of pre-cross-linking.

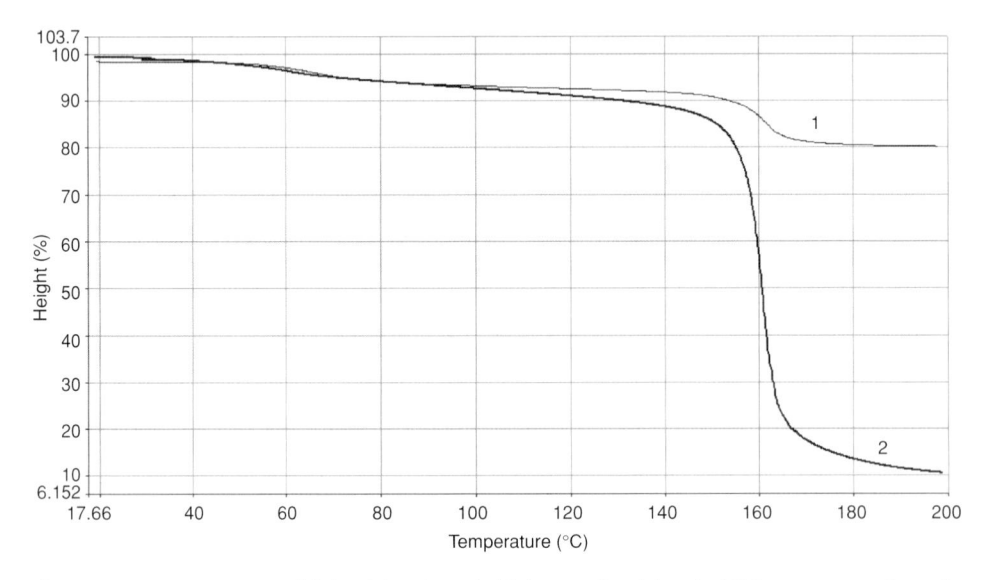

Figure 5.10 TMA response of defect (1) and matrix (2) for samples giving the DSC response, as shown in Figure 5.7.

5.3.2 Cross-linking

The cross-linking of materials such as polyethylene, ethylene propylene rubber and some other rubbers used for cable insulation and sheathing is conventionally measured for quality purposes by means of a test of elongation under a specified load at elevated temperature of the permanent set measured when the load is removed and the sample allowed to cool. This hot-set test is very good when it is possible to cut appropriately sized tensile test dumb-bells for the measurement. Where samples are irregular and small, as may be the case in work of a more investigative nature, it is much more difficult to apply, and TMA has been applied, albeit in a fairly qualitative way, to the problem of finding out if a specimen is cross-linked.

The data of Figure 5.10 are suitable to illustrate this point; the uncross-linked matrix yielded completely in this experiment, whereas that containing the pre-cross-linked defect did not yield completely. Under the experimental conditions described, well-cross-linked materials used for cable manufacture yield only slightly at the crystalline melting point and so the TMA measurement yields a useful qualitative result, with some possibility of commenting on the gross degree of cross-linking. The ability to observe this yield point in the TMA is however useful on occasion, and TMA experiments on material known to be cross-linked, carried out with greater loadings, have proved extremely useful in this respect.

It would be possible to obtain more quantitative data by measurement of gel fraction by solvent extraction of the material in order to effect calibration but this is a time-consuming process and for a single measurement it would render the TMA measurement redundant. If the TMA data are considered likely to prove fit for purpose, they can be obtained easily and quickly, an advantage in an environment where information may be required on an urgent basis.

5.3.3 Material identification

Some of the materials commonly used in cable making do not give very clear crystalline melting points in the DSC experiment. This was remarked upon with respect to the data of Figures 5.7, 5.8 and 5.10. It will be observed that the softening points in Figure 5.10 are very clear, so penetration measurements have proved very useful for melting point determination of materials with low crystallinity and with heavy filler loadings and also for gaining additional confidence with very small samples of the type described earlier.

As remarked earlier, greater loadings are required for cross-linked samples than for the corresponding uncross-linked material in order to achieve sufficient penetration to adequately detect the crystalline melting point with confidence. TMA data would typically be used in conjunction with some or all of FTIR spectra, DSC responses and also data from thermogravimetric analysis (TGA) as appropriate, in order to more fully characterise an unknown.

A knowledge of the properties of the materials available, or provision of reference samples where some material fault is suspected, greatly facilitates this type of work. It is a great advantage, therefore, to log all thermal analysis experiments and to archive all data for future reference. These will, at some time, amply repay the analyst in interpretation of another investigation and may save much time in collecting reference thermograms, a factor of considerable importance in matters of production or customer support.

5.3.4 Extrusion studies

Cable production involves extrusion of insulation and sheaths in long, continuous lengths. For technological reasons, both electrical and mechanical, as well as aesthetic considerations, these extrusions have to be of good quality and smoothness is important. TMA studies have been valuable into providing insights into the differences in behaviour of material as a support to extrusion trials and rheological measurements. As an example consider the data illustrated in Figure 5.11, which were obtained from two fire-retardant cable sheaths.

It is clear that these materials are likely to have differing extrusion properties and will, to achieve satisfactory finishes, require different processing conditions, such as extrusion head pressure. A higher viscosity may affect output since it will need to be run at a slower speed than the lower viscosity material if increasing extrusion pressure is ineffective or not possible. It should be noted that the extrusion temperature will need to be kept well below the onset temperature of any polymer or fire-retardant decompositions, otherwise there is a risk that even a small degree of decomposition will produce gas bubbles and hence porosity into the extrudate. Such porosity is extremely deleterious to the final product properties and cannot be tolerated.

This type of TMA experiment can be valuable in screening out experimental formulations which are unlikely to succeed, prior to moving to other material evaluations and, more importantly, to production trials. Such screening can lead to time saving and considerable cost reductions for the manufacturing plant, since any failed trial produces only scrapped

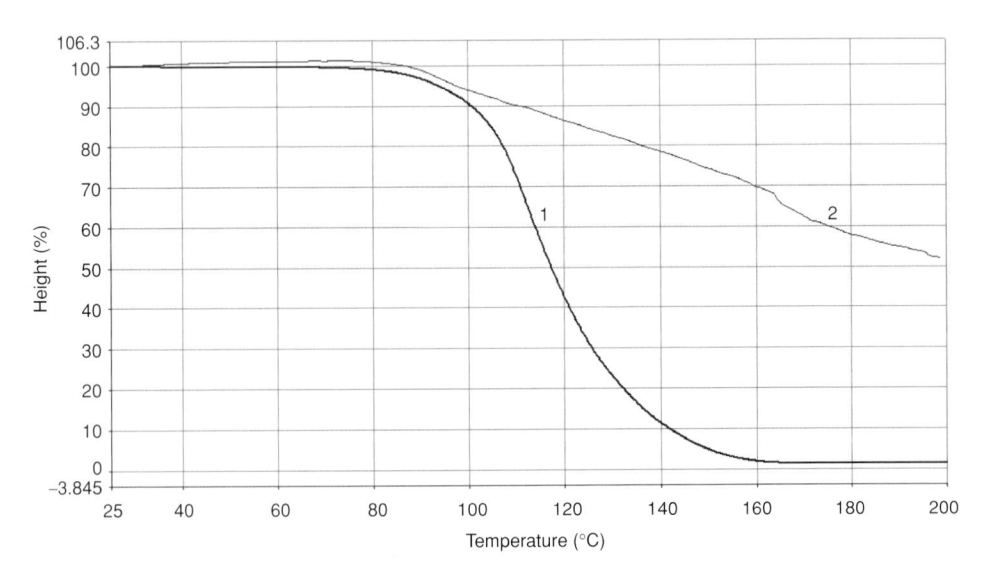

Figure 5.11 Penetration data for two grades of cable sheath using 1-mm-diameter probe and 110-mN loading. Heating rate 10 K/min with nitrogen purging throughout.

material. This approach has also been used with some success in combined evaluations and will be discussed in more detail later.

5.3.5 Fire-retardant mineral insulations

Some types of cables are designed to give continued service under fire conditions, so that services such as fire-detection systems and emergency lighting are maintained until evacuation is completed. Common approaches to the provision of such a cable include that of using silicone rubber, which burns to leave an insulating residue, or mica paper backed with woven glass fabric and bound in resin ('mica tape') as at least one layer of the insulation. The behaviour of these materials at elevated temperatures is therefore of considerable interest.

One investigation centred on the comparative performance of different grades of mica tape and the melting and sintering temperatures of the mica and glass was of particular interest.

To facilitate the investigation the mica tapes were solvent extracted to remove the binder and then dried. The mica and glass were separated by hand and cut up into squares approximately 4 mm on each side. Small piles of these squares were placed into copper DSC pans and a second, smaller, lid placed over the samples. These precautions were necessary to avoid adhesion of the sample to the quartz TMA probe and sample support. The height reading for the TMA probe was zeroed with the pans in place but without the sample, and a blank experiment, without sample, was run to verify that the pans would not contribute spurious transitions to the experiment. The TMA was run from ambient to 900°C with an air purge and heating rate of 10 K/min; the applied load was 100 mN. The curves given in Figure 5.12 are representative of those obtained.

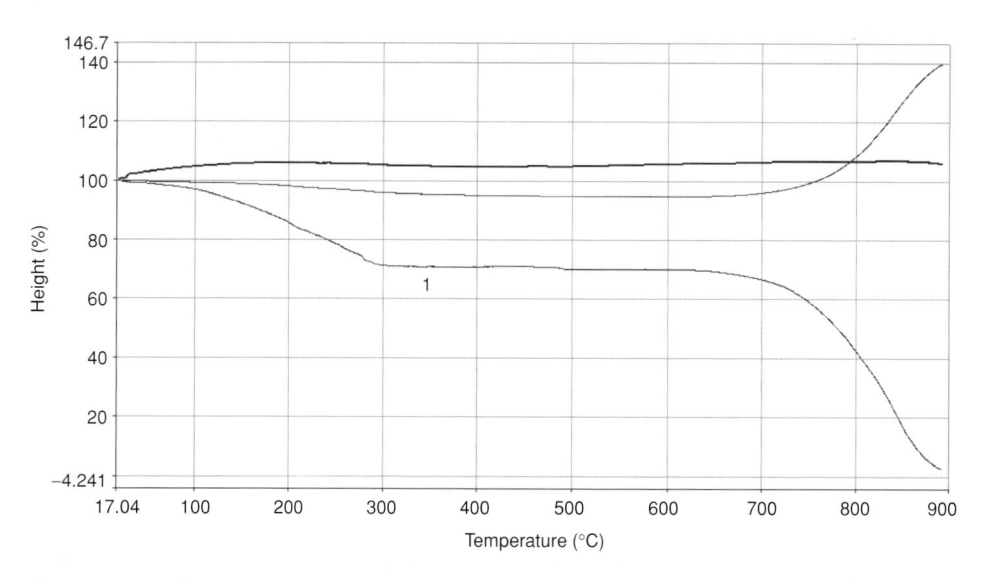

Figure 5.12 Deformation curves for glass-woven tape (1) and two different varieties of mica paper.

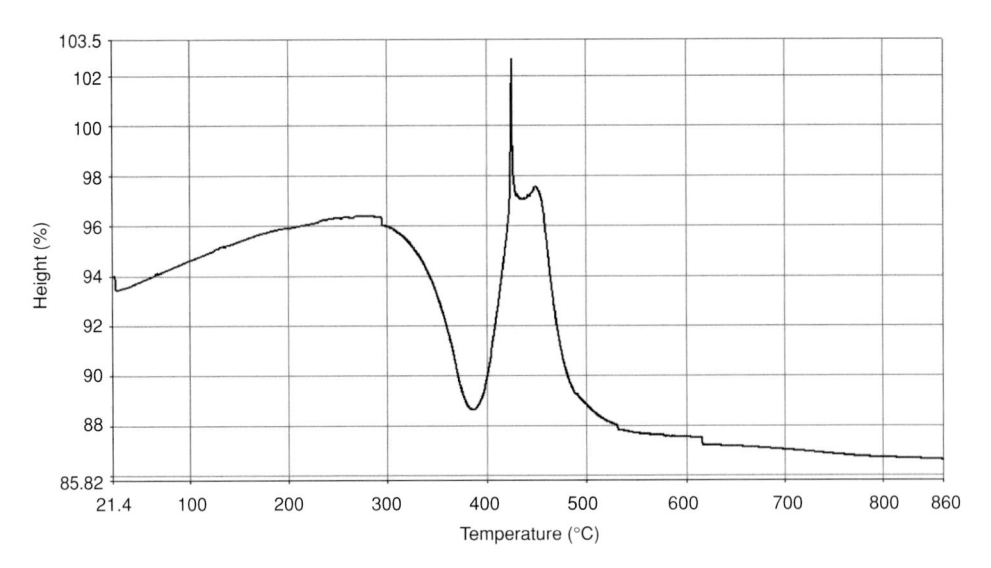

Figure 5.13 Burning of silicone rubber showing sample movement as the temperature increases.

Inspection of Figure 5.12 shows that the glass tape yields somewhat up to 300°C and begins to melt at 650–700°C. The two micas, which have different compositions, do not melt but one begins to expand at around 700°C, whist the other expands minimally and then begins to contract at about 850°C. These temperatures are those at which the two types of micas are reported to sinter and are therefore of interest in terms of the fire process because it is at these temperatures that they will ceramify well, further stabilising the high-temperature insulation.

Silicone rubber burns in air and, if the formulation includes appropriate fillers, will produce a hard insulating ash. A TMA experiment to simulate burning yields interesting results when carried out with a 1-mm-diameter probe, a load of 100 mN, air purge and a heating rate of 10 K/min. Figure 5.13 shows the result of one such experiment.

As expected, the rubber expands up to about 300°C and then begins to contract until, at about 400°C, there is a rapid expansion which can be correlated with the onset of decomposition processes observed by thermogravimetry. As the rubber continues to decompose and burn, it eventually contracts again and then slowly stabilises. This behaviour is of interest in terms of understanding the interaction of the silicone with other materials in the cable as it begins to burn and also in terms of the eventual development of the mineralised insulation required for fire service.

5.4 Thermogravimetric analysis

5.4.1 Practical comments

Thermogravimetry offers the opportunity to examine polymeric formulations by way of a weight-loss curve and also the first derivative of the weight loss. From experience in these laboratories the value of this derivative curve is difficult to overstate. Peak onset, position and end data are all of great value in interpreting the meaning of the weight-loss steps observed in the curve and are usually characteristic of particular formulations, provided that a number of conditions are met.

It is important that sample geometry and weight are kept consistent because a thin planar sample can lose volatiles more rapidly than a cubic one of the same mass, because the surface area available for volatile loss is greater and the mean path to the outside of the sample smaller for the planar sample. Similarly, a small sample can lose each weight-loss step more quickly than can a large sample, partly due to the effect of path length to the outside but also because a large sample requires more heat and is longer to reach the programme temperature. For filled polymers this is important because the samples are usually good thermal insulators. It therefore follows that adequate validation of temperature calibration is also attended to for the same reason. Generally, for most applications, a 10-mg sample, cut as close as possible to a cube, has proved satisfactory.

It has also been found important to carry out separate standard runs for materials which may be presented in both uncross-linked and cross-linked conditions. Cross-linked materials do not melt and flow, and so escaping gases cannot escape so rapidly as those from a fully molten uncross-linked sample. This leads to relative delays in the onset of each weight-loss step in the weight-loss curve and in the derivative peaks for the cross-linked sample as compared to the uncross-linked sample. This observation has proved useful for some formulations in verifying cross-linking or indeed its lack, where this is a specification requirement.

In general, the heating rate must also be carefully specified, since the thermal insulation effect of the sample will be exaggerated if the heating rate is increased. It follows, therefore, that standard reference runs should be carried out if it is desired to heat at high rates. In addition purge gas rate should be standardised, as purge rate will affect removal of the gases liberated from the test sample.

It has proved necessary on many occasions to specify the instrument for use in TGA, as test samples have given slightly different behaviour in instrument of different designs, so that derivative peak onsets, shapes and positions vary very slightly, although the quantitative analysis is consistent between instruments. These variations are presumably due to differences in gas flows and thermal profiles between the designs and do mean that comparison of data obtained on different designs of TGA should be approached with caution.

5.4.2 Investigation of composition

Many cable sheaths are manufactured from heavily filled polymers, and TGA has been found to be of immeasurable value in studies of such materials and some of their ingredients. The combination of the weight-loss curve, for quantitative analysis, with the derivative, for a qualitative interpretation, has been found to be particularly powerful for these applications particularly when taken together with data obtained by the other thermal methods already discussed but also with that obtained by FTIR spectroscopy. In addition the vinyl acetate content of ethylene–vinyl acetate (EVA) copolymer used in the preparation of some sheathing formulations has been measured and used as a confirmation of grade.

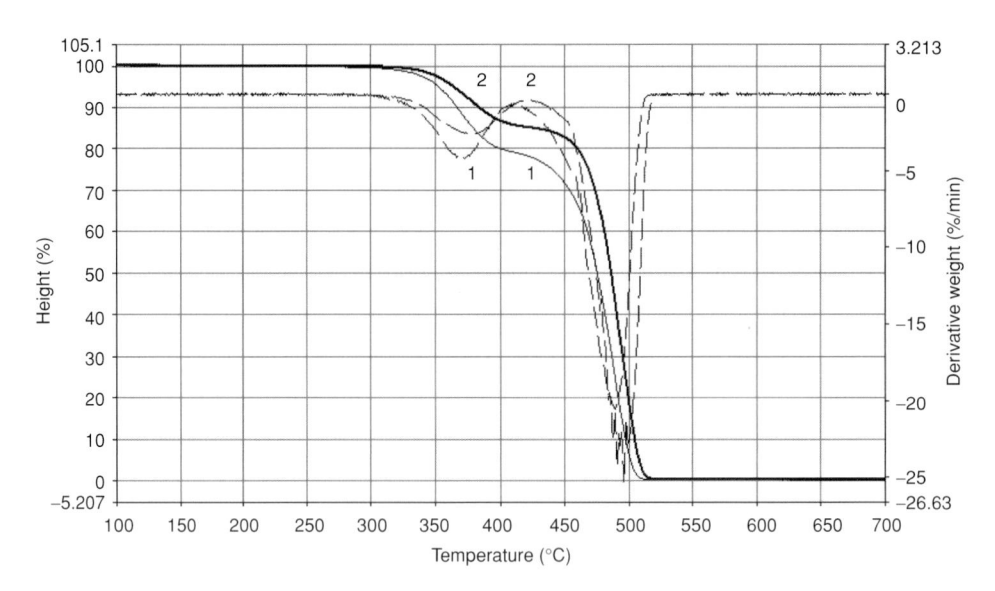

Figure 5.14 Weight-loss (solid) and derivative (dashed) curves for two EVA copolymers.

Figure 5.14 shows the weight-loss curves and derivatives for a pair of EVA grades; they are readily distinguished. The first weight-loss step is due to loss of acetic acid from the vinyl acetate residues in the polymer and so the vinyl acetate concentration can be calculated from the stoichiometry of the decomposition reaction. The turning point (minimum) between the peaks in the derivative curve is used to define the end of the decomposition of the vinyl acetate for purposes of defining the weight-loss end.

It is also, of course, possible to observe mineral filler decompositions, such as those of aluminium and magnesium hydroxide, which are frequently used in cable sheathing as fire retardants (see Figure 5.15). Both materials quantitatively decompose to liberate water and their oxides; for example

$$2Al(OH)_3 = Al_2O_3 + 3H_2O$$

The data of Figures 5.14 and 5.15 were obtained with nitrogen purging at 20 cm^3/min and a heating rate of 10 K/min.

Examination of Figure 5.16 shows that, although it is reasonable to conclude that there is much the same amount of aluminium hydroxide present in both samples, based on both decomposition and final ash content, the polymer matrix is different because the escape rate for the water liberated by the aluminium hydroxide is noticeably different in the weight-loss curve and the corresponding derivative curve peaks have different shapes. Furthermore, the inflection regions between the decompositions of the aluminium hydroxide and the polymer are also different for the two formulations, suggesting differences in copolymer. The actual polymer decompositions, whilst superficially similar, are different in detail, and finally curve 1 shows a response due to burning of some residual char when the purge gas is changed to air above 700°C.

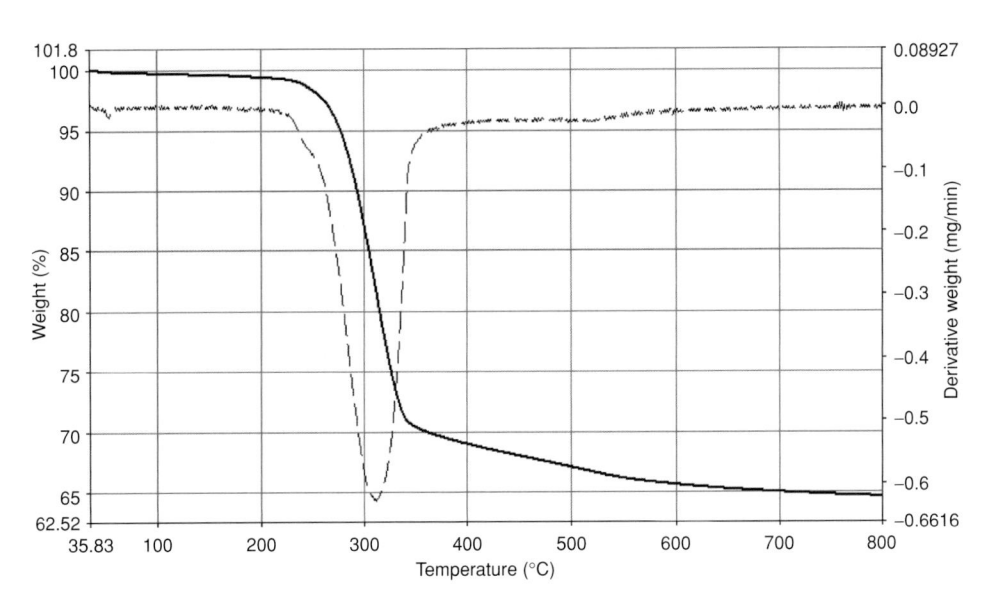

Figure 5.15 Decomposition of aluminium hydroxide.

Data from FTIR spectroscopy were used to corroborate the interpretation of the first weight-loss peak for each sample as due to aluminium hydroxide fire retardant. Data from both techniques were combined with crystalline melting point data obtained by DSC and TMA, as already discussed to form conclusions about the polymers present. Usually, such

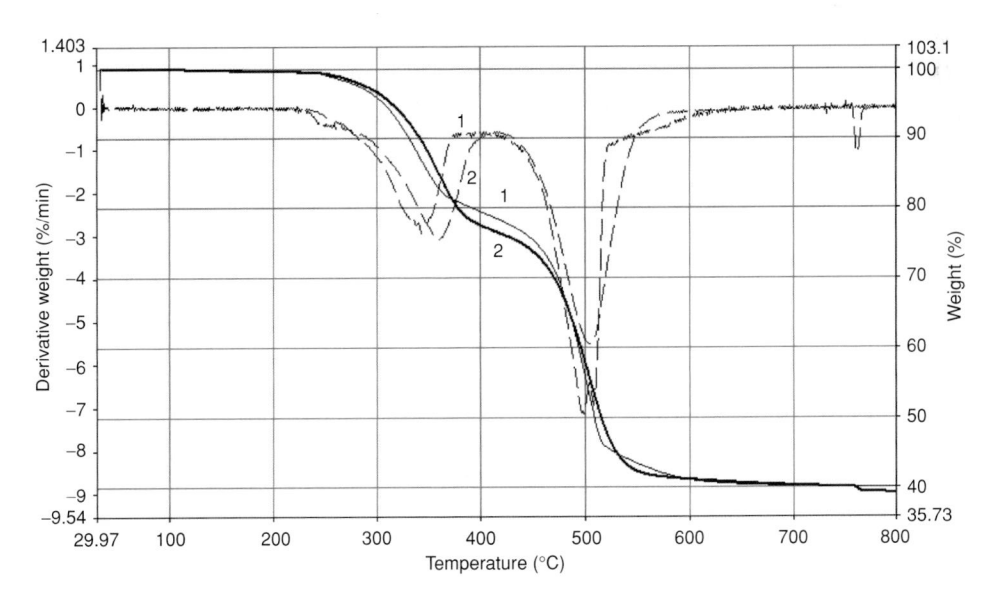

Figure 5.16 Thermogravimetric data for two alternative formulations of fire-retardant cable sheath. Heating rate 10 K/min and nitrogen purging at 20 cm^3/min until 750°C and then change to air.

conclusions will also be drawn and will be consistent, with known formulation information or with knowledge of production plant conditions leading to the requirement for an investigation.

Thermogravimetry on polymer recovered by solvent extraction has also been found valuable for testing initial interpretations, although sample preparation can prove somewhat time consuming. A second approach is to prepare small experimental samples with which to test the analytical interpretation. The data from these samples should verify a correct analysis or provide further insight for a second iteration of interpretation. The effort to be given to these more laborious procedures will obviously depend on whether a simple survey, or a very detailed analysis, is deemed most fit for the purpose towards which the original analysis was directed.

5.4.2.1 PVC insulation and process investigation

For many years much cable for house wiring and domestic appliance flexible cords has been manufactured using plasticised and filled polyvinyl chloride (PVC). Calcium carbonate has commonly been used as a filler, with plasticisers chosen from a range to suit specific applications, phthalate esters being a frequently selected solution. PVC itself begins to dehydrochlorinate at just above 200°C whereas calcium carbonate is stable up to about 600°C after which it slowly liberates carbon dioxide and leaves a calcium oxide ash. Thus PVC insulation is amenable to examination by TGA, either as a quality check or for purposes of investigating process problems during blending of the ingredients prior to pelletising ready for subsequent extrusion.

A suitable experiment has been found to consist of using nitrogen as a purge and a heating rate of 10 K/min together with a sample size of 10 mg. The data in Figure 5.17 are typical. The final step of weight loss in both curves is that of calcium carbonate decomposition, which is shown to vary in temperature and derivative peak shape from sample to sample, with concentration having an effect. This illustrates the point that data obtained from mixtures will not always exactly correspond to a proportional sum of curves for the various unmixed ingredients and that care should be taken to select appropriate reference mixtures for confirmation of peak identity. However, in this case a further effect is that arising from the insulation being a compressed cube, with a small surface area, and the dust having a much larger surface area per unit weight so that the weight-loss peaks occur at lower temperature.

For investigative work one can arrive at an approximate formulation (since the assumptions made ignore minor ingredients such as stabilisers) by assuming that the final derivative peak is due to filler decomposition and that the second weight-loss step is due to evaporation of the polymer backbone of the PVC. The first weight-loss step then consists of PVC dehydrochlorination and evaporation of plasticiser. It will be appreciated that it is then possible to calculate filler and PVC concentrations from the decomposition stoichiometries (see Section 3.6.2) and to deduce the plasticiser concentration. For quality control it has been found to be preferable to utilise carefully weighed and mixed standards of known concentration so that a specification based on percentage mass loss for each step can be given and standard derivative curves retained for comparison since

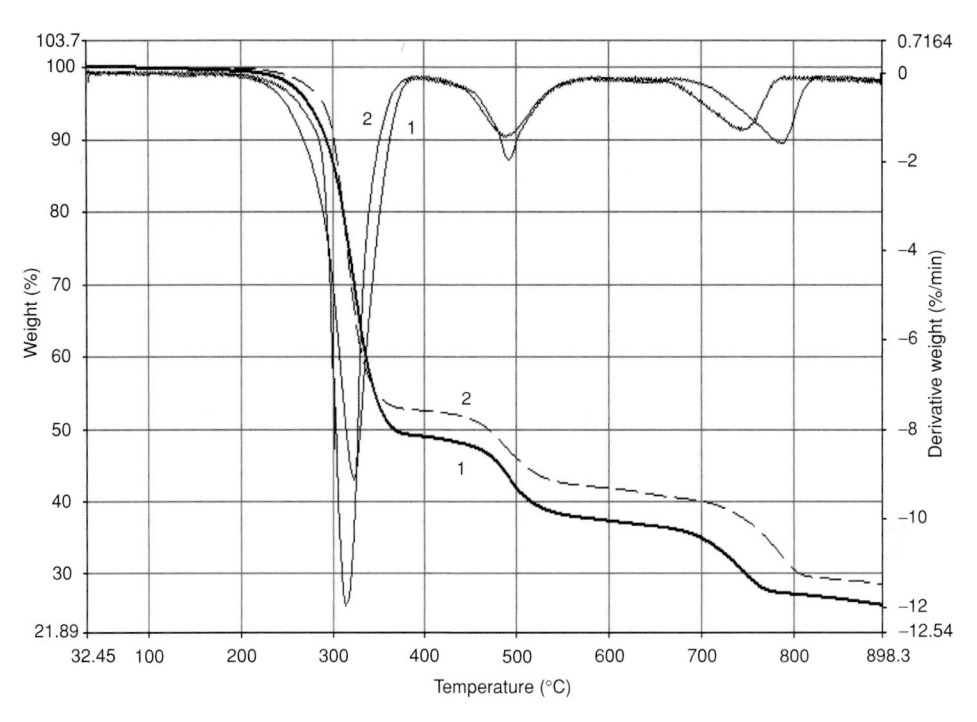

Figure 5.17 Typical TGA data for PVC insulation (2) and dust found in part of the mixing plant (1).

such standardisation takes account of stabiliser and other additions as well as the major ingredients.

Using these strategies it would be possible to examine Figure 5.17 and confirm that the insulation was satisfactory, given reference data, but that the dust is an unrepresentative sample of the overall formulation since it is appreciably filler deficient. Such information can prove valuable to plant engineers in the course of investigations to rectify the process problem.

5.4.3 Carbon content

Carbon loadings are used in extruded cable screens in concentrations up to almost 40%, to give appreciable electrical conductivity for management of electric fields so that measurement of carbon content is of some interest. Carbon is also incorporated into polyethylene sheathing for protection against ultraviolet (UV)-radiation-induced degradation, but at much lower levels around 2.5%.

In principle the TGA measurement of carbon is straightforward. The sample is heated in nitrogen until the polymer is completely decomposed and then the purge is changed to air or oxygen and the carbon burnt off. Use of a pre-programmed gas-switching valve simplifies the gas change. A heating rate of 10–20 K/min is usually found suitable. It is advisable to install an oxygen trap on the nitrogen line or else some of the carbon may be burnt off by any impurity oxygen in the nitrogen before the oxidising purge is introduced and also to be scrupulous in eliminating leaks in the purge gas system, which may admit air (see Section 3.4.4). At the end of the experiment the carbon burning step is calculated.

In practice it is necessary to consider some additional factors. First, the polymer may leave some carbon residue on decomposition, and this must be checked separately. Second, some of the carbon grades used are blacks produced by combustion and not graphitic as such. Some of these have been found to have appreciable volatile fractions in nitrogen in the temperature range 550–750°C so that false low results have been obtained. This difficulty can be overcome by introducing an isothermal phase in the decomposition at sufficient temperature to complete the polymer decomposition but not to volatilise the black. At the end of the isothermal period the oxidising purge is admitted and the temperature ramp resumed so that the black combustion can be observed and evaluated. These precautions have been found particularly useful for the analysis of UV-protected sheathing grades.

5.4.4 Rapid scanning with TGA

At first sight the idea of rapid heating over a wide temperature range for a TGA experiment would be expected to result in weight-loss steps shifted to higher temperature and broadening of derivative peaks with attendant loss of qualitative resolution. These objections can be overcome by use of standards to calibrate the range for each weight loss and derivative curve peak shape if one is required to validate many samples, and the writer has used this technique on occasions when validation of all polymeric components in a multicable system was required.

On another occasion it was desired to investigate the effect of changed heating rate on the data obtained from some fire-retardant material. Temperature lags for scans carried out at 100 K/min were found to be only of the order of 11°C in the Perkin Elmer TGA 7, when determined using Curie point standards, whereas the analytical samples gave greater lags, suggesting kinetic limitations to the decomposition or burning rate, depending on purge gas, of the samples.

Recently, this investigation also resulted in the realisation that the rate of weight loss for a given step is much greater for a high heating rate than for a low one and so the sort of sensitivity improvement already discussed for high-speed DSC might also be accessible by TGA where analysis of very small samples has proved difficult at conventional heating rates, because signal to noise is poor. Initial results show some promise and are reproduced in Figures 5.18–5.20.

The improvement in sensitivity in the derivative curve is considerable, although the quantitative data presented are rather inconsistent between the four samples. Variability of final ash content has not proved a problem with 10-mg samples and the inconsistency is believed to be a function of the small sample size. Baseline subtraction has been found valuable in allowing for this, as some buoyancy effects will always be inescapable in the TGA experiment. The qualitative data are a great improvement at the higher heating rate

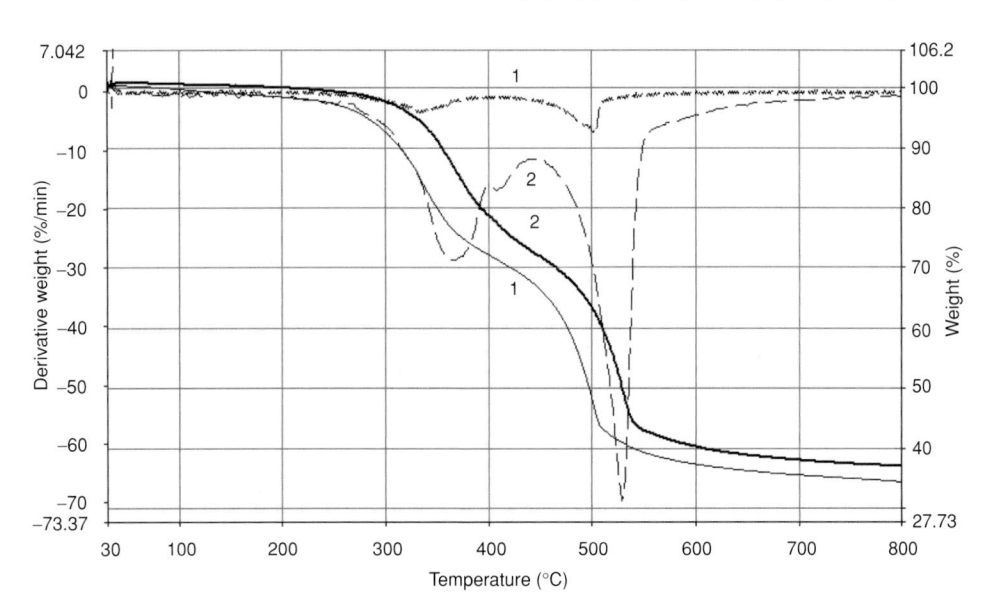

Figure 5.18 TGA data for cable sheath: 1.0-mg run at 10 K/min (1) and 1.2-mg run at 100 K/min (2).

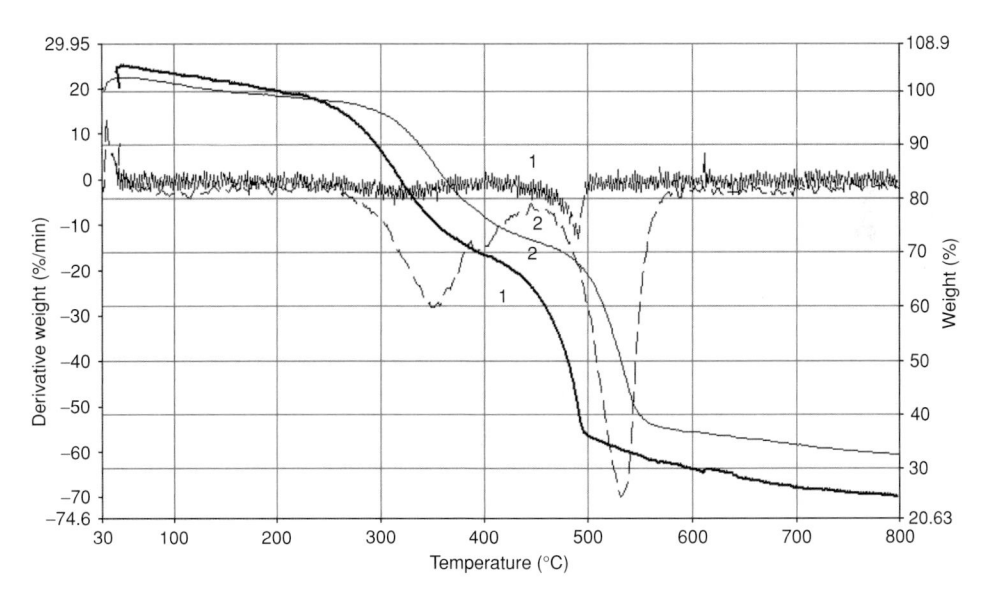

Figure 5.19 TGA data for same cable sheath: 0.14-mg run at 10 K/min (1) and 0.2-mg run at 100 K/min (2).

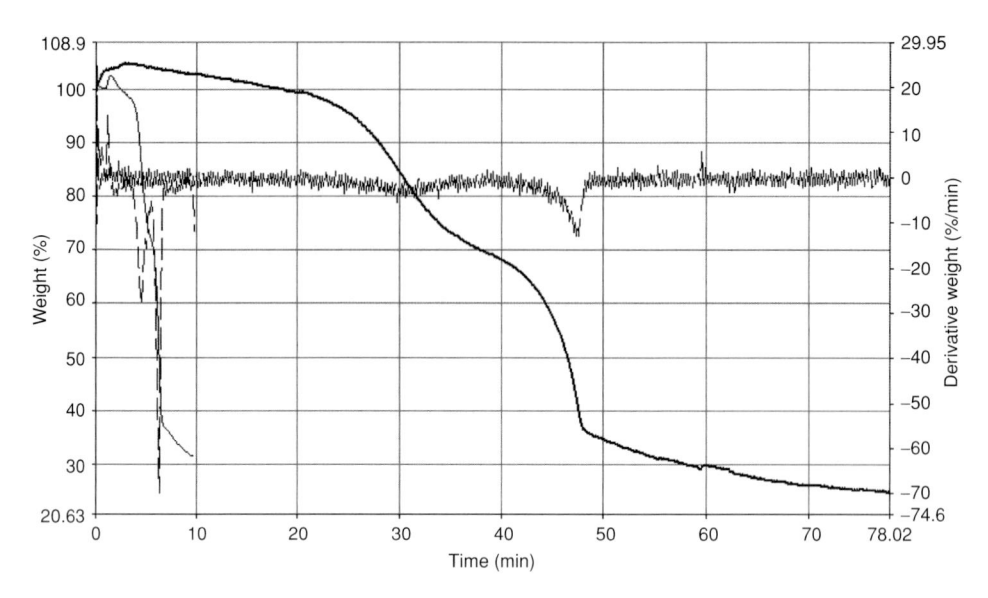

Figure 5.20 Data from Figure 5.19 presented on a time-scale. The faster scan took around 10 min and the slower scan around 80 min.

for the small samples and therefore leads to an opportunity to carry out sequential FTIR, TMA, DSC and TGA investigation of small difficult samples, which would otherwise be inaccessible.

5.5 Combined studies

The advantage of using several analytical methods to attack the identification of unknowns, particularly when small in size and contaminated by the matrix, has already been emphasised.

A second possibility, that of investigating the mutual compatibility of the various layered components of cables, has also been exploited from time to time as follows.

During a fire the components of a cable will variously melt, burn, deform and decompose at differing temperatures. The possibility of a molten layer interpenetrating a deformed or decomposed layer has, on occasion, been demonstrated in experimental designs, with resultant electrical failure under simulated fire conditions. Identification of fire-damaged material is often ambiguous using only a single technique and, on occasion, it has proved necessary to combine date from several thermal analyses with IR spectroscopy and light microscopy in order to understand the process taking place.

To avoid incompatible materials being placed together in costly prototypes and so avoid this mode of failure, a strategy of material screening was adopted in which interpretations of data from DSC, TMA and TGA experiments were plotted on a temperature axis so that overlaps of different domains, e.g. material A foams whilst material B melts, could be examined. This strategy led to design improvements for some products whilst achieving cost savings in terms of avoiding the production of non-viable prototypes.

5.6 Concluding remarks

The preceding discussion has covered the application of several thermal analysis techniques to a number of problems in polymer science, as it is concerned with electric cable manufacture, but many of these are readily generalised and should be transferable to other technologies. The strategy of attacking a problem with more than one technique, particularly for investigative work, has proved to be invaluable in some of the work described. Additional confidence is obtained from this approach in the investigation of unknowns. This is particularly true of very small, possibly irreplaceable, samples, where sequential analysis in order of increasing destructiveness has proved useful. With these very small samples the use of high heating rates has proved valuable in obtaining an analysis which would have proved impossible at conventional rates.

References

1. Billing J. Investigation of thermal history of cable insulation using differential scanning calorimetry. ERA Report 87-0138, 1987.
2. Pijpers Thijs FJ, Mathot Vincent BF, Goderis B, Scherrenberg Rolf L, van der Vegte Eric W. High-speed calorimetry for the study of the kinetics of (de)vitrification, crystallisation, and melting of macromolecules. *Macromolecules* 2002;35:3601–3613.

Chapter 6
Application to Thermoplastics and Rubbers

Martin J. Forrest

Contents

6.1	Introduction	191
6.2	Thermogravimetric analysis	192
	6.2.1 Background	192
	6.2.2 Determination of additives	194
	6.2.3 Compositional analysis	199
	6.2.4 Thermal stability determinations	206
	6.2.5 High resolution TGA and modulated TGA	208
	6.2.6 Hyphenated TGA techniques and evolved gas analysis	210
6.3	Dynamic mechanical analysis	212
	6.3.1 Background	212
	6.3.2 Determination of polymer transitions and investigations into molecular structure	215
	6.3.3 Characterisation of curing and cure state studies	218
	6.3.4 Characterisation of polymer blends and the effect of additives on physical properties	220
	6.3.5 Ageing, degradation and creep studies	224
	6.3.6 Thermal mechanical analysis	227
6.4	Differential scanning calorimetry	228
	6.4.1 Background	228
	6.4.2 Crystallinity studies and the characterisation of polymer blends	231
	6.4.3 Glass transition and the factors that influence it	236
	6.4.4 Ageing and degradation	238
	6.4.5 Curing and cross-linking	241
	6.4.6 Blowing agents	243
	6.4.7 Modulated temperature DSC	244
	6.4.8 HyperDSC^{TM}	245
	6.4.9 Microthermal analysis	246
6.5	Other thermal analysis techniques used to characterise thermoplastics and rubbers	247
	6.5.1 Dielectric analysis	247
	6.5.2 Differential photocalorimetry (DPC)	248

	6.5.3 Thermally stimulated current (TSC)	248
	6.5.4 Thermal conductivity analysis (TCA)	249
6.6	Conclusion	249
References		250

6.1 Introduction

The polymeric nature of thermoplastics and rubbers means that they are comprised of very large macromolecules. This dictates that, in contrast to their low molecular homologues, they have broad melting points and molecular weight distributions, and glass transition temperatures (the temperature at which the molecules start to undergo segmental rotation and the material becomes flexible). Being polymers, these materials also display viscoelastic properties and energy loss upon deformation due to the effects of hysteresis (steric hindrance within the molecules due to the interaction of side groups and chain branches). In the case of rubbers, curing (also know as vulcanisation) is also widely employed to achieve satisfactory products. In the case of both thermoplastics and rubbers, degradation can occur due to harmful influences from agencies such as ultraviolet light, heat, radiation and specific environmental chemicals such as ozone. The irregular nature of polymers (e.g. the orientation of the repeat units – head to head, tail to tail, etc. – and the presence of short and long chain branching) also ensures that the matrix is inhomogeneous containing, for example, both crystalline and amorphous regions.

The thermal analysis group of techniques, which provide information on materials by observing structural and property changes that occur with variations in temperature, are therefore ideally suited to these materials and they enable the polymer analyst to obtain a wealth of information on the composition and physiochemical properties of thermoplastics and rubbers materials.

Some of the more important properties of thermoplastics and rubbers that can be measured using these techniques are listed below:

(a) Chemical composition
(b) Cross-link density and cross-linking reactions
(c) Thermal conductivity
(d) Glass transition temperatures
(e) Crystalline melting temperatures
(f) Heats of reaction (e.g. curing or degradation)
(g) The influence of additives and thermal history (e.g. processing temperature) on a number of the above.

Although there are a growing number of thermal analysis techniques that are commercially available, this chapter will concentrate on the three principal ones that are used for the analysis of the thermoplastics and rubbers. These are as follows:

1. Thermogravimetric analysis (TGA)
2. Dynamic mechanical analysis (DMA)
3. Differential scanning calorimetry (DSC).

Mention will be made of the other techniques that are commercially available (e.g. dielectric analysis – DEA, and thermal mechanical analysis – TMA) within this chapter when it is appropriate.

In terms of structure, this chapter has been divided into sections dealing with each of the principal techniques in turn. Within each section there is a description of how the technique is used for the analysis of plastics and rubbers, a review of the application of the technique, the data that are obtained, and how it can be used in problem solving and compositional analysis work. Throughout the chapter the terms thermoplastic and its abbreviation, plastic, can be regarded as interchangeable. It is also the case that as thermoplastics and rubbers are members of the group of materials known as polymers, this word is often used in the chapter to refer to them as a single class.

6.2 Thermogravimetric analysis

6.2.1 Background

Thermogravimetric analysis measures the amount and rate of weight change in a material as a function of temperature, or time, in a controlled atmosphere. These measurements are mainly used to determine the composition of polymer products and precursors (e.g. additive master batches), but can also be used to measure thermal stability and study the effects that additives such as antidegradants have on this property.

In the analysis of polymers, temperatures from ambient to 1000°C can be used in either oxidising (e.g. oxygen or air) or non-oxidising (e.g. nitrogen or helium) atmospheres, and either isothermal or dynamic heating rates can be employed.

A typical TGA programme that can be used for the analysis of thermoplastics and rubbers is shown below.

6.2.1.1 Stage 1

The sample is heated in a nitrogen atmosphere at 20°C/min from 40 to 550°C and held at that temperature for 10 min to achieve constant weight.

6.2.1.2 Stage 2

The temperature is then reduced to 300°C, the atmosphere changed to air, and then the sample heated again at 20°C/min to 850°C.

In common with the other thermal analysis techniques, in TGA work it is important to standardise the analysis conditions that are used as much as possible to enable meaningful comparisons to be made between samples. In addition to ensuring that the same heating programme is used, it is also important to use similar sample masses (around 10 mg, for example) and, where a sample is in the solid, as opposed to powder or liquid state, a similar number and size of fragments.

A representative TGA trace of a polymer product (a nitrile rubber compound) is shown below in Figure 6.1.

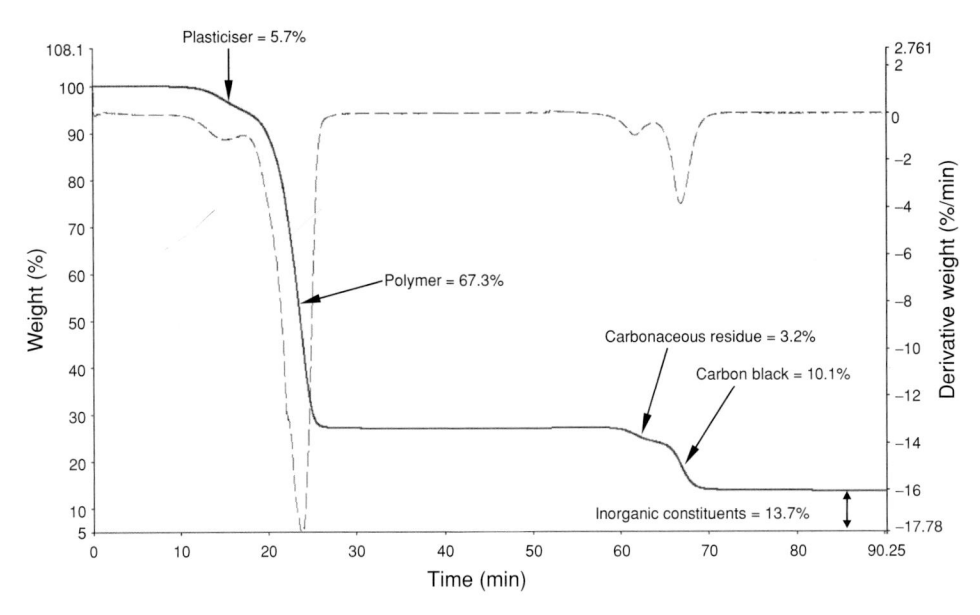

Figure 6.1 Typical TGA trace for a rubber compound. The weight loss is shown as a full curve and the derivative curve is dashed. The derivative curve helps to define suitable calculation limits.

The data in Figure 6.1 show the primary weight loss events that are present in such traces, with both the weight loss and derivative weight loss curves shown. In particular, the following information can be obtained:

Plasticiser content	5.7%
Polymer content	67.3%
Carbonaceous residue from the polymer*	3.2%
Carbon black filler content	10.1%
Inorganic residue	13.7%

Figure 6.1 shows the TGA trace plotted in the form of weight loss against time, which is the standard format as it makes the information present easier to see by eliminating the effect of the two independent heating steps (i.e. stages 1 and 2). However, it is often useful to obtain temperature data on the weight loss events, for example the onset temperature of weight loss, the peak degradation temperature for the polymer component, etc. Figure 6.2 shows how the data in Figure 6.1 would look if displayed in the form of weight loss against temperature.

---

* Polymers that contain atoms such as nitrogen, oxygen, sulfur and a halogen give carbonaceous residues due to incomplete combustion during the nitrogen atmosphere region (stage 1 above). This residue oxidises to carbon dioxide when heated in an oxidising atmosphere (i.e. in stage 2 above).

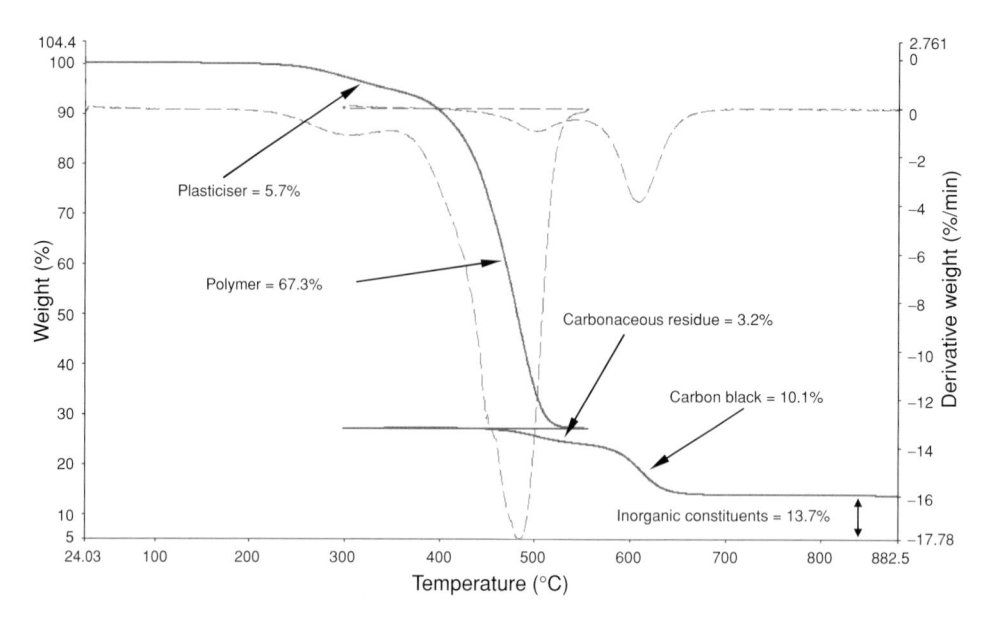

Figure 6.2 The data in Figure 6.1 shown in the form of weight loss against temperature.

A number of developments have improved the ease of use, flexibility and accuracy of these instruments over the last ten years. These include the following:

(i) Self-loading/unloading onto the hang down wire.
(ii) Multi-position autosamplers.
(iii) Sealed pans for the use of samples that have volatile components.
(iv) High-resolution running programmes to improve the resolution of weight loss events in complex samples.

The main application areas of TGA can be summarised as follows:

(a) Quality control and failure diagnosis via the production of compositional fingerprints.
(b) Research and development, e.g. behaviour of antidegradants and flame retardants.
(c) Reverse engineering, e.g. determination of the amounts of the principal ingredients (e.g. plasticiser, polymer and inorganic filler) within a polymer product.

6.2.2 Determination of additives

Polymeric compounds, particularly rubber compounds, contain a number of functional additives. TGA can be used to identify the presence of these additives, to evaluate the effect that they have on certain properties of the product (e.g. oxidative stability in the case of antioxidants) and, in the case of the major additives (e.g. plasticisers and fillers), it can be used to quantify them.

6.2.2.1 Antioxidants

A number of different types of antidegradants are incorporated into plastics and rubbers, e.g. antioxidants, UV stabilisers and antiozonants. The last two require the presence of a specific degradation agent (i.e. UV light and ozone) in order to display their effectiveness and so the use of TGA, where the experiment is carried out in the dark and in a controlled environmental, is not an appropriate choice for their study. The first class, the antioxidants, however, can be studied using TGA and the approach used to comparatively determine their effectiveness is similar to the oxidative inductive time measurements made by DSC (see Section 6.4.4), with isothermal experiments being run at an elevated temperature (e.g. 100°C) and the time for the sample to lose weight monitored. The earlier and faster that a sample loses weight, the less stable it is to oxidative attack, the result of which is to break up the molecular structure of the polymer. The situation with using weight loss measurements as opposed to changes in specific heat (using DSC) is not as straightforward, though, as there is always competition between cross-linking and depolymerisation-type reactions as a polymer oxidises, with only the latter generating volatile, low molecular weight fragments. The relative proportions of these two events are determined by the type of polymer present in a sample, e.g. the predominant mechanism with polymethyl methacylate is depolymerisation, but in the case of polyolefins it is cross-linking. This is why the standard methods tend to use the DSC approach, where only the fact that a reaction of some kind is occurring is important.

6.2.2.2 Plasticisers and process oils

These types of additives include man-made compounds such as phthalates and adipates, and naphthenic hydrocarbon mineral oils. Being viscous liquids they have relatively low molecular weights and so are usually volatile within the temperature range 150–300°C. This means that they are lost from a sample during a TGA analysis by volatilisation, rather than pyrolysis or oxidation, and this mode of loss is the same irrespective of the atmosphere type being used.

Due to this behaviour, it is possible from the peak of the weight loss derivative to obtain some qualitative information on the type of plasticiser present. As always with TGA, the rate of heating will affect the precise temperature obtained for a given compound – the faster the rate of heating the higher the temperature – but the type of data that can be obtained are given in Table 6.1 using a series of monomeric ester plasticisers.

Table 6.1 Derivative peak temperatures for a series of monomeric ester plasticisers

Ester plasticiser	Derivative peak temperature (°C)
Dibutyl phthalate	220
Di(2-ethylhexyl) adipate	250
Di(2-ethylhexyl) phthalate	265
Di(2-ethylhexyl) sebacate	280

It can be seen that the order is independent of molecular weight; the ease with which the plasticiser volatilises being of overriding significance.

More complicated additives in this class (e.g. polymeric plasticisers and multi-component mineral oils) do not lend themselves to this type of interpretive work with, in the case of the higher molecular weight examples, some overlap occurring with the pyrolysis breakdown of the polymer component in the sample, but the relative position of the derivative peak can still be used to obtain comparative data on two or more compounds. Knappe [1] has described how operating a TGA under vacuum conditions can improve the separation of additives such as plasticisers from the polymer component in a rubber. The resolution improvements that are possible using a rate-controlled mass loss heating method are also discussed as is the coupling of a top-loading TGA to an infrared spectrometer to identify volatile components.

Often the quantification of plasticisers by TGA is more accurate than a quantitative solvent extraction step. This is because a solvent is never completely selective in what it extracts from a polymer and any extract will always contain a certain amount of other material, particularly low molecular weight oligomers from the polymer, and so the extract value obtained is always higher than the actual plasticiser level that is present.

6.2.2.3 Carbon black and other carbon-based fillers

Carbon black, acetylene conducting black, graphite and carbon fibres can be quantitatively determined in samples by their characteristic oxidative weight loss in the oxidising atmosphere (e.g. air) region. The reasonably impure carbon blacks oxidise at a lower temperature (500–600°C) than the others (i.e. acetylene black, etc.) which are much purer and as a consequence oxidise at around 700°C. This difference in behaviour enables the two classes to be differentiated from each other.

It is possible to use oxidation temperature to identify the generic type of carbon black filler present in a rubber. Research carried out in the 1970s [2] showed that a more accurate result is obtained if the partial (e.g. 15%) weight loss position is used rather than the derivative peak temperature. The examples given in Table 6.2 reports how the oxidation temperature varies with different types of carbon black. Oxidation temperature is reduced as the particle size of the black reduces, with a resulting increase in surface area.

Table 6.2 Oxidation temperatures for different types of carbon black

Carbon black type	Temperature (°C)
Medium thermal	600
Semi-reinforcing furnace	575
High-abrasion furnace	550
Medium process channel	520

It is important, however, to appreciate that there can be a number of problems in assigning the carbon black type, such as the following:

(a) Blacks can vary in 'structure' (i.e. degree of particle agglomeration) as well as particle size. This can blur the distinction between grades as there is a large degree of overlap.
(b) The analysis conditions used can affect the data and so all the samples and controls must have been analysed in the same way.
(c) The type of polymer in the sample can affect the data because carbonaceous residues (see Section 6.2.3) can overlap with the carbon black.

The result is that TGA can only be an indicative technique and other more accurate techniques (e.g. surface absorption of nitrogen – BET method, or iodine adsorption method) should be used to obtain a more definitive identification of the black present in a sample after its recovery by non-oxidative pyrolysis.

6.2.2.4 Inorganic fillers and other inorganic additives

Inorganic fillers, such as silica, silicates, barium sulphate and calcium carbonate, are used extensively in plastics and rubbers and to be effective these are usually incorporated at a level well in excess of 5%. This means that they can easily be quantified by TGA. Other inorganic compounds are also used in most products, and these include zinc oxide (a common co-agent used in the cure system of rubbers) and titanium dioxide (a popular weight pigment). In contrast to fillers, these additives are usually only added at levels below 5% but, given that the detection limit of TGA is around 0.5%, they can still be detected and a reasonably accurate quantification performed.

The majority of the inorganic compounds used for either filler or pigmentation are stable up to 1000°C (the maximum temperature usually employed in TGA work of polymers) and the residue remaining at the end of a TGA will enable them to be quantified. Unfortunately though, this means that TGA is only capable of determining the total amount of inorganic material in sample – other tests (e.g. XRF and infrared spectroscopy) are required to identify the inorganic compounds present. The exceptions to this are compounds such as calcium carbonate; this decomposes into calcium oxide and carbon dioxide at around 700°C. This characteristic decomposition enables this species to be detected within a sample and, because the breakdown is quantitative with a 40% weight loss occurring, the amount present can also be calculated whether other inorganic compounds are present or not.

The use of nanofillers, such as organo-clays and silica, in polymers is a very active area for research and TGA has been used by a number of workers to assist in their characterisations of the resulting nanocomposites [3–5].

Another group of inorganic compounds used are polymers that are certain flame retardants and these are dealt with in the next section.

6.2.2.5 Flame retardants

The flame retardants used in plastics and rubbers are usually of two principal types: involatile liquids (e.g. chlorinated paraffins, brominated aromatics or phosphate esters) or inorganic compounds (e.g. antimony dioxide, zinc borate or hydrated alumina). The exceptions to this are compounds which are organic solids, such as urea derivatives and melamine materials.

The presence of flame retardants in polymer compounds alters their overall thermal stability and changes their pyrolysis breakdown chemistry, although the precise changes will depend upon the polymer present in the sample [6, 7]. It is also the case that liquid fire retardants will volatilise in a similar way as plasticisers, although certain examples (e.g. polybrominated compounds) have very high molecular weights and will merge with the polymer weight loss event. Overall, these effects mean that it is relatively easy to detect the presence of a flame retardant additive by TGA, and it can also be used to obtain a measure of the additional stability that it is inferring even though the conditions that a sample experiences during a TGA run are very different to a fire situation. This increase in stability is complicated and, as mentioned above, tends to be specific for a given polymer system. For example, flame retardants in ABS compounds produce a lower initial temperature of decomposition for the sample, create a two-stage pyrolysis weight loss pattern for the polymer itself (there being only a single stage in the non-flame retardant material), and increase the level of carbonaceous residue that is produced.

Certain inorganic flame retardants also have characteristic fingerprints. A classic example of this is hydrated alumina. This compound gives off a quantitative amount of water in the early stages of a run (35%), with the residue that remains being aluminium oxide. As long as no other constituents mask or interfere with this weight loss (e.g. volatilisation of a plasticiser or dehydrohalogenation of a polymer) it can be used to quantify this flame retardant.

6.2.2.6 Miscellaneous additives and the limitations of the TGA technique

TGA is a technique that is very useful to identify the principal constituents in a polymer sample, but it has limitations in terms of its detection limit. This means that a number of important, but minor, additives with respect to the amounts that are added to a particular product cannot be detected or quantified using the technique. In addition to being present at low levels (e.g. <1% w/w), a number of these additives interact with the sample in a complex way and so they will not be present in an easily definable section of the TGA trace. This also contributes to the difficulty in detecting and quantifying them.

A good example of this situation exists in the case of the curatives (e.g. sulphur) and accelerators that are used in the large number of rubber products. These cure system species undergo the following types of reaction with the sample:

(a) Reaction with the cure co-agents to produce intermediates.
(b) Reaction with the polymer to produce cross-links.
(c) Reaction products from the above two and thermal breakdown products react to produce an additional series of compounds.

In addition to the above, because the efficiency of the reactions is below 100%, some of the initial, unreacted additive(s) will also remain in the sample.

A similar problem is encountered by other reactive species such as antidegradants. At least these can be detected in an indirect way by measuring the effect that they have on the stability of the sample (see Section 6.2.4).

A summary of the most important additives that cannot be directly determined by TGA is listed below:

(a) Stearic acid (rubber cure co-agent) and stearate process aids (calcium stearate) used in plastics.
(b) Curatives used in rubbers (e.g. sulphur, peroxides and amines).
(c) Accelerators used in rubber cure systems (e.g. thirams and sulphonamides).
(d) Cross-linkers used in plastics (e.g. peroxides).
(e) UV stabilisers used in plastics.
(f) Antistatic additives used in plastics.
(g) Antiozonants used in rubbers.
(h) Slip additives used in rubbers and plastics (e.g. molybdenum disulphide).

For all of the above, techniques that are more suited to the analysis of additives at the parts per million level (e.g. GC-MS, LC-MS and ICP) have to be used. The use of such techniques for the analysis of rubbers and plastics has been reviewed by Forrest [8, 9].

6.2.3 Compositional analysis

The use of TGA to identify and quantify individual classes of additive has been covered in Section 6.2.2. It is often the case that the purpose of an analytical experiment is to find out as much compositional information on a sample as possible. It is then the case that the knowledge of how the individual elements (i.e. additives) of a polymer system perform during a TGA experiment can be used to interpret the entire trace and produce what is often referred to as a 'bulk composition' – i.e. a list of the proportions of the major ingredients present. It has already been mentioned in Section 6.2.2 that additives at low levels (i.e. <1%) cannot be detected by TGA. This section will build upon the discussion given in Section 6.2.2 to enable the analyst to use TGA effectively for complete compositional analysis work.

Firstly, when using the TGA to undertake compositional analysis work on rubbers and plastics, it is important to use the two-stage analysis programme given in Section 6.2.1 because it will be likely that the weight loss events for certain ingredients will overlap otherwise. An obvious example of this is in the case of rubbers where the polymer pyrolyses at a similar temperature to that at which carbon black fillers oxidise – around 400–500°C (see Section 6.2.2). Performing the analysis under only air with a single temperature ramp will therefore produce a single, composite weight loss whenever these two are present. Given that carbon black is easily the most common filler used in rubbers (and it is also used as a pigment in plastics) this is a very likely occurrence. Using the two-stage approach, the polymer is pyrolysed first in the nitrogen atmosphere, leaving the carbon black untouched and available to be quantitatively lost due to oxidisation in the subsequent air atmosphere.

The TGA trace for a typical plastic, a glass reinforced polyamide 6, is shown below in Figure 6.3. The bulk compositional analysis data that can be obtained from such a trace are

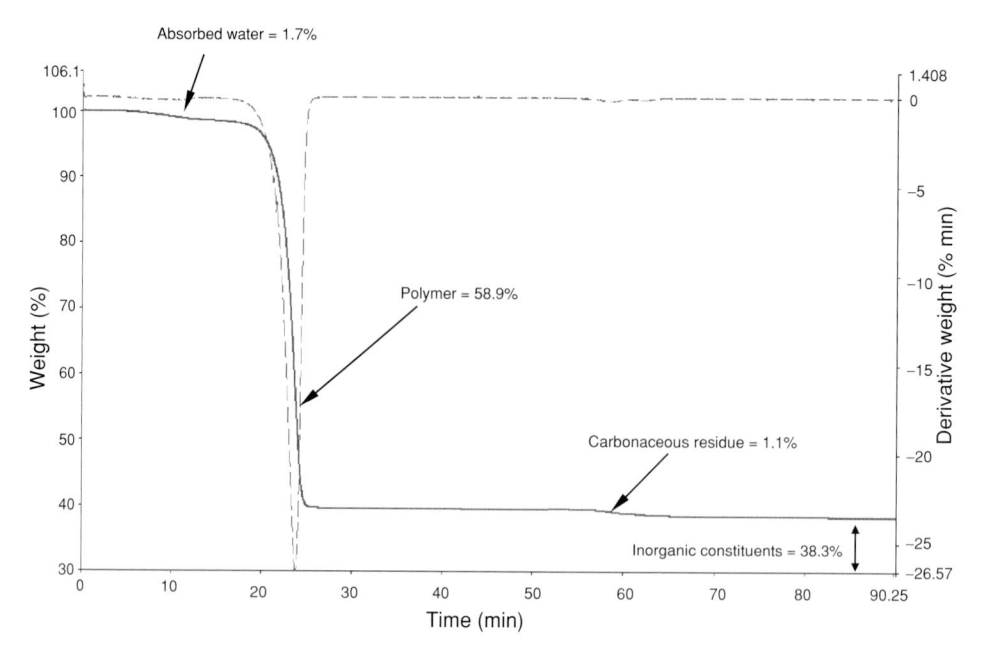

Figure 6.3 TGA trace for a typical plastic sample. The weight loss is shown as a full curve.

as follows:

Absorbed water	1.7%
Polymer	58.9%
Carbonaceous residue	1.1%
Inorganic constituents[†]	38.3%

6.2.3.1 Solvents and monomers

The use of TGA to determine the low molecular weight, non-volatile liquid additives that are referred to as plasticisers and process oils has already been covered in Section 6.2.2. Solvents (including water) and monomers are similar to plasticisers and oil in that they appear early on in a TGA trace, but they are usually of a lower molecular weight and are not regarded as additives. Solvents can be present in a sample due to absorption in service, a classic example being the absorption of water by polyamides, or because they are used as a 'carrier' for an additive to aid dispersion within the polymer matrix during the mixing stage. A certain amount of residual, unreacted monomer is always left in a polymer after the polymerisation stage, and sometimes the levels are sufficient for it to be quantified by TGA, e.g. polymethylmethacrylate produced by a casting process. Another reason for the

---

[†] Mainly the glass fibre reinforcement.

presence of monomer is if the polymer has undergone some thermal degradation in service and the material has 'unzipped' to regenerate the monomer.

Due to their volatility solvents and monomers will volatilise very early on in the TGA trace and they are usually reasonably well resolved from any weight loss due to the polymer. The exception to this is in the case of polymers that undergo a two-stage weight loss where a small molecule (e.g. acetic acid or halogen halide) is generated at a low temperature) and is lost early on. Usually, though, good quantification data are obtained and, if further accuracy is required, Forrest has described how the use of high resolution TGA can be used to improve the quantification data that are obtained on these types of volatile species [10].

It is important to appreciate that one problem often encountered in obtaining an accurate quantification of volatile, low molecular weight solvents and monomers is that they can be lost by volatilisation from the sample in the time that elapses between the being prepared and the analysis stage. The manufacturers have addressed this problem, and all modern TGA instruments now have the ability to use sealed pans up to the point of analysis. Care still needs to be taken in the storage (i.e. use of a refrigerator) and preparation of samples though. It is also good practice to start the run as soon as possible.

6.2.3.2 Polymers

Polymers, irrespective of their chemical composition, are lost from a sample during a TGA experiment by a pyrolysis reaction – and this will be the case irrespective of the atmosphere type, i.e. oxidising or non-oxidising. This occurs because of their high molecular weight, which means that they will pyrolyse before the temperature can reach a level that would enable them to volatilise. It is the case though that, in addition to residual monomers (dealt with above), some polymers contain a sufficient quantity of low molecular weight oligomers for these to register in a TGA trace. The presence of any plasticiser, water, solvent or any other low molecular weight species will mask these however.

When using the type of TGA programme given in Section 6.2.1, enough data have been generated using standard materials to enable some qualitative identification to be made from the pyrolysis derivative peak maximum. For example, the following temperature–polymers matches have been made for some polymer rubbers and plastics (see Table 6.3).

Table 6.3 Derivative peak maxima temperatures for a selection of rubbers and plastics

Polymer type	Derivative peak maximum (°C)
Natural rubber	370
Butyl rubber	390
SBR, polybutadiene or EPDM	460
Fluorocarbon rubber	480
Polymethylmethacrylate	325
Polyethylene	400
PTFE	520
Polyamide	600

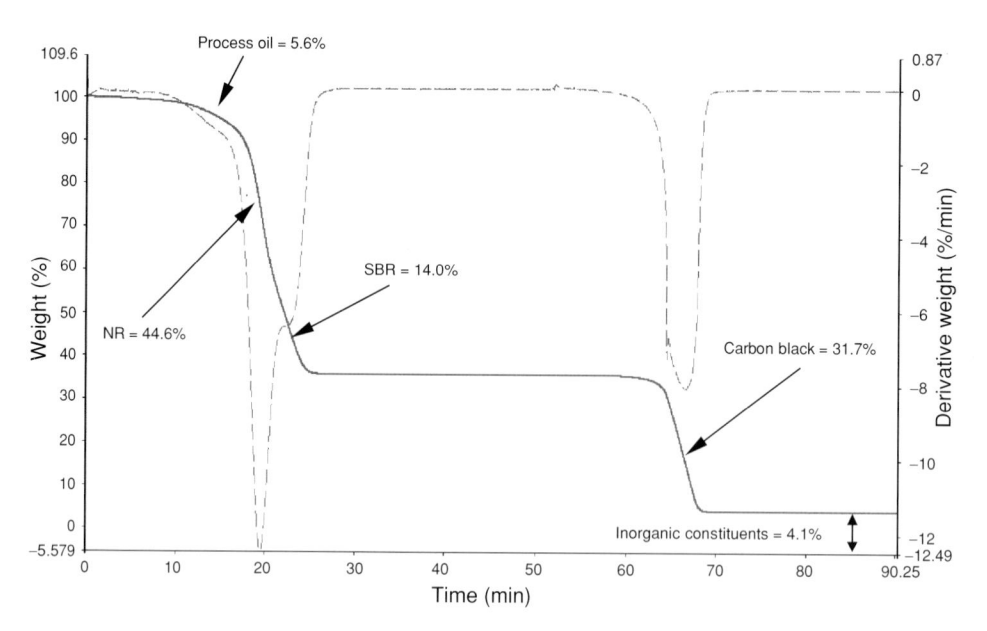

Figure 6.4 TGA trace for a tyre tread compound containing both NR and SBR rubber. This illustrates how the amounts of polymer in a material can be obtained by TGA. The weight loss derivative trace has been crucial in this application to determine the crossover point of the weight losses.

The differences shown above can be of use in enabling the analyst to detect the presence of a blend of polymers in a sample. If there is sufficient resolution between the two polymer weight losses, for example between natural rubber (NR) (peak temperature 370°C) and SBR (peak temperature 460°C), it is possible to obtain a reasonably accurate quantification. An example of this is shown in Figure 6.4, which illustrates how the amounts of NR and SBR in a truck tyre can be obtained by TGA. The weight loss derivative trace has been crucial in this application to determine the crossover point of the weight losses.

Unfortunately, though the use of high resolution modes of operation makes the detection of a blend easier, they yield less accurate quantification data due to the generation of complex multi-peak weight loss profiles for each of the individual rubbers and these overlap in the blend thermogram. This problem is discussed further in Section 6.2.5.

In addition to the pyrolysis breakdown mechanism, a number of polymers also undergo an earlier breakdown step where a small molecule is lost from the polymer backbone due to, for example, a dehydrohalogenation reaction [11]. This additional weight loss provides the advantage of making certain polymers easier to detect in a TGA trace, but usually this is outweighed by the fact that it overlaps and combines with any early weight loss event due to a plasticiser or oil additive, thus making the quantification of both this additive and the polymer itself impossible. The approach that has to be taken to get round this is to carry out a quantitative solvent extraction to obtain a value for the plasticiser and use this value to adjust the TGA data to enable the amount of polymer to be calculated.

Table 6.4 Examples of polymers that undergo an initial low-temperature weight loss during a TGA experiment

Polymer	Molecule lost
Polychloroprene rubber	Hydrogen chloride
Chlorinated rubber	Hydrogen chloride
Chlorosulphonated rubber	Hydrogen chloride
PVC	Hydrogen chloride
Ethylene-vinyl acetate (EVA)	Acetic acid
Vinyl acetate	Acetic acid

Examples of polymers that undergo an initial weight loss at a relatively low temperature are given in Table 6.4.

In some cases, and EVA is the classic example, the loss is quantitative and the amount of acetic acid generated can be used to estimate the amount of vinyl acetate monomer in the EVA copolymer. The use of TGA for this purpose is covered by Affolter and Schmid in a paper reporting the results of interlaboratory trials involving thermal analysis techniques [12].

In a recent paper, Agullo and Borros [13] have reported on an easy and rapid way of using TGA to determine qualitatively and quantitatively the type of polymer or polymer blend used in rubber compounds. The technique was applied successfully to a range of commonly used, commercially important polymers, including NR, EPDM, chlorosulphonated polyethylene and chloroprene. It was found that TGA alone could not differentiate between NBR and SBR, with another technique capable of detecting either styrene or acrylonitrile (pyrolysis GC-MS) being required to complement the data.

In some cases, it can also be possible to use TGA to obtain a measure of the relative proportions of two monomers in a copolymer. For example, Shield and Ghebremeskel [14] have obtained a quantitative determination of the amount of styrene in tyre tread grade, SBR rubbers. A methodology was established for determining the styrene content from the magnitude of the shifts in the polymer pyrolysis region of the TGA curve.

6.2.3.3 Carbonaceous residues

This substance is formed by a number of heteroatom-containing polymers as they pyrolyse during the nitrogen atmosphere region (see Figures 6.1 and 6.3). This residue then oxidises to carbon dioxide during the subsequent heating in the air atmosphere region. The oxidation temperature of these residues is usually lower than carbon black, and much lower than the other carbonaceous materials (carbon fibres or graphite), due to their impure, heterogeneous nature and this means that, although some overlap can occur, they are usually well resolved and easily identified.

In order to obtain an accurate quantification of the amount of polymer in a sample, the value of the carbonaceous residue must be added to the weight loss, or losses (see above), that the polymer has undergone in the nitrogen atmosphere region.

Table 6.5 Relationship of carbonaceous residue to the level of acrylonitrile monomer present in a series of polymers

Polymer	Level of acrylonitrile (%)	Carbonaceous residue (%)
Polybutadiene	0	0
NBR	20	2
NBR	25	3
NBR	35	5.5
NBR	40	6
NBR	50	12
Polyacrylonitrile	100	45

Examples of rubbers and plastics that produce carbonaceous residues include the following:

1. Nitrogen-containing polymers, e.g. polyamides and nitrile rubbers.
2. Oxygen-containing polymers, e.g. polycarbonate, PEEK and polyesters.
3. Halogen-containing polymers, e.g. PVC and polychloroprene rubbers.

In addition to the diagnostic value of these residues in identifying the presence of one of the above types of polymers, as their production is consistent and quantitative, the level of carbonaceous residue can also be of value. The prime example of this is the range of polymers referred to as nitrile rubbers. These materials are copolymers of butadiene and acrylonitrile. It is the acrylonitrile units that give the residue and, the more there are of these in the polymer molecule, the greater the level of residue. Some indicative values for a range of pure (i.e. uncompounded) nitrile rubbers (NBRs), together with the two homopolymers, are given below in Table 6.5.

The approximate amounts of carbonaceous residue generated by other commercially important rubbers and plastics are given below in Table 6.6. The ranges given reflect the variation that has been found between the grades available for a particular polymer. In common

Table 6.6 Amounts of carbonaceous residue obtained from a range of commercially important polymers

Polymer	Carbonaceous residue (%)
Polychloroprene rubber	20–25
Chlorosulphonated polyethylene rubber	2–4
Fluorocarbon rubber	3–10
Epichlorohydrin rubber	5–15
Ethyl acrylate rubber	6–8
PVC	15–20
PET	10–15
PBT	5–8
Polyamide 6	1–3
Polyamide 6,6	2–4

A range is given because the actual amount is dependent on the grade of polymer.

with the nitrile rubber group, a range exists often because the materials are copolymers or modified homopolymers, and the level of a particular monomer, or the degree of chemical modification, can be varied.

The presence in a rubber compound of one of the polymers listed above can result in problems due to its carbonaceous residue overlapping with the weight loss due to carbon black. Usually the carbonaceous residue oxidises at a temperature below that of the carbon black, but in the case of small particle size blacks, problems can be encountered and it can then be difficult to obtain accurate values for both the polymer and the black. This situation is exacerbated by polymers (e.g. polychloroprene) that produce a large carbonaceous residue. Sometimes a high resolution programme can assist in separating the two, but these in themselves can cause problems (see Section 6.2.5).

6.2.3.4 Blends of additives in samples

It has already been mentioned above how individual additives (e.g. process oils, carbon black and inorganic) can be detected and quantified in samples. The areas that can prove problematical (e.g. the presence of an oil in a polymer that undergoes a dehydrohalogenation reaction) have also already been discussed in their respective sections. As the majority of the additive types have their own technology associated with them, a range is commercially available and so it is possible to find that two have been used in a compound to achieve certain end product performance goals.

This type of scenario is particularly common with flame retardants and carbon black fillers. In the case of the latter, it has already been discussed how the variation in particle size relates to changes in the oxidation temperature. If a standard TGA analysis is performed (high resolution operation can cause problems – Section 6.2.5) and two weight loss events are observed in the 'carbon black region', it is then reasonably safe to assume that two different particle sizes of carbon have been used in the compound. The split will also be quantitative and so the relative proportions of the two blacks can be calculated. It is very unusual for three types of carbon black to be used in a single sample and so more than two weight losses in this region usually indicate that a problem has occurred with the analysis. An easier case to interpret results when a carbon black and a carbonaceous material having a much higher oxidation temperature (e.g. acetylene black or graphite) have been used in the same material.

With respect to flame retardants, these are often used in complex formulations and so the TGA data are hard to interpret. In the absence of any other analytical data, this complexity alone is sometimes a good guide that a flame retardant formulation is being examined. The combination of a large, early weight loss and the presence of a significant inorganic residue is also diagnostic as it is common practice to use both organic and inorganic flame retardants in a synergistic blend (e.g. halogenated organic with antimony trioxide). In reality, though, a TGA trace in isolation cannot reveal much detail on the composition of such samples (see the example given below in Table 6.7) and a full suite of tests (including infrared spectroscopy and elemental analysis) is required to compliment the TGA data to produce a full picture.

A number of polymer additives are inorganic in nature and it is also the case that the majority, with the notable exception of the some carbonates (e.g. calcium carbonates), are stable to very high (i.e. $>1000°C$) temperatures. The stable inorganic residue obtained by TGA is therefore a good guide as to the total level of inorganic constituents present, although not as accurate as a quantitative ashing experiment due to the relatively small

Table 6.7 TGA weight loss event position of the components present in a flame retardant nitrile rubber/PVC compound

Constituent/function	Position in TGA trace
Pentabromodiphenylether (flame retardant)	1st weight loss
Phosphate ester (flame retardant)	1st weight loss
Phthalate ester (plasticiser)	1st weight loss
Nitrile rubber (polymer)	2nd and 3rd weight loses
PVC (polymer)	1st, 2nd and 3rd weight loses
Carbon black (pigment)	4th weight loss
Hydrated alumina (flame retardant)	1st weight loss and final residue
Antimony trioxide (flame retardant)	Final residue
Zinc oxide (cure co-agent)	Final residue

The 3rd weight loss is the carbonaceous residue.
Other additives such as a pre-vulcanisation inhibitor and a sulphur-based cure system with accelerators were also present but at low levels, i.e. <1–2%.

amount of sample used (e.g. 10 mg), but further analysis work (e.g. a combination of infrared spectroscopy and an elemental technique) has to be carried out on the ash to identify and quantify the inorganic compounds in the blend.

The complexity that can be present in the TGA thermogram of an extensively compounded sample is given in Table 6.7 using the example of a flame retardant rubber based on a nitrile rubber/PVC blend [15].

6.2.4 Thermal stability determinations

TGA is often used to determine the relative stability of materials. This is carried out by observing the onset temperature, in a given atmosphere, of the decomposition of the polymer. Since the majority of polymers are used in real life scenarios, the atmosphere usually chosen is air. It is also possible though to assess relative thermal stability of materials that are used in service under adiabatic conditions, for example rubber seals that are immersed in oil. It is always important to remember in these cases though that the actual lifetime of the material will also be influenced by the media that it contacts and not just the temperature that it is exposed to.

Although it can be possible to conduct these experiments using a temperature programme with a ramped heating rate (for example, Denardin et al [16] have shown how the dynamic, non-isothermal Kissinger and Osawa methods can be used to determine the activation energies of the degradation processes that occur with polychloroprene rubber, and Jana and Nando have used non-isothermal TGA to study the thermal stability of LDPE/silicone blends containing compatibilisers [17]) it is sometimes preferable to use isothermal conditions at a relatively high temperature (e.g. 100°C) where a finer discrimination between materials can be obtained.

In this way an accurate ranking of relative stability of different polymers can be obtained. For example, the polymers polyetherimide (PEI), polycarbonate (PC), polyethylene terephthalate (PET) and polyvinyl chloride (PVC) differ in their thermal stabilities due to the chemical make-up of their backbones (in the order PEI > PC > PET > PVC) and the order in which they will show the onset of weight loss in a TGA experiment reflects this correctly. However, an important point to make with thermoplastics such as these is that the experimental conditions have to be chosen carefully as too high a test temperature (isothermal analysis), or too fast a heating rate, will mean that the data will have limited use in terms of assessing the final product's life span since the materials will be molten, and hence useless, before their onset temperatures. Saccani et al [18] have used both isothermal and non-isothermal experiments to investigate the short-term thermal endurance of a number of thermoplastic polyesters, including PET and PBT. The results obtained were used to calculate the activation energy of the degradation processes for these materials. The situation is slightly different in terms of rubbers, which are thermosets, but some care is needed and the best approach is to use an isothermal temperature that accelerates the degradation that will take place in service without creating a completely unrepresentative environment.

Skachkova et al [19] used isothermal TGA experiments, with subsequent atomic force microscopy investigations of the sample surfaces, to investigate the thermo-oxidative stability of the commercially important polypropylene–EPDM thermoplastic rubbers; the rubbers analysed contained a hydrocarbon oil plasticiser as did control samples of polypropylene and EPDM rubber. The results obtained indicated that the oil degraded and became integrated into the oxidised polymer surface at high temperatures forming a multi-layer which inhibited the ingress of oxygen and hence provided protection to the elastomer. Pruneda et al [20] have used TGA in conjunction with a number of other analytical techniques (e.g. DSC and SEM) to determine the activation energies associated with the degradation processes associated with NBR/PVC blends that contained the antioxidant TMQ. In a comprehensive study, involving a wide range of EPDM rubber compositions, Gamlin et al [21] used isothermal TGA experiments in the range 410–440°C to evaluate the kinetic parameters of degradation using mathematical models that were based on various proposed degradation mechanisms. One of the results was that no observable trend was found between the ethylene content of the EPDM polymer and the activation energy of degradation.

It has been shown above that TGA is used in a number of degradation studies to determine the kinetics of decomposition, and this can involve conducting experiments on a particular sample at different heating rates, or at different isothermal temperatures. In the first case, a graph of the log of the heating rate against the reciprocal temperature for a specific weight loss is plotted, and in the second the log of the time required to achieve the specified weight loss is plotted against the reciprocal of the test temperature. In both cases, the data obtained are used to obtain the time for decomposition to occur at a lower, usually nearer to ambient, temperature. There are some problems in using TGA data to determine these quantitative relationships; for example, in identifying the crucial weight loss mechanism (it has already been mentioned that some polymers have more than one, and this is certainly true of compounded products) and equating a specific weight loss value to the end of a material's useful life [22, 23]. It is also the case that small errors in the calculation of the activation energy can result in large uncertainties in the predicted lifetime if the range of temperature extrapolation is large [24].

For a given polymer type, e.g. polyethylene, TGA can show how thermal history (e.g. heat ageing) can affect the oxidative stability of the material, as well as the effect of additives such as antioxidants (both types and levels), and the influence that manufacturing impurities such as residual metal catalysts can have. High-temperature studies carried out at temperatures such as 600°C can show how fillers and pigments can affect stability [25, 26].

6.2.5 High resolution TGA and modulated TGA

In order to improve the resolution that can be obtained between the weight loss events for a given sample, there have always been a number of possible approaches, such as the following:

(i) Using a very slow heating rate (e.g. 2°C/min).
(ii) Increasing the surface area of a sample by grinding.
(iii) Using a helium (as opposed to nitrogen) atmosphere to improve heat transfer properties.

Unfortunately, the first of these significantly increases experimental run times (e.g. from 2 to 20 h), and the last two increase the cost of the analysis work.

Another option, which has been available in the form of software packages that enable TGA instruments to be operated in high resolution modes, has been commercially available for over 10 years. High resolution TGA was developed as a technique to enhance the resolution of multiple components in complex systems. It therefore offers particular potential advantages in the analysis of complex rubber formulations and complex thermoplastic products such as flame retardant PVC compounds for treating upholstery fabrics.

The principle of its operation is that rather than have a constant heating rate (e.g. 20°C/min) throughout the analytical run, the TGA furnace is actively controlled and the rate of heating determined by the rate that the sample loses weight. The faster the sample loses weight, the slower the furnace temperature rises. In the cases of very fast weight loss the furnace operates in an isothermal mode. The operator has a large degree of control over how the high resolution mode is applied by choosing parameters, such as sensitivity settings, which determine at what weight loss rate the mode 'kicks in'. Another feature of the high resolution operation is that a high default heating rate (e.g. 50°C) is used so that when no weight loss is occurring there is an opportunity to regain time lost during the high resolution stage. In theory then, with the correct choice of experimental settings it should be possible to obtain better resolution without the penalty of very long experimental times.

A review of the use of high resolution TGA has been published by Forrest [27]. This shows that, in practice, high resolution has been found to offer distinct advantages in the case of some samples, but it is also necessary to be cautious in its application to others. A summary of these two instances is given below.

Cases where high resolution TGA offers advantages:

1. Quantification of the plasticiser and polymer components in rubbers and plastics.
2. Quantification of low molecular weight species (e.g. residual monomers and solvents) in plastics.
3. Differentiation between different grades of some complex plastics such as ABS.
4. More detailed TGA compositional fingerprints of very complex samples.

Cases where high resolution TGA can cause problems:

1. On some occasions the weight loss event due to the carbon black filler in rubbers can be split giving the impression that two grades of carbon black (i.e. having different oxidation temperature maxima) are present in the compound.
2. Less accurate quantifications are obtained of two polymers in a blend compared to the data obtained using the conventional approach.

These disadvantages are thought to be due to additional weight loss events being 'pulled out' of these heterogeneous materials and complicating the data. For example, natural rubber gives one weight loss in a conventional TGA experiment, but four in a high resolution operation [28].

It is also the case that incorrect choice of the operating parameters can lead to long periods of isothermal operation and consequently very long experimental times. To mitigate the possibility of this, it is highly advisable to record a conventional TGA trace on a sample to observe its weight loss profile before choosing which high resolution settings to employ. Once the optimum analysis conditions have been established, the advantages that high resolution TGA can bring to compositional analysis work are highlighted by comparing the data shown in Figures 6.5 and 6.6. Figure 6.5 shows the data obtained on a flame retardant, heat shrink tubing product using conventional TGA. In addition to a number of additives, such as the flame retardant plasticiser, the product also contained two polymers (a polyester and a fluoropolymer) and it can be seen from Figure 6.5 that the conventional TGA

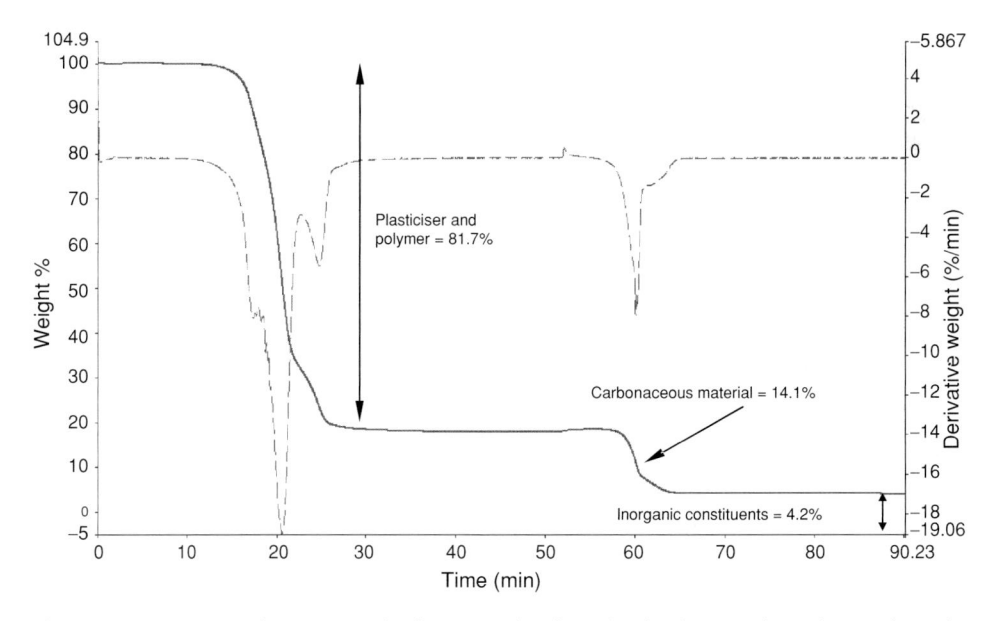

Figure 6.5 Conventional TGA trace of a flame retardant heat shrink tubing product. The initial weight loss shows a number of overlapping events which are not well separated.

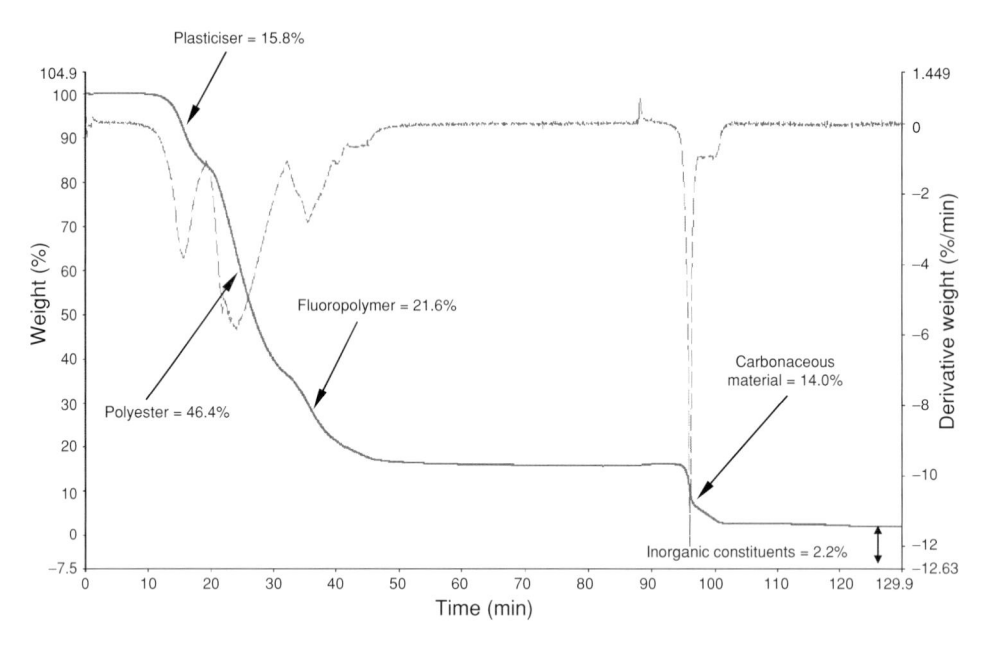

Figure 6.6 High resolution TGA trace of a flame retardant heat shrink tubing product. This shows better separation of events.

experiment was unable to resolve these constituents from each other. However, repeating the analysis using a high-temperature programme produced the data shown in Figure 6.6 and, from this, it was possible to quantify the plasticiser and the two polymers present in the product.

Another development of TGA designed to improve the accuracy of the data obtained is modulated TGA. This technique was developed by TA Instruments [29, 30] and operates using a similar principle to modulated DSC (see Section 6.4.7), with a sinusoidal modulation overlaid upon the linear heating rate to produce a situation in which the average sample temperature changes continuously with time in a non-linear way. Gamlin et al [31] have used a combination of high resolution TGA and modulated TGA to study the effects of the ethylene content and the maleated EPDM content on the thermal stability and degradation kinetics of EPDM rubbers. A comparison was made between the data generated from isothermal and non-isothermal TGA experiments, and other values reported in the literature.

6.2.6 Hyphenated TGA techniques and evolved gas analysis

The option of connecting the outlet of a TGA instrument, via a heated transfer line, to a spectroscopic or chromatographic technique has been available to analysts for over 15 years. This is an attractive proposition because of the much greater amount of information that can be accumulated on a sample in essentially one analytical operation and it addresses the main limitation of TGA – the lack of a significant amount of qualitative data. For example, by this route, not only can the amount of plasticiser or oil in sample be determined but also its chemical type. This alleviates the need to carry out a separate analytical step, such as a solvent extraction of the sample followed by infrared analysis of the extract.

There are a number of instrumental combinations that can be set up, including the following:

(a) TGA-IR
(b) TGA-MS
(c) TGA-GC-MS.

These three techniques vary in the fundamental mechanism being applied. In the case of the IR and MS continuous sampling may be carried out, and with GC-MS intermittent sampling is conducted where the gas evolved over a particular temperature range, or time (particularly in isothermal work) is collected in a intermediate trap and then analysed.

The TGA-IR combination is the most cost effective and can provide information on both low molecular weight additives volatilising off at the beginning of a run as well as higher molecular weight constituents such as polymers (and blends of these), resins, etc. In the latter example, the compounds will be pyrolysed to low molecular weight fragments and so it is useful to run standards to enable accurate identifications to be performed.

The other two techniques have an additional capability to provide very detailed information on a sample. This includes identification of low level addition additives such as antioxidants. This is particularly true in the case of TGA-GC-MS because of the intermediate chromatographic step which enables the evolved gas produced at a given point in time to be resolved into its component parts.

It is also possible to analyse the gases produced during a TGA experiment by a wide variety of other, mostly indirect, routes. Some of the more useful for polymer compounds are as follows:

(i) Draeger tube.
(ii) Trapping onto an adsorption tube (e.g. one packed with tenax) and then analysis using thermal desorption GC–MS.
(iii) Absorption into a solution and then analysis by chromatographic (e.g. LC-MS) and wet chemistry (e.g. titrimetric) techniques.

The first example is limited, but the other two have a much wider capability, although they are relatively time consuming and are better suited to research projects than contract analysis work.

Although it can be seen that there are distinct advantages to these hyphenated combinations, these only really extend over the portion of the TGA thermogram that deals with the organic components in a polymer sample (i.e. stage 1 in Section 6.2.1). Once all of the organic components have been removed by volatilisation or pyrolysis, only carbonaceous materials (carbon-type fillers or carbonaceous residue) and/or inorganic compounds are left and these only tend to produce simple gases (e.g. carbon dioxide), or are completely inert.

A typical example of the applications that hyphenated-TGA can be used for has been published by Pielichowski and Leszczynska [32]. They have shown how a combination of TGA and IR can be used to characterise the most abundant volatile products of polyoxymethylene/thermoplastic polyurethane blends and, having done so, have gone on to discuss the mechanism of their formation. Gupta *et al* [33] have shown how high resolution TGA-MS can be used to characterise the properties of a range of flame retardant bromobutyl rubber formulations.

Although the most commonly available, TGA is not the only thermal technique for which evolved gas analysis can be performed. It can also be found in conjunction with DSC, for example, to discover if thermal events are associated with mass loss. It is also the case that a commonly encountered combination of techniques is TGA-DSC, with a number of manufacturers offering instruments with this dual capability.

6.3 Dynamic mechanical analysis

6.3.1 Background

DMA, or DMTA (dynamic mechanical thermal analysis) as it can also be called, measures the modulus (stiffness) and damping (energy dissipation) properties of polymers as a function of temperature as they are deformed under a periodic stress. This latter criterion distinguishes this technique from the closely related one of TMA, which is a static technique. Because DMA is able to provide information on the viscoelastic properties of polymers, it has a greater capability than TMA and this is why the latter will not be included in this section, but is covered in Section 6.3.6.

In common with most thermal analysis techniques, DMA analyses can be performed under both ramped and isothermal temperature programmes. The outputs from an experiment are elastic (storage) modulus and viscous (loss) modulus, and tan δ, which is the ratio of viscous modulus over the elastic modulus.

The operator in a DMA experiment has a number of variables that can be set prior to running an experiment, namely the following:

1. Clamping force that is applied to the end(s) of the sample.
2. Temperature programme, i.e. heating rate or isothermal conditions.
3. Frequency of deformation.
4. Start and finish temperature.

In addition to this, the instrument can be operated using a number of different configurations to enable data to be collected on a variety of polymeric samples:

(a) In tension
(b) Flexure – single and double cantilever/three-point bend

(c) Compression
(d) Shear
(e) Torsion.

In addition, instruments can be operated in constant stress (creep) and constant strain (stress relaxation) modes.

As with the other thermal analysis techniques, the data that are obtained by a particular DMA experiment will be very dependent on the experimental conditions used and so it is important to keep as many of the potential variables constant if the data are going to be used for comparative purposes or the study of unknowns. A typical example of this type of experiment is the identification of the polymers in a blend by use of the tan δ peak temperatures and comparison with reference data obtained on standard materials.

A typical set of running conditions that have proved useful at the Rapra for a wide range of plastics and rubbers experiments are given below:

Deformation mode	Flexible (double cantilever)
Sample dimensions	2 mm × 10 mm × 25 mm
Heating rate	3°C/min
Frequency of deformation	1 Hz
Start temperature	−150°C
Finish temperature	Dependent on material[‡]
Amplitude of deformation	Dependent on material[§]

Typical DMA traces for a rubber and a plastic sample are shown in Figures 6.7 and 6.8. Both of these contain modulus and tan δ plots. Figure 6.7, the DMA trace for a polychloroprene rubber, shows the extensive loss of modulus as the rubber moves from the glassy region (−100–50°C), through its glass transition (tan δ peak at −30°C), to the rubbery plateau (0–250°C). Figure 6.8 shows a DMA trace for a semi-crystalline plastic (LLDPE) and illustrates how the glass transition can be determined (tan δ peak at −113°C) and how the modulus (and tan δ) varies as the melting temperature is approached. The loss of data after 120°C is due to the sample melting in the grips of the instrument.

DMA can be used for determining a large range of material properties and in several characterisation options. The most important of these are listed below:

1. Identification of polymers in isolation and in blends, and obtaining a measure of the degree of mixing in blends.

[‡] Rubbers: +250°C; plastics: +100 to +400°C, depending upon crystalline melting point.
[§] Rubbers: 50 μm; plastics: 20 μm.

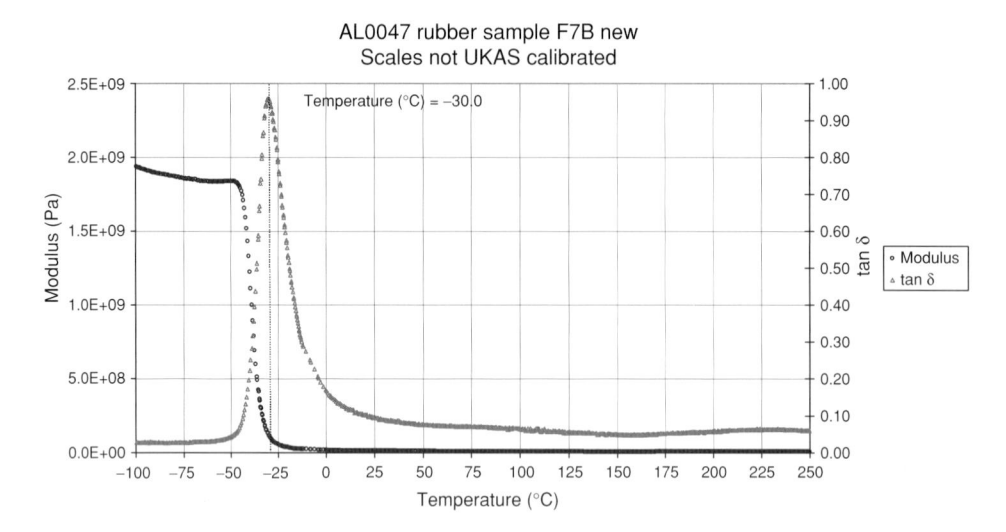

Figure 6.7 Typical DMA trace for a rubber sample (a polychloroprene rubber). The modulus shows a significant reduction at T_g, whilst tan δ shows a peak.

2. Determination of the modulus of a polymer at a given temperature.
3. Cure studies and investigations in the degree of cross-linking.
4. Evaluation of long-term creep behaviour.
5. High-temperature performance and thermal stability studies.

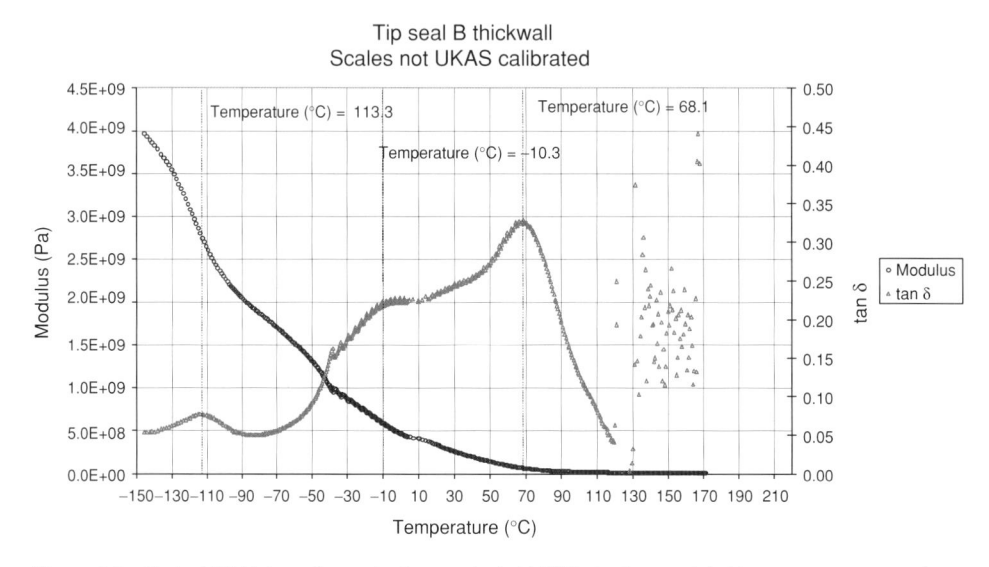

Figure 6.8 Typical DMA trace for a plastic sample (a LLDPE plastic sample). Measurement proceeds up to the melt region.

Table 6.8 Typical polymer-related applications for DMA

Temperature range (°C)	Event	Typical application areas
−150 to +150	Glass transition	Determination of T_g of rubbers and plastics/identification of blends
		Compatibility of blends and the degree of mixing of polymers
		Effect of additives/ageing on T_g
	Changes in tan δ	Effect of polymer structure/additives on viscoelastic properties
		Effect of degree of cross-linking on viscoelastic properties
	Changes in modulus	Long-term creep studies
+150 to +250	Curing/vulcanisation	Assessment of degree of cure and studies of cure kinetics
		Evaluation of curing agents/catalysts
		Quality control
	Melting behaviour	Identification of plastics/resin modifiers
> +250	Thermal degradation	Determination of thermal stability
		Evaluation of antidegradants
	Melting behaviour	Identification of high-performance plastics

These applications are discussed in more detail in Sections 6.3.2–6.3.5.

The information above is presented in Table 6.8 in terms of the region of a DMA thermogram where the events occur.

6.3.2 Determination of polymer transitions and investigations into molecular structure

DMA is the most sensitive thermal technique for the determination of the glass transition temperature (T_g) of plastics and rubbers (see Figures 6.7 and 6.8), although it has long been recognised that the experimental method used to determine the T_g of a polymer will influence the value of the result obtained. Recently, Sevcik et al [34] have discussed the contribution of the analysis methods used to determine the T_g of commercially important rubbers and concluded that, in fact, DSC, and not DMA, gave the best agreement with published results. This greater degree of sensitivity is due to the large change in modulus that these materials undergo as they pass through their T_g's and segmental rotation takes place in the polymer backbone. This change is obvious in the rapid decline in the modulus as the material moves from a glassy to a rubbery state and a value for the T_g can be obtained by reference to the peak (often referred to as the α peak) in the tan δ (viscous modulus/elastic modulus) plot.

Its sensitivity is such that it can also detect the more minor transitions that occur in plastics and rubbers at lower temperatures. These transitions are referred to as β and γ peaks in the tan δ plot and are due to movements within the structure of the polymer

Table 6.9 Motional and structural transitions in semi-crystalline polymers

	Relaxation temperature (°C)			
		Amorphous phase		
Polymer	γ	β	α (i.e. T_g process)	Crystalline phase
Linear polyethylene	–	–	−130	+60 and +120
Branched polyethylene	nd	nd	−20	+60 and +120
Polypropylene	−253	−80	+5	+90 and +170
Polyoxymethylene	–	−100	−60	−and +180
PTFE	–	−97	+130	+320 and +400
PCTFE	−240	−20	+80	+150 and –
Polyvinylidene chloride	–	–	+15	+80 and +190
Polyvinylidene fluoride	–	−110	−40	+130 and +180
Polyamide 6,6	−130	−60	+75	−and +250
Polyamide 6	−125	−45	+65	−and +225
PET	–	−60	+90	−and −

–, an event is present but no reliable data are available; nd, no event detected.
All the data were recorded using a measurement frequency of 1 Hz.

backbone that required less energy than segmental rotation. For example, in the case of polystyrene, the β transition is due to the rotation of the pendant phenyl group relative to the chain and the lower energy γ transition is due to a 'wagging' movement of the same group in and out of the plane of the main chain. Because they require less energy than the T_g (or α transition) these minor transitions always occur at a lower temperature and they are less intense due to the fact that the molecule is still effectively in the glassy state and so they have less effect on the moduli. As demonstrated above, these additional peaks in the tan δ plot are understood for well-studied polymers such as polystyrene, but they have not been elucidated for all commercially available polymers. This means that it is not possible to obtain as much structural information from a DMA experiment as it is from the use of a spectroscopic technique such as NMR. Although a complete understanding of all of the transitions in the tan δ peak has not been obtained, it is still possible to use DMA as a fingerprinting quality control tool to differentiate between different polymers, for example between low-density polyethylene (LDPE) and high-density polyethylene (HDPE). The complex differences observed between these polymers are due to the branching, of differing chain lengths, present in the low-density version. Table 6.9 above lists some of the transitions that have been identified for a number of commercially important polymers.

DMA was used by Khonakdar et al [35] to study miscibility and the α, β and γ transitions in blends of ethylene-vinyl acetate (EVA) with LDPE and HDPE. The results showed that the presence of the EVA reduced the temperature of all three types of transition. Increasing the level of the EVA was also found to broaden the peak of the tan δ and this was thought to be due to a reduction in the level of crystallinity of the materials.

The sensitivity of DMA to the glass transition temperature of polymers can be exploited for a number of different applications:

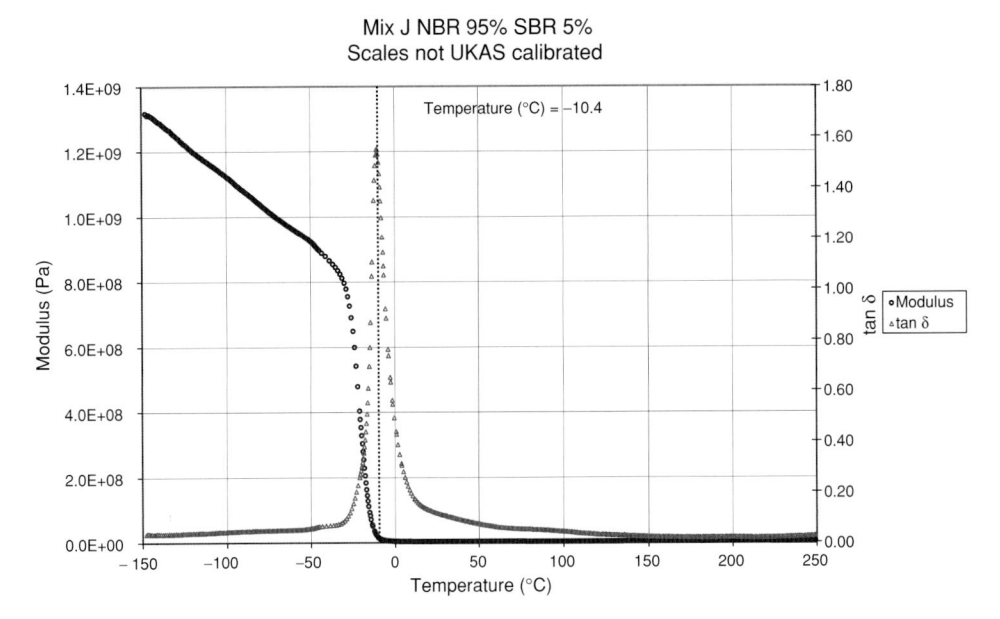

Figure 6.9 DMA trace showing the presence of 5% SBR within an NBR matrix; see the small peak in tan δ just above −50°C.

1. The identification of two polymers in a blend, even when one of the components is present at a low, e.g. 5%, level.
2. The determination of the degree of mixing of two polymers in a blend.
3. Differentiation between copolymers having different monomer ratios, e.g. styrene–butadiene rubber and styrene–butadiene thermoplastic resin.
4. Effect of additives (e.g. plasticiser and fillers) on the softening point of polymers.

The first of these applications is illustrated in Figures 6.9 and 6.10 where low levels of SBR rubber (5% in the case of Figure 6.9 and 15% in the case of Figure 6.10) have been detected within a nitrile rubber (NBR) matrix due to the presence of the T_g for the SBR component in the tan δ trace at around −37°C. The main peak in the tan δ trace (at around −10°C) is due to the T_g of the NBR component.

Rana et al [36] have used DMA to carry out investigations into the viscoelastic behaviour, plastic deformation behaviour, molecular orientation, and the orientation relaxation behaviour of an alternating styrene–methylmethacrylate copolymer. The experiments were carried out at a number of isothermal temperatures, covering the glassy state, the T_g region and the rubbery plateau; with a range of frequencies used for each temperature. The data obtained were compared to those produced by a random copolymer of styrene and methylmethacrylate, and were interpreted in relation to the role of polar–polar intermolecular interactions between the methylmethacrylate units in the two polymers.

Mina et al [37] used DMA in conjunction with other analytical techniques to relate the micromechanical behaviour of rubber-toughened polymethylmethacrylate with its glass

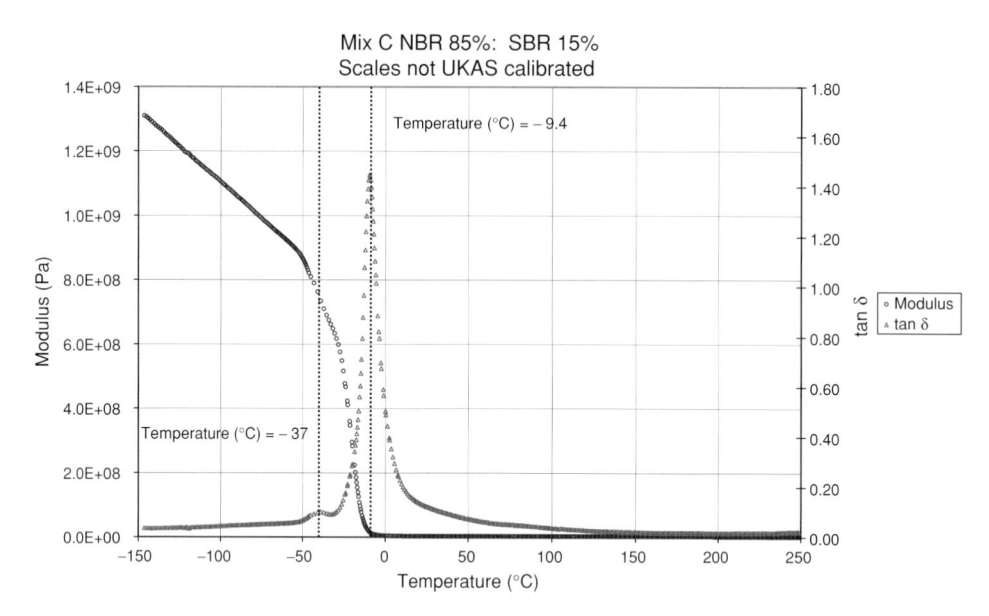

Figure 6.10 DMA trace showing the presence of 15% SBR within an NBR matrix; see the small peak in tan δ just above −50°C.

transition temperature. The plastic matrix had been toughened by the incorporation of rubber-coated core shell particles (CSPs), but the effect of the presence of these was to reduce the hardness of the material, e.g. by 40% of the initial value at a loading of 35% by volume. The results obtained showed that, contrary to expectations, the hardness was shown to decrease with increasing glass transition temperature.

6.3.3 Characterisation of curing and cure state studies

DMA is an effective tool for investigating the cure of rubbers, because the cure state of a sample has an effect on both the tan δ plot (reducing its height) and the final modulus value of the compound (increasing its value). It is, however, less sensitive than DSC (see Section 6.4.5), which measures the exotherm that results from a curing reaction.

The ability to use the height of the tan δ peak originates from the fact that the reduction in storage modulus as the rubber passes through its T_g is less severe the higher the cross-link density, and hence hardness, of the rubber. The reduction in the height of the tan δ peak is dependent on the type of rubber that is being studied; for example, the difference between a fully cured and unvulcanised natural rubber sample is 15%, whereas the difference for synthetic polyisoprene rubber is much higher at 76% [38]. By the use of standard compounds, it is therefore possible to use the height of the tan δ peak to gauge the state of cure of an unknown.

Schubnell and Schawe [39] used DMA to investigate the effect that differing degrees of cross-linking had on the viscoelastic properties of SBR rubbers that had been cured using different levels of sulphur. A relationship between the degree of cure and the loss factor at a characteristic frequency was established for compounds that had low cross-link densities. Vijayabasker and Bhowmick [40] used DMA to study nitrile rubber vulcanisates that had been cured using different levels of sulphur and then irradiated using an electron beam. The results showed that there were significant changes in the tan δ peak temperature and the storage modulus after irradiation. The vulcanisates containing a higher level of sulphur formed intense cross-link networks and cross-link rearrangements leading to an increase in the storage modulus and a higher tan δ peak temperature. An increase in the height of the tan δ peak due to the occurrence of chain scission and subsequent plasticisation was also recorded.

In the case of very highly cross-linked polymers (e.g. epoxy resins and phenol–formaldehyde resins) it is possible to observe an increase in the glass transition temperature of the polymer matrix as cross-linking occurs by DMA. This effect, however, is rarely apparent with conventional rubbers as the actual degree of cross-linking is relatively small; there still remaining long segments, upwards of 5000 units of molecular weight in size, of polymer molecule between each cross-link. This is insufficient to have a significant influence (i.e. a restriction) on segmental rotation and so the temperature at which this energy is sufficient for this event to occur is essentially unaltered. The one exception to this rule is ebonite (mainly derived from natural rubber) where very high loadings of curative (usually sulphur) are used to achieve a very highly cross-linked matrix. This material retains some rubbery properties, but is essentially a chemically resistant thermoset used in tank linings for its good chemical resistance, the majority of the double bonds in the natural rubber having reacted with the sulphur atoms.

With respect to the modulus, the degree of change moving from the uncured to cured state will be influenced by the composition of the rubber compound; for example the amount of filler that is present, as this will have a significant influence on the modulus in itself. For a highly filled compound, the difference in modulus between a fully cured and uncured sample will be less than the difference observed for an unfilled rubber. These factors underline the fact that the data produced by DMA for this, and a number of other applications, are best used in a comparative way.

This section has so far been concerned with rubbers. Though thermoplastics (as the name suggests) are not inherently three-dimensional cross-linked systems, there are some applications for which some degree of cross-linking is imparted to a thermoplastic system. A classic example of this is the partial cross-linking of polyethylene cable covering compounds. This is done by the use of peroxides and peroxide co-agents and the cross-linking processes that take place in this system, and any other thermoplastic example can be monitored and characterised in the same way as described here for rubber-based compounds.

In addition to evaluating the degree of cross-linking in finished products, it is also possible to use DMA to directly monitor a curing process. This is possible because storage and loss modulus data can be used to calculate viscosity. Although this type of study is usually achieved with parallel plate methods, which are better suited for polymers in the form of powders and pastes, materials that have an adequate level of reinforcement (e.g. due to the presence of fillers or fibres) can be self-supporting enough to be analysed using standard, solid sample, techniques. The key measurements that are useful for these types of applications are the shear storage modulus (G'), the shear loss modulus (G'') and the

Table 6.10 Glass transition temperatures for a selection of rubbers

Rubber type	Glass transition temperature (°C)
Polybutadiene	−100
Polyisoprene	−70
Styrene–butadiene (~25% styrene content)	−60
Polychloroprene	−45
Acrylonitrile–butadiene (~40% ACN content)	−20

complex viscosity – the complex shear modulus divided by the phase angle. Tan δ can also be useful as the gellation point of a polymer product is where $G' = G''$ and so tan δ is equal to 1. This type of approach is useful for characterising the cure of rubbers that originate from liquid precursors, e.g. some silicone- and polysulphide-based products. It is also possible to use DMA to monitor the cure of products which are very important in rubber technology, such as the agents used in rubber to metal bonding applications. Such a study has been carried out by Persson, Goude and Olsson [41] using proprietary bonding products and an NR/polybutadiene rubber blend. They concluded that DMA was a promising technique for increasing the available knowledge of current bonding agents and for their optimisation through development work in the future.

6.3.4 Characterisation of polymer blends and the effect of additives on physical properties

Mention has already been made in Section 6.3.2 of how the high sensitivity of DMA to the glass transition of polymers can be used to characterise polymer blends and investigate the effect of additives on physical properties.

DMA has been found to be particularly useful in studying blends of rubbers, particularly those that are structurally similar, and so are difficult to characterise by the use of other techniques such as infrared spectroscopy. A good example of this is the diene rubbers (polybutadiene, polyisoprene, styrene–butadiene rubber, etc.), which are often blended together to produce a wide range of commonly encountered products, e.g. car and truck tyres, seals and gaskets. The difference in the glass transition temperatures of these rubbers (see Table 6.10 below) is sufficient for them to be resolved from each other in a DMA experiment (see Figures 6.9 and 6.10).

This not only enables the presence of these rubbers to be identified in a blend but, by the use of standard samples of known blend composition, it is also possible to estimate the relative amounts of the two rubbers; although if an accurate quantification of these rubbers is required it is necessary to use nuclear magnetic resonance spectroscopy. Because rubbers are amorphous materials, there is a large drop in modulus as they pass through their glass transition and so DMA is more sensitive than techniques such as infrared spectroscopy. Although it depends upon the rubbers present in the blend, levels as low as 5% by weight of one rubber can be detected (see Figure 6.9). A large difference in the T_g of the two rubbers is necessary to achieve this level of detection.

Pandey et al [42] used DMA in conjunction with a number of other analytical techniques (e.g. DSC and AFM) to assess the compatibility of blends of NBR and EPDM. Data were obtained using various blend ratios, with some of these containing either chlorinated polyethylene or chlorosulphonated polyethylene as a compatibiliser. The viscoelastic properties of oil-extended polybutadiene/EPDM blends containing various levels of carbon black filler as a function of strain amplitude have been investigated by Ibarra-Gomez et al [43] using DMA. Among other things, they found that the storage modulus was dependent on strain amplitude, as described by the Payne effect, and that the linear viscoelastic region was dependent on the level of carbon black. The blends also exhibited high hysteresis above the percolation threshold.

In addition to the diene rubbers, it is also possible to use DMA to identify blends of the other commercially available rubbers, provided that there is a sufficient difference in their glass transition temperature, and to characterise their properties. For example, Carlberg, Colombini and Maurer [44] mixed ethylene-propylene rubber and silicone rubber in a number of blend ratios and studied their morphology and viscoelastic properties. The results obtained experimentally by DMA and DSC were compared to theoretical data produced from self-consistent models, both indicating that the silicone rubber was the dispersed phase in a continuum of ethylene-propylene rubber.

It is also possible to use DMA to detect the presence of rubbers that are used as toughening agents within plastic matrixes. A good example of this is the detection of polybutadiene within polystyrene in high impact polystyrene products. Although it is regarded as a plastic, styrene–acrylonitrile–butadiene (SAN) is in fact comprised of two phases: polybutadiene and styrene–acrylonitrile. The two phases are grafted (i.e. chemically combined) together via covalent bonds during the manufacturing process and so technically are not blends as such, but the relative amounts of the two can be investigated via the size of their respective glass transition tan δ peaks at $-100°C$ and $+120°C$, respectively. Zhang, Li and Huang [45] have used DMA in conjunction with a number of physical property tests, to investigate the toughening mechanism of polypropylene filled with both rubber particles and calcium carbonate particles. The effects of the phase concentrations, the particle size of the toughening agents and the nature of the rubber were all studied.

Figure 6.11 shows the DMA trace for 50:50 mixture of butyl rubber and LDPE. The T_g for the butyl rubber can be clearly seen at $-55.8°C$ (in the tan δ trace), and in the region over which the storage modulus of the sample has reduced. The melting of the LDPE phase is best seen in the tan δ trace. The sample remained in the grips of the instrument due to the influence of the butyl rubber phase, which was still within its rubbery plateau region at this temperature.

Shivakumar et al [46] used DMA to investigate the compatibility of ternary blends of acrylic rubber, PET and a liquid crystalline polymer (LCP). The work concentrated on evaluating the effect that the LCP content of the mixes had on the compatibility of the acrylic rubber and the PET, and the crystallinity and thermal properties of the resulting products.

DMA can also be used effectively to characterise blends of plastics, although in this case there is an important distinction between amorphous and crystalline plastics. For a number of crystalline plastics, the T_g is well below the crystalline melting point and the presence of the supporting influence of the crystalline matrix has a significant effect in reducing the fall in modulus as the material passes through the T_g. This means that the T_g for these materials

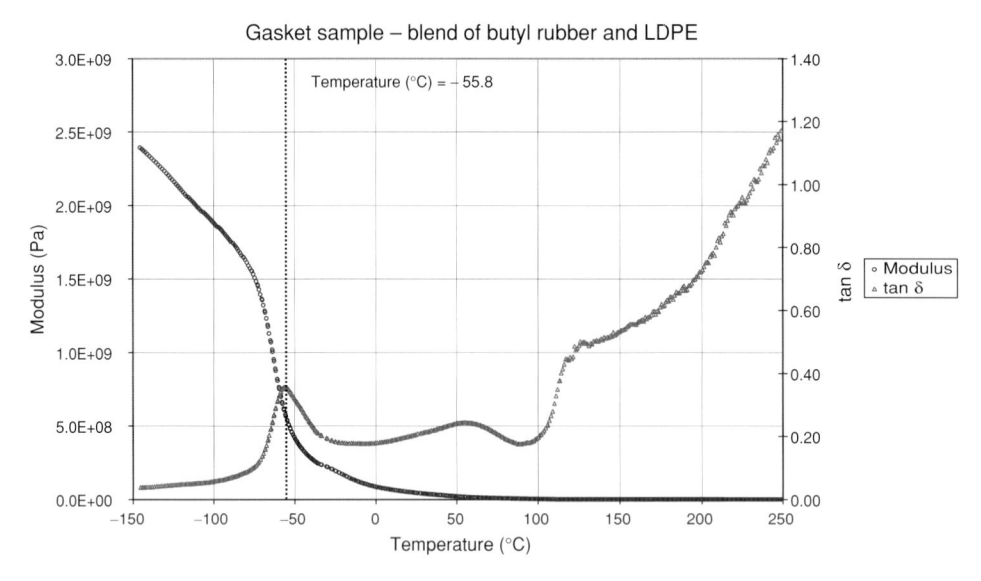

Figure 6.11 DMA trace of a 50:50 blend of butyl rubber and LDPE. The butyl rubber supports the material through the melting range of the LDPE.

can be difficult to detect and so is not as diagnostic as it is for rubbers. It is also the case that a T_g in a material due to the presence of an amorphous plastic blended with a crystalline one will also be reduced in intensity and hence harder to detect.

The loss of modulus that occurs when a crystalline plastic reaches its melting point means that this can sometimes be used to detect and identify a blend of two crystalline plastics – the one with the lowest melting point producing a step change in modulus, but the material will not become completely molten due to the supporting effect of the plastic that is still below its T_m.

The DMA trace (containing both modulus and tan δ plots) for a car mat is shown below in Figure 6.12. This product was a blend of a thermoplastic block copolymer rubber (SBS) and polypropylene. The T_g for the butadiene phase within the SBS is apparent at −68°C. The transitions above this temperature are complex, being due to a combination of the contributions from the T_g of the styrene phase within the SBS and the T_g of the polypropylene. It was not possible to collect data above 150°C due to the sample melting as the melting point of the polypropylene component in the sample was reached.

DMA has been used to obtain an empirical measure of the degree of blending of two polymers in a material. For the reasons described above, this type of investigation works best in the case of two amorphous polymers (rubbers or plastics) where there is a reasonable degree of separation in their T_g's. In the case where two polymers are mixed but there is a considerable degree of separation between the two in the matrix (i.e. two different phases exist), two distinct T_g's for the material will result and these will have values consistent for the original polymers in the unblended state. As the degree of mixing of the two polymers improves, the tan δ peaks broaden, due to the influence of the two polymers on each other

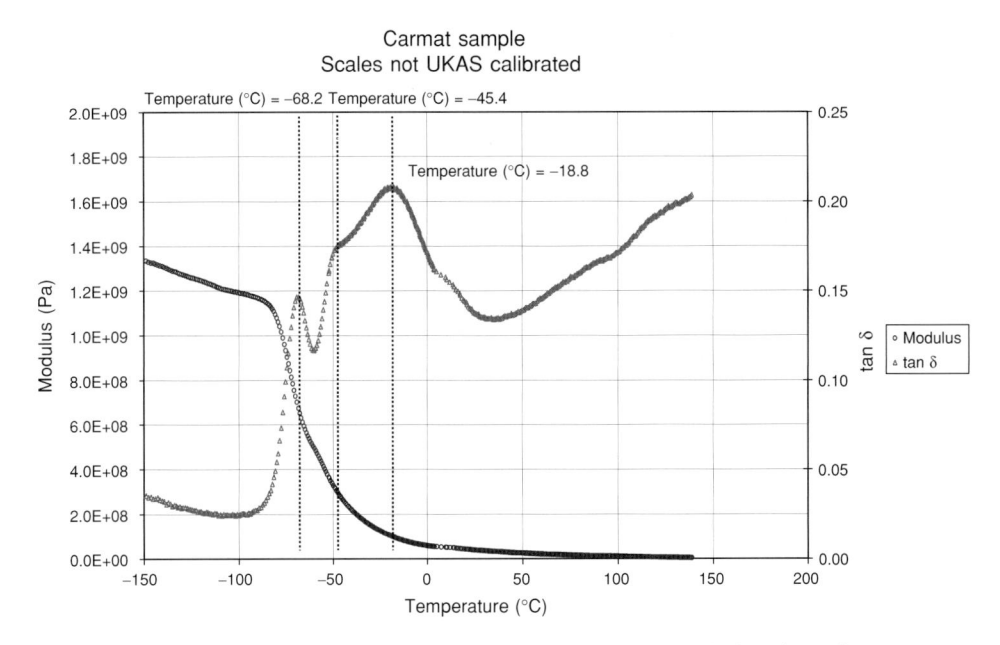

Figure 6.12 DMA trace for a car mat product. Transitions can be seen from the butadiene phase at −68°C and from the styrene and polypropylene components at higher temperatures.

and the 'bridge' in the tan δ plot between the two rises. In the ideal situation where very intimate mixing occurs (i.e. mixing of the two polymers at the molecular level) the material will only have one broad T_g which will be a rough average of the two T_g's of the original polymers. This type of study is often of value for, in order for polymer blends to have the desired physical properties, a certain degree of mixing is required. As usual, standards which display the required degree of mixing are required in order to provide useful comparative data. This type of study has been carried out by McConnell et al [47]. They produced blends of different grades of PVC with polyether and polyester types of polyurethane (PU). The results obtained with DMA showed that the two polymer types were partially miscible and, overall, the impact performance of the blends was found to be more dependent on the PVC/PU ratio than the type of PVC or PU used.

With respect to studying the effect of additives, it is a well-established fact that low molecular weight liquids that have a high degree of compatibility with a particular polymer become intimately mixed within the polymer matrix due to the effects of intermolecular forces such as van der Waals forces and hydrogen bonding. This intimate mixing means that they come between the polymer molecules, effectively creating additional free volume and hence reducing the energy required for segmental rotation to occur and therefore the glass temperature of the whole sample.

DMA can, therefore, be used to follow and monitor the effects of this plasticisation. Practical examples that can be studied include the effect of adding various amounts of dioctyl phthalate plasticiser to PVC and the absorption of water by polyamides. In addition to reducing the α peak on the tan δ trace, the values of the other transitions, e.g. β transition,

Table 6.11 Use of DMA to illustrate the plasticisation of polyamide 6,6 by the absorption of water

Sample	T_g (α transition)	β transition
Dried at 120°C for 3 h	80°C	−42°C
Boiled in distilled water for 2 h	7°C	−63°C

are also reduced. To show the data that can be obtained, the results obtained on polyamide 6,6 prepared in two different ways are given in Table 6.11.

The effect of water absorption on the viscoelastic properties of glass fibre reinforced polyester and polyvinyl ester composites has been investigated using DMA by Fraga et al [48]. These workers contrasted the water absorption behaviour of the pure, unfilled polymers with composites at two temperatures (40°C and 80°C) and evaluated the relative degree of hydrothermal degradation that occurred.

DMA has also been used to investigate the effect of reinforcing fillers in rubbers. Lopez et al [49] used the technique in a study involving the use of mesoporous silica, with and without a coupling agent, in SBR rubber. The degree of chemical interaction between the filler and the rubber was evaluated, and the results obtained compared to those obtained using a commercial precipitated silica filler.

The potential and limits of DMA for the evaluation of the fracture resistance of isotactic polypropylenes, random copolymers of isotactic polypropylene with ethylene, and isotactic polypropylene/ethylene-propylene rubber blends have been evaluated by Grein, Bernreitner and Gahleitner [50]. They showed that a minimum amount of information about the samples is required in order to interpret the DMA traces correctly. For example, in the case of the isotactic polypropylene/ethylene-propylene rubber blends, the rubber relaxation was affected by both the molecular weight and the amount of the rubber and, for isotactic polypropylene, the catalyst system had an effect. In general, the results showed that DMA is a powerful tool for ranking materials in terms of toughness, but that care should be taken with nucleated materials.

Nanofillers are increasingly being used as high-performance additives for plastics and Finegan et al [51] have used DMA to qualitatively and quantitatively study the fibre–matrix adhesion of various carbon nanofibres that had been subjected to several different types of surface treatments in a polypropylene matrix. Lee, Hsieh and McKinley [52] have characterised a nanocomposite of poly(methylmethacrylate) containing 1–7.5% of modified montmorillonite clay. They established a number of relationships, which included that the storage and loss moduli were strongly dependent upon the clay content, and that the presence of the clay increased the distortion temperature of the plastic.

6.3.5 Ageing, degradation and creep studies

In the majority of cases, polymers in service will encounter both heat and oxygen. It is the norm therefore that these materials will undergo oxidative thermal decomposition and

DMA is an ideal technique to monitor the important changes that occur during this process. In order to provide predictive data on these properties, a number of isothermal experiments are performed and the change in the modulus of the sample determined with respect to time. Time–temperature superposition is then used to predict long-term performance (i.e. oxidation, creep, cross-linking, etc.) at a given temperature. Examples of this type of study include those carried out by Patel, Morrell and Evans [53] on silica-filled, cross-linked silicone rubbers, and by Patel [54] on polystyrene samples of differing molecular weight. They found that the underlying degradation chemistry appeared to be more temperature dependent at higher temperatures, to be independent of the applied strain, and also independent of the atmosphere type, with experiments conducted in air producing similar results to those conducted in nitrogen.

The chemical changes that can occur in a rubber or plastic when it undergoes oxidative thermal degradation can vary according to the particular polymer concerned. There are two principal mechanisms involved:

(a) Chain scission – leading to a reduction in molecular weight, and hence a loss in modulus, and a reduction in T_g due to 'self plasticisation' effects.
(b) Cross-linking – resulting in the development of a three-dimensional structure with an increase in both modulus and T_g due to an elimination of free volume and inhibition of segmental rotation.

In reality, both of these mechanisms are usually present, but one will predominate over the other and this is determined by the chemical structure of the polymer backbone and the availability of hydrogen (or other reactive species) for the free radicals to react with. Once a free radical has formed on the polymer molecule by hydrogen abstraction (a typical route in oxidative degradation), it becomes stable either by a rearrangement reaction (often leading to chain scission) or by reacting with another radical (cross-link formation).

Increases in cross-linking will reduce the impact resistance of a plastic due to the loss of mechanical damping properties and this will be apparent from a reduction in the height of the tan δ peak, or another of the important sub-ambient transition (β or γ) peaks. Since rubbers are usually cross-linked in the first place, ageing of this sort will increase the effective cross-link density (although the chemical nature of these additional cross-links will usually be different to those that were originally present), again reducing the damping properties of the material.

The effect of ageing on the modulus of a rubber component, in this case a polychloroprene rubber impeller used in a food production line, is illustrated by the change in modulus shown in Figure 6.13 (new sample), Figure 6.14 (450-h service) and Figure 6.15 (750-h service). Due to the relatively low service temperature (ambient), the T_g of the three samples (see the peak in the tan δ plot) has not been affected, but the modulus of the component has been, with the following modulus values at 25°C (in Pa $\times$ 10^{-7}): new sample 1.5, 450-h sample 2.0 and 750-h sample 2.6. This increase in modulus due to ageing is typical for both rubber and thermoplastic products and reflects the chemical changes that are taking place within the polymer matrix due to a combination of both the physical stresses and the heat build-up in service.

In amorphous plastics, ageing will be apparent by a narrowing of the tan δ peak associated with the T_g of the material. Chemical ageing has been discussed above, but physical ageing,

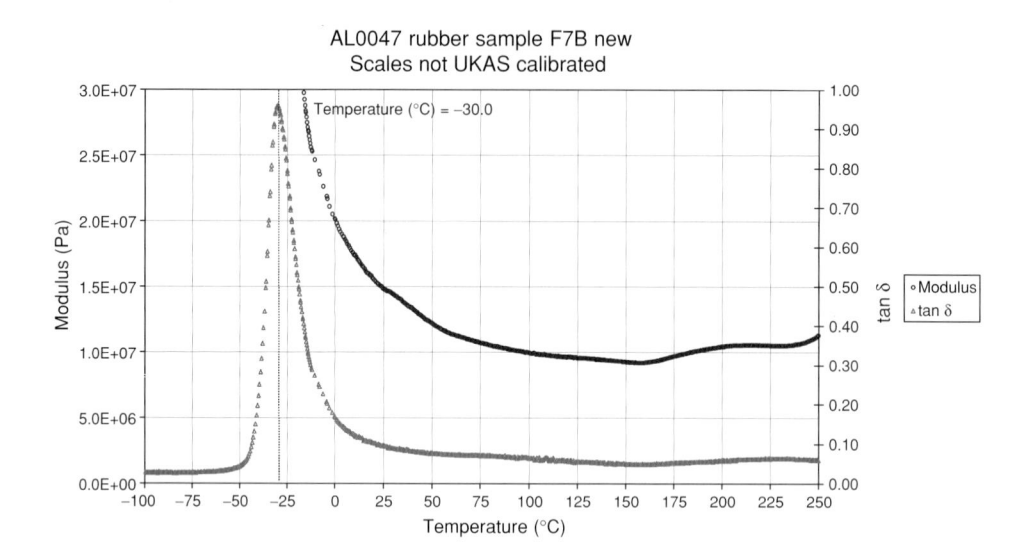

Figure 6.13 DMA trace for the new rubber impeller. Note the modulus trace, particularly the value at 25°C.

in which molecular rearrangements within the glassy region of these materials occur, is also possible and this affects the relaxation time, and hence a sample that has undergone physical ageing will have a higher storage modulus than an unaged specimen. This effect reduces as the temperature is increased and the sample approaches its T_g. The distance from the T_g

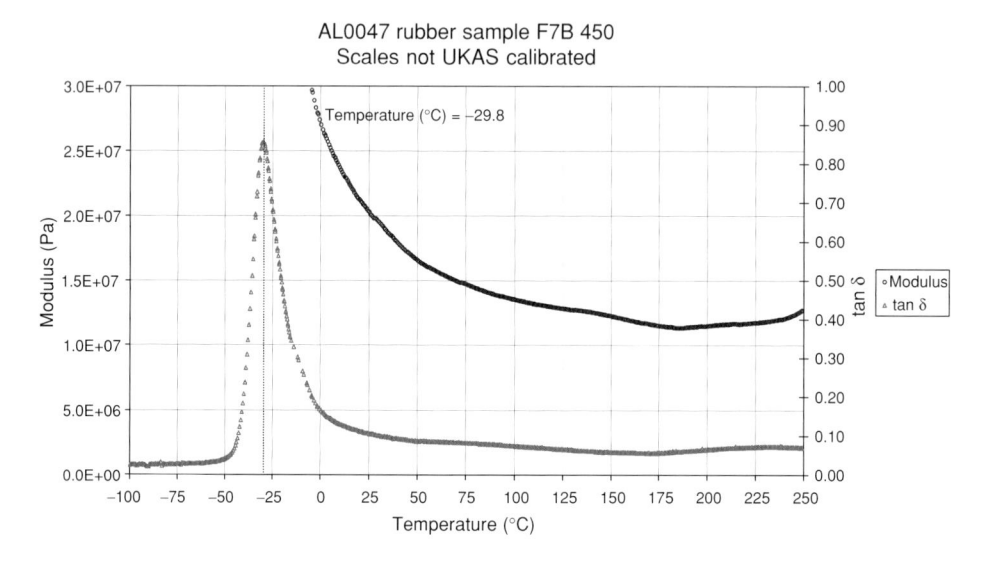

Figure 6.14 DMA trace for the rubber impeller which had seen 450 h of service. Note the modulus value at 25°C, which has increased with length of service.

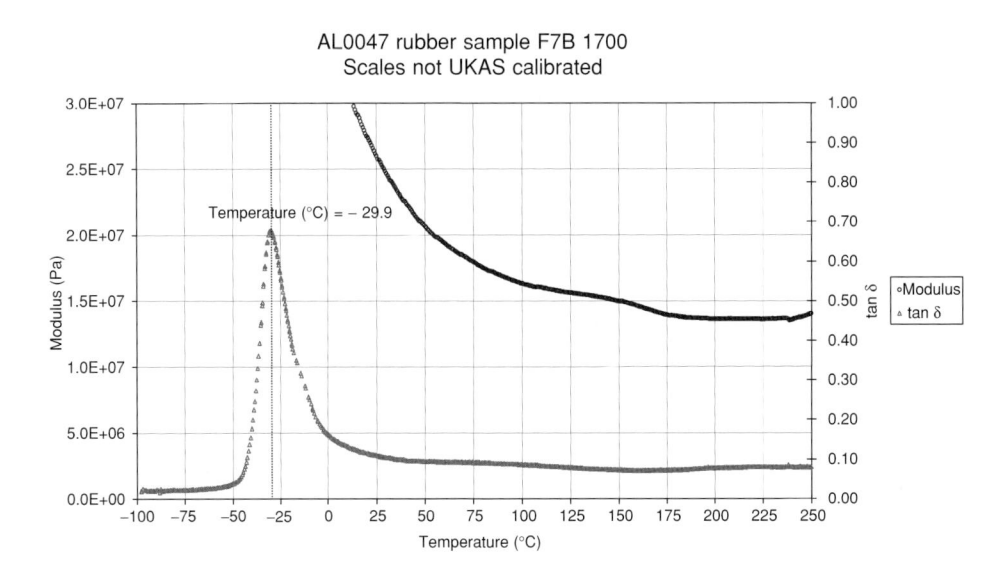

Figure 6.15 DMA trace for the rubber impeller which had seen 750 h of service. Note the modulus value at 25°C, which has increased with length of service.

that the aged and unaged curves cross over is an empirical measure of the degree of physical ageing that a sample has undergone – the higher the temperature the greater the degree of ageing.

It can also be possible for a cross-linked rubber to undergo ageing that involves attack by a chemical agent (i.e. a type of chemical ageing) at either ambient or elevated temperature. This is often partially or completely anaerobic as it is immersed in the attacking medium. This type of attack can destroy the cross-link structure and result in a reduction in T_g and, in the early stages, increase the damping properties of the material.

How DMA can be used in the characterisation of blends has already been mentioned in Section 6.3.4. It is a possibility that the thermal ageing of some partially miscible blends (e.g. PBT and PC) can result in an increase in the separation of the two phases due to changes to the interface. This is apparent by the widening of the difference between the two T_g's for the materials and continued ageing will result in T_g's returning to the value of the original polymers [55].

6.3.6 Thermal mechanical analysis

TMA is a method for measuring linear or volumetric changes in polymers as a function of temperature, time or force and is often used in conjunction with DSC to investigate the structure and properties of polymers. This combination is often found in commercial microthermal analysers (see Section 6.4.9). While the DSC is concerned with the energetics of physical and chemical changes, TMA measures the dimensional effects associated with these changes and so it can be used to obtain the coefficient of thermal expansion for polymer samples.

TMA can also measure properties such as T_g and T_m, and determine the modulus of polymers and is similar to DMA in this regard.

6.4 Differential scanning calorimetry

6.4.1 Background

DSC measures the heat flow into or out of a material as a function of time or temperature. What is being measured during a DSC experiment on a polymer sample is changes to its heat capacity (C_p) as a result of either endothermic or exothermic events. Examples of events that bring about the two types of changes are shown below:

Endothermic changes:

(a) Melting of a crystalline structure within the matrix.
(b) Loss of a low molecular weight substance (e.g. water or a solvent) from the sample due to its volatility.
(c) Stress relaxation.

Exothermic changes:

(a) Formation of a crystalline structure within the sample.
(b) Formation of cross-links due to a chemical reaction.
(c) Decomposition of a sample due to anaerobic or oxidation degradation.
(d) Decomposition of a sample due to chemical degradation.

In addition to these events, DSC can also be used to determine the T_g of rubbers and plastics as these polymers will undergo a step change in their heat capacity as they pass through their T_g. Unfortunately though, the T_g is not always easy to detect for a number of polymers, particularly semi-crystalline plastics and, in general, DSC is estimated as being up to 1000 times less sensitive than DMA for this purpose (see Table 6.14) as it measures different parameters. The HyperDSC™ technique has been reported as offering significant improvements in sensitivity for the determination of T_g (see Section 6.4.8).

A summary of the polymer-related applications that DSC can be used for is given below:

1. Identification of crystalline polymers and polymer blends by their melting points.
2. Semi-quantitative analysis of crystalline polymer blends by the areas of their melting point endotherms and reference to standard materials.
3. Identification of polymers by their T_g.
4. Investigations into factors affecting both T_g and T_m (e.g. molecular weight, additives and thermal history).
5. Thermal stability testing.
6. Cure studies on rubber compounds and cross-linking studies on thermoplastics.
7. Crystallisation kinetic studies on crystalline plastics.

Table 6.12 Typical polymer-related applications for DSC

Temperature range (°C)	Event	Typical application areas
−150 to +25	Glass transition	Determination of T_g
		Blend compatibility
		Phase heterogenicity of copolymers
		Analysis of copolymers
	Crystallisation	Nature of crystallinity
+25 to +150	Melting of additives	Analysis and purity of raw materials
+150 to +250	Curing/vulcanisation	Characterisation of curing kinetics
		Analysis of curatives
		Quality control
> +250	Thermal degradation	Identification of polymers
		Determination of thermal stability
		Evaluation of antidegradants
	Oxidative degradation	Determination of oxidative stability
		Evaluation of antioxidants

Semi-crystalline polymers melt over a large temperature range (e.g. 50–400°C) and so the use of DSC to study this event covers the majority of these regions. Some typical melting points are shown in Table 6.13.

The information above is presented in Table 6.12 in terms of the region of a DSC thermogram where the event occurs.

It is possible to carry out a DSC experiment on rubbers and plastics under both isothermal and dynamic heating rate conditions. The former is used extensively for stability studies, where the time for the onset of the exotherm due to degradation is determined, and the latter for compositional, polymer identification and quality control purposes. It is also possible to use both oxidising (for oxidation stability tests) and inert atmospheres, e.g. nitrogen. An inert atmosphere is the preferred choice for compositional work as it helps ensure that the data obtained are consistent and, hence, representative of the material. In common with TGA, relatively small samples sizes (5–10 mg) are used and semi-open aluminium pans are the norm, with sealed pans capable of holding the internal pressures only being used for curing studies (e.g. on rubbers), where loss of volatiles from the sample would create an endothermic response that would interfere with the curing exotherm.

A typical two-stage, dynamic heating programme used with a nitrogen atmosphere to investigate the composition of a rubber or plastic is a follows:

1. Two minutes at −100°C.
2. Heating from −100 to +250°C at 20°C/min.
3. Isothermal at +250°C for 2 min.
4. Cool from +250 to −100°C at 20°C/min.
5. Isothermal at −100°C for 2 min.
6. Heat from −100 to +250°C at 20°C/min.

Note. The top temperature should be below the decomposition temperature of the polymer. TGA can be used to determine the safe temperature range for DSC work.

230 *Principles and Applications of Thermal Analysis*

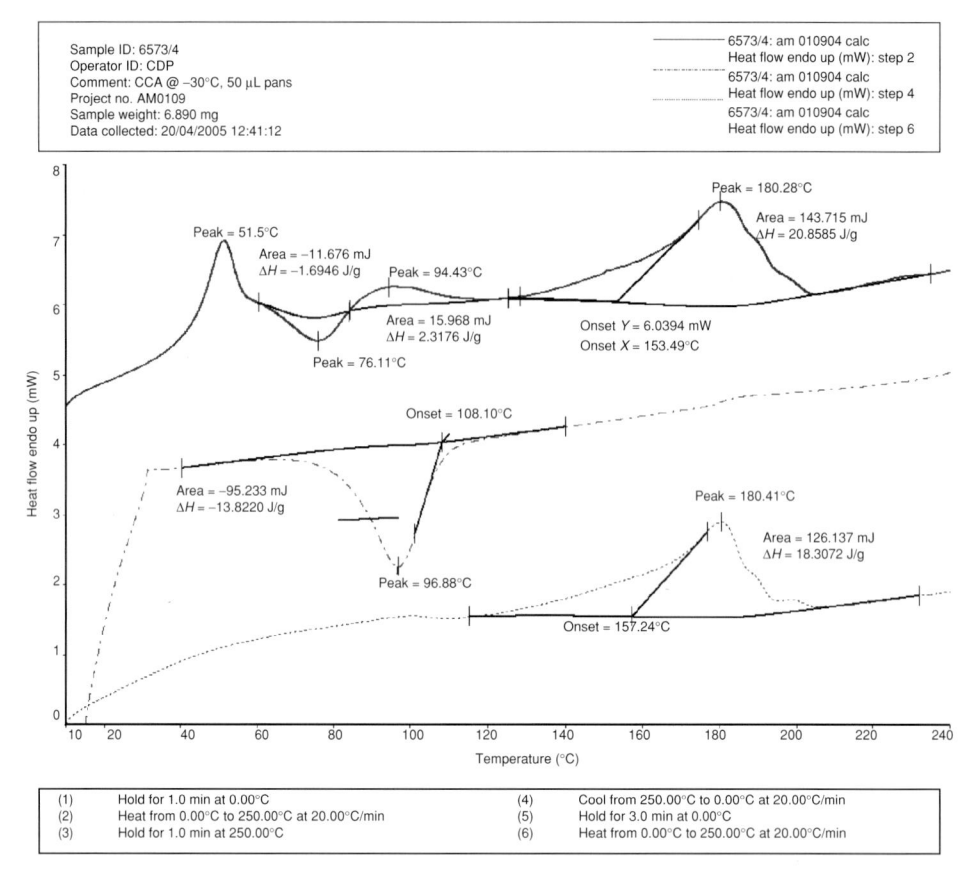

Figure 6.16 As-received, (full curve) cool down and reheat DSC traces for a thermoplastic polyurethane tube product, illustrating the effects of thermal history on the sample.

The first heating stage is referred to as the 'as-received' run and this produces a composite thermogram consisting of features that are representative of the material and those that are due to its thermal history (e.g. the processing and storage temperatures). This thermal history is removed by this initial run and the second heating experiment then gives data that only relate to the material, i.e. it is sample specific. In addition to ensuring that all samples are analysed in a consistent way, this initial run can also be used as a diagnostic tool in problem solving etc. (see Section 6.4.4). In semi-crystalline plastics for example, if the material has been cooled too quickly during processing this can lead to embrittlement and failure in service and this 'frozen in stress' can be detected by DSC. In effect, the as-received and reheat thermograms look quite different due to the amount of reorientation that the crystalline region undergoes once it is above its melting point. This is illustrated in Figure 6.16, which shows the as-received (top trace), cool down (middle trace) and reheat (bottom trace) DSC runs for a thermoplastic PU tubing product. It can be seen that there is a significant amount of exothermic and endothermic activity in the early part of the as-received trace due to the thermal history that the product has experienced. Once this has been removed from the sample, the reheat trace provides data (e.g. crystalline melting point and melting range) that are representative of the product. These types of investigations are mainly qualitative in nature since most crystalline polymers show some thermal history, and the DSC data have to be compared to information produced from another test (e.g. impact strength in this instance) to enable an assessment of where good and bad samples fall.

Because it monitors the change in the heat capacity of a polymer as it is heated, DSC can be used to determine the specific heat of polymers. For example, Changwoon Nah *et al* [56] have used DSC to determine the specific heat of a number of rubber ingredients and compounds that contained them over the temperature range 45–95°C. The values obtained were compared with previously published results for NR, SBR, polybutadiene and butyl rubber. The influence of the level of carbon black and effect of vulcanisation on specific heat was also evaluated.

6.4.2 Crystallinity studies and the characterisation of polymer blends

DSC is a very useful tool for providing complementary data to techniques such as infrared spectroscopy when it comes to identifying unknown plastic samples [57]. A significant number of plastics are semi-crystalline and have characteristic melting points. Some of the most popular examples are shown below in Table 6.13. These and many other examples are given in a collection of DSC thermograms published by Rapra Technology Ltd [58].

It can be seen from Table 6.13 that there is a reasonable difference between the melting point values, and this means that DSC is very useful for identifying blends of plastics. This is particularly useful in cases such as a blend of polyamide 6 and polyamide 6,6 where techniques such as infrared spectroscopy are of little use due to the similarity of the spectra

Table 6.13 DSC melting point peak maxima for a range of commercially important polymers

Material	Melting point – peak maxima (°C)*
LDPE	110
HDPE	135
PP	160
Polyacetal homopolymer	170
Polyacetal copolymer	160
Polyamide 6	220
Polyamide 6,6	260
Polyamide 12	175
PBT	220
PET	250
Ethylene-vinyl acetate (4% VA)	105
Ethylene-vinyl acetate (18% VA)	80
PP/EPDM rubber blend	150[†]

* As determined by DSC using a heating programme similar to that shown in Section 6.4.1.
[†] A thermoplastic elastomer.

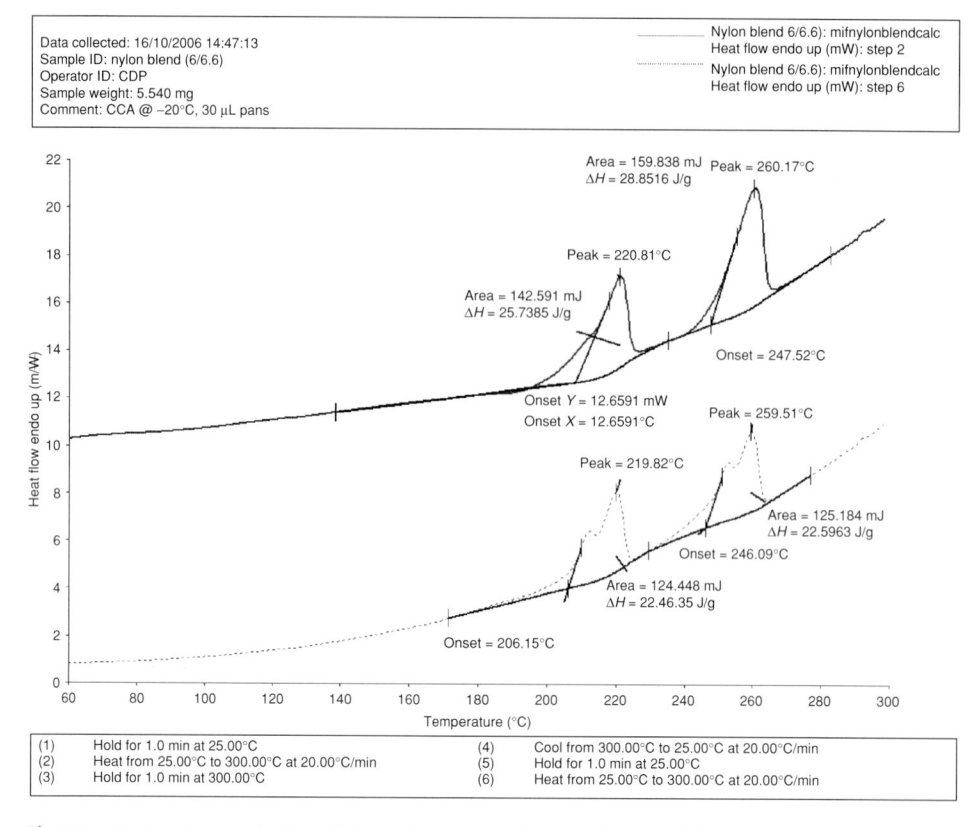

Figure 6.17 DSC trace of a 50:50 blend of polyamide 6 and polyamide 6,6.

of these two materials. An example of this type of application is shown in Figure 6.17, where a sample has been shown to be comprised of a 50:50 blend of polyamide 6 (melting point 220°C) and polyamide 6,6 (melting point 260°C). Another example where DSC is often employed is in the characterisation of blends where a small amount of polyethylene (e.g. 5%) has been added into a polypropylene material to toughen it. It is very effective at this type of investigation due to its high sensitivity to energetic changes within polymers. Because they retain the characteristic melting points of their homopolymers, DSC can also be used to identify block copolymers of ethylene and propylene.

DSC is often used to identify and approximately quantify two, or more, polymers in a blend. As shown in Figure 6.17, identification is by reference to the peak melting temperatures (i.e. those given in Table 6.13), as the presence of other crystalline polymers in the matrix does not usually have a significant effect on these, and semi-quantification data are obtained using the area under the melting endotherms and comparing this with that obtained for pure (i.e. unfilled, un-plasticised) reference materials that have been analysed under similar experimental conditions. Examples where this type of exercise has proved very useful include blends of different polyamides, blends of polypropylene with polyethylene, and blends of different types of polyethylene (i.e. LDPE, HPDE, etc.) with EVA. In this latter case, it is also possible to estimate the amount of vinyl acetate monomer in the EVA copolymer as

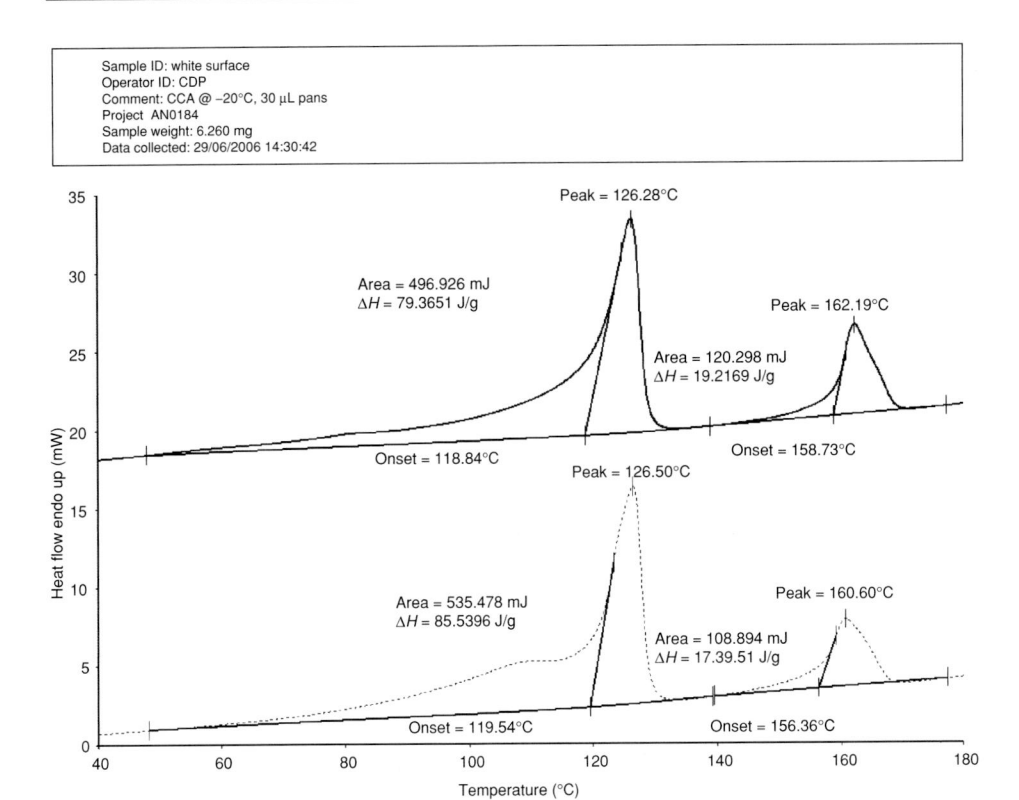

Figure 6.18 DSC trace for an 80:20 blend of polyethylene and polypropylene.

increasing amounts of this monomer successively reduce the crystallinity of the copolymer and, hence, its melting point.

The use of DSC to provide an approximation of the amount of polypropylene and polyethylene in a sample is shown in Figure 6.18. The areas for the polyethylene melting peak (at 127°C) and the polypropylene melting peak (at 160°C) show that the proportion of the two polymers is approximately polyethylene 80/polypropylene 20.

Although it is the crystalline melting temperature that is traditionally used in these types of applications, it is also possible to use the crystalline formation temperature by monitoring the heat capacity change that occurs as the sample cools. Table 6.13 shows that it is possible for two commercially important plastics (polypropylene and polyacetal) to have very similar melting peaks, but fortunately it is possible to differentiate between these materials by examining the crystalline formation peaks, which are usually different. In the case of polypropylene and polyacetal, for example, the crystallisation peaks are at 142°C and 105°C, respectively. This type of capability is particularly useful in failure diagnosis work where one polymer is suspected of having contaminated another. Once again, reference to standard data will also enable the area under the crystalline formation curve to be used to estimate the proportions of the two polymers.

Because DSC is such a sensitive tool for characterising the crystallinity of a plastic sample [59], it can also be used to monitor the effects that certain parameters have on the degree of crystallinity of a sample [60] and its crystalline melting point. One example of its use for this type of investigation, the determination of the proportions of the two monomers in a copolymer (EVA), has been given above. Other important examples are as follows:

(a) A qualitative measure of the thermal history that a sample has been subjected to.
(b) Within a series of plastics (e.g. LDPEs), the molecular weight of the polymer.
(c) Effect of additives.

With respect to the first example, the thermal history of a plastic is evident in the difference in the heat capacity trace produced by the initial experimental run given in the typical heating programme shown in Section 6.4.1 and the second, reheat run. Although thermal history will also be apparent in heat capacity changes around the T_g of a plastic (see Section 6.4.3), it often has the greatest influence on the data generated during the crystalline melting region. This is because it will have disrupted the formation of the crystalline domains within the polymer and affected both the degree of regularity (i.e. melting point) and the extent of crystallinity (i.e. the area under the melting endotherm). Sometimes this degree of change will be sufficient to affect the physical properties of the product (e.g. toughness, tensile strength) and failure in service will ensue. Because a DSC experiment can be performed using a relatively small amount of sample, it is also possible to investigate the change in thermal history through a plastic sample from the outside to the centre. If a material has been cooled too rapidly during an injection moulding operation a skin can form on the outside which will have different properties to the inner layers and so come away in service. This type of problem can also affect the extrusion of plastics where the extrudate is run through a cooling bath, and in plastic compounding operations where pelletisation occurs at the end of the process.

In the case of the second example in the list, although it cannot be used as a primary tool for determining molecular weight, in contrast to techniques such as GPC for example, DSC can be used to determine if the molecular weight of a series of samples varies to a significant degree and so it can be used as a quality control tool in this respect. Although work carried out at Rapra has not revealed any obvious trend, for example the T_m of a series of LDPE materials does not follow a pattern for decreasing molecular weight, there are obvious differences in the melting endotherms overall, i.e. in the onset temperature, the peak temperature and the area. Similar to this phenomenon is the effect that additives will have on the ability of plastics to crystallise and so DSC can also be used as a quality control tool to assess their presence and level.

An extension of the use of DSC for the above specific applications is its use as a general quality control tool and the production of 'fingerprint' thermograms. Differences between samples, such as the molecular weight of the polymer, the level of additives, both 'passive' (e.g. plasticisers and fillers) and reactive (e.g. curatives and antidegradants), will all be apparent by changes in the fingerprint obtained. Knowledge of the effect that components of the sample can have on the transitions present within the specific heat trace of the sample will indicate the nature of the differences between samples, and these can then be targeted by the

Operator ID: CDP
Comment: CCA @ −30°C, 50 μL pans
Project no. AN0056
Sample weight: 6.740 mg
Data collected: 02/03/2006 11:42:32

(1)	Hold for 1.0 min at 20.00°C	(4)	Cool from 200.00°C to 20.00°C at 20.00°C/min
(2)	Heat from 20.00°C to 200.00°C at 20.00°C/min	(5)	Hold for 3.0 min at 20.00°C
(3)	Hold for 1.0 min at 200.00°C	(6)	Heat from 20.00°C to 200.00°C at 20.00°C/min

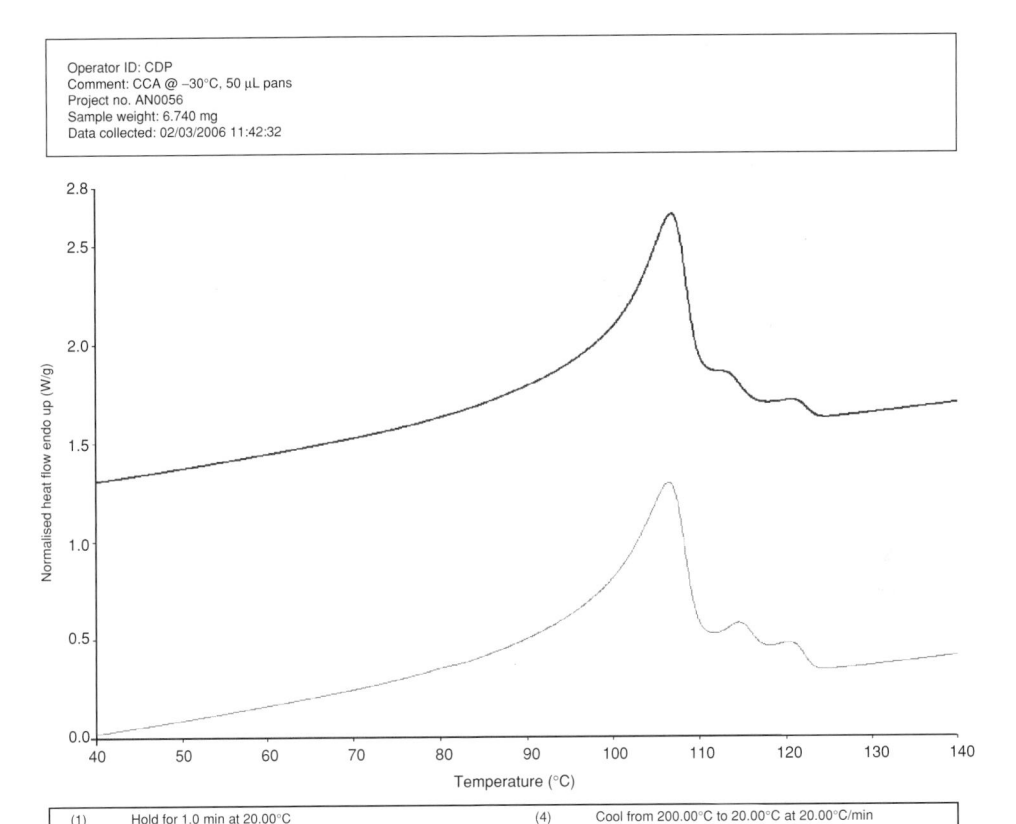

Figure 6.19 DSC trace showing the reheat traces for two LDPE products – sample A (top trace) and sample B (bottom trace). Small differences in the melting region 110–125°C can be attributed to differences in the material.

use of other analytical techniques (e.g. infrared spectroscopy, GPC, etc.). The use of DSC as a quality control tool is illustrated in Figure 6.19, which shows the reheat traces for two LDPE samples. It can be seen that there are differences in the shape of the melting endotherm in the region 110–125°C for the two samples – sample A (top trace) and sample B (bottom trace). The thermal history has been removed from both samples by the 'as-received' runs and so these differences must be related to the LDPE polymers in the samples and indicate that they are different grades having different molecular weights and/or polymer chain characteristics (e.g. degree of branching).

In addition to these product characterisation and differentiation type applications, DSC is also used extensively for fundamental studies into crystallisation kinetics. A typical example is the work published by Li *et al* on the isothermal and non-isothermal crystallisation kinetics of elastomeric polypropylene [61].

In common with a number of the other thermal techniques (e.g. DMA and DEA), DSC can be used to investigate the compatibility of polymers in a blend. Examples of the use of

Table 6.14 Relative sensitivity of a range of thermal analysis techniques to the determination of the glass transition of polymers

Technique	Measured parameter	Relative sensitivity*
DSC	Heat capacity	1
TMA	Coefficient of expansion	5
DMA	Modulus (damping modulus)	10
DEA	Permittivity (loss factor)	10

* 10 represents the most sensitive.

DSC for this type of study have been reported by Abou-Helal and El-Sabbagh [62] (EPDM and NR), Wang, Luo and Wang [63] (thermoplastic PU and aliphatic polyester), Li et al [64] (LLDPE and polyethylene butene), and Qixia et al [65] (polypropylene and EPDM rubber).

In addition to compatibility studies, DSC is often used to characterise the properties of a blend. For example, Van dyke et al [66] have used it to investigate a number of variables (e.g. rubber type, rubber content and degree of cure) on the properties of polyamide 12/butyl rubber blends. More examples of the application of DSC to the characterisation of blends are given in Section 6.4.3.

6.4.3 Glass transition and the factors that influence it

When a rubber or plastic goes through its glass transition, there is a change in the heat capacity of the material due to the segmental rotation within the polymer molecules which can be detected by DSC. DSC is less sensitive to T_g than DMA because the change in heat capacity that results from this phenomenon is far less than the change in modulus that the material undergoes. This is illustrated above in Table 6.14, which gives a comparison of the relative sensitivities of the principal thermal analysis techniques for the measurement of the T_g of polymers.

The advantage that DSC has, though, is that no specific sample geometry is required and the amount of material required to perform an analysis is much less. As already mentioned, DSC detects the T_g as a step change in the heat capacity of a polymer. This change, however, is not instantaneous and occurs over a temperature range (e.g. 5°C) and the midpoint or inflection point is often taken as the T_g temperature. This is shown in Figure 6.20 where the as-received (top) and reheat (bottom) traces are shown for a PET sample, the T_g being found to be 73.4°C and 80.4°C, respectively. The second value resulting from the reheat trace is regarded as the more accurate as any heat history has been removed from the sample. This method of using the inflexion point to determine the T_g of a polymer is only one parameter that can be used. Others, such as the onset temperature of the heat capacity change, can be used; see Section 1.5.5 for further comments on T_g calculations.

In the case of rubbers, all of which have T_g's that are sub-ambient and, in some cases as low as 120°C (e.g. silicone rubber), the experiments have to start at very low temperatures, e.g. −150°C [67, 68].

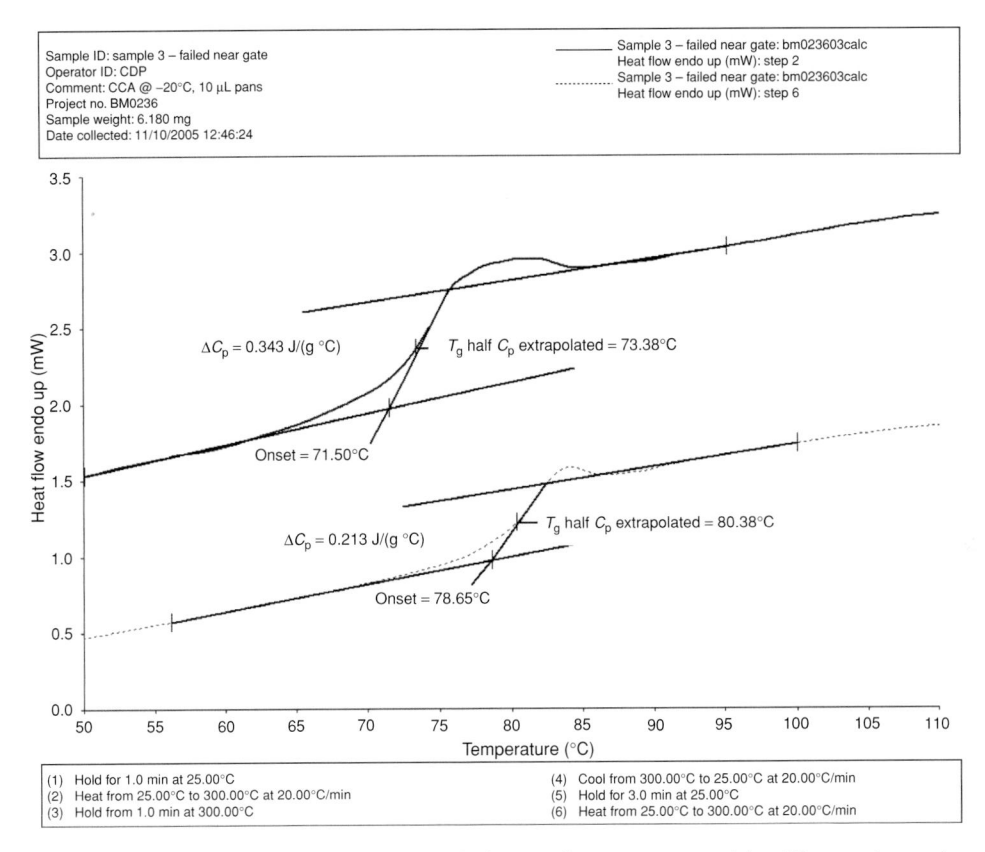

Figure 6.20 Heat capacity trace for a PET sample showing the T_g transition and the difference that results from effects of thermal history.

In addition to the relatively low sensitivity of DSC to T_g, the situation is often complicated by the fact that a large number of polymer compounds contain a significant quantity of additives which can reduce the polymer content to 50%, or less, of the sample by weight. As the response for the T_g is solely reliant on the polymer content of a sample the presence of additives reduces the change in heat capacity that the material undergoes and hence makes the detection of T_g even harder. Other properties of polymer systems that reduce the ability of DSC to detect the T_g are the presence of cross-links and the crystallinity of certain plastics, particularly high levels of both of these. In the former case, the stiffening effect of the cross-links reduces the heat capacity change as the material passes through its T_g and, in the latter, it is only the amorphous regions within the sample that contribute as the polymer chains within the crystalline regions cannot undergo a glass transition. An additional problem which can occur in the case of crystalline plastics is if other transitions, such as recrystallisation and melting, occur within a similar temperature range to the T_g, thus totally or partially obscuring it. However, these problems are far from insurmountable and, in addition to the T_g or, as it is sometimes referred to, the α transition, DSC has also been used to characterise β relaxations in polymers such as amorphous and semi-crystalline polyester plastics [69].

With respect to the effects that additives have, ones that are not soluble in the polymer (glass fibres, inorganic fillers, carbon black, etc.) do not increase free volume and so do not reduce the energy required for main chain motion; i.e. the T_g is not reduced in temperature. It is possible, though, for such additives to increase the temperature of T_g if they are intimately mixed with the polymer molecules and inhibit their ability to rotate.

Additives that are soluble in the polymer (e.g. process oils and plasticisers) do reduce the T_g of polymer samples by increasing free volume. It is therefore possible to use DSC to investigate the suppression effect that particular loadings of a given plasticiser can have on T_g. This is very important when a material is being compounded to enable it to perform satisfactorily (i.e. without hardening) at sub-ambient temperatures. Examples of this type of application include that reported by Sengers *et al* [70], who monitored the plasticising effects of paraffinic oil on the glass transition of both the plastic and rubber phases within blends of polypropylene/EPDM rubber and polypropylene/SEBS rubber, and the work carried out by Marais *et al* [71] who investigated the plasticisation effect that water had on EVA films and blends of PVC and EVA.

Cassel and Li [72] have described how developments in DSC have improved its ability to characterise semi-crystalline thermoplastics and mixed recyclates. The techniques covered show the influence of impurities on the glass transition, crystallinity and crystallisation. It is also possible to use this effect to investigate the degree to which two compatible polymers are blended – the greater the degree of mixing the nearer to merger the T_g's for the individual polymers become [73].

Other factors that can influence T_g and which DSC can, therefore, be used to investigate include the following:

(a) Molecular weight
(b) Molecular weight distribution
(c) Degree of crystallinity
(d) Orientation in fibres
(e) Stereoregularity of the polymer
(f) Cross-link density.

With respect to the last two examples, Bukhina *et al* [74] investigated, via T_g measurements, the long-term low-temperature properties of polybutadiene rubbers with differing cis-1,4 unit contents, and then made a comparison of the behaviour of blends of these rubbers with cis-1,4-polyisoprene, and SBR, and Cook, Groves and Tinker [75] demonstrated that a linear relationship existed between cross-link density and T_g for NR, polybutadiene and SBR gum (i.e. unfilled and un-plasticised) vulcanisates.

6.4.4 Ageing and degradation

DSC is a useful technique for detecting the chemical and morphological changes that occur in conjunction with ageing and degradation. For example, in semi-crystalline polymers heat ageing can cause a process referred to as solid state crystallisation to occur. When such a sample is heated in the DSC, a secondary melting endotherm having a lower melting point to the principal one will be apparent. The crystals formed in this way are small and their

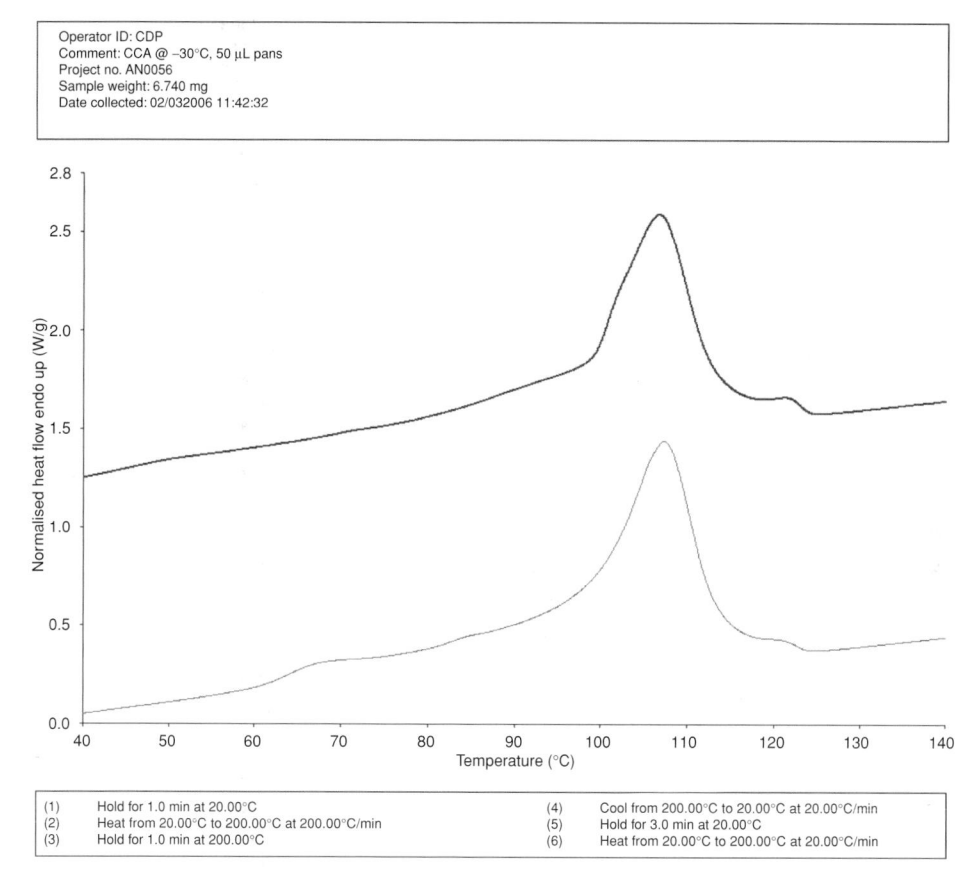

Figure 6.21 Comparison of the DSC data obtained on two samples, one of which (sample B) has undergone heat ageing in service and has an additional low-temperature endotherm as a consequence.

melting point is related to their thickness, which in turn is related to the time of ageing. It is therefore possible to develop correlations between ageing time and temperature and degree of secondary crystallisation. This effect is illustrated in Figure 6.21, where a small endotherm at a relatively low temperature (65–70°C) is apparent in the as-received data for sample B (bottom trace); a sample that had been heated in service and had failed as a consequence. This endotherm is absent in the as-received data for sample A (top trace), which had not been exposed to any heat in service.

In other cases, the principal endotherm due to the melting of crystals formed from the melt is affected by ageing; an increase in both the heat of fusion and the melting point being observed. In these cases no small, secondary crystals are formed.

In amorphous plastics, ageing can be detected by changes in the glass transition region of the DSC thermogram. Ageing below the materials' T_g causes a slow relaxation to occur in the polymer glass and, when the sample is heated in the DSC, this relaxation produces an enthalpic event that is apparent as an endotherm at the end of the glass transition. The greater the degree of ageing the greater this endotherm and it is possible, by carrying out controlled ageing experiments at known temperatures and times, to establish the thermal history of an unknown sample [76].

Degradation of a polymer, whether by heat [77], ultraviolet light [78], irradiation [79] or chemicals, results in irreversible changes that can be detected by DSC. In the case of semi-crystalline plastics, it is detected by a reduction in both the melting and recrystallisation temperatures, and the energy associated with these changes [80]. When rubbers undergo oxidative degradation, this is apparent in a broader glass transition that also occurs at a higher temperature; the effect being enhanced with rubbers that undergo cross-linking (e.g. the diene rubbers) under these circumstances. Diene rubbers are prone to ozonolysis reactions due to the double bond in the polymer backbone being attacked by ozone in the atmosphere. This type of attack leads to a high degree of chemical modification, and Rakovsky and Zaikov [81] have used results obtained by DSC to propose reaction mechanisms for the ozonolysis of a wide range of diene rubbers, including polychloroprene, polyisoprene and acrylonitrile–butadiene rubber.

The onset of the oxidative exotherm due to thermal degradation can be used to obtain a measure of the stability of a polymer. An oxidative atmosphere is used and the sample is either heated at a constant rate to a given temperature, or heated isothermally at a specified temperature. In the former case the temperature at the onset point is used as the endpoint, and in the latter the time is used. A common test to determine the stability of plastics is the OIT (oxidation induction time) test. The method to be used for plastics in general is described in ISO 11357-6 (2002), and methods for polyolefins are described in the standards EN 728 (1997) and ASTM D3895 (2003). Schmid and Affolter [82] have reported on the results of an interlaboratory study investigating the use of DSC for the determination of OIT.

The use of DSC to determine the relative stability of two polyethylene samples is shown in Figure 6.22. Both of these HDPE samples had been exposed to sea water at 85°C, but the sample that had been exposed for 48 h (bold line) has considerably reduced stability compared to the sample that had been exposed for only 2 h (dotted line).

Resolution of this oxidation exotherm can be improved by the use of the high-pressure pan technique, which suppresses the release of additives and degradation by-products. The other advantage is that lower test temperatures and shorter times can be used. This ability to obtain data at lower temperatures makes it possible to determine the kinetics of oxidative degradation of the plastics in the solid state. This circumvents one of the main drawbacks associated with ambient pressure experiments; that of measurable oxidation not occurring until the plastic is above its melting point. It is also the case that kinetic studies are most accurate when isothermal runs are performed (e.g. the OIT test – see above) but tests performed above the melting point of a plastic cannot be extrapolated reliably to temperatures below it. Pressure DSC, therefore, allows reasonable test times in the solid state and provides good experimental data for extrapolation to expected lifetime at lower temperature [83].

Because of its ability to investigate the stability of polymers, DSC is often used to evaluate the effectiveness of stabilisers (e.g. antioxidants and antiozonants) and to establish the energetics of their protective action. Examples of this type of application are the work carried out by Klein *et al* [84] and Cibulkova *et al* [85] who studied the effect of p-phenylene diamine antidegradants with polyisoprene rubber, and Parra and Matos [86] who investigated the synergistic effects of a range of different antioxidants in natural rubber compounds. Burlett

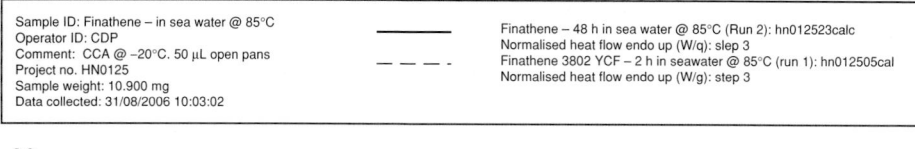

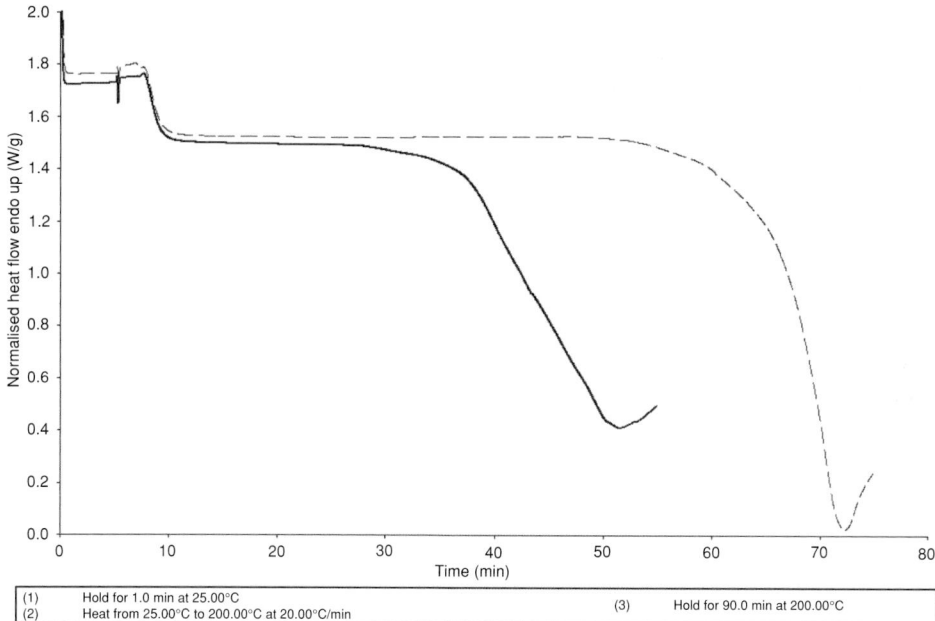

Figure 6.22 Comparative DSC traces for two HDPE samples that had been exposed to sea water at 85°C for two different periods – 48 h (full curve) and 2 h (dotted curve). The lowering of thermal stability with exposure time is illustrated.

[87] has described how DSC can be used to determine the apparent activation energies that are associated with the oxidation process.

6.4.5 Curing and cross-linking

The vast majority of rubbers are cured (or vulcanised) and a wide range of curing systems can be used, such as the following:

(a) Accelerated sulphur systems
(b) Sulphur donor systems
(c) Peroxide systems
(d) Amine and metal oxide cures.

All of these systems have the shared characteristic of a strong curing exotherm due to the wide range of chemical reactions that accompany the cross-linking process. This exotherm,

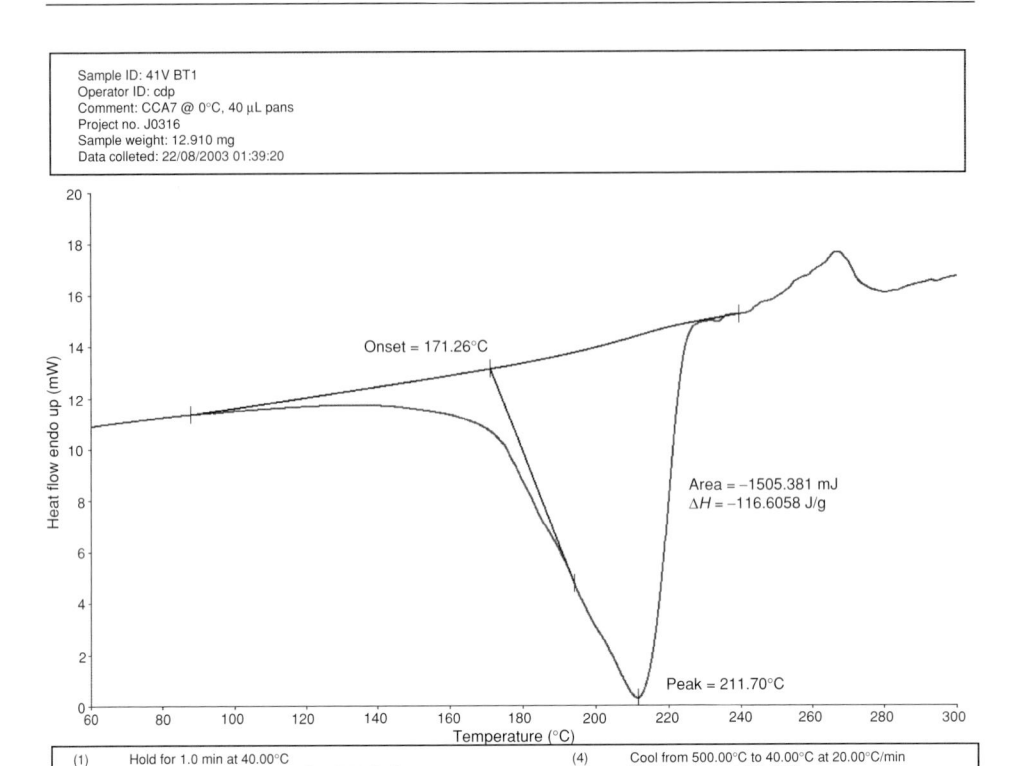

Figure 6.23 DSC trace showing the curing exotherm of a peroxide-cured fluorocarbon rubber.

the size and position (i.e. temperature range) of which is dependent on a number of factors such as the rubber type, the proportion of rubber in a sample, as well as the amount and type of cure system present, can be used for a number of purposes, for example,

(i) determination of the amount of curative in a sample;
(ii) qualitative assessment of the type of curative or cure system present;
(iii) characterisation of a cure in terms of the temperature that the cure begins, or the cure time at a given temperature.

An example of the cure data that can be provided by DSC is shown in Figure 6.23. This is a DSC trace of a peroxide-cured fluorocarbon rubber. The onset temperature of the cure is 172°C in this case and the area of the curing exotherm is 1421 mJ. The size of the sample was 12.9 mg and so this gives a value of 110 J/g of sample. The negative sign in front of the area values in Figure 6.23 only signifies the direction that the heat capacity curve has moved relative to the reference sample – endothermic changes being regarded as positive in this case and exothermic changes negative. The endotherm peak from 240 to 280°C is due to a loss of volatiles from the sample after the curing reaction has taken place. Some of these volatiles will be low molecular weight breakdown products of the curing reaction that has just taken place.

Although it is desirable to carry out these analyses using sealed pans (see also Section 6.4.6 below), to ensure that no volatile breakdown products can escape and produce an endotherm in the heat capacity curve work can be carried out using the standard semi-open pans if the objective is to identify unknowns, determine the onset temperature of cure, and produce comparative data on a range of closely related samples.

There are some cases in which peroxides are used in plastic compounds to produce a cross-linked matrix. A classic example of this is in the production of polyethylene cable covering materials. These types of systems can be studied in the same ways, and with the same considerations, as those that apply to rubbers.

Although it can be seen from the above that DSC has proved to be a useful technique for studying and comparing curing behaviour, it is also a popular technique for characterising the cure kinetics of rubbers and plastics. This quantitative work is important as the degree of cure of a polymer product is vital in terms of its properties and performance in service. However, interlaboratory trials have shown that a number of experimental conditions, as well as instrumental effects, can interfere with achieving a reliable value for the heat of reaction [88, 89]. In particular, it has been shown that even with curing reactions that do not involve volatile by-products, reactants with low vapour pressures can cause problems with accurate measurement if volatiles are not properly contained. Sepe has discussed other considerations that can prevent DSC from being used as a reliable method for the absolute determination of the degree of cure [90].

Wong-on and Wootthikanokkan [91] have used DSC to determine the degree of cross-linking of the acrylic rubber phase in blends of PVC and acrylic rubber that had been prepared by using a twin-screw extruder and then dynamically vulcanised in a compression mould at a temperature of 170°C. Baba et al [92] have used a novel experimental approach to evaluate the cross-link density of EPDM and polybutadiene rubbers that had been cross-linked to varying degrees either by photo-oxidation or dicumyl peroxide.

It is also possible to use DSC to determine the influence that additives have on the curing of polymer products. For example, Karbhari and Kabalnova [93] have investigated the curing kinetics of elastomer-modified, carbon fibre reinforced vinyl ester resins. The influence that the loading of carbon fibre and the percentage of sizing on the fibre have on the cure rate, degree of cure, time to maximum cure and the rate constant were all evaluated.

Workers at Port Elizabeth University have produced an important series of papers addressing the fundamental chemistry involved in the vulcanisation of different types of rubber compounds. DSC has been used extensively in this body of work to study the behaviour of accelerators and other essential cure ingredients in isolation [94] and in the presence of rubber molecules [95].

Simon and Kucma [96] have studied the vulcanisation of rubber compounds by DSC using both isothermal and non-isothermal conditions and, among other things, have described the temperature dependence of the induction period using an Arrhenius-type equation.

6.4.6 Blowing agents

Blowing agents are added to a number of plastic and rubber compounds to produce sponges and foams. There are two main types of blowing agents: chemical and physical. The chemical types (e.g. azo compounds, hydrazides and nitroso compounds) decompose at high processing temperatures to produce a gas, which is usually nitrogen. The physical types are low boiling point simple organic compounds such as pentane and dichloromethane which volatilise during processing to produce a cellular structure within the material.

It is possible to use DSC to study the performance of both these types, as the chemical ones will produce an exotherm in the heat capacity trace and the physical ones an endotherm as they remove heat from the system to volatilise. The trace obtained in both cases will enable the type of blowing agent present to be deduced, by the position of the event and by reference to traces produced from standard materials, as well as the amount of blowing agent, again by comparison with data obtained on standards.

However, an important consideration with the chemical types is that in addition to an exotherm resulting from the chemical breakdown of the compound, unless steps are taken to seal the system there will also be an endotherm due to the liberation of the gas. This loss of heat from the system will cancel out some of the exotherm and interfere with the data. The use of high-pressure, sealed pans is therefore very important in studies involving chemical blowing agents as these ensure that no volatiles can escape during the analysis.

A further complication occurs in the case of rubbers as the chemical blowing agents used will breakdown within the same temperature range that the curatives in the rubber react to cure it. The exotherm obtained in these cases is therefore a composite of the two processes, and control samples that do not contain any blowing agent have to be analysed to evaluate the contribution of the blowing agent. Again, sealed pans are necessary if accurate data on the exotherms are to be obtained as both the blowing and curing processes produce volatile low molecular weight products.

Moulinie and Woefle [97] have described how DSC can be used to study the kinetics of the decomposition of azodicarbonamide (AZDC) in polyolefin foams. Activation energies and rate constants were computed from the results that were obtained using variable heating rate experiments.

6.4.7 Modulated temperature DSC

The advent of modulated temperature DSC (also described as alternating or oscillating DSC) over 15 years ago offered advantages to polymer analysts, in that its ability to separate the heat flow into its reversing (heat capacity related) and non-reversing (kinetic) events improved the quality of the data that could be obtained by DSC and hence increased its usefulness [98, 99].

The major reversing events that occur in plastics and rubbers include the T_g in both cases, and the melting of the crystalline region in the case of certain semi-crystalline plastics. There are a number of non-reversing events including recrystallisation processes occurring during the melting process, and the enthalpic relaxation that occurs in the region of the T_g for amorphous polymers that have undergone physical ageing. In addition, with crystalline plastics, the situation is complicated by the fact that the heating process can alter the crystalline structure of the material, with imperfect crystals melting and becoming absorbed into larger crystals prior to these melting. Conventional DSC cannot separate these two events as they often happen simultaneously, but modulated DSC can resolve the reversing heat flow of crystal melting and the non-reversing heat flow of crystallisation. Modulated DSC is therefore capable of providing more accurate data on the latent heat of fusion for the crystal-melting event. This is important as the study of the crystallinity of plastics is one of the principal uses of the DSC technique.

Other advantages that have been claimed for the use of modulated DSC in the analysis of plastics and rubbers include the following:

(a) Improvements in the accuracy of T_g measurements in plastics due to the greater separation of the T_g (reversing) and enthalpic relaxation event (non-reversing) which occurs just after the T_g [100].
(b) The inherent greater sensitivity of the technique makes it easier to detect the T_g in highly filled/reinforced polymers, or highly crystalline plastics, where these features have reduced the change in the heat capacity that occurs as the material passes through this transition. HyperDSC™ also offers this advantage over conventional DSC.
(c) It can be useful in the analysis of certain polymer blends. For example, in the case of a PC/PBT blend, the PC T_g may overlap with the recrystallisation of parts of the PBT phase.
(d) It provides, along with power compensation DSC, a means of directly measuring the heat capacity of a material and it has been demonstrated that the method can also be used to produce thermal conductivity data that agree well with values established by more conventional techniques. This and other applications are covered by a TA Instruments guide to modulated DSC [101].
(e) Provided that the calibration has taken into account the heat loss through the purge gas surrounding the sample, the thermal conductivity of polymers can also be measured in the range from 0.1 to 1.5 W/°C
(f) Its ability to generate an instantaneous heating rate during quasi-isothermal experiments allows cure measurements on rubbers to be made which are not possible by conventional DSC. This can assist in obtaining a greater understanding of the cure mechanism.

Hourston and Song [102] have proposed a modulated DSC method for the measurement of the weight fraction of the interface in a wide range of rubber–rubber blends. A quantitative analysis using the differential of the heat capacity versus the temperature signal from the modulated temperature DSC allowed the weight fraction of the interface to be calculated. The data obtained showed how essentially miscible the different rubber types were, as well as revealing the influence of properties such as stereoregularity (with BR blends) and the proportions of co-monomers (with SBR blends) on the extent of mixing.

6.4.8 HyperDSC™

This variant on DSC involves recording the heat flow cure of a sample as it is heated very rapidly (e.g. at 500°C/min), as opposed to a standard 20°C/min heating ramp. This technique can only be used with power compensation systems, as an analysis chamber having a low thermal mass is essential; and the ability to record data very quickly is another prerequisite. Although the technique of HyperDSC™ is relatively new, there are a number of publications available which provide further information on its use and application [103, 104].

The advantages that this technique offers in the analysis of polymers result due to its twin attributes of very low short analysis times and increased accuracy when it comes to measuring crystallinity. In the former case, the instrument has been used effectively on the production lines of plastic materials to record data, such as T_g and T_m, almost as soon as they are produced; a full characterisation being possible within 60 s. This offers considerable practical benefits to a manufacturer, as any quality control problems can be identified very quickly and rectified before too much production has been lost. Although the samples are heated very rapidly, it has been demonstrated that no significant increase in T_g, for example, results from this. The data obtained from these HyperDSC™ runs can therefore be compared to library standards that have been run using more traditional conditions.

With respect to the latter case, many semi-crystalline polymers exist in a number of potentially unstable states and a fast heating rate enables data to be recorded on them before they change due, for example, to micro-crystalline regions melting and then recrystallising during the analysis. If this type of effect occurs under a slow heating rate, it effectively means that the extent of crystallinity of the polymer can change during an experiment and so the data obtained are not totally representative of the original material.

Two other advantages of HyperDSC™ are as follows:

(i) The increased scan rate provides much higher sensitivity due to the resulting higher heat flow and this enables the analyst to identify transitions which would not be apparent using the standard conditions.
(ii) The ability to have fast heating and cooling rates means that it is possible to closely mimic the conditions that some samples experience in their manufacture and production.

6.4.9 Microthermal analysis

TA Instruments were one of the first companies to produce a commercial microthermal analysis instrument [105, 106]. This device is useful for the analysis of polymers in certain applications because it can obtain DSC and TMA data on very small (i.e. micrometre-sized) areas. Examples of these are presented below:

1. *Determination of the homogeneity of polymer blends.* The surface of a product can be mapped and occurrence of the events which are characteristic of individual polymers can be obtained. The relative intensity of these will indicate whether some areas of the sample are richer in one polymer than the others. For example in a gasket, which is a blend of polyisobutylene rubber (PIB) and LDPE, the T_g of the PIB (TMA trace) and the T_m (DSC trace) of the LDPE will be focused on.
2. *Studying surface modification processes.* A number of rubber and plastic products are modified after production by such processes as corona discharge and sterilisation by radiation. These processes will change the physical properties (e.g. modulus) of the surface layer of the polymer due to cross-linking and/or oxidation reactions. The microthermal analyser can be used to determine how uniform across the surface this modification is.
3. *Characterisation of localised thermal history.* During the processing of plastics, stresses can become 'frozen' into products due to their being cooled too quickly. This is undesirable as when these stresses relax in service the product can become distorted. In certain manufacturing processes, such as the extrusion of thin walled and small bore tubing,

these can be very localised and hence not easy to detect by standard thermal analysis techniques. A microthermal analyser is able to map the heat capacity of the extrudate surface and see if frozen irregularities in the structure can be detected, e.g. by a reduction in crystallinity in certain regions. The processing conditions can then be changed to eliminate the problem.

A thorough review of the techniques and applications of microthermal analysis has been published by Pollock and Hammiche [107].

6.5 Other thermal analysis techniques used to characterise thermoplastics and rubbers

There are a large number of thermal analysis techniques that can be applied to the analysis of plastic and rubber systems, and the scope of this chapter has only allowed three of the principal ones to be covered. Some of the more important techniques that it has not been possible to include are mentioned briefly below.

6.5.1 Dielectric analysis

DEA has proved to be very useful over the years and measures a polymer's response to an applied alternating voltage signal. It therefore measures two fundamental electrical characteristics of a polymer – capacitance and conductance – as a function of time, temperature and frequency. In practice a polymer sample is placed in contact with a sensor (or electronic array for samples in the form of pellets or powders) and the oscillating voltage signal applied at frequencies between 0.001 and 100 000 Hz. Whilst important in themselves, capacitance and conductance acquire more significance when they are related to changes in the molecular state of a polymer:

Capacitance, the ability of a polymer to store an electrical change, dominates at temperatures below T_g or T_m.
Conductance, the ability of a polymer to transfer an electrical charge, dominates at temperatures above T_g or T_m.

The actual polymer properties monitored by DEA are as follows:

E' (permittivity): a measure of the degree of alignment of the molecular dipoles to the applied electrical field (analogous to the elastic modulus in DMA).
E'' (loss factor): it represents the energy required to align the dipoles (analogous to the viscous modulus in DMA).

The tan δ in DEA is the ratio of these two, i.e. the same as in DMA.
　　DEA is worthy of mention because, although it can be used for a very similar range of applications to DSC and DMA, it has a number of practical advantages over those techniques.

These are as follows:

(a) The data obtained are not affected by the evolution of volatile species during such events as curing reactions, the breakdown of additives such as blowing agents and degradation reactions.
(b) Data can easily be obtained on samples in a wide range of physical forms (e.g. solid, liquid, pellets or powder).
(c) It is more sensitive than DSC and so it can be used to obtain superior data on samples where there is only a small polymer fraction, e.g. highly extended or filled polymer products.
(d) As data can be obtained by probe samplers situated a long distance from the instrument, it can be used in industry for the 'in-process' monitoring of events such as curing/polymerisation reactions.

There are a number of application notes on the use of DEA to characterise polymers available from instrument suppliers [108, 109].

6.5.2 Differential photocalorimetry (DPC)

DPC is an adaptation of DSC which involves the use of an ultraviolet or visible light source to study the behaviour or photoreactive systems and has specific application to light curable coatings, films, inks, adhesives and dental materials. In particular, it can be used to study the effect of different initiator systems on the cure behaviour of these products, and to investigate the influence that parameters such as the other additives in the products and the cure conditions have on the cure rate etc. The technique is therefore particularly useful for studying the curing behaviour of coatings and inks.

Early instrumentation involved the modification of existing DSC instruments, but a commercial instrument was introduced by TA Instruments in 1987. As with DSC, kinetic studies of photoreactive systems can be performed through the use of isothermal experiments performed at different temperatures, or multiple dynamic heating runs utilising different heating rates [110]. The majority of studies are still carried out isothermally, though, as modern equipment still relies on modified versions of existing equipment due to cost.

6.5.3 Thermally stimulated current (TSC)

TSC, which is also referred to as thermally stimulated depolarisation (TSD) in the literature, is a technique with capabilities similar to those of DMA and DEA. However, TSC is superior to these two in its sensitivity to the investigation of the fine molecular structure of polymers and hence its ability to find relationships between this and physical properties that may have eluded the other techniques.

The principle of TSC is to orient the dipoles in molecules by applying a high voltage field to a material at a high temperature and then quenching the material to freeze in the orientation created by the electric field. The external field is then removed and the sample heated at a constant rate. At temperatures that are associated with specific relaxations the sample will depolarise; a process which produces a depolarisation current, which is plotted as dynamic conductivity (or current density) against temperature. This plot is analogous to the tan δ plot against temperature plot in DMA and DEA. The ability of the technique to work at very low frequencies (e.g. 10^{-2}–10^{-4} Hz) means that it has the potential to resolve molecular transitions that may overlap at the higher frequencies used by DMA and DEA.

Because of this high sensitivity, TSC is ideally suited to investigating the detailed structure in blends and copolymers that have not been characterised satisfactorily by DMA and DEA [111, 112].

A significant increase in the precision of TSC was achieved when Lacabanne [113] developed the method of windowing polarisation, and this in turn enables the process known as relaxation map analysis to be applied to polymer systems [114, 115]. This also enables fine detail within a polymer to be elucidated and a thorough description of the technique and its capabilities has been provided by Ibar [116].

6.5.4 Thermal conductivity analysis (TCA)

Accurate thermal conductivity data are important in assessing the suitability of a particular polymer for a given application. It is also being used extensively in the application of computer modelling packages to predict the flow of plastics and rubber compounds during processing operations.

The ability to use modulated DSC to determine such data has already been mentioned (see Section 6.4.7). More traditional methods involve the use of the standard guarded hot plate device that is described in ASTM F-433. In addition to this, Lobo and Cohen [117] have described an in-line method that can be applied to polymer melts, and Oehmke and Wiegmann [118] describe an apparatus that can perform thermal conductivity measurements as a function of temperature as well as pressure.

6.6 Conclusion

The application of thermal analysis techniques to thermoplastics and rubbers is of major interest, being one of the principal uses for this group of analytical techniques, and unfortunately it has only been possible to give a brief overview within the confines of this chapter. It should be evident from the information that has been provided that for any polymer laboratory or facility to be fully effective, and provide a comprehensive service, it must possess as a minimum requirement at least the three thermal techniques that have been showcased here.

This area of research and development is very active, with a healthy number of quality publications being produced each year from the many key workers in the field. The instrument manufactures have also continued to provide improvements in both the software and in the fundamental design of their instruments and their capability (e.g. modulated TGA and HyperDSC™).

Another relatively recent innovation which is showing increasing promise and usefulness to researchers and contract analysts alike is the micro-TA technique. When combined with an atomic force microscope, the ability of this technique to provide topographical, DSC and DMA/TMA data on very small areas of a sample is proving invaluable in fundamental characterisation and failure diagnosis work, and it complements the data which can be recorded by the standard versions of these techniques.

The micro-TA instrument is a prime example of the fact that, in addition to providing a wealth of data on the chemical structure, morphology and chemical properties of a polymer sample, the ability of these techniques to be combined with each other and other analytical techniques (e.g. IR, GC and MS) enables their capability to be enhanced significantly.

Thermal analysis techniques will continue to play a vital role in assisting polymer scientists in all of their endeavours: fundamental characterisation work and failure diagnosis studies; the development of better polymers, polymer blends and compounds; the addressing of pressing current environmental concerns, such as the recycling of used tyres and the polymer components of electrical devices; and the substitution of renewable raw materials (e.g. fibres such as hemp and natural rubber) for synthetic materials derived from petroleum products.

References

1. Knappe S. Rubber analysis made more accurate and informative by special TGA techniques. *Rubber World* 2002;January:33–34.
2. Loadman MJR. *Analysis of Rubber and Rubber-like Polymers*, 4th edn. Dordrecht: Kluwer Academic Press, 1998; Chapter 11, pp. 279–283.
3. Wang S, Peng Z, Zhang Y, Zhang Y. Structure and properties of BR nanocomposites reinforced with organoclay. *Polym Polym Compos* 2005;13(4):371–384.
4. Mahaling RN, Jana GK, Das CK. Modified nanofiller epichlorohydrin elastomer composite. *Compos Interfaces* 2005;8(9):701–710.
5. Fibiger R, Garces JM, Palmieri J, Traugott T. Nanocomposite reinforced polypropylene. In *SPE Automotive TPO Global Conference*, 29 September to 1 October, 2003; pp. 25–33.
6. Ramesan MT. Thermogravimetric analysis, flammability and oil resistance properties in natural rubber and dichlorocarbene modified styrene butadiene rubber blends. *React Funct Polym* 2004;59(3):267–274.
7. Vlad S. Effects of flame retardants on thermal and mechanical behaviour of some thermoplastic polyurethanes. *Mater Plast* 2003;40(3):112–115.
8. Forrest MJ. Rubber analysis – polymers, compounds and products. *Rapra Review Report*, Number 139. Rapra Technology Ltd, 2001.
9. Forrest MJ. Analysis of plastics. *Rapra Review Report*, Number 149. Rapra Technology Ltd, 2002.
10. Forrest MJ. The use of high resolution TGA in the analysis of polymer systems. In *Polymer Testing'97*, day 3: chemical analysis, paper 5, Rapra Technology Ltd, 7–11 April, 1997.
11. Torosyan KA, Yutudzhyan KK, Sarkisyan V Yu, Voskanyan ES. Heat Stability of chlorinated rubbers. *Int Polym Sci Technol* 1999;26(2):38–41.
12. Affolter S, Schmid M. Interlaboratory tests on polymers: thermal analysis. *Int J Polym Anal Charact* 2000;6:35–57.
13. Agullo N, Borros S. Qualitative and quantitative determination of the polymer content in rubber formulations. *J Therm Anal Calorim* 2002;67:513–522.
14. Shield SR, Ghebremeskel GN. Determination of the styrene content of styrene-butadiene rubber using TGA. *Rubber World* 2000;223(2):24–66.
15. Forrest MJ. *Rapra Review Report*, Number 139. Rapra Technology Ltd, 2001.
16. Denardin ELG, Samios D, Janissek PR, de Souza GP. Thermal degradation of aged chloroprene rubber studied by thermogravimetric analysis. *Rubber Chem Technol* 2001;74(4):622–629.

17. Jana RN, Nando GB. Thermogravimetric analysis of blends of low density polyethylene and poly(dimethyl siloxane) rubber: effects of compatibilisers. *J Appl Polym Sci* 2003;90(3):635–642.
18. Saccani A, Motori A, Montanari GC. Short-term thermal endurance evaluation of thermoplastic polyesters by isothermal and non-isothermal thermogravimetric analysis. *J Appl Polym Sci* 2005;98(5):968–973.
19. Skachkova VK, Erina NA, Chepal LM, Prut EV. Thermooxidative stability and morphology of polypropylene-rubber-paraffin oil blends. *Polym Sci Ser A* 2003;45(12):1220–1224.
20. Pruneda F, Sunol JJ, Andreu-Mateu F, Colom X. Thermal characterisation of nitrile butadiene rubber (NBR)/PVC blends. *J Therm Anal Calorim* 2005;80(1):187–190.
21. Gamlin CD, Dutta NK, Choudhury NR. Mechanism and kinetics of the isothermal degradation of ethylene-propylene-diene (EPDM) elastomers. *Polym Degrad Stab* 2003;80(3):525–531.
22. Flynn J. Lifetime prediction for polymeric materials from thermal analytical experiments – problems, pitfalls and how to deal with them. In *Antec'88, Proceedings of the 46th Annual Technical Conference*, Atlanta, 18–21 April,1988; 930–932.
23. Schneider HA. Thermogravimetric kinetics of polymer degradation: science or fiction ? *Polym Eng Sci* 1992;32(17):1309–1315.
24. Flynn JH, Dickens B. Durability of macromolecular materials. In Eby RK (ed.), *ACS Symp. Ser. No 95.* Washington, DC: American Chemical Society, 1979; p. 97.
25. Scheirs J, Bigger SW, Billingham NC. Effect of chromium on the oxidative pyrolysis of gas-phase HDPE as determined by dynamic thermogravimetry. *Polym Degrad Stab* 1992;38(2):139–145.
26. Sepe MP. Use of thermal analysis in polymer characterisation. *Elastomer Technology Handbook.* Boca Raton, FL: CRC Press, 1993; pp. 105–258.
27. Forrest MJ. The use of high resolution TGA in the analysis of polymer systems. In *Polymer Testing'97*, day 3: chemical analysis, paper 5, Rapra Technology Ltd, 7–11 April, 1997.
28. Forrest MJ. The use of high resolution TGA in the analysis of polymer systems. In *Polymer Testing'97*, day 3: chemical analysis, paper 5, Rapra Technology Ltd, 7–11 April, 1997.
29. TA Instruments Application Notes TA237.
30. TA Instruments Application Notes TA251.
31. Gamlin C, Markovic MG, Dutta NK, Choudhury NR, Matisons JG. Structural effects on the decomposition kinetics of EPDM elastomers by high resolution TGA and modulated TGA. *J Therm Anal Calorim* 2000;59:319–336.
32. Pielichowski K, Leszczynska A. TG-FTIR study of the thermal degradation of polyoxymethylene(POM)/thermoplastic polyurethane (TPU) blends. *J Therm Anal Charact* 2004;78(2):631–637.
33. Gupta YN, Chakraborty A, Pandey GD, Setua DK. High resolution thermogravimetry/coupled mass spectroscopy analysis of flame retardant rubber. *J Appl Polym Sci* 2003;89(8):2051–2057.
34. Sevcik A, Klepal V, Betakova S, Sionova Z. Contribution to method for measurement of glass transition temperature and brittle temperature in rubber materials. In *IRC Conference*, Prague, 1–4 July, 2002; p. 9. Proceeding Paper 69.
35. Khonakdar HA, Wagenknecht U, Jafari SH, Haessler R, Eslami H. Dynamic mechanical properties and morphology of polyethylene/ethylene vinyl acetate copolymer blends. *Adv Polym Technol* 2004;23(4):307–315.
36. Rana D, Mounach H, Halary JL, Monnerie L. Differences in the mechanical behaviour between alternating and random styrene-methyl methacrylate copolymers. *J Mater Sci* 2005;40(4):943–953.
37. Mina MF, Ania F, Huy TA, Michler GH, Balta Calleja FJ. Micromechanical behaviour and glass transition temperature of poly(methyl methacrylate) – rubber blends. *J Macromol Sci B* 2004;4(5):947–961.
38. Forrest MJ. Rubber analysis – polymer, compounds and products.*Rapra Review Report*, Number 139. Rapra Technology Ltd, 2001; p. 31.

39. Schubnell M, Schawe JEK. Influence of cross-link density on the mechanical behaviour of elastomers. In *International Rubber Conference*, Institute of Materials, 7–9 June, 2005; pp. 471–472.
40. Vijayabasker V, Bhowmick AK. Dynamic mechanical analysis of electron beam irradiated sulphur vulcanised nitrile rubber network – some unique features. *J Mater Sci* 2005;40(11):2823–2831.
41. Persson S, Goude M, Olsson T. Studying the cure kinetics of rubber-to-metal bonding agents using DMTA. *Polym Test* 2003;22(6):671–676.
42. Pandey KN, Setua DK, Mathur GN. Determination of the compatibility of NBR-EPDM blends by an ultrasonic technique, modulated DSC, dynamic mechanical analysis and atomic force microscopy. *Polym Eng Sci* 2005;45(9):1265–1276.
43. Ibarra-Gomez R, Marquez A, Mendoza-Duarte M. Dynamic mechanical analysis of conductive BR/EPDM/carbon black blends. *Rubber Chem Technol* 2004;77(5):947–954.
44. Carlberg M, Colombini D, Maurer FHJ. Ethylene-propylene-rubber (EPR)/polydimethylsiloxane (PDMS) binary polymer blends: morphology and viscoelastic properties. *J Appl Polym Sci* 2004;94(5):2240–2249.
45. Zhang L, Li C, Huang R. Toughness mechanism of polypropylene/elastomer/filler composites. *J Polym Sci Polym Phys Ed* 2005;43(9):1113–1123.
46. Shivakumar E, Srivastava RB, Pandey KN, Das CK. Compatibility studies of blends of acrylic rubber (ACM), poly(ethyleneterephthalate) (PET), and liquid crystalline polymer (LCP). *J Macromol Sci A* 2005;9(42):1181–1195.
47. McConnell DC, McNally GM, Murphy WR. The performance of polyvinylchloride/thermoplastic polyurethane blends In *Antec'04; Proceedings of the 62nd Conference*, 16–20 May, 2004; pp. 2753–2757.
48. Fraga AN, Alvarez VA, Vazquez A, de la Osa O. Relationship between dynamic mechanical properties and water absorption of unsaturated polyester and vinyl ester glass fiber composites. *J Compos Mater* 2003;37(17):1553–1574.
49. Lopez BL, Perez LD, Mesa M, et al. Use of mesoporous silica as a reinforcement agent in rubber compounds. *E-Polym* 2005;18:1–13.
50. Grein C, Bernreitner K, Gahleitner M. Potential and limits of dynamic mechanical analysis as a tool for fracture resistance evaluation of isotactic polypropylenes and their polyolefin blends. *J Appl Polym Sci* 2004;93(4):1854–1875.
51. Finegan IC, Tibbetts GC, Glasgow DC, Ting J-M, Lake ML. Surface treatments for improving the mechanical properties of carbon nanofibre/thermoplastic composites. *J Mater Sci* 2003;38(16):3485–3490.
52. Lee H, Hsieh AJ, McKinley GH. Rheological properties and thermal characteristics of clay/PMMA nanocomposites In *ACS Polymeric Materials: Science and Engineering, Fall meeting*, Boston, 18–22 August, 2002; pp. 19–20.
53. Patel M, Morrell PR, Evans J. Load bearing property testing of a silica filled room temperature vulcanised polysiloxanes rubber. *Polym Test* 2004;23(5):605–611.
54. Patel M. Viscoelastic properties of polystyrene using dynamic rheometry. *Polym Test* 2004;23(1):107–112.
55. Sepe MP. Thermal analysis of polymers. *Rapra Review Report*, Number 95. Rapra Technology Ltd, 1997; p. 31.
56. Changwoon Nah, Je Hwan Park, Choon Tack Cho, Young-Wook Chang, Shinyoung Kaang. Specific heats of rubber compounds. *J Appl Polym Sci* 1999;72(12):1513–1522.
57. Bertucci M. Practical examples of IR and DSC analysis. *Macplas International* 2004;October: 95–97.
58. Price C. *The Rapra Collection of DSC Thermograms of Semi-Crystalline Thermoplastic Materials.* Rapra Technology Ltd, 1997.
59. Kong Y, Hay JN. The measurement of the crystallinity of polymers by DSC. *Polymer* 2002;43: 3873–3878.

60. Koszkul J. Study of the degree of crystallinity of thermoplastics after injection moulding and annealing. *Polymer* 1999;44(4):255–261.
61. Li J, Zhou C, Wang G, Tao Y, Liu Q, Li Y. Isothermal and non-isothermal kinetics of elastomeric polypropylene. *Polym Test* 2002;21(5):583–589.
62. Abou-Helal MO, El-Sabbagh SH. A study of the compatibility of NR-EPDM blends using electrical and mechanical techniques. *J Elast Plast* 2005;37(4):319–346.
63. Xiaodong Wang, Xin Luo, Xinfeng Wang. Study on blends of thermoplastic polyurethane and aliphatic polyester: morphology, rheology and properties as moisture vapour permeable films. *Polym Test* 2005;24(1):18–24.
64. Li C, Zhao J, Zhao D, Fan Q. Linear low-density polyethylene/poly(ethylene-ran-butene) elastomer blends: miscibility and crystallinity. *J Polym Res* 2004;11(4):323–331.
65. Zhang Qixia, Fan Hong, Bu Zhiyang, Li Bogeng. Blending modifications of PP by EPDM. *China Synth Rubber Indust* 2004;27(3):1000–1255.
66. Van Dyke JD, Gnatowski M, Koutsandreas A, Burczyk A. Effect of butyl rubber type on properties of polyamide and butyl rubber blends. *J Appl Polym Sci* 2004;93(3):1423–1435.
67. Xueqin Z, Yang L, Hongde X. Glass transition temperature of integral rubber SBR-IR-SBR and SBR-BR-SBR. *China Synth Rubber Indust* 2004;27(2):69–72.
68. Abdullah A, Long OE. Glass transition temperature of standard Malaysian rubbers. In *IRC'97*, Kuala Lumpur, 6–9 October, 1997; pp. 1102–1107.
69. Kattan M, Dargent A, Grend J. Relaxations in amorphous and semi-crystalline polyesters. *J Therm Anal Calorim* 2004;76(2):379–394.
70. Sengers WGF, Wubbenhorst M, Picken SJ, Gotsis AD. Distribution of oil in olefinic thermoplastic elastomer blends. *Polymer* 2005;46(17):6391–6401.
71. Marais S, Bureau E, Gouanve F, et al. Transport of water and gases through EVA/PVC blend films – permeation and DSC investigations. *Polym Test* 2004;23(4):475–486.
72. Cassel RB, Li L. Thermodynamic and kinetic analysis of semicrystalline recyclates by DSC. In *Antec'00*, Orlando, 7–11 May, 2000. Proceedings Paper 441.
73. Bukhina MF, Zorina NM, Severina NL, Morozov Yu L. Physical problems of low temperature behaviour of elastomeric materials based on copolymer and blends. In *IRC Conference*, Prague, 1–4 July, 2002; p. 5. Proceedings Paper 42.
74. Bukhina MF, Zorina NM, Morozov YL, Kuznetsova EI. Low temperature behaviour of butadiene rubbers of different cis-1,4 unit content. *Elastomery* 2003;7(5):3–8.
75. Cook S, Groves S, Tinker AJ. Investigating crosslinking in blends by differential scanning calorimetry. *J Rubber Res* 2003;5(3):121–128.
76. Wunderlich B. *Thermal Characterisation of Polymeric Materials*, Turi EA (ed.). New York: Academic Press, 1981; Chapter 2.
77. Giurginca M, Popa L, Zaharescu T. Thermo-oxidative degradation and radio-processing of ethylene vinyl acetate elastomers. *Polym Degrad Stab* 2003;82(3):463–466.
78. dos Santos KAM, Suarez PAZ, Rubim JC. Photodegradation of synthetic and natural polyisoprenes at specific UV radiations. *Polym Degrad Stab* 2005;90(1):34–43.
79. Stevenson I, David L, Gauthier C, Arambourg L, Davenas J, Vigier G. Influence of silica fillers on the irradiation ageing of silicone rubbers. *Polymer* 2001;22:9287–9292.
80. Pages P, Carrasco F, Saurina J, Colom X. FTIR and DSC study of HDPE structural changes and mechanical properties variation when exposed to weathering ageing during Canadian winter. *J Appl Polym Sci* 1996;60(2):153–159.
81. Rakovsky S, Zaikov G. Ozonolysis of polybutadienes. *J Appl Polym Sci* 2004;91(3):2048–2057.
82. Schmid M, Affolter S. Interlaboratory tests on polymers by DSC: determination and comparison of oxidation induction time and oxidation induction temperature. *Polym Test* 2002;22:419–428.
83. Sepe MP. Oxidation induction time : an old method for a new era. In *Antec'95*, Boston, 7–11 May, 1995; pp. 2468–2474.

84. Klein E, Cibulkova Z, Lukes V. A study of the energetics of antioxidant action of p-phenylene diamines. *Polym Degrad Stab* 2005;88(3):548–554.
85. Cibulkova Z, Simon P, Lehocky P, Balko J. Antioxidant activity of phenylene diamines studied by DSC. *Polym Degrad Stab* 2005;87(3):479–486.
86. Parra DF, Matos JR. Some synergistic effects of antioxidants in natural rubber. *J Therm Anal Calorim* 2002;67(2):287–294.
87. Burlett DJ. Studies of elastomer oxidation via thermal analysis. In: *153rd ACS Rubber Division Meeting*, Indianapolis, 5–8 May, 1998. Paper 11.
88. Richardson MJ. Characterisation of the cure of resins by DSC. *Pure Appl Chem* 1992;64(11):1789–1800.
89. Affolter S, Schmid M. Interlaboratory tests on polymers: thermal analysis. *Int J Polym Anal Charact* 2000;6:35–37.
90. Sepe MP. Use of thermal analysis in polymer characterisation. In: *Elastomer Technology Handbook*. Boca Raton, FL: CRC Press, 1993; pp. 105–258.
91. Wong-on J, Wootthikanokkan J. Dynamic vulcanisation of acrylic rubber blended with PVC. *J Appl Polym Sci* 2003;88(11):2657–2663.
92. Baba M, George SC, Gardette JL, Lacoste J. Evaluation of crosslinking in elastomers using thermoporometry, densimetry and differential scanning calorimetry analysis. *Rubber Chem Tech* 2002;75(1):143–154.
93. Karbhari VM, Kabalnova L. Effect of sizing and loading levels on the cure kinetics of carbon fibre vinyl ester composites. *J Reinf Plast Compos* 2001;20(2):90–104.
94. Gradwell MHS, Grooff D. Comparison of tetraethyl and tetramethylthiuram disulphide vulcanisation. I. Reactions in the absence of rubber. *J Appl Polym Sci* 2001;80(12):2292–2299.
95. Hendrikse KG, McGill WJ, Reedijik J. Vulcanisation if chlorobutyl rubber. I. The identification of crosslink precursors in compounds containing zinc oxide and zinc chloride. *J Appl Polym Sci* 2000;78(13):2290–2301.
96. Simon P, Kucma A. DSC analysis of the induction period in the vulcanisation of rubber compounds. *J Therm Anal Calorim* 1999;56(3):1107–1113.
97. Moulinie P, Woefle C. Investigation of the reaction kinetics within expandable mixtures used for preparing injection moulded polyolefin films. In *Antec'99*, 2–6 May, 1999; Vol. 2, pp. 2031–2034.
98. TA Instruments Application Notes TA211.
99. TA Instruments Application Notes TA210.
100. Hutchison JM. Studying the glass transition by DSC and TMDSC. *J Therm Anal Calorim* 2003;72:619–629.
101. TA Instruments. *Modulated DSC Compendium: Basic Theory and Experimental Considerations*. 1996.
102. Hourston DJ, Mo Song. Quantitative characterisation of interfaces in rubber-rubber blends by means of modulated-temperature DSC. *J Appl Polym Sci* 2000;76(12):1791–1798.
103. Pijpers TFJ, Mathot VBE, Goderis B. High speed calorimetry for the study of kinetics of (re) vitrification, crystallisation and melting of macromolecules. *Macromolecular* 2002;35:3601–3613.
104. Gabbott P, Clarke P, Mann T, Royall P, Shergill S. A high sensitivity, high speed DSC technique: measurement of amorphous lactose. *American Laboratory* 2003;August.
105. TA Instruments Application Notes TS58.
106. TA Instruments Applications Notes TA257.
107. Pollock HM, Hammiche A. Micro-thermal analysis: techniques and applications. *J Phys D Appl Phys* 2001;34:R23–R53.
108. TA Instruments. *Thermal Analysis Technical Literature (Theory and Applications)*. 1994.
109. PerkinElmer Applications Notes PETAN-54 and PETAN-55.
110. Groves IF, Lever TJ, Hawkins NA. Use of differential photocalorimetry (DPC) for the study of photoinitiator systems *Polym Paint Color J* 1993;43:8–16.

111. Hsiao BS, Sauer BB. Broadening of the glass transitions in blends of poly(aryl ether ketones) and a poly(ether imide) as studied by thermally stimulated currents. *J Polym Sci Polym Phys* 1993;31(8):917–932.
112. Sauer BB, Avakian P, Starkweather HW. Cooperative relaxations in semi-crystalline fluoropolymers studied by thermally stimulated currents and AC dielectric. *J Polym Sci Polym Phys* 1996;34(3):517–526.
113. Lacabanne C. PhD Thesis. University of Toulouse, France, 1974.
114. Liu S-F, Lee Y-D. *J Polym Sci Polym Phys* 1995;33(9):1333–1341.
115. Sauer BB, Avakian P, Starkweather HW. Cooperative relaxations in semi-crystalline fluoropolymers studied by thermally stimulated currents and AC dielectric. *J Polym Sci Polym Phys* 1996;34(3):517–526.
116. Ibar JP. *Fundamentals of Thermally Stimulated Current and Relaxation Map Analysis*. New York: SLP Press, 1993.
117. Lobo H, Cohen C. Measurement of thermal conductivity of polymer melts by the line-source method. *Polym Eng Sci* 1990;30(2):65–70.
118. Oehmke F, Wiegmann T. Measuring thermal conductivity under high pressures. In *Antec'94*, San Francisco, 1–5 May, 1994; pp. 2240–2242.

Chapter 7
Thermal Analysis of Biomaterials

Showan N. Nazhat

Contents

Abbreviations	257
7.1 Biomaterials	257
7.1.1 Introduction	257
7.2 Material classes of biomaterials	260
7.2.1 Metals and alloys	260
7.2.2 Ceramics and glasses	260
7.2.3 Polymers	261
7.2.4 Composites	262
7.3 The significance of thermal analysis in biomaterials	262
7.4 Examples of applications using dynamic mechanical analysis (DMA) in the development and characterisation of biomaterials	263
7.4.1 Particulate and/or fibre-filled polymer composites as bone substitutes	264
7.4.2 Absorption and hydrolysis of polymers and composites	270
7.4.3 Porous foams for tissue engineering scaffolds	271
7.5 Examples of applications using DSC in the development and characterisation of biomaterials	275
7.5.1 Thermal history in particulate-filled degradable composites and foams	275
7.5.2 Plasticisation effect of solvents	276
7.5.3 Thermal stability and degradation	278
7.5.4 Setting behaviour of inorganic cements	278
7.6 Examples of applications using differential thermal analysis/thermogravimetric analysis (DTA/TGA) in the development and characterisation of biomaterials	280
7.6.1 Bioactive glasses	280
7.6.2 Thermal stability of bioactive composites	282
7.7 Summary	283
References	283

Abbreviations

BCP: Biphasic calcium phosphate.
HA: Hydroxyapatite.
In vitro: In an experimental situation outside the organism. Biological or chemical work done in the test tube (*in vitro* is Latin for 'in glass') rather than in living systems.
In vivo: Biological or chemical work done in a living organism.
PDLLA: Poly(DL-lactic acid).
T_c: Crystallisation temperature.
T_g: Glass transition temperature.
TiTa30: Titanium–tantalum alloy
T_m: Melting temperature.

7.1 Biomaterials

7.1.1 Introduction

The field of biomaterials is currently recognised as a multidisciplinary field that encompasses the expertise of materials scientists, engineers, chemists, biologists and clinicians working together to solve problems in medicine through the use of materials. Although the use of biomaterials dates back to more than 2000 years ago, for example through the use of metals as teeth or collagen as sutures, it was not until the turn of the twentieth century that saw the expansion of the biomaterials field, particularly through the development of ceramics and synthetic polymers. Nowadays, biomaterials originate from a wide range of material classes including metals, ceramics, glasses, polymers and composites, which before use must undergo rigorous examination in order to be acceptable for use in the body.

Over the years there have been numerous definitions for biomaterials and the most quoted definition states that 'a biomaterial is a nonviable material used in a medical device, intended to interact with biological systems' [1]. However, within the last decade research into biomaterials has moved towards regenerative medicine and the engineering of tissues where 'organoids' are developed through the combination of viable cells and materials.

Any material can be a biomaterial as long as it serves the stated medical and surgical requirements, and in fact materials from all classes have been used as biomaterials. Biocompatibility is a general term that describes the biological performance of materials and refers to a number of processes that occur at the interface between biomaterials and tissues. It has been defined as 'the ability of a material to perform with an appropriate host response in a specific application' [1]. Classically, biomaterials needed to be inert in that they needed to be corrosion resistant, non-toxic and biocompatible. This means that while the material is in the body it must not elicit a specific interaction with the host. However, it has long been recognised that no material is totally inert in the body and while all biomaterials need to be non-toxic, they can also be bioactive or biodegradable. Bioactive materials are designed to elicit a specific biological response at the tissue/material interface, which results in the formation of a bond between the tissues and the material. In contrast, biodegradable materials should safely breakdown within the body over a period of time, and be replaced by the surrounding tissues. In addition to not producing an inflammatory reaction, a biomaterial must meet certain performance requirements, which include mechanical performance, durability and physical properties. These originate from the need to perform a physiological function consistent with the physical (bulk) properties of the material. The implanted material is expected to perform mechanically during its planned lifetime: it must be able to withstand any anticipated physiological loading without substantial dimensional change, catastrophic brittle fracture, or fracture in the longer term due to creep, fatigue or stress corrosion [2].

The success or failure of a biomaterial device is related to the effect of the biomaterial on the host and also the effect of the host tissues on the biomaterial. These consequences will be significantly dependent on both the bulk and the surface properties of the material. At the surface, the interaction of the host tissue or cells with the biomaterial is affected by the chemical, biological or topographical properties of the material. Topography, or surface roughness, of materials has been shown to have an important role in increasing adhesion strengths; for example, porous surfaces on titanium implants enhance bone ingrowth. This adhesion is more biomechanical rather than of a chemically bioactive nature.

Table 7.1 provides a list of some of the many examples of materials used as biomaterials [3]; see also Figure 7.1 [4]. Depending on the type of application, these materials are moulded

Table 7.1 Some examples of biomaterials

Application	Types of materials
Skeletal system	
Joint replacement (hop, knee)	Titanium, Ti–Al–V alloy, stainless steel, polyethylene
Bone plate for fracture fixation	Stainless steel, cobalt–chromium alloy
Bone cement	Poly(methylmethacrylate)
Bony defect repair	Hydroxyapatite
Artificial tendon and ligaments	Teflon, Dacron
Dental implants	Titanium, Ti–Al–V alloy, stainless steel
Cardiovascular systems	
Blood vessel prosthesis	Dacron, Teflon, polyurethane
Heart valve	Reprocessed tissue, stainless steel, carbon
Catheter	Silicone rubber, Teflon, polyurethane
Organs	
Artificial heart	Polyurethane
Skin repair template	Silicone–collagen composite
Artificial kidney	Cellulose, polyacrylonitrile
Heart lung machine	Silicone rubber
Senses	
Cochlear replacement	Platinum electrodes
Intraocular lens	Poly(methylmethacrylate), silicone
Contact lens	Silicone–acrylate, hydrogel
Corneal bandage	Collagen, hydrogel

Reprinted from [3], Copyright (2004), with permission from Elsevier.

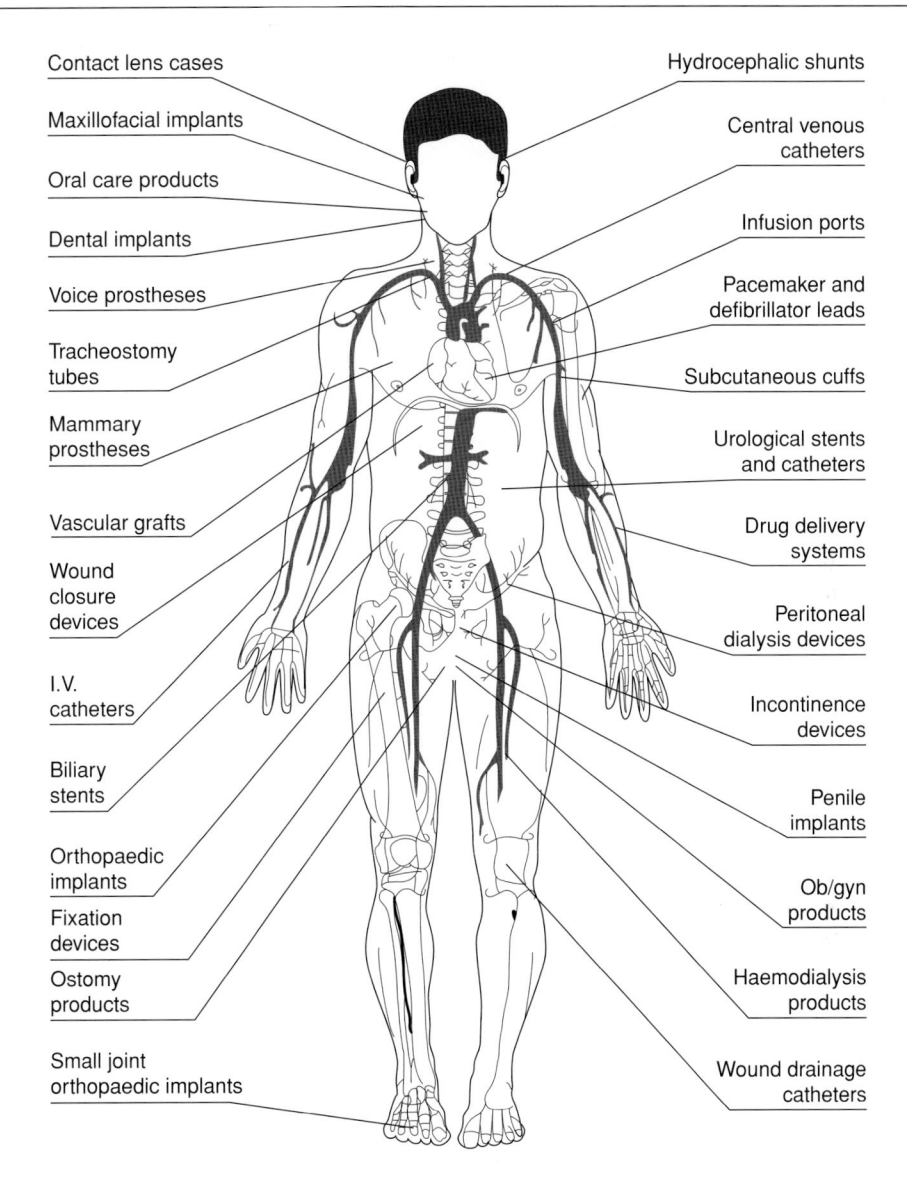

Figure 7.1 Anatomical range of medical devices. (From [4]. Copyright 1998. Copyright John Wiley & Sons Limited. Reproduced with permission.)

or machined to form the biomaterial device, or used as coatings, fibres, films, foams and fabrics. Therefore in many cases, the processing route will put the material through large temperature changes, which in turn may lead to possible changes in the material's physical structure, morphology and properties. Furthermore, a biomaterial must also be sterilised before use, which may involve heat or radiation and may result in further changes to the properties of the biomaterial. Therefore, the properties of a biomaterial must always be considered in its final fabricated sterilised form.

Thermal analysis is a useful tool for the characterisation of biomaterial properties and has routinely been used for the characterisation of biomaterials during their processing, as well as during clinical use. This chapter focuses on the applications of the different types of thermal analysis used in biomaterials characterisation.

7.2 Material classes of biomaterials

7.2.1 Metals and alloys

Metals and metal alloys tend to have significantly greater mechanical strength and fatigue properties than other classes of biomaterials. Therefore they have been used mainly in load-bearing applications for skeletal replacements which include hip and knee prostheses, plates, screws and nails. Non-load-bearing applications of metals include valves, heart pacemakers and cages for pumps. Metals have also been used as dental materials. For example, mercury-based amalgams have been used as restorative materials as they are easily processed at room temperature, and TiTa30 bonded to alumina are used in dental implants due to their similar thermal expansion coefficients.

Metallic biomaterials can be inert or bioactive. Stainless steel and cobalt–chromium are classic examples of inert metallic biomaterials, their inertness being due to a passive oxide layer on their surface. Titanium and their alloys fall into the bioactive metallic biomaterial group and have good bone-bonding abilities. As they also have favourable physical and mechanical properties, they have found increasing applications as orthopaedic and dental implants. Typically, metals and alloys are assessed thermally with differential scanning calorimetry (DSC) and differential thermal analysis (DTA) for T_m.

7.2.2 Ceramics and glasses

Bioceramics and bioactive glasses are biomaterials which enjoy use in bone healing. A common characteristic of bioactive ceramics or glasses (or glass–ceramic materials) is that their surface develops a biologically active hydroxyl carbonate apatite layer which bonds with collagen fibrils. This interface can resist substantial mechanical force. Hydroxyapatite (HA) is a natural component of bone that can also be synthetically processed into powders, solids and porous scaffolds. Bioglass®, developed by Larry Hench (1967), is one of the first completely synthetic materials with high bone-bonding abilities. These materials are generally composed of Na_2O–CaO–P_2O_5–SiO_2 and result in a single-phase amorphous materials that have low mechanical strength and toughness. Therefore these materials are used as particulates and coatings or in low load bearing applications. Bioactive glass–ceramics, on the other hand, are multi-phased materials, i.e. contain amorphous and crystalline regions, with better mechanical properties, but have a relatively lower bone-bonding ability. Due to their higher transition temperatures (T_g, T_c and T_m) ceramics and glasses are typically assessed for their thermal properties using high-temperature DTA equipment.

7.2.3 Polymers

Polymers constitute the largest group of biomaterials and can be of either synthetic or natural base. Since polymers are made of long-chain molecules, they can have a wide range of properties and therefore greater potential for biomedical applications. However, their properties tend to be more complex than the other classes of biomaterials. Generally synthetic polymers can be divided into three main classes, thermoplastics, thermosets and elastomers, all of which have found applications as biomaterials. Polymers can be fabricated with a range of properties and into numerous shapes such as fibres, sheets, rods, tubes, screws and foams. They can be processed into end-use shapes starting from monomers or low molecular weight pre-polymers where the final polymerisation process is carried out within the casting mould and once completed, the end product should be ready for use.

The morphology of polymers can be either amorphous or semi-crystalline, where the tendency of a polymer to crystallise depends on small side chains and chain regularities. Poly(methylmethacrylate) is an example of an amorphous thermoplastic biomaterial which is used in the applications of bone cement, artificial teeth, dentures, dental fillings, bone prostheses, intraocular lens, membrane for dialysis or ultrafiltration. Crystallisation in polymers leads to changes in their mechanical and thermal properties. Examples of semi-crystalline polymers used as biomaterials are low-density polyethylene (catheters, film substrate, tubing, connectors, bottles, plastic surgery implants, packaging materials), ultra high molecular weight polyethylene (acetabular cup in total hip prosthesis, tibial plateaux in artificial knee prostheses), polypropylene (disposable syringe, finger joint prosthesis, non-resorbable sutures), polytetrafluoroethylene (vascular graft prostheses) and polyethylene terephthalate (vascular drafts, heart valves, sutures). These are tough thermoplastics with relatively high modulus and low strain to break, whereas amorphous elastomeric (rubbery) polymers tend to have the opposite mechanical properties. At high temperatures, thermoplastics possess enough thermal energy for long segments to move randomly, while at low temperatures these long segments cease. They can therefore be melt-processed and their properties are sensitive to their processing history. Thermosets on the other hand cannot be softened on reheating due to cross-linking. Silicones (shunts, catheters, artificial skin for burns, plastic surgery implants, artificial heart, heart assist pump materials, drug release systems, facial and ear prostheses, denture lining materials, tendon and finger joint repair, tracheal and bladder prostheses) and epoxy resins are examples of thermosets used as biomaterials.

Copolymers are a combination of two monomers that are either polymerised randomly or exist in block form. The properties of random block polymers approximate to the weighted average of the two monomers whereas block copolymers tend to phase-separate into two phases exhibiting properties that are rich to each of the monomer phases. Degradable sutures made of α-polyesters of poly(glycolides-co-lactides) are an example of random copolymers used as biomaterials. Polyurethanes, on the other hand, are an example of block copolymers used as biomaterials as they contain hard and soft segments. They are tough with good blood-containing properties and consequently are used in vascular applications such as grafts, and heart assist balloon pumps. Polymers can also be blended, to create new polymers that combine properties of the individual polymers.

Unlike synthetic polymers, natural polymers have the advantage of being very similar to macromolecular substances which the biological environment recognises. Numerous types of natural polymers, from plant or animal origin, have been researched. Examples of plant-originating polymers are cellulose, starch and natural rubber. Collagen, silk, fibrin, hyaluronan, elastin, chitin, and deoxyribonucleic acid (DNA) are examples of animal-derived polymers. Natural polymers have the ability to produce biomaterials with biological function at the molecular level rather than at the macroscopic level. These polymers degrade enzymatically, which is a process that can be controlled through chemical cross-linking. However, since they are structurally more complex than synthetic polymers, their technological manipulation is much more elaborate.

Collagen is the most frequently used protein-based biomaterial. It originates from mammalian connective tissue extracellular matrix and has been used in many different forms including sponges, membranes, fibres and gels. These biomaterials cannot be thermally processed as they decompose below their melting temperature. The characteristic triple helix structure of collagen is converted to randomly coiled gelatine by heating above the helix–coil transition temperature, which is approximately 37°C for bovine collagen. This will have clinical implications as gelatine is more rapidly degradable than collagen and is therefore used in systemic drug delivery devices.

7.2.4 Composites

A composite is a new complex of two or more materials put together in order to combine the properties of the separate components. They are generally classified on the basis of their structure; for example, they are classed as fibrous when composed of fibres in a matrix, laminar when composed of layers of materials, and particulate when composed of particles in a matrix. The properties of composites are strongly influenced by their constituent materials, their content and distribution, as well as the nature and quality of their interactions. Composites can be of metal, ceramic and polymer-based matrices and have been developed to be bioactive, non-degradable and degradable. Composites comprising a matrix and a reinforcing agent are common.

The main reinforcing agents used in biomaterials research are carbon and polymer fibres, ceramics and glasses. Carbon fibres have been used to reinforce PTFE for soft tissue augmentation. Bioactive ceramic and glass particle reinforced polymeric matrices have been developed to produce biomaterials with improved mechanical and biological properties, e.g. HAPEX^{TM}, which is HA particulate reinforced high-density polyethylene that has been clinically used for middle ear implants. Fillers have also been used in bone cements to increase the modulus and the bioactivity as well as the radiopacity since the radiological detection of these cements is clinically important. Dental restorative composites were first developed as an aesthetic alternative to amalgam. These composites are based on particles of glass, quartz, filled with methacrylates such as Bis-GMA and have the ability to undergo either chemical or light activated curing.

7.3 The significance of thermal analysis in biomaterials

In any biomaterial application, the composition and structure of a candidate biomaterial relevant to the properties under consideration should be evaluated. Generally, the biomaterial specification is set at the beginning of the design cycle and the properties of the candidate biomaterials evaluated at that stage. At this initial phase, the thermal properties of a biomaterial should be considered as important as its physical, electrical, mechanical, chemical and biological properties. Thermal characterisations will provide general information on stability, shrinkage, expansion, effect of sterilisation methods, insulation and storage. Furthermore, the thermal behaviour of a material forms an important part of characterising a material's properties as it is intrinsically linked to the morphology and therefore has a direct effect on its clinical performance. An evaluation of the thermal behaviour of biomaterials post *in vivo* or *in vitro* testing or explanted can also provide a valuable insight into the behaviour of the biomaterial in a biological environment, e.g. its oxidation/crystallisation.

It is a misconception to suggest that tests should not be carried out at low or high temperatures since materials only need to perform at ambient or physiological temperatures. In fact, thermal analyses should be carried out across a wide temperature range as they will provide a wealth of information on the material's amorphous versus crystalline phases, thermal transitions, mechanical properties, heat of reaction in the body, and residual cure as well as working and setting time. For example, before a polymer is used as a biomaterial, it must be processed into the desired shape which can induce physical, thermal and mechanical changes. High molecular weight polymers often require additives, such as antioxidants, UV stabilisers, reinforcing agents, lubricants, mould release agents and plasticisers, which can alter the properties of the polymers and also result in the formation of new compounds. Furthermore, prior to clinical use, biomaterials will need to be sterilised to reduce the risk of infection. Sterilisation methods include the application of either wet (steam) or dry heat, chemical sterilisation or irradiation. Exposing polymers to heat, chemicals or ionising radiation may affect their properties, for example because of chain scission or cross-linking. Therefore, it is vital to evaluate the effect of sterilisation on the properties of the biomaterial.

The thermal behaviour of materials can also provide important information about the structure and morphology of a material. For example, while most synthetic polymers have a glass transition temperature (T_g) associated with amorphous structure in the material, only polymers with regular chain architecture can crystallise and so have a melting temperature (T_m). These in turn can have a direct effect on the mechanical performance of the materials since below T_g polymers tend to be glassy and become more rubbery above T_g. These thermal properties can also be used to identify or verify the nature of the composition. For example, random copolymers will only exhibit one T_g that will be somewhere in between the T_g's of the individual homopolymers, whereas block copolymers will exhibit T_g's characteristic of each homopolymer but will be slightly shifted due to imperfect phase separations. Similarly this can be applied to polymeric blends, which are essentially two polymeric systems mixed together.

7.4 Examples of applications using dynamic mechanical analysis (DMA) in the development and characterisation of biomaterials

The following sections will highlight some examples of how thermal analysis techniques have been applied in the characterisation of biomaterials to demonstrate the versatility of these complementary techniques.

Over the past two decades, DMA has proven to be a useful technique for the characterisation of biomaterials since it not only gives a quantitative assessment of material properties such as stiffness and damping, but also provides structural information. This is because the dynamic mechanical properties of materials are sensitive to all kinds of transitions, relaxation processes, structural heterogeneity and morphology of multi-phase systems such as crystalline polymers, polyblends and composites. DMA can also pinpoint thermal transitions; for example, typical output of tan δ versus temperature will display a peak at T_g. Above T_g, peaks correspond to the crystalline regions and eventually T_m. As a technique, DMA is also sensitive for the characterisation of polymers of similar chemical compositions, as well as detecting the presence of moderate quantities of additives such as plasticisers or leachable materials. For example, PVC is very stiff, however, with the addition of plasticisers it can be made more flexible.

Generally, the viscoelastic behaviour of polymeric systems can be used to classify their thermal behaviour. In all polymers, the modulus is relatively high below the T_g (see Section 4.1.3) and they are glassy in nature. For linear amorphous polymers (thermoplastics), the onset of T_g induces a significant change in the modulus and the material becomes leathery in nature in this region. For some amorphous polymers, the region that occurs beyond T_g is of relatively constant modulus and is described as the rubbery plateau region where long-range segmental motion occurs but the thermal energy is insufficient to overcome entanglement interactions that inhibit flow. As the temperature is increased the material will begin to flow, and a further decrease in modulus is seen over a narrow temperature range. This temperature range is important in terms of processing the polymer into shapes such as orthopaedic screws and plates, to suit the end application. The plateau temperature region of semi-crystalline polymers tends to be longer so that the modulus is more stable over a longer temperature range. This is due to the reinforcing effect of the crystals. Chemically cross-linked polymers (thermosets) on the other hand do not display flow behaviour as the cross-links inhibit flow at all temperatures below degradation temperature. Therefore, these cross-linked polymers cannot be melt-processed.

7.4.1 Particulate and/or fibre-filled polymer composites as bone substitutes

DMA is a particularly useful tool for the characterisation of bone substitute materials. Bone is viscoelastic within the range of physiological loading and this should be recognised as an essential feature of an optimised prosthetic material for bone [5]. Pure polymers generally have insufficient mechanical properties for applications as bone fracture fixation devices in load-bearing applications. Consequently, various forms of composite technologies have been used to enhance their mechanical properties, including particulate and fibre reinforced polymer composites. Reinforcing elements of different bioceramics and bioactive glasses have been applied in particulate form. The addition of these fillers into the polymer matrix is also used to enhance their osteoconductivity and potential bone-bonding purposes. As these materials are of relatively high stiffness, flexural, three-point bend analyses have been undertaken on rectangular specimens. The sample rests on two supports (e.g. 20 mm apart) and is loaded by means of a knife-edge probe tip midway between the supports. The aspect

Thermal Analysis of Biomaterials 265

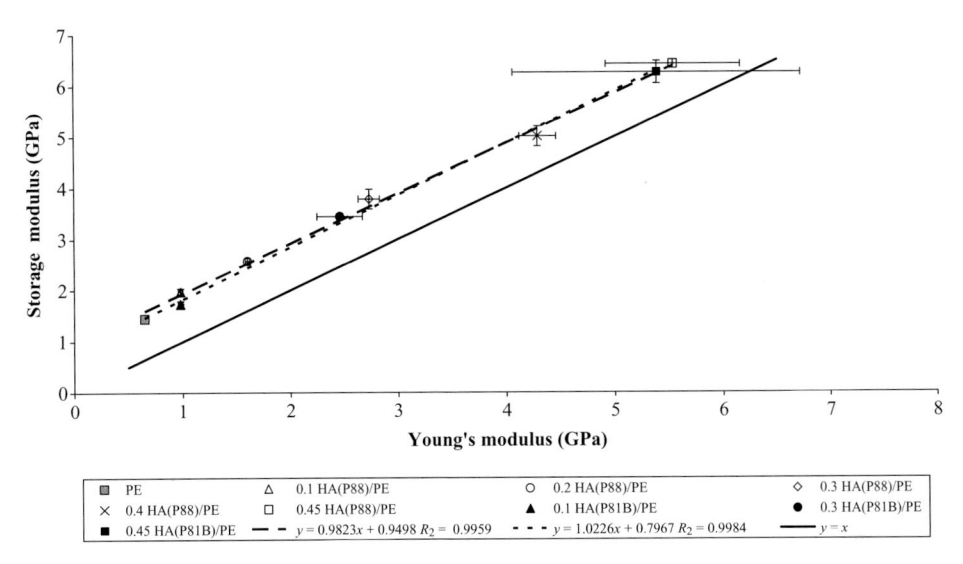

Figure 7.2 Storage modulus versus Young's modulus for HA(P88)/PE and HA(P81B)/PE composites. The HA/PE samples were prepared from moulded plates, which were sectioned into rectangular beams (1 × 3 × 24 mm³). The specimens were lightly dry polished using silicon paper to improve the surface finish and to remove cutting marks. Five specimens were prepared for each variable. Analyses were carried out in three-point bend, where the beam rests on two supports and is loaded by means of a knife-edge probe tip midway between the supports 20 mm apart. A temperature scan mode of analyses was carried out using the conditions as given in Table 7.2, but with a temperature range of 20–100°C. E' measured at 20°C versus E measured at room temperature for both composites are linearly related with a gradient within 2% of unity but with an offset. The observed increase in E' with increasing HA volume fraction is comparable to Young's modulus observed in quasi-static analysis. (From [6], with kind permission of Springer Science and Business Media.)

ratio of the samples should ideally be greater than 14 to ensure a flexural action (smaller aspect ratios lead to a greater shear influence). However, DMA studies on particulate systems have shown that an aspect ratio of 12 is still acceptable. In a study on particulate composites of HA reinforced semi-crystalline matrices of high-density polyethylene (HAPEX™) developed for permanent bone replacement, the effects of volume fraction and HA particle size on the dynamic mechanical properties were investigated. As can be seen in Figure 7.2 [6] the observed increase in E' (measured at 20°C and 1 Hz) with increasing HA volume fraction is comparable to Young's modulus observed in quasi-static analysis. Higher E' values are expected due to the higher strain rate and lower strain levels used in DMA compared to quasi-static testing. This confirmed the accuracy of the DMA technique in measuring the modulus of materials compared to those obtained from standard tensile testing.

The properties of polymeric composites depend primarily on the relative amounts and properties of the filler and matrix as well as the quality of interaction or interface between them. The shape of the particles and efficiency of dispersion and packing of the filler also influence the magnitude of the enhancement. To analyse the efficacy of the filler, relative modulus (E_r^I) values could be calculated according to Equation (7.1) [7] where E_c^I and E_m^I

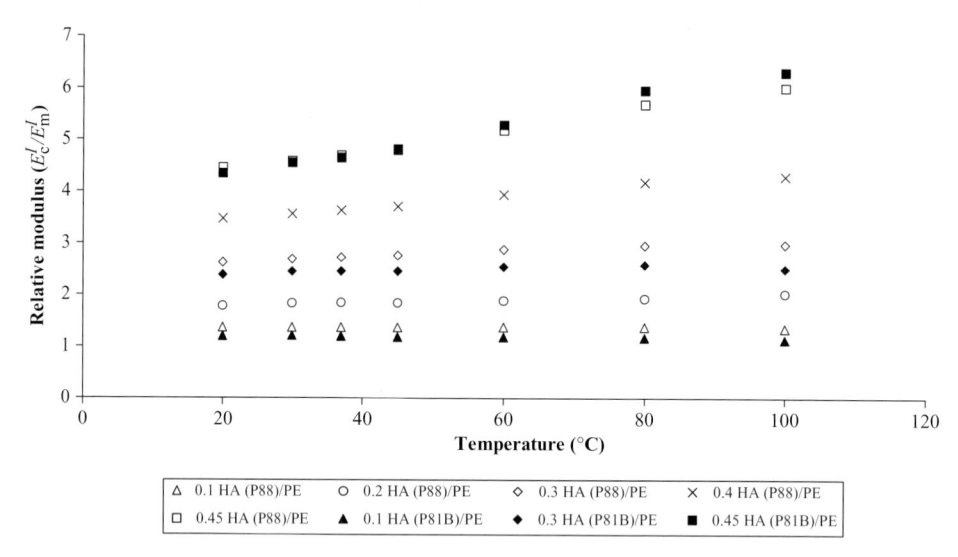

Figure 7.3 Relative modulus versus temperature for HA/PE composites (Equation (7.1)). The samples were prepared and tested as described below. It can be seen that E_r^I is always >1, confirming the reinforcing effect of HA, namely as the HA volume fraction increases, so does the relative modulus. At lower volume fractions, 0.1 and 0.3, the smaller P88 particles (median HA particle size = 4.53 μm) gave higher moduli at an equivalent volume fraction than P81B particles (median HA particle size = 7.43 μm). This trend was not reflected with the 0.45 volume fraction where above 60°C there was a crossover between the two composites, when the larger particles (P81B) gave higher values than the smaller (P88) particle reinforced composite. It can also be seen that for HA(P88)/PE the values of relative modulus for 0.1 and 0.2 volume fractions do not appear to vary significantly with temperature, whereas those obtained at 0.3, 0.4 and 0.45 volume fractions increase with increasing temperature. The slope of relative modulus versus temperature increased with higher HA volume fraction, indicating that the decrease in storage modulus was dependent on the amount of HA in the composites. The increase in relative modulus with temperature suggests that HA at higher volume fractions tends to control the values of E^I. (From [6], with kind permission of Springer Science and Business Media.)

are the storage moduli of the composite and matrix, respectively, at a specific temperature:

$$E_r^I = \frac{E_c^I}{E_m^I} \qquad (7.1)$$

Figure 7.3 [6] shows the relative modulus values as a function of temperature indicating that the fillers with smaller particle size have a greater effect in increasing the modulus compared with the similar composites of the same volume fractions.

Degradable three-component composite systems consisting of the amorphous degradable poly(DL-lactic acid) (PDLLA) self-reinforced with unidirectional semi-crystalline poly(L-lactic acid) fibres and/or particulate HA have also been investigated by DMA in three-point bend. Testing was performed using a temperature scan mode with parameters as given in Table 7.2, with the temperature measured with a thermocouple positioned approximately 1 mm away from the sample. Helium- or oxygen-free nitrogen gas can be used as purge in

Table 7.2 Typical test conditions applied in DMA temperature scan methods

Parameter	Conditions
Temperature range	−50–100°C
Heating rate	4°C/min
Static control – tension	120%*
Dynamic control – strain	0.02%
Frequency	1 Hz

*The static stress applied was maintained at 120% of the value of dynamic stress.
From [8]. Copyright 2001. Copyright John Wiley & Sons Limited. Reproduced with permission.

the furnace and a liquid nitrogen bath maintained an isothermal environment outside the furnace. A heating rate of 4°C/min is usually chosen since it gives the best compromise between accuracy and speed of testing. The acceptable range using this method should be 1–5°C/min. Figure 7.4a [8] shows the storage modulus values of the various composites as a function of temperature. At sub-ambient temperatures, below the T_g of the matrix, the materials displayed a fairly constant storage modulus value. T_g can be defined in a number of ways including the peak in loss modulus (E'') tan δ and onset of storage modulus. In this case, the former was used and the PDLLA T_g was calculated at 45°C, which increased with filler addition as can be seen by the broadening of the reduction in E'. The loss modulus curves showed two peaks in the composites containing PLA fibres. The second peak results from T_g of the amorphous phase of the fibres (Figure 4b).

Similar effects of fillers onto the matrix T_g were also observed in polymer films prepared by solvent casting. Polymer and composite films are useful for characterising the effect of filler on the matrix in terms of mechanical properties, bioactivity and thermal properties. Particulate fillers have also been predicted to have a controlling effect on the degradation of the matrix. In films, DMA tests can be carried in tension on specimen strips. Figure 7.5 shows that the T_g (measured by using the peak in loss modulus) of the matrix increased with increasing amounts of biphasic calcium phosphate (BCP) in PDLLA films [9]. An increase in T_g of the polymer matrix has been observed by reinforcing with bioceramic particles (e.g. HA and BCP) and is attributed to the filler–polymer interaction and partial immobilisation of the polymer and the physical blocking of a number of molecular configurations as a result of adsorption onto the filler surface. This would cause a change in the density of the packing of polymer chains and a modification of conformity and orientation in the neighbourhood of the surface. In turn, this would lead to the adsorbed layer being more rigid than the polymer matrix that is away from the particle, thus broadening the transition region and increasing the range of the transition temperature. This has either been achieved through mechanical interlocking which can be dependent on the size and surface roughness of the particulate filler or when there is a chemical bond as achieved through surface treatment. Glass particulates in general have smoother surfaces than ceramics and have lower interfacial adhesion between filler and matrix. Therefore the reinforcing effect of glasses without surface treatment would be lower than that of ceramics.

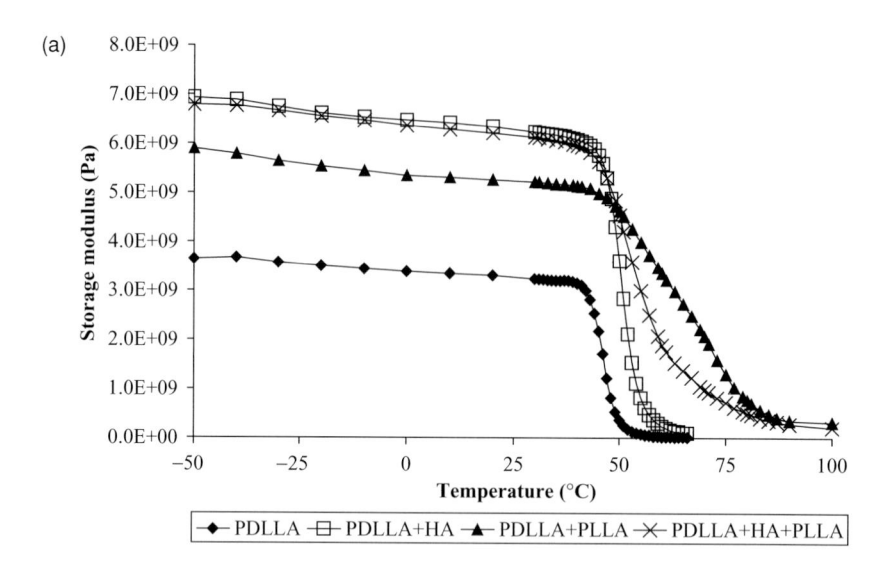

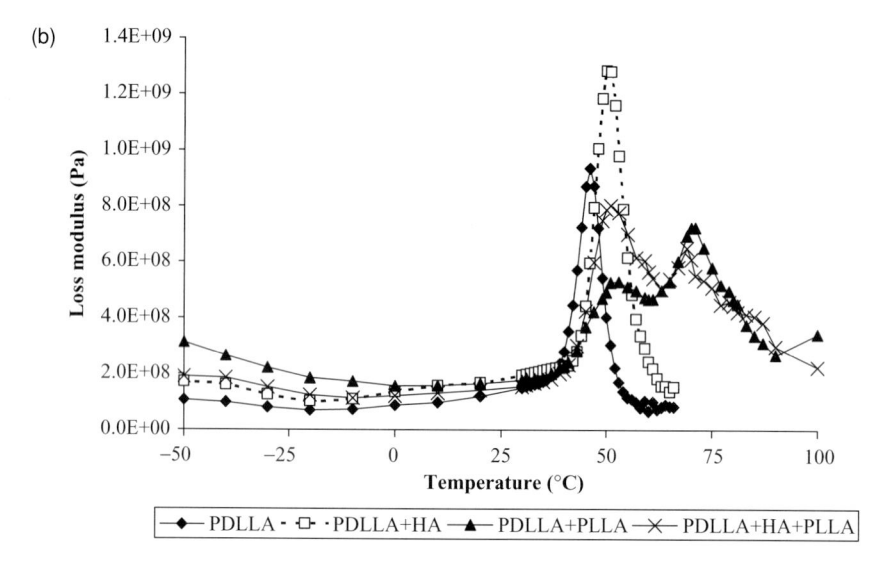

Figure 7.4 (a), (b) Mean storage modulus and loss modulus, respectively, of PLA fibre and/or HA particulate reinforced PDLLA matrix as a function of temperature. Tests were carried out using the conditions defined in Table 7.2. (From [8]. Copyright 2001. Copyright John Wiley & Sons Limited. Reproduced with permission.)

In particulate composites, the energy loss mechanism (damping) is complex. As well as the loss mechanisms induced by the polymer matrix, there is the possibility of additional loss mechanisms occurring at the filler–matrix interface and this is dependent on the extent of adhesion between the phases. The filler contribution to damping is extremely low compared to the matrix. If damping is only dependent on the volume fraction of the matrix, the

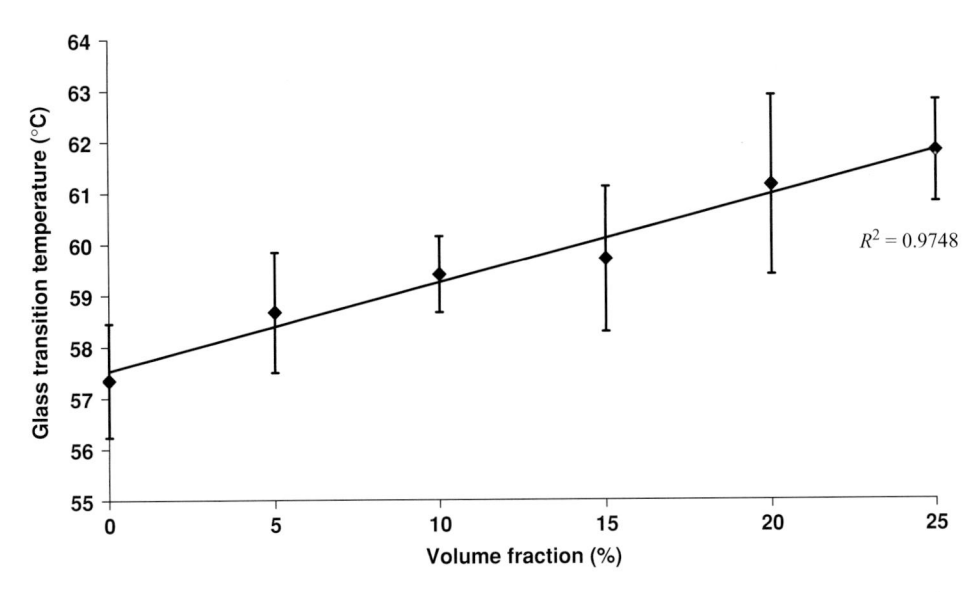

Figure 7.5 Glass transition temperature of PDLLA versus BCP particulate filler fraction investigating the filler on polymer films. BCP particles addition increased T_g of the amorphous PLA matrix and resulted in the broadening of the peak obtained in E''. The increase in T_g-PLA was linearly related to the volume fraction of BCP. The particle size distribution was monomodal with $d_{0.1} = 1.85$ μm, $d_{0.5} = 4.62$ μm and $d_{0.9} = 17.37$ μm and specific surface area was 59.79 m^2 g^{-1}. Strip specimens of approximately 15 × 4 mm^2 were cut and tested in tension. The grips were set 10 mm apart and this was used as the specimen length. Five specimens were cut for each V_f BCP. Testing was performed in the temperature scan mode heating from 20 to 80°C at a rate of 4°C/min. The static and dynamic controls were set at 0.2% strain and 0.1% strain, respectively, and the sample was tested at a frequency of 1 Hz. Helium gas was used as a purge in the furnace and a water bath was frozen with liquid nitrogen to maintain an isothermal environment outside the furnace. T_g of the composite was defined as the temperature where the loss modulus was a maximum. (Reprinted from [9], Copyright (2002), with permission from Elsevier.)

theoretical relative is given by Equation (7.2) where V_m is the volume fraction of the matrix in a composite, tan δ_c is the damping of the composite, and tan δ_m is the damping of the unfilled polymer:

$$V_m = \frac{\tan \delta_c}{\tan \delta_m} \tag{7.2}$$

If the damping in a filled polymer results from the same mechanism which produces the damping in the unfilled matrix then the ratio holds. However, it is possible that new damping mechanisms may be introduced that are not present in the unfilled polymer, which include particle–particle friction where particles are in contact with each other in the weak agglomerates that occur at high volume fractions.

DMA can also be used to investigate the effect of sterilisation on the structural and thermal properties of polymers and composites intended for biomedical use. Due to the sensitivity of some polymers to heat and solvent attack, one option to sterilise would be to γ-irradiate. In a study investigating the properties of composite materials consisting of bioceramic fillers,

Table 7.3 Effects of γ-irradiation on molecular weights and glass transition temperatures

	After processing			After radiation		
Sample	SEC M_w (g/mol)	MWD	DMA T_g (°C)	SEC M_w (g/mol)	MWD	DMA T_g (°C)
PEU-BDI	170 100	2.3	54.2 ± 0.2	102 000	2.2	53.5 ± 0.2
PEU20HA	147 900	2.1	56.4 ± 0.2	99 700	2.1	55.8 ± 0.2
PEU40HA	123 100	2.2	57.6 ± 0.5	86 500	2.1	56.4 ± 0.6
PEU20BCP	153 300	2.4	56.3 ± 0.3	99 300	2.2	55.0 ± 0.5
PEU40BCP	136 500	2.4	57.3 ± 0.2	99 300	2.2	57.3 ± 0.2

Average molecular weights as measured through size exclusion chromatography (SEC) and T_g values, determined from the peak of E'', before and after γ-irradiation. Rectangular bar specimens were tested using three-point bending over a temperature range of 20–80°C at a rate of 4°C/min, at 1 Hz. The presence of the fillers induced some degradation, i.e. loss of molecular weight, during processing, which may be due to higher shear forces in the twin-screw mixer. Thus, adding 40 vol % of filler produced more degradation than 20 vol %. In addition, HA caused slightly more degradation than BCP. More significant degradation occurred during the sterilisation of the composites. Polymer without any filler present suffered the greatest loss of molecular weight, about 40%. Glass transition temperatures recorded after irradiation were marginally lower than in the native materials due to the loss of molecular weight.
From [10]. Copyright 2002. Copyright John Wiley & Sons Limited. Reproduced with permission.

either HA or BCP, and biodegradable lactic acid based poly(ester-urethane) (PEU-BDI), the effects of processing and γ-irradiation were evaluated using size exclusion chromatography and DMA [10]. The composite components were mixed while the polymer was in the melt, followed by compression moulding of the mix. As can be seen from Table 7.3, the presence of the fillers induced some degradation, i.e. loss of molecular weight, during processing, probably due to higher shear forces in the twin-screw mixer. Thus, adding 40 vol % of filler produced more degradation than 20 vol %. In addition, HA caused slightly more degradation than BCP. More significant degradation occurred during the sterilisation of the composites. The pure polymer suffered the greatest loss of molecular weight, about 40%. The average molecular weights post sterilisation were virtually the same for all materials although they were different for the non-sterilised materials post processing. Irradiation did not cause cross-linking in PEU-BDI, which was confirmed by the gel content measurements. Glass transition temperatures recorded after irradiation were marginally lower than in the native materials due to the loss of molecular weight.

7.4.2 Absorption and hydrolysis of polymers and composites

DMA has been used to investigate the hydrolysis of degradable composites by immersing rectangular specimens in saline at 37°C for up to 5 weeks [10]. Tests were carried out on dried samples before and after incubation, to assess the degradation of lactic acid based poly(ester-urethane) incorporated with fillers of either HA or BCP. Initial three-point bend analysis, using similar test conditions outlined in Table 7.2, showed that as has been previously demonstrated, there was an increase in the storage modulus of the materials

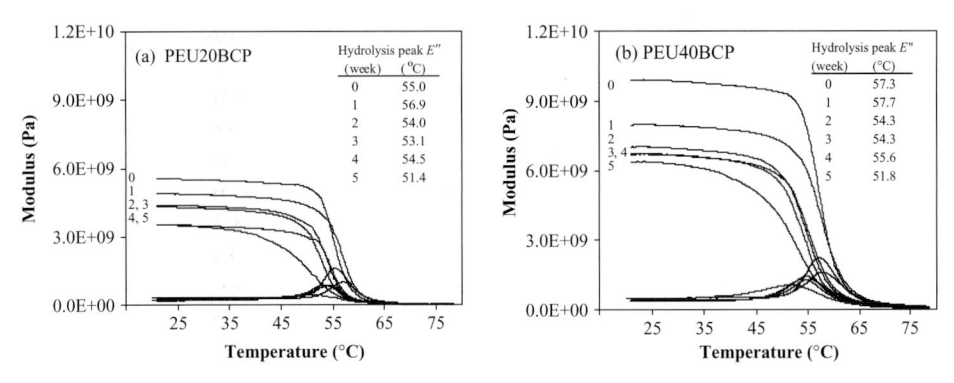

Figure 7.6 Dynamic mechanical properties of composites (a) PEU20BCP and (b) PEU40BCP over 5-week hydrolysis. Rectangular bar specimens were tested using three-point bending over a temperature range of 20–80°C at a rate of 4°C/min, at 1 Hz. Five repeat specimens (20 × 4 × 1.5 mm^3) were immersed in saline at 37°C for set periods and specimens were also vacuum-dried for 6 days before testing. Glass transition temperatures, measured through the peak in E'', for each material are also tabulated. Reduction in the storage modulus values is most prominent with the composites containing 40 vol % BCP and after 1 week the glass transition decreases in all materials. Some deformation of the amorphous polymer samples was observed over the hydrolysis time, whereas the dimensions of the composites were not significantly affected during the hydrolysis. (From [10]. Copyright 2002. Copyright John Wiley & Sons Limited. Reproduced with permission.)

with an increase in filler volume fractions. There was also an increase in T_g as there were good interactions between filler and matrix. Any changes in dynamic mechanical properties through polymer degradation can be monitored along with water absorption, mass loss and changes in molecular weights and crystallinity. Figure 7.6 shows the ageing behaviour of the composites containing 20 and 40 vol % BCP. As can be seen, over the 5-week period, it was found that the polymer matrix alone retained its stiffness compared to the composites containing 20 vol % as they lost their stiffness in a gradual manner. However, the reduction in the E' was most prominent with the composites containing 40 vol %, along with the matrix T_g which was lowered considerably.

7.4.3 Porous foams for tissue engineering scaffolds

DMA of porous foams developed for tissue engineering has been a focus of interest during the last decade. Tissue engineering scaffolds, generally made of natural or synthetic-based materials, should provide mechanical support for cells while at the same time be porous to allow for cell growth and the migration and transport of nutrients and oxygen. Numerous forms of scaffolds have been developed through gas foaming, salt leaching, thermally induced phase separation (TIPS) and the electrospinning of nanofibres into porous structures. However, unlike in non-porous materials, the modulus of composites foams is affected by a number of factors such as filler–matrix interaction, apparent density of the foam, porosity, and polymer cell wall size and shape as well as the orientation effects of the gross foam.

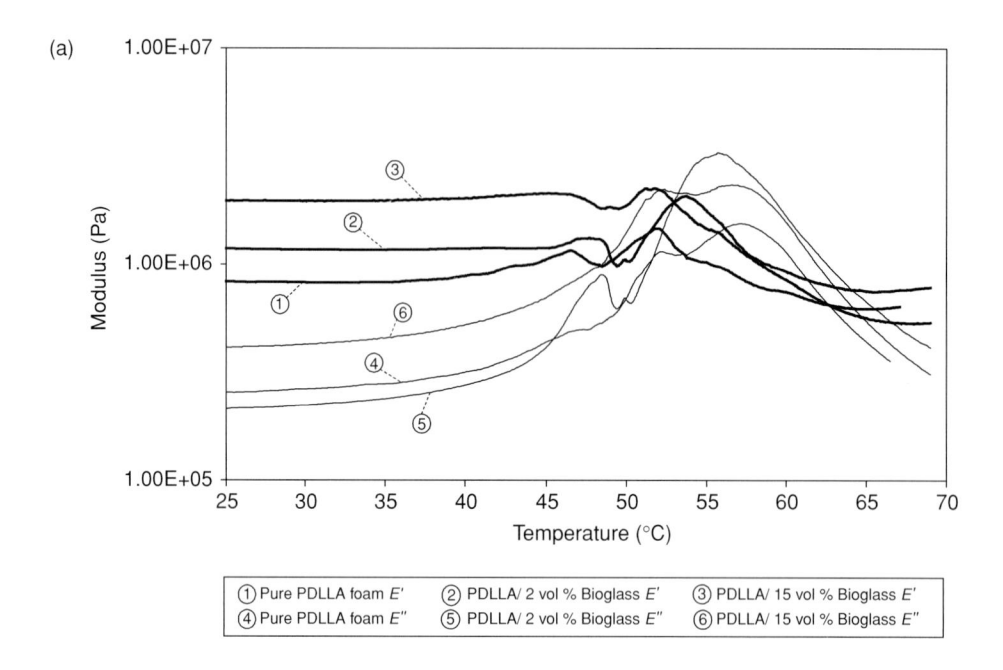

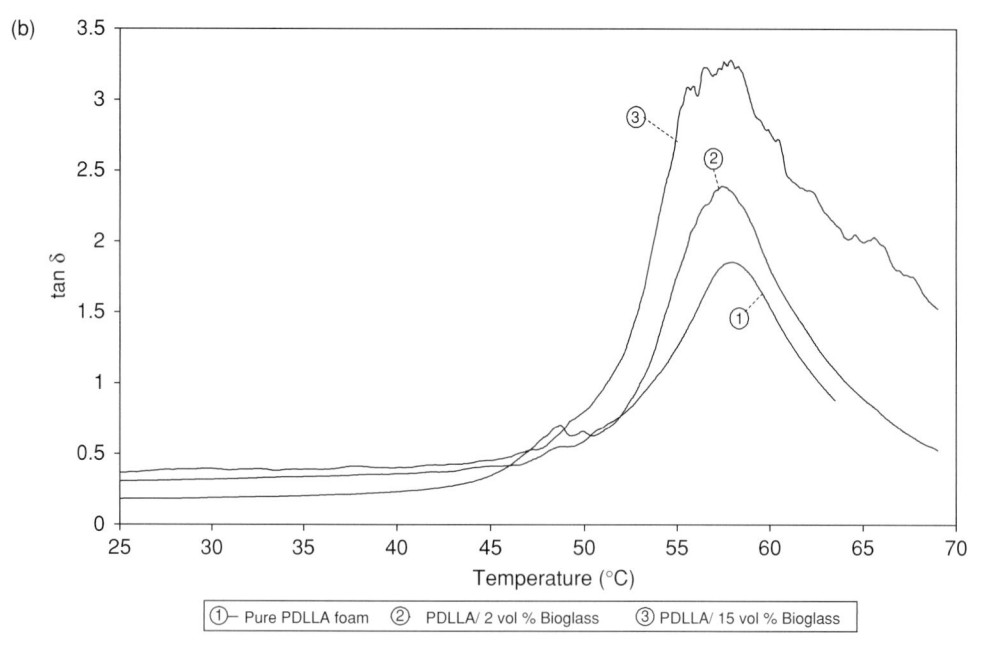

Figure 7.7 (a) DMA of as-fabricated pure PDLLA foam (1), PDLLA/2 vol % Bioglass® foam (2), PDLLA/15 vol % Bioglass® (3), showing the storage modulus (E') and loss modulus (E'') variation with increasing testing temperature; (b) DMA of as-fabricated pure PDLLA foam (1), PDLLA/2 vol % Bioglass® foam (2), PDLLA/15 vol % Bioglass® (3), showing the variation of mechanical loss tangent (tan δ) response with increasing temperature. Tests were carried out on cubic samples (measuring 5 × 5 × 5 mm³) obtained from the homogeneous region of each of the three foam systems. Five repeat specimens were tested in

Figure 7.7 [11] shows the DMA properties of highly porous (>90% porosity) foams of Bioglass® incorporated PDLLA that were produced through TIPS and subsequent solvent sublimation. The resultant foams are highly anisotropic with continuous tubular macro-pores of ~100 μm that are ordered in parallel and interconnected micro-pores of 10–50 μm. Due to the anisotropic nature of the foams and processing route, DMA was carried out in compression on cubic specimens. As can be seen, there was no effect of temperature on the viscoelastic parameters up to the onset of the matrix T_g. The onset of softening above 45°C, which was defined by an immediate apparent increase in E', may be due to immediate densification (increase in density) of the foams due to the onset of T_g. This behaviour was a result of the materials' porous nature and as the temperature approached the transition there was an apparent collapse of the structure resulting in an increase in density and possible increase in contact area with the compressional plates causing a temporary increase in E'. This illustrates the need for care when interpreting data generated in compressional modes of analysis. These data are in contrast to non-foamed samples which demonstrate a reduction in E'. As the temperature increased during the testing of the foams, this reduction in E' is observed and is accompanied by peaks observed in E'' and tan δ. The glass transition temperature in these foams did not increase in the same way for composites containing ceramics due to the lack of thermal processing and the smoother surface finish of the glasses.

Foams of elastomeric polymer blends resulting in rubber-toughened heterocyclic methacrylates developed for the repair of soft tissues have also been assessed through DMA. Here the foaming was carried out through supercritical carbon dioxide [12]. DMA was carried out to investigate the effect of blend composition on the viscoelastic properties and locate the thermal transitions of the various foams. These were based on styrene–isoprene–styrene (SIS) (with 15 wt % styrene) and tetrahydrofurfuryl methacrylate (THFMA) compositions. Figures 7.8a, 7.8b and 7.8c show E', E'' and tan δ versus temperature for 30/70, 50/50 and 70/30 SIS/THFMA respectively. As can be seen, the dynamic mechanical properties of the materials showed distinct behaviours with the modulus and thermal transition features generally moving to higher temperatures with an increase in THFMA content. There was an increase in the initial storage modulus (measured at 25°C) with decreasing SIS content. In each case these values remained relatively constant, until the onset of thermal transition. Since these are essentially polymer blends, two transition regions were observed, as demonstrated by a peak obtained in E'' and the onset of increase in the tan δ values; one at 65°C, which is associated with poly(THFMA) and another at around 90°C which is associated

Figure 7.7 (*Continued*) compression under the stress control mode, where a static stress of 30 kPa was initially applied followed by a dynamic stress of 24 kPa, at 1 Hz. A temperature scan from 25 to 75°C and a heating rate of 4°C/min was applied. There was a trend observed towards an increase in the storage modulus with higher volume fractions of Bioglass®. There was no effect of temperature on the parameters up to the onset of softening above 45°C, which was defined by an immediate apparent increase in E' that may have been due to the immediate collapse and densification (increase in density) of the foams as a result of the onset of the PDLLA glass transition. As the temperature increased further there was a significant decrease in E' in all the materials, which was accompanied by peaks in E'' and tan δ that are associated with the glass transition. The broadening of the peak in tan δ with increasing amounts of Bioglass® is likely to be due to regions of order within the polymer chains in close proximity to the glass particulates. The noisy appearance of tan δ above T_g can be attributed to the foams and compression nature of the tests. (Reprinted from [11], Copyright (2005), with permission from Elsevier.)

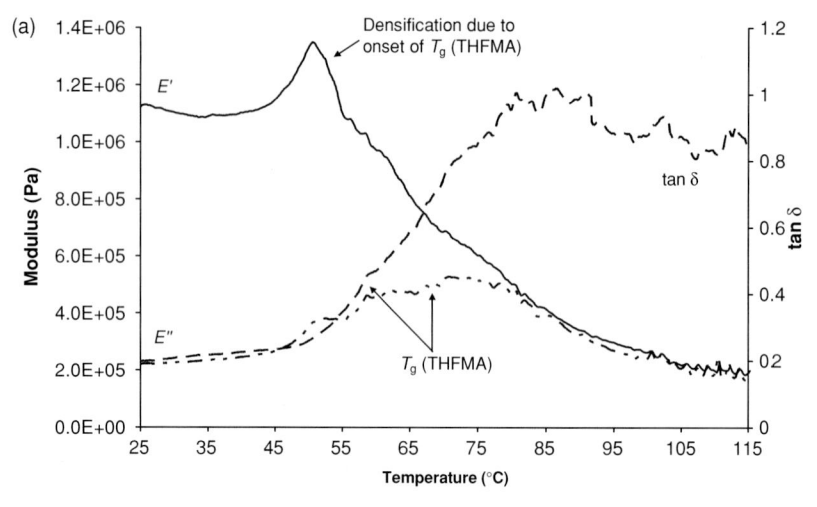

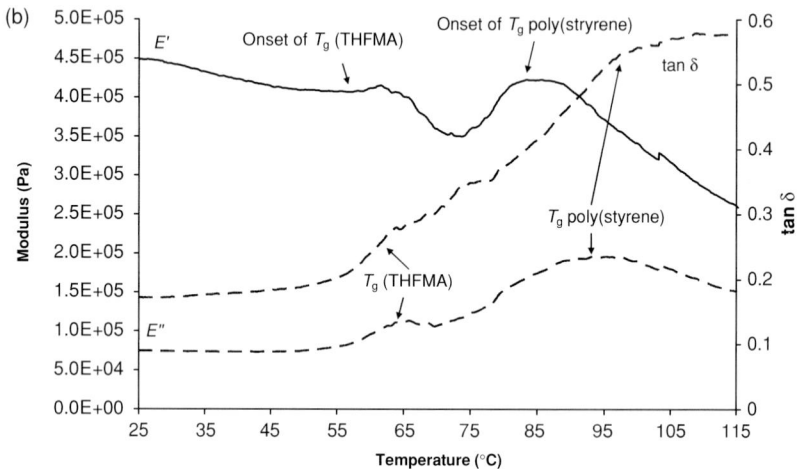

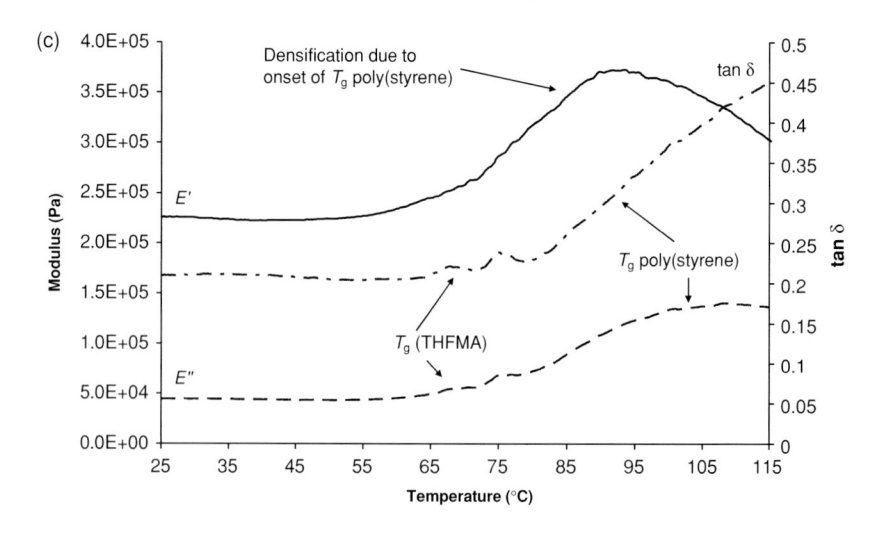

with poly(styrene). Poly(isoprene), the third component in these blends, has previously been shown to have a transition in the region of −50°C [13] and is outside of this temperature range. As the test temperature approaches T_g of poly(THFMA), there was an apparent increase in the storage modulus, in particular for the 30/70 composition (Figure 7.8a). This apparent increase in storage modulus was demonstrated in two temperature regions for the 50/50 composition (as indicated in Figure 7.8b) and tended to shift to a higher temperature with increasing poly(styrene) content (forming 15 wt % in SIS) as demonstrated in the case of 70/30 SIS/THFMA (Figure 7.8c). This densification region is due to the porous nature of the scaffolds as described above. In all cases, beyond the densification region, there was a decrease in the storage modulus, a result of the T_g poly(THFMA) and poly(styrene), respectively, which were accompanied by a peak obtained in the loss modulus. The tan δ also underwent an increase as a consequence of the glass transition temperatures of the different components. The magnitude of this increase was found to be relative to the components undergoing the transition; i.e. the greatest change in tan δ in the 30/70 SIS/THFMA blend took place in the T_g poly(THFMA) temperature region as indicated in Figure 7.8a, whereas in the 70/30 blend it took place in the T_g poly(styrene) temperature region as indicated in Figure 7.8c.

7.5 Examples of applications using DSC in the development and characterisation of biomaterials

As a technique, DSC has been used extensively for the characterisation of the thermal properties of polymers and composites used as biomaterials. Typically, resultant graphs of the energy flow versus temperature or time can be used to identify a number of endothermal or exothermal transitions occurring in materials and parameters can be identified such as glass transition temperature (T_g), crystallisation temperature (T_c), melting temperature (T_m) and the heat of cure. In its most common use, DSC has been used to identify the T_g of polymeric biomaterials. For this a number of methods are possible: a constant heating rate (typically at 10 or 20°C/min), a modulated temperature (which overall is a much slower rate of heating) and recently HyperDSC™ (which operates at very high heating rates >100°C/min).

7.5.1 Thermal history in particulate-filled degradable composites and foams

Locating the polymer T_g and how this would change as a consequence of filler addition or the thermal history due to processing would be of obvious relevance to its physical and mechanical properties. Table 7.4 [14] gives the glass transition temperature of degradable composite foams developed for bone tissue engineering consisting of a semi-crystalline PLA matrix

---

Figure 7.8 The storage modulus (E'), loss modulus (E'') and mechanical loss tangent (tan δ) of the foamed SIS/THFMA blends (a) 30/70, (b) 50/50, (c) 70/30 versus temperature. Cylindrical samples were tested in compression on five repeat specimens which were sinusoidally loaded at a frequency of 1 Hz. A temperature scan from 25 to 120°C at 4°C/min was applied while nitrogen gas was used as purge in the furnace. A clear trend can be seen in the behaviour of the blends as the percentage of SIS decreases. (From [12]; reproduced by permission of The Royal Society of Chemistry.)

Table 7.4 DSC data of compression-moulded and foamed PLA-PG composites comparing the PLA-T_g taken from the first and second heating ramps

	Moulded composites		Foamed composites	
Material	1st heat T_g/°C	2nd heat T_g/°C	1st heat T_g/°C	2nd heat T_g/°C
PLA	60.3 ± 0.5	62.5 ± 0.2	68.2 ± 3.0	62.2 ± 0.1
PLA-5G	62.2 ± 0.1	62.1 ± 0.2	69.1 ± 6.1	62.3 ± 0.2
PLA-10G	61.7 ± 0.2	62.4 ± 0.1	70.6 ± 0.1	61.9 ± 0.5
PLA-20G	62.1 ± 1.0	63.0 ± 0.6	–	–

DSC was carried out on ∼5-mg samples taken through heat/cool/heat ramps from 0 to 200°C at a rate of 20°C/min. T_g values were calculated from the onset point on the heat flow versus temperature curve. From [14]. Copyright 2007. Copyright John Wiley & Sons Limited. Reproduced with permission.

reinforced with phosphate-based glass particulates and compared to compression-moulded samples. The measurements were carried out using a standard constant heating rate of 20°C/min of heat/cool/heat. The DSC data obtained through the first heating run showed an increase in PLA-T_g for all compositions when compared to the T_g of the compression-moulded samples. However, when all the thermal history that may have been introduced as a consequence of foam processing of high temperature and pressure was removed during the second heating run, similar T_g values were obtained for PLA and PLA-PG composites. This suggested that the apparent increase may have been due to the combined heat and foaming conditions. These findings were also confirmed by complementary DMA data.

7.5.2 Plasticisation effect of solvents

Poly(methylmethacrylate) has been used extensively as a biomaterial in dental and orthopaedic applications as it has the ability to chemically cure at room or physiological temperatures and forms a glassy tough polymer. As a technique, DSC has proven to be very convenient and versatile in the characterisation of the thermal properties of these materials. It has been used to investigate the polymerisation kinetics, the glass transition temperature and the estimation of residual cure using both isothermal and constant rate methods of testing [15–19]. DSC can also be used to determine the effect of immersing these polymeric bone cements in a wet environment since it has been found that the T_g of bone cements can be up to 20°C lower than dry samples when they have been aged in water at 37°C and are therefore water saturated. The T_g of water-saturated higher methacrylate cements such as butyl methacrylate can therefore drop to the physiological temperature range as water has a plasticisation effect. This in turn can lead to dimensional changes due to a cold flow effect, which may be clinically catastrophic [20]. The recent introduction of HyperDSC^{TM} may be useful in accurately assessing the plasticisation effects of solvents on T_g since the rapid heating rates would prevent the drying out of samples due to slower rates of heating.

In a similar manner to the plasticising effect of water, any residual monomer present due to incomplete reactions within polymeric bone cements is expected to influence the thermal properties [21]. Figure 7.9a compares the thermograms of the first heating cycles for bone cements cured at room temperature and at 37°C obtained by the constant rate method of

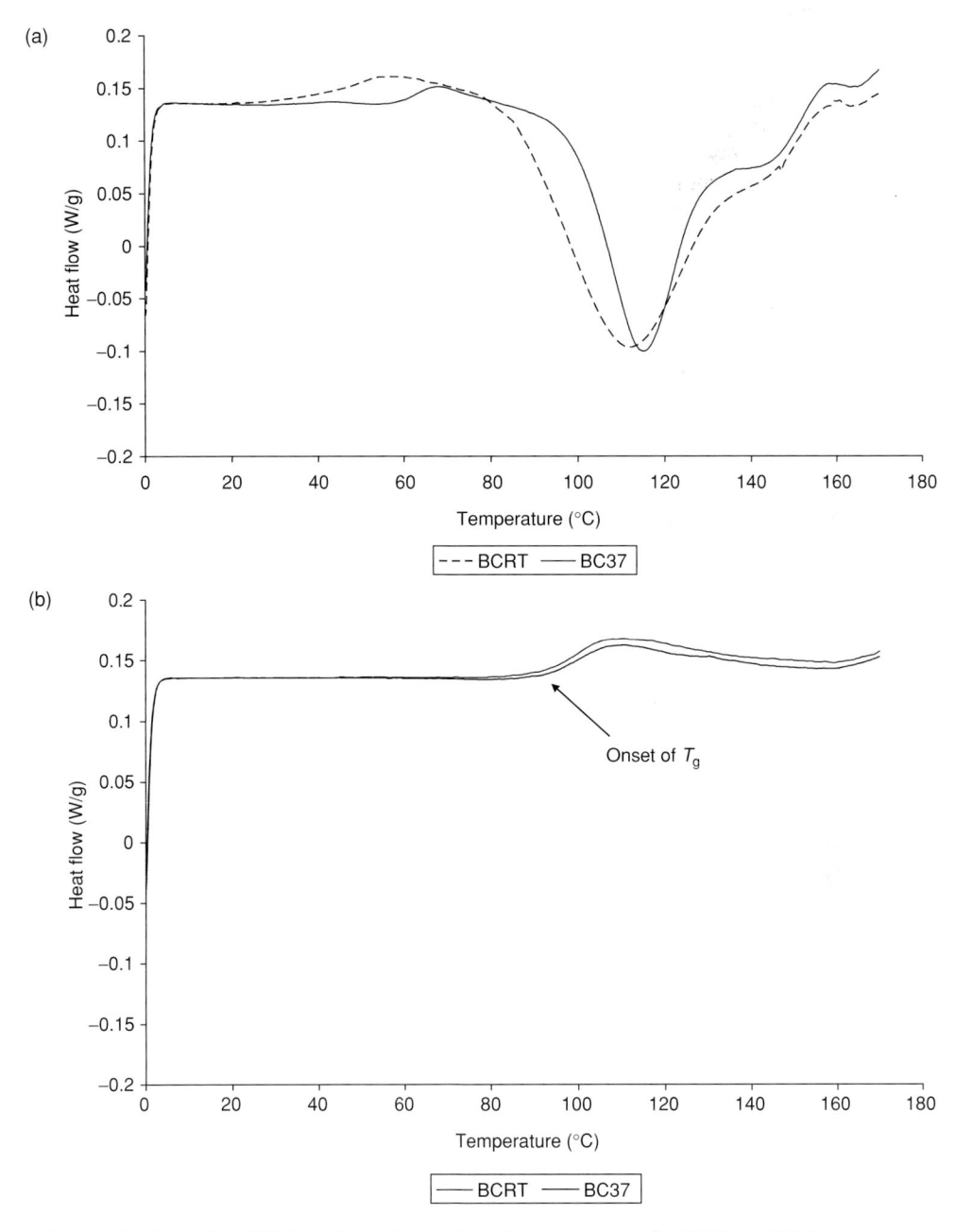

Figure 7.9 Parts (a) and (b) show first and second heating thermograms for BCRT and BC37 respectively. Tests were carried out under constant rate analysis consisting of three cycles, heating/cooling/heating, at a rate of 10°C/min from 0 to 170°C under nitrogen purge. T_g was characterised by a change in the upward direction of the heat flow and the value was calculated by the onset of change in the heat flow. (From [21], with permission.)

testing. As can be observed, in both the BCRT and BC37 there appears to be a low glass transition that is in the range of 40–60°C indicated by an endothermic step (upwards) in both thermograms. T_g was calculated using the onset of endothermic inflection. At temperatures above the transitions, both materials underwent exothermic peaks (characterised by a downward inflection in the thermograms) that reached a maximum at around 112 and 115°C for BCRT and BC37 respectively. If the curing is not complete during the initial setting of the bone cements, unreacted residual monomers trapped within the polymerised chains are released as the temperature increases above T_g, and therefore undergo further reactions. Figure 7.9b shows the second heating curve and, as can be seen, in all materials there was an absence of the residual exothermic peak and a clear T_g inflection showing that the heating process during the first cycle allowed the unreacted residual monomer to undergo further curing. This illustrates the fact that both residual exotherm values and the temperature of T_g can be used to estimate the degree of cure of a material. Generally, T_g of methacrylate-based bone cements is reported using the endothermic change in the thermogram obtained during the second heating cycle. This may be due to a number of overlapping events taking place during the first heating cycle, which make it difficult to calculate the glass transition temperature. These include the T_g of the bone cement, the overlapping of T_g of the pre-polymerised PMMA powder, and residual monomer reactions. Figure 7.10a shows the first heating cycles of PMMA-based cements with different amounts of triphenyl bismuth (TPB) that were added to render them radiopaque. As with the unmodified cements, these materials tended to not show a clear glass transition region. Also, it was observed that crystals of TPB appeared in the 25% TPB-containing specimens evidenced by a sharp melting peak at 80°C. Higher concentrations of TPB in the system would lower the net amount of TPB in solution and both rapid vitrification and decreasing solubility result in precipitation. TPB cements showed that there was a limit to the amount of TPB that can be dissolved in the methacrylate monomer. Figure 7.10b shows the thermograms obtained for the second heating cycle for the same material. As was also observed for the cements without TPB, a clear T_g transition region was obtained for these materials indicating the removal of any residual monomer as a consequence of further reactions. However, there was a reduction in T_g with TPB inclusion.

7.5.3 Thermal stability and degradation

The biodegradability of collagen films and sponges can be controlled by cross-linking, which can be either chemical or physical in nature. The level of cross-linking can be evaluated by the thermal stability of collagen and can be related to the susceptibility of the material to collagenase degradation [22]. Studies have shown that there is a direct link between the thermal stability as measured by a transition observed through DSC, and the time required for collagen to degrade.

7.5.4 Setting behaviour of inorganic cements

DSC has also been used to investigate the kinetics of fast setting inorganic based calcium phosphate cements (CPC). Conventionally, the setting times of CPC have been determined

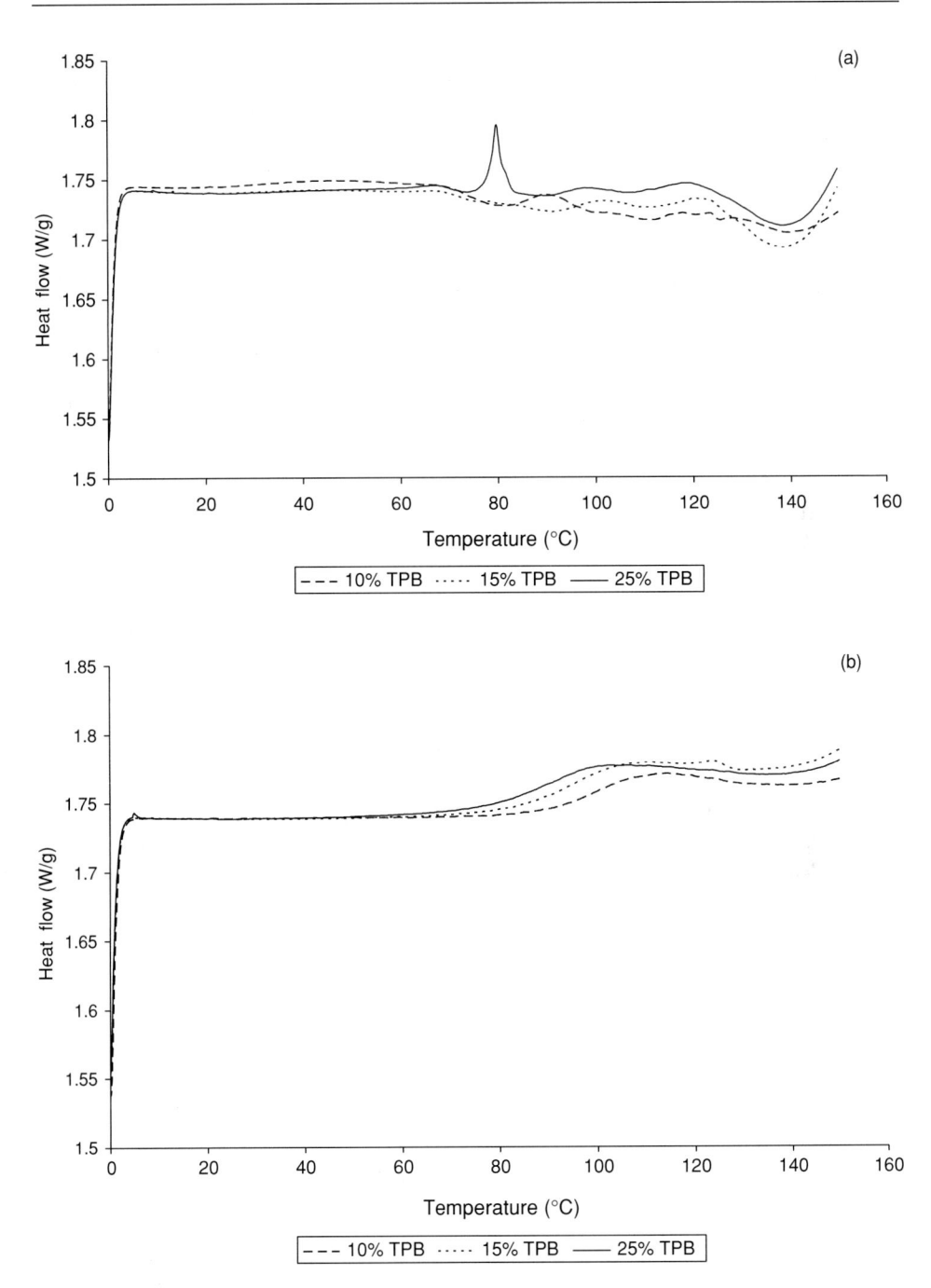

Figure 7.10 Parts (a) and (b) show first and second heating thermograms of B-TPB cements respectively. Tests were carried out under constant rate analysis consisting of three cycles, heating/cooling/heating, at a rate of 10°C/min from 0 to 170°C under nitrogen purge. T_g was characterised by a change in the upward direction of the heat flow and the value was calculated by the onset of change in the heat flow. (From [21], with permission.)

using indentation techniques, such as the Gilmore needles and the Vicat needles methods, or based on the time after which no disintegration occurs after immersion in water. Although these techniques are attractive as they are cheap and easy to use they suffer from being somewhat subjective, i.e. user dependent, and the results obtained may be of little relevance to the clinical application of the cements. Isothermal DSC in which the heat generated during the setting is monitored can be a viable alternative method for determining the setting time of an exothermic brushite forming system. Brushite cement may be formed by mixing the components β-tricalcium phosphate (β-TCP) and monocalcium monophosphate (MCPM) in the presence of water using the reaction described in (7.3) [23]:

$$\beta\text{-}Ca_3(PO_4)_2 + Ca(H_2PO_4)_2 \cdot H_2O + 7H_2O \rightarrow 4CaHPO_4 \cdot 2H_2O \qquad (7.3)$$

In CPC systems, a number of parameters can be changed including the powder/liquid ratio (PLR) and the use of a retardant, e.g. citric acid. In a test to investigate the isothermal setting behaviour of these cements [24], after an initial mixing period of 30 s, 30–50 mg of the paste is placed in a standard aluminium pan and lid, with the lid covering by slight pressing (but not sealed) in order to obtain an even distribution of the paste in the cup and to maximise the contact area with the furnace. The pan is then placed into the DSC at 20°C, which was then increased to 37°C (a process taking approximately 20 s), at which point the measurement of the heat flow was initiated. The calorimetry measurements were recorded until the thermal curve was parallel to the time axis and maintained a stable baseline, which was usually after 20–30 min depending on the mixture composition, which indicated the completion of the reaction. Three typical DSC measurements showing the normalised heat flow of the exothermic setting reaction against reaction time for a PLR of 2.0 g/mL and three different citric acid concentrations are shown in Figure 7.11a. Typical measurements for a fixed retardant concentration of 800 mM citric acid solution and different PLR are presented in Figure 7.11b. These results showed that while the citric acid concentration had an effect on the setting times, the PLR did not. The time at which maximum heat flow took place increased with citric acid concentration and this strongly correlated with both the initial and final setting time measured by conventional indentation methods. The course of the DSC curves was typical of an autocatalytic-like reaction [25]; i.e. the growth of brushite crystals within the cement paste being simultaneously produced and seeded for the crystal precipitation and growth defined the speed of the reaction.

7.6 Examples of applications using differential thermal analysis/thermogravimetric analysis (DTA/TGA) in the development and characterisation of biomaterials

7.6.1 Bioactive glasses

DTA equipment has been used in the characterisation of glasses and ceramics (see Chapter 10) as it has the ability to measure higher temperature parameters compared to DSC equipment, though modern high-temperature analysers are now capable of offering a heat flow signal. As for DSC analysis, a typical heating run can measure the T_g, T_c and T_m of

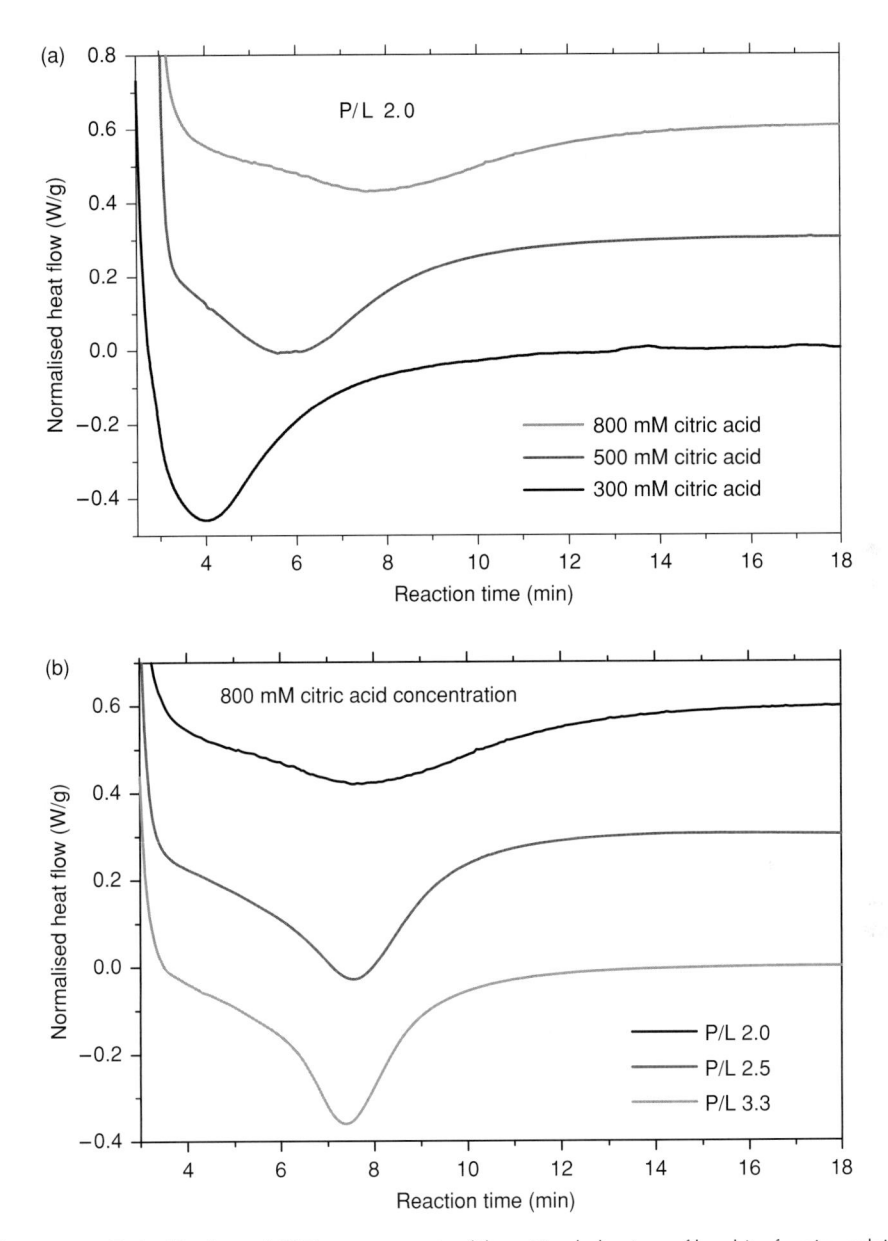

Figure 7.11 Typical isothermal DSC measurements of the setting behaviour of brushite-forming calcium phosphate cements based on cement (β-tricalcium phosphate/monocalcium monophosphate) at 37°C, showing (a) the normalised heat flow of the exothermic setting reaction with time for fixed PLR 2.0 g/mL and (b) three different citric acid retardant concentrations and fixed retardant concentration of 800 mM citric acid and three different PLRs. While the citric acid concentration had an effect on the setting times, in terms of the time at which the maximum heat of reaction took place, the PLR did not. (From [24]. Copyright 2006. Copyright John Wiley & Sons Limited. Reproduced with permission.)

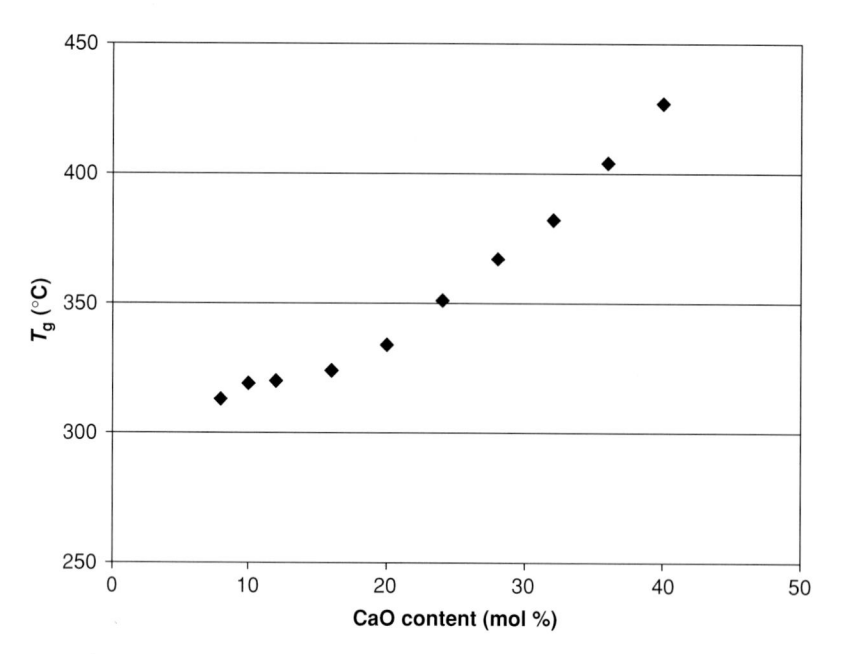

Figure 7.12 The variation of T_g with CaO content in degradable ternary CaO–Na$_2$O–P$_2$O$_5$ based glass system. This was carried out by heating samples from room temperature up to 1000°C at a heating rate of 10°C/min, under nitrogen purge. T_g was calculated at the midpoint of inflection during transition. There was an increase in T_g with CaO content, which has a direct effect on the degradability rate of the glasses. (Reprinted from [26], Copyright (2001), with permission from Elsevier.)

materials through the energy of a reaction. Figure 7.12 [26] shows how the T_g's of ternary phosphate glasses based on CaO–Na$_2$O–P$_2$O$_5$ were varied as a consequence of changing the formulation. Phosphate glasses have the potential to be used as scaffold materials since they are degradable due to the abundance of the easily hydrated P–O–P bonds, where their degradation products are natural components and are tolerated by the body. Furthermore, the degradation rate is strongly dependent on the glass composition, and accordingly a wide range of materials with different degradation rates can be obtained. For example, by substituting Na$_2$O with CaO glass systems that were less degradable are formed, since Ca^{2+} ions have much stronger field strengths than Na$^+$ and a chelating structure could be formed with ionic bonding between two adjacent [PO$_4^{3-}$] tetrahedron to strengthen the P–O–P bond. The increase in T_g of the glasses can be related to the increase in its stability and therefore degradation. Similar studies based on quaternary phosphate glasses based on CuO–CaO–Na$_2$O–P$_2$O$_5$ developed as antimicrobial glasses found comparable T_g values when comparing DTA and DSC (obtained through rapid rate DSC of 100°C/min) [27].

7.6.2 Thermal stability of bioactive composites

TGA can be used to verify the composition of composites as well as to analyse the effect of fillers on the thermal stability of the matrix. Figure 7.13 shows the TGA profile of foamed PDLLA on Bioglass® particles and on the two PDLLA/Bioglass® composite foams [11].

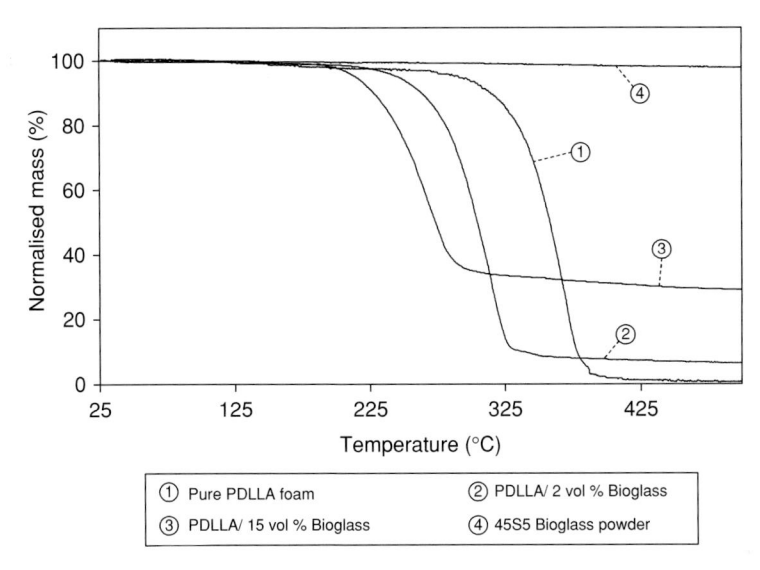

Figure 7.13 TGA profiles of a pure PDLLA foam (1), PDLLA/2 vol % Bioglass® foam (2), PDLLA/15 vol % Bioglass® (3) and 45S5 Bioglass® powder (4). (Reprinted from [11], Copyright (2005), with permission from Elsevier.)

PDLLA foam exhibited no significant change in mass above 425°C and therefore this temperature was chosen to record the residual mass for all foam systems. It was also shown that below this temperature there is no significant weight loss occurring in the glass. TGA tests conducted on three samples of the composite foams demonstrated that there was slight variability in the amount of Bioglass® present throughout the material, which may have been due to partial sedimentation of the Bioglass® particles during foam fabrication. Another interesting finding from this study was that Bioglass® had an effect on the degradation temperature of the PDLLA matrix, which significantly decreased with increasing amounts of filler. This effect has not been demonstrated with other types of bioceramic fillers.

7.7 Summary

This chapter has highlighted the significance of thermal analysis for the characterisation of biomaterials. Through experience it has become apparent that the thermal properties of materials are clearly linked to their structure and can have a significant effect on their clinical performance. Any of the techniques highlighted here can be used as a complementary tool in the general characterisation of materials. However, for a full understanding of the thermal behaviour of materials it is often much more useful to use a combination of the above techniques.

References

1. Williams DF. Definitions in biomaterials. In *Proceeding of a Consensus Conference of the European Society for Biomaterials*, Vol. 4, Chester, England, 3–5 March, 1986. New York: Elsevier, 1987; Vol. 4.

2. Bonfield W. *Engineering Applications of New Composites.* In: Paipetis SA (ed.), Oxford: Omega Scientific, 1988; pp. 17–21.
3. Ratner BD, Hoffman AS, Schoen FJ, Lemons JE. *Biomaterials Science: An Introduction to Materials in Medicine*, 2nd edn. New York: Elsevier Academic Press, 2004.
4. Hill D. *Design Engineering of Biomaterials for Medical Devices.* New York: John Wiley and Sons, 1998.
5. Bonfield W. *Biocompatibility of Tissue Analogues*, Vol. 2. Boca Raton, FL: CRC Press, 1985; p. 89.
6. Nazhat SN, Joseph R, Wang M, Smith R, Tanner KE, Bonfield W. Dynamic mechanical characterisation of hydroxyapatite particulate reinforced polyethylene: effect of particle size. *J Mater Sci, Mater Med* 2000;11:621–628.
7. Nielsen LE, Landel RF. *Mechanical Properties of Polymers and Composites*, 2nd edn. New York: Dekker, 1994.
8. Nazhat SN, Kellomäki M, Tanner KE, Törmälä P, Bonfield W. Dynamic mechanical characterisation of biodegradable composites of hydroxyapatite and polylactides. *J Biomed Mater Res (Appl Biomater)* 2001;58:335–343.
9. Bleach NC, Nazhat SN, Tanner KE, Kellomäki M, Törmälä P. Effect of filler content on mechanical and dynamic mechanical properties of particulate biphasic calcium phosphate polylactide composites. *Biomaterials* 2002;23:1579–1585.
10. Rich J, Tuominen J, Kylmä J, Seppälä J, Nazhat SN, Tanner KE. Lactic acid based PEU/HA and PEU/BCP composites: dynamic mechanical characterization of hydrolysis. *J Biomed Mater Res (Appl Biomater)* 2002;63:346–353.
11. Blaker JJ, Maquet V, Jérôme R, Boccaccini AR, Nazhat SN. Mechanical properties of highly porous PDLLA/Bioglass® composite foams as scaffolds for bone tissue engineering. *Acta Biomater* 2005;1:643–652.
12. Barry JJA, Nazhat SN, Rose FRAJ, Hainsworth AH, Scotchford CA, Howdle SM. Supercritical carbon dioxide foaming of elastomer/heterocyclic methacrylate blends as scaffolds for soft tissue engineering. *J Mater Chem* 2005;15:4881–4888.
13. Nazhat SN, Parker S, Patel MP, Braden M. Isoprene-styrene copolymer elastomer and tetrahydrofurfuryl methacrylate mixtures for soft prosthetic applications. *Biomaterials* 2001;22:2411–2416.
14. Georgiou G, Mathieu L, Pioletti DP, et al. Polylactic acid-Phosphate glass composite foams as scaffolds for bone tissue engineering. *J Biomed Mater Res (Appl Biomater)* 2007;80B:322–331.
15. Yang JM, You JW, Chen HL, Shih CH. Calorimetric characterization of the formation of acrylic type bone cements. *J Biomed Mater Res* 1996;33:83–88.
16. Yang JM, Shyu JS, Chen HL. Polymerization of acrylic bone cement investigated by differential scanning calorimetry: effects of heating rate and TCP content. *Polym Eng Sci* 1997;37:1182–1187.
17. Vallo CI, Montemartini PE, Cuadrado TR. Effect of residual monomer content on some properties of a poly(methyl methacrylate)-based bone cement. *J Appl Polym Sci* 1998;69:1367–1383.
18. Vallo CI. Residual monomer content in bone cements based on poly(methyl methacrylate). *Polym Int* 2000;49:831–838.
19. Borzacchiello A, Ambrosio L, Nicolais L, Harper EJ, Tanner KE, Bonfield W. Isothermal and non-isothermal polymerization of a new bone cement. *J Mater Sci, Mater Med* 1998;9:317–324.
20. Ege W, Kuhn K-D, Tuchscherer C, Maurer H. Physical and chemical properties of bone cements. In Walenkamp GHIM (ed.), *Biomaterials in Surgery.* New York: Thieme, 1998.
21. Abdulghani S, Nazhat SN, Behiri JC, Deb S. Effect of triphenyl bismuth on glass transition temperature and residual monomer content of acrylic bone cements. *J Biomater Sci, Polym Ed* 2003;14:1229–1242.
22. Chevallay B, St. Roche, Herbage D. Collagen based biomaterials and tissue engineering. In Walenkamp GHIM (ed.), *Biomaterials in Surgery.* New York: Thieme, 1998.
23. Lemaitre J, Mirtchi A, Mortier A. Calcium phosphate cements for medical use: state of the art and perspectives of development. *Silicates Ind* 1987;9–10:141–146.

24. Hofmann MP, Nazhat SN, Gbureck U, Barralet JE. Real-time monitoring of the setting reaction of brushite-forming cement using isothermal differential scanning calorimetry. *J Biomed Mater Res (Appl Biomater)* 2006;79B:(2006) 360–364.
25. Tenhuisen KS, Brown PW. The effects of citric and acetic acid on the formation of calcium deficient hydroxyapatite at 38°C. *J Mater Sci, Mater Med* 1994;5:291–298.
26. Franks K, Abrahams I, Georgiou G, Knowles JC. Investigation of thermal parameters and crystallisation in a ternary $CaO–Na_2O–P_2O_5$ based glass system. *Biomaterials* 2001;22:497–501.
27. Abou Neel EA, Ahmed I, Pratten J, Nazhat SN, Knowles JC. Characterisation of antibacterial copper releasing degradable phosphate glass fibres. *Biomaterials* 2005;26:2247–2254.

Chapter 8
Thermal Analysis of Pharmaceuticals

Mark Saunders

Contents

8.1	Introduction	287
8.2	Determining the melting behaviour of crystalline solids	288
	8.2.1 Evaluating the melting point transition	288
	8.2.2 Melting point determination for identification of samples	289
8.3	Polymorphism	290
	8.3.1 Significance of pharmaceutical polymorphism	291
	8.3.2 Thermodynamic and kinetic aspects of polymorphism: enantiotropy and monotropy	292
	8.3.3 Characterisation of polymorphs by DSC	293
	8.3.4 Determining polymorphic purity by DSC	297
	8.3.5 Interpretation of DSC thermograms of samples exhibiting polymorphism	302
8.4	Solvates and hydrates (pseudo-polymorphism)	303
	8.4.1 Factors influencing DSC curves of hydrates and solvates	304
	8.4.2 Types of desolvation/dehydration	305
8.5	Evolved gas analysis	308
8.6	Amorphous content	310
	8.6.1 Introduction	310
	8.6.2 Characterisation of amorphous solids: the glass transition temperature	311
	8.6.3 Quantification of amorphous content using DSC	313
8.7	Purity determination using DSC	315
	8.7.1 Types of impurities	315
	8.7.2 DSC purity method	316
	8.7.3 Practical issues and potential interferences	317
8.8	Excipient compatibility	320
	8.8.1 Excipient compatibility screening using DSC	321
	8.8.2 Excipient compatibility analysis using isothermal calorimetry	321
8.9	Microcalorimetry	323
	8.9.1 Introduction	323
	8.9.2 Principles of isothermal microcalorimetry	324

8.9.3	High-sensitivity DSC	324
8.9.4	Pharmaceutical applications of isothermal microcalorimetry	325
References		327

8.1 Introduction

Pharmaceutical drug development is defined by many companies as the process of taking a new chemical lead compound identified from the discovery phases through a plethora of preclinical testing activities, with the data generated allowing for the production of a suitable and efficacious formulation to be tested in human clinical trials. On average, it takes 12 years and $800 million for an experimental drug to travel from laboratory to the pharmacist's shelf, with 5 in 5000 compounds that enter preclinical testing making it to human trials. Subsequently, 1 of these 5 tested in humans is ultimately approved for final use.

A raft of tests are needed to ensure that the physiochemical properties of promising new chemical entities (NCEs) are fully understood. In addition to screening for *in vivo* activity and safety, drug development is required to establish properties such as chemical composition, stability, intrinsic solubility, physical form, etc. Table 8.1 shows some common testing procedures carried out during the preclinical stages to understand the physicochemical properties of the lead compound.

In Table 8.1 a range of very familiar techniques are listed, and all are important. This is especially true during preformulation stages where it is best to use as wide a range of characterisation technologies as possible, to build up a comprehensive physicochemical profile of the molecule under study. In today's R&D environment where both budgets and time frames are extremely limited, not all laboratories possess the instruments that may be needed, and increasingly test projects are outsourced to specialist facilities possessing such instrumentation. However, one range of equipment that is common to most R&D laboratories is that of thermal analysis (TA) and in particular calorimetry which is used to

Table 8.1 Some techniques commonly employed during the preformulation testing phases to characterise drug substances

Technique	Applications
Scanning electron microscopy	Particle shape, size and morphology
Differential scanning calorimetry	Melting point, physical form identification, investigation of recrystallisation behaviour and identification of glass transitions
Thermogravimetric analysis	Identification of residual solvents, solvates, stability and general characterisation
Gravimetric vapour sorption analysis	Hygroscopicity testing (hydrate conversion)
Raman spectroscopy	Physical and chemical form identification
X-ray diffraction	Crystal form identification and/or verification
Intrinsic solubility	Kinetic and thermodynamic solubility testing
Isothermal microcalorimetry	Excipient compatibility screening, amorphous content screening and stability analysis

understand the physical properties of drug substances, drug substance/excipient interactions and final drug product stability. TA is the term used to describe all analytical techniques that measure the physical and chemical properties of a sample as a function of temperature.

In practice, calorimetry is the quantitative measurement of the heat required or evolved during a chemical process; a calorimeter is an instrument for measuring the heat of a reaction during a well-defined process. Therefore, calorimetry is not just confined to differential scanning calorimetry (DSC) but also includes numerous isothermal methods, which have become popular for the rapid assessment of stability/compatibility [1] and for quantification of small amounts of amorphous content in processed materials [2]. This chapter will therefore highlight the use of DSC thermogravimetric analysis (TGA) and, where appropriate, isothermal calorimetry in the investigation and quantification of various physicochemical parameters important during the preformulation stages of pharmaceutical product development. Where necessary, examples from the literature will be given to emphasise the methodology.

8.2 Determining the melting behaviour of crystalline solids

Perhaps the most common parameter obtained from a DSC experiment is the determination of the melting point of a test compound. The melting point of a crystalline solid is the temperature at which it changes state from a solid to liquid. (When considered as the temperature of the reverse change from liquid to solid, it is referred to as the freezing point.) During melting, the standard enthalpy change measured (the heat of fusion) is the amount of thermal energy which must be absorbed or lost for 1 g of a substance to change states from a solid to a liquid or vice versa. For most substances, the melting and freezing points are equal; for example the melting point and freezing point of the element mercury is 234.32 K ($-38.83°$C or $-37.89°$F). However, certain substances possess differing solid–liquid transition temperatures; for example agar melts at 85°C (185°F) and solidifies from 32 to 40°C (89.6–104°F). In practice most substances show a degree of undercooling (supercooling) when cooled in a DSC, making accurate measurement of freezing points very difficult. It is also for this reason that no recognised standards currently exist for the calibration of DSC analysers upon cooling.

From a thermodynamics point of view, at the melting point, the change in Gibbs free energy (ΔG) of the material is zero, because both the enthalpy (H) and the entropy (S) of the material are increasing ($\Delta H, \Delta S > 0$). Melting phenomenon happens when the Gibbs free energy of the liquid becomes lower than that of the solid for that material.

8.2.1 Evaluating the melting point transition

Prior to conducting any DSC experimentation it is useful to run a corresponding TGA in order to identify the exact onset point of degradation. Exceeding the degradation onset temperature in a DSC is not recommended, as this can lead to erroneous results due to the numerous exothermic and endothermic events associated with the transition. Also, condensation of the desorbed organic volatiles from the sample pan back into the DSC instrumentation can occur and can lead to contamination of the system, which will affect subsequent analysis.

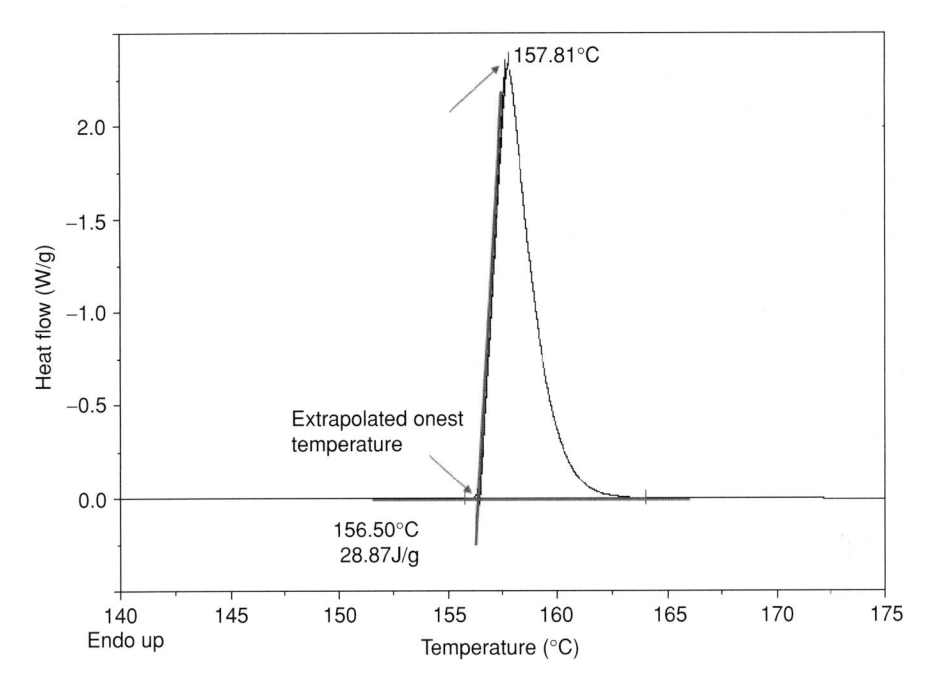

Figure 8.1 The melting point (T_m) of a single crystal melt is determined from the extrapolated onset of melting as shown. The heat of fusion is obtained from the area under the curve.

Because of the nature of the measurement, during a DSC experiment, there will always be some degree of broadening of the peak associated with any phase change. In theory the melt of a pure single crystal at an infinitely slow scan rate should result in a peak that is infinitely narrow, but broadening of the peak occurs as a direct result of the thermal gradients that occur across a sample. Thermal gradients are caused by the time it takes for energy to transfer through the sample. As a result, the only accurate method of measurement of the melting point of a pure crystalline material is to quote the extrapolated onset temperature which is where melting begins. The peak maximum or half height transition point will vary as a function of particle size and sample mass, and it does not reflect the true melting point. In contrast, the onset temperature should not vary with changes in any of these parameters, nor with changes in heating rate, provided calibration has been performed correctly (see Section 1.5.2 for a more detailed discussion of melting).

For determination of the onset temperature, a line is extrapolated from the slope of the leading edge of the peak to the x-axis. The point at which the line bisects the x-axis is denoted the extrapolated onset temperature (Figure 8.1).

8.2.2 Melting point determination for identification of samples

Determination of the melting point of an investigational test compound allows us to establish its identity (if we have suitable reference standards for comparison) and it can also be used to distinguish between different physical forms (i.e. polymorphs) [3], different isomers [4] and differing salts of the same compound. This is of extreme importance during early product development stages, as polymorphs and salts of the same compound may possess not only different melting points (generally, the more thermodynamically stable polymorph possesses a higher melting point) but also different dissolution rates, 'apparent' solubilities and in some cases bioavailability (BA) [5].

DSC is ideally suited for determination of melting points, as it allows a direct measurement of not only the melting point but also the enthalpy of fusion of the substance. Enthalpy measurements are obtained from the area of the melting peak, as shown in Figure 8.1. The further benefit of using DSC is that it requires only small quantities of material (typically only a few milligrams) – a particular advantage as the amount of sample generally available during the preformulation stages is extremely limited. Note, though, that if small sample sizes are used then a very accurate balance will be needed to measure the weight of the sample. This will typically need to be to six-figure (microgram) level.

For any pure compound the melting point is a thermodynamically fixed point and so it can be used to identify the material. Many pharmaceuticals exhibit polymorphism (see Section 8.3) where each crystal structure has a different specific melting point. Measuring the melting point by DSC therefore gives the analyst the ability, in principle, to distinguish between different crystal forms. This provides one of the largest application areas for TA in the pharmaceutical industry and is considered in more detail in the following section.

8.3 Polymorphism

Polymorphism is the ability of a solid material to exist in more than one form or crystal lattice structure and can potentially be found in any crystalline material, including polymers and metals, and is related to allotropy, which refers to packing variations in elemental solids. (For example, both graphite and diamond are allotropes of carbon.) Together with polymorphism, the complete morphology of a material is described by other variables, such as crystal habit, amorphous fraction or crystallographic defects.

In principle, all organic molecules can exist in more than one distinct crystal form (i.e. polymorph). In reality, approximately 70% of all pharmaceutical drugs on the market have been shown to be able to exist in more than one discrete anhydrous crystalline lattice arrangement or polymorphic form. (McCrone's law states that every compound has different polymorphic forms and that, in general, the number of forms known for a given compound is proportional to the time and money spent in research on that compound.) In fact modern computer programs predict the possibility of multiple crystal forms for most pharmaceuticals, only some of which are usually discovered. In the past, the term 'polymorph' was generally used to describe the crystal packing of an anhydrous form of a crystalline solid; however, polymorphism, as defined in the International Conference on Harmonization (ICH) Guideline Q6A [6], now includes solvation/hydration products and amorphous forms.

When the melting profile of a polymorphic material is obtained from a DSC, it is often found to contain a number of endothermic peaks corresponding to the melting of different crystal forms, possibly separated by an exothermic peak as molten material recrystallises into a more stable form. Figures 8.5, 8.6(b) and 8.7 (discussed later in the chapter) show examples of this type of behaviour. Sometimes the melting temperatures of different forms occur so close together that shoulders and unusual peak shapes are observed as different events overlap, and a range of scan rates are useful to help determine what is happening. The choice of scan rate should take into account the desired resolution, sensitivity and kinetics of events in question; more details about these choices are given in Section 8.3.3. TGA should also be obtained to confirm whether or not any endothermic peaks correspond to a weight loss and are therefore the result of mass loss.

8.3.1 Significance of pharmaceutical polymorphism

Polymorphic forms of a drug substance can have different chemical and physical properties, including melting point, chemical reactivity, apparent solubility, dissolution rate, optical and mechanical properties, vapour pressure and density [6–9]. These properties can have a direct effect on the ability to process and/or manufacture the drug substance and the drug product, as well as on drug product stability, dissolution, bioavailability (BA) and bioequivalence (BE). Thus, polymorphism can affect the quality, safety and efficacy of the final drug product formulation and therefore must be tightly controlled and monitored throughout all phases of the drug development process.

Depending on the stability relationship between the different polymorphic forms, the drug substance can undergo phase conversion when exposed to a range of manufacturing processes, such as drying, milling, micronisation, wet granulation, spray drying and compaction. Samples obtained for analysis therefore should not be ground or milled prior to analysis, as this can change the sample. Exposure to environmental conditions such as humidity and temperature can also induce polymorph conversion (such as hydrate formation or solid-state phase transformations). The extent of conversion generally depends on the relative stability of the polymorphs, kinetic barriers for phase conversion and applied stress. Nonetheless, phase conversion generally is not of serious concern, provided that the conversion occurs consistently and as a part of a validated manufacturing process where critical manufacturing process variables are well understood and controlled and where drug product BA/BE has been demonstrated.

8.3.1.1 Influence on stability

The most thermodynamically stable polymorphic form of a drug substance is often chosen during development, based on the minimal potential for conversion to another polymorphic form and on its greater chemical and physical stability. A less stable (metastable) form can be chosen for various reasons (including BA enhancement); however, it must be noted that this form is unstable with respect to the thermodynamically stable state, and, depending on the processing and storage conditions employed, conversion to the most thermodynamically stable polymorphic form may occur. Therefore, it is the stable form that is generally considered the most suitable form to take forward the development in order to avoid any unexpected late-stage product failures due to polymorphic conversion during manufacture and/or storage. In the liquid and gaseous states, polymorphs of the same compound behave identically since the crystal lattice history has been completely erased.

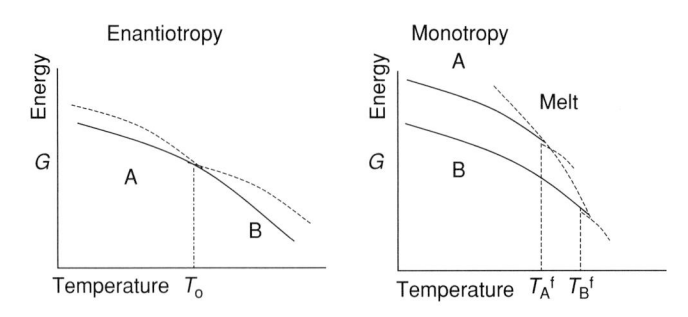

Figure 8.2 The relationship between Gibbs energy (*G*) and temperature for two modifications in the cases of enantiotropy (reversible) and monotropy (irreversible) transition between forms.

8.3.2 Thermodynamic and kinetic aspects of polymorphism: enantiotropy and monotropy

When polymorphism is detected in any compound selected for pharmaceutical product development, a precise knowledge of the thermodynamic stability relationship between the different solid phases is essential in order to fully understand the crystallisation process and specific solid-form stability. During crystallisation of a material from a chosen solvent system, depending on the extent of supersaturation and the temperature–solubility curves of each polymorph or pseudopolymorph, generally the first nucleating crystals or metastable forms are the kinetically favoured. Subsequently, as the thermodynamic equilibrium is re-established due to changes in the solubility product as the sample crystallises (i.e. the solution become less saturated or concentrated), the solid sample undergoes a solvent-mediated phase conversion to the more stable state. However, depending on factors such as crystal growth, temperature and solubility, this conversion may not occur and the metastable form will prevail in the solid phase. In the case of enantiotropy, the stable forms are different above and below the reversible equilibrium temperature, and conversion to that specific form under the conditions employed will occur. The term monotropy applies in the case of an irreversible transition from one form to another. Knowing the relationship among the thermodynamic quantities H (enthalpy), G (free energy), S (entropy) and T (temperature), it is often simple to represent equilibrium states by plotting the free energy G as a function of the temperature for each form. If the two curves intersect before the melting point, there is reversibility, that is, enantiotropy, and if the reverse is true, there is monotropy (Figure 8.2).

In the case of monotropy, the higher melting form is always the thermodynamically stable form. In the case of enantiotropy, the lower melting form has a higher heat of fusion compared to the higher melting form and is the thermodynamically stable form at temperatures below the transition point. (The higher melting form is the thermodynamically stable form at temperatures above the transition point.) The relationship between the melting enthalpies of two solid phases, A and B, AH_f^A and AH_f^B, and the heat of transition, AH_t, is

$$AH_t = AH_f^A - AH_f^B$$

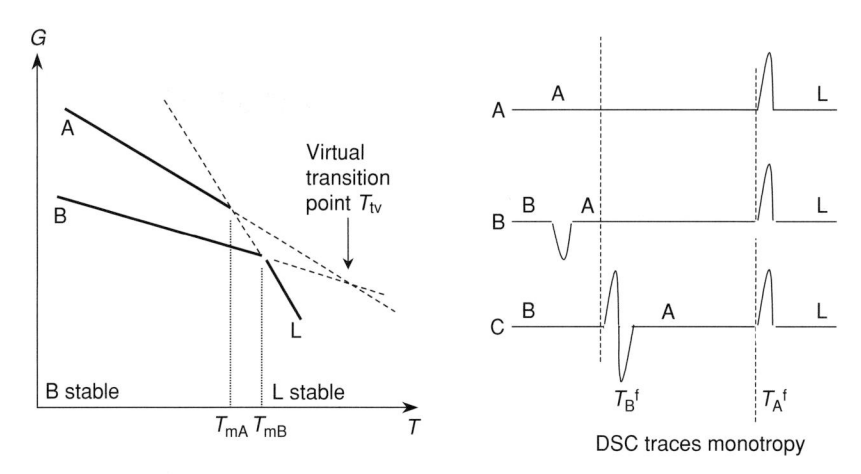

Figure 8.3 Energy diagrams showing H (melting enthalpy) and G (free energy) for monotropic polymorphism and the corresponding DSC curves: T_o is the temperature of the transition A to B, T_A^f is the melting temperature of A and T_B^f is the melting temperature of B. DSC scans: A, the thermodynamically high melting form A melts; B, the low melting form undergoes an exothermic transition into A; C and B, melt and A crystallises from the melt and then A melts.

The transition point can be measured by TA, solubility measurements or a combination of measurements of solubilities and melting enthalpies. The form which is thermodynamically stable at the temperature and pressure of measurement is that which has the lowest free energy and apparent solubility. For each modification, the following equation is valid:

$$\mathrm{Log}C = -\frac{\Delta H_\mathrm{diss}}{RT} + K$$

where C is solubility, T is the gas constant, ΔH_diss is the heat of dissolution (or heat of solution) in the solvent and K is a constant. In the case of enantiotropy both modifications have the same solubility at the transition point. From DSC measurements, the melting point, the melting enthalpy and perhaps the transition point may be measured. The plots of the thermodynamic quantities H (enthalpy) and G (free energy) clarify the relationships between the polymorphs. Figures 8.3 and 8.4 show such plots and also show the DSC curves which may be obtained in monotropy or enantiotropy, with a stable state or a metastable state. Table 8.2 is a summary of the thermodynamic rules established by Burger and Ramburger [10] in order to help distinguish between monotropic and enantiotropic transitions.

8.3.3 Characterisation of polymorphs by DSC

There are a number of methods that can be used to characterise polymorphs of a drug substance. Demonstration of a non-equivalent structure by single-crystal X-ray diffraction [11] is currently regarded as the definitive evidence of polymorphism. X-ray powder diffraction (XRPD) can also be used to support the existence of polymorphs. However supporting

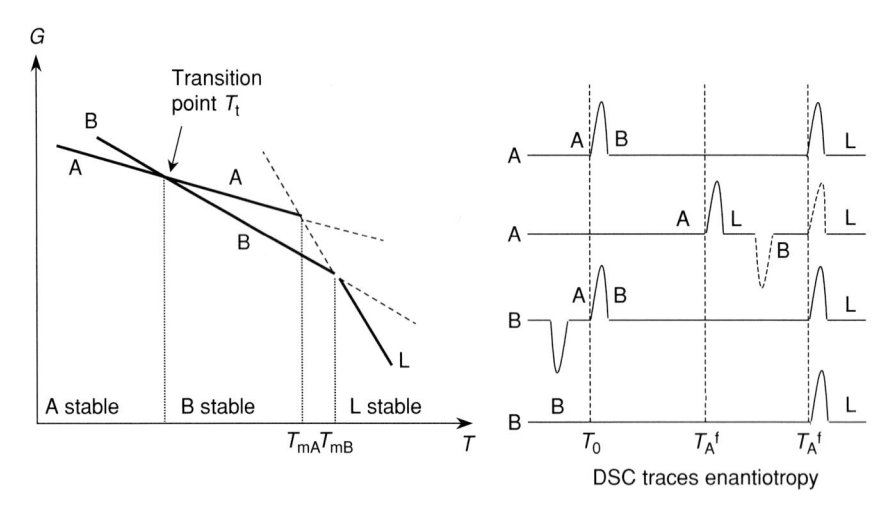

Figure 8.4 Energy diagrams showing H (melting enthalpy) and G (free energy) for enantiotropic polymorphism and the corresponding DSC curves: T_0 is the temperature of the transition A to B, T_A^f is the melting temperature of A and T_B^f is the melting temperature of B. DSC scans: A, endothermic solid–solid transition into B and then B melts or A melts and eventually B crystallises from the melts. B is at room temperature and a spontaneous exothermic transition into A occurs or B melts.

evidence is normally sought from a range of other complementary methods, including hot-stage microscopy, spectroscopy (e.g. infrared (IR), Raman [12], solid-state nuclear magnetic resonance [13]) and TA methods, most notably DSC.

Many examples exist in the literature of studies using DSC as a method for predicting the presence of polymorphs of a compound [14]. A recent example is that of carbamazepine, a drug used in the treatment of epilepsy. It has been shown successfully using DSC that three distinct polymorphs of carbamazepine exist. These findings were backed up by results obtained from Fourier transform infrared (FTIR) and XRPD experiments [15]. However,

Table 8.2 Thermodynamic rules for polymorphic transitions according to Burger and Ramburger [10]

Enantiotropy	Monotropy
Transition < melting I	Transition > melting I
I stable > transition	I always stable
II stable < transition	
Transition reversible	Transition irreversible
Solubility I higher < transition	Solubility I always lower than II
Solubility I lower > transition	
Transition II ▶ I endothermic	Transition II ▶ I endothermic
$\Delta H_f^I < \Delta H_f^{II}$	$\Delta H_f^I > \Delta H_f^{II}$
Density I < density II	Density I > density II

I is the higher melting form.

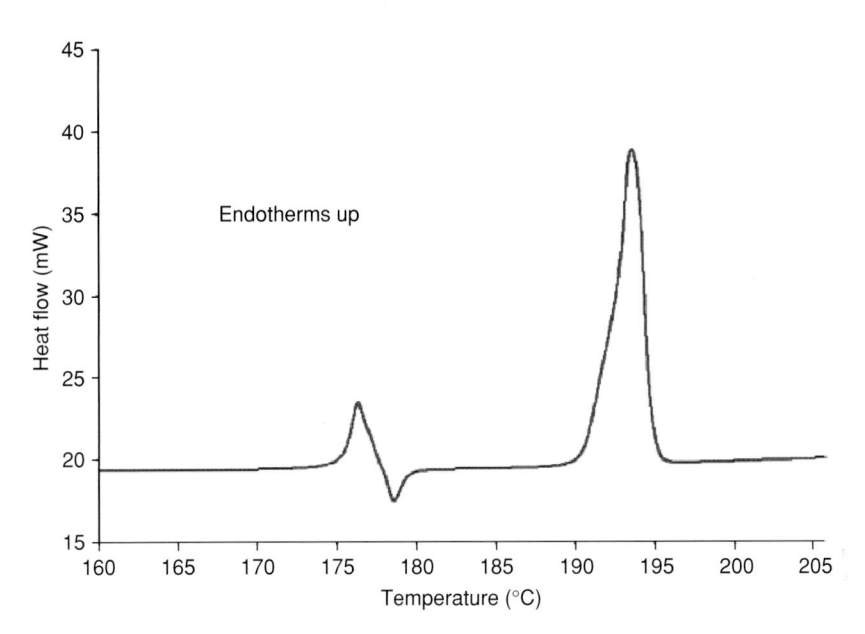

Figure 8.5 Typical polymorphic behaviour as seen with DSC. The initial crystal form melts upon heating (at about 175°C), giving the initial endotherm, and recrystallises into a second form which shows as an exotherm. This then melts at a higher temperature. It is probable that melting of the initial form and subsequent crystallisation are fully separated, so in this case energy measurements of these events are not possible.

although DSC has proved its worth in identifying the presence of multiple polymorphs via differences in melting point, it is not always possible to characterise the lower melting (metastable) form of a compound. This is because the material will often spontaneously recrystallise into a more stable form during the melt process, resulting in a concurrent exothermic response during the endothermic melt. (An example is shown in Figure 8.5.) At the heating rates employed in standard DSC experiments (usually around 10°C/min), carbamazepine form III is seen to melt and then recrystallise, the two events occurring simultaneously and so the peaks are not separated. Neither the heat of fusion nor the heat of recrystallisation can be accurately measured in this case. This is followed by the melting of a new, more stable form. It should be noted that this more stable form is not necessarily the most thermodynamically stable form of the material (see Section 8.3.2). There are also methods using isothermal calorimetry (see Section 8.6.2) that provide this information.

The fact that two peaks are observed in Figure 8.5 separated by an exotherm is useful information, and the trace can be easily understood. All too frequently it seems that the separation of melting events is not so large and just one peak may be observed, with a shoulder indicating that more than one event is happening. The higher melting peak in Figure 8.5 is an example. A further example is shown in Figure 8.6(a) where a very small peak can just be observed in the tail of the main melt. To discover further information about what is occurring it is useful to vary the scan rate. In Figure 8.6(b) the same sample is run with the scan rate reduced from 10 to 3°C/min. The events are now more clearly understood since it can be seen that one form is converting to second form. The existence of

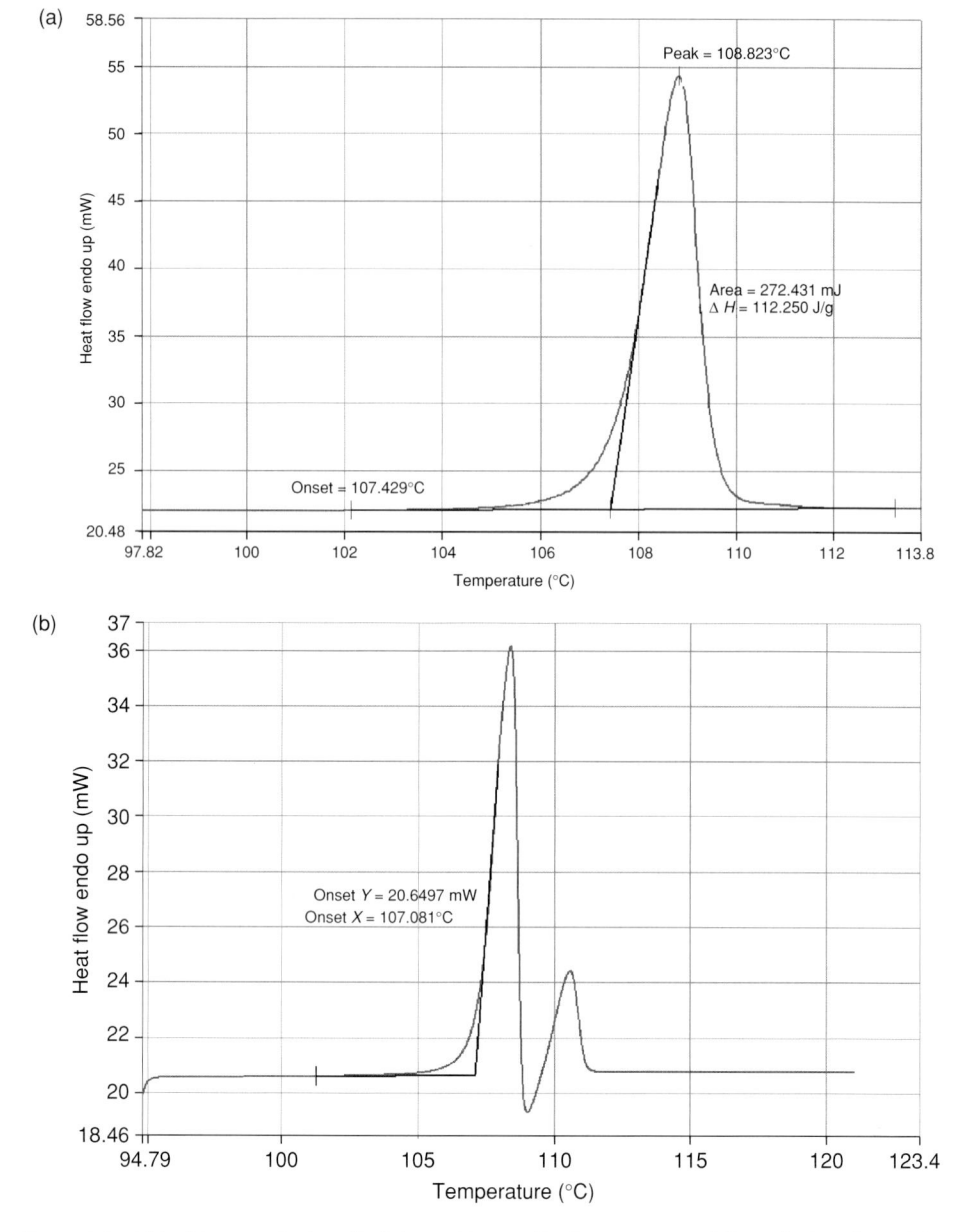

Figure 8.6 (a) In this example the tail of the melting profile does not return fully to the baseline, giving evidence of a possible transition in the region above 110°C. This can be clarified by reducing the scan rate. (b) At a slower rate the separation of events is observed. The initial form is melting and recrystallising, and the resulting form melting at a slightly higher temperature. Often resolution of such events can be a challenge. Slower heating rates and reduced sample size help to improve resolution. In some cases higher heating rates can completely prevent the recrystallisation, allowing accurate characterisation of the initial melt (see Section 8.3.4).

the exotherm is clear evidence of recrystallisation. Reducing the scan rate further may also be helpful; frequently, the use of a range of different scan rates can be helpful to interpret the events that are occurring. The use of the second derivative can also aid understanding of a trace (see Section 1.2.5). The effects of increasing the scan rate to very high rates are discussed in Section 2.5.

Unusual peak shapes may also result from poor practical technique. For example, if a sample collapses during melting, the change in thermal contact may induce a change in peak shape. For this reason it is generally helpful to examine melting phenomena on relatively small samples, typically around 3 mg that have been well compressed in the sample pan to give good thermal contact. If there is doubt about a trace, it is best to repeat it.

8.3.4 Determining polymorphic purity by DSC

8.3.4.1 Confirming polymorphic purity using fast scan rates

The recrystallisation behaviour observed at relatively low scan rates impacts upon the capacity to characterise the material placed into the pan, since the observed changes in the material affect the measurements. In particular it is not possible to determine the polymorphic purity of a given sample, either from the enthalpy of fusion of the metastable state or by measurement of the higher melting peaks since these may, or may not, have formed during the scan. If we return to the carbamazepine example described in Figure 8.5, quantification of the enthalpy of fusion of pure form III was not possible because of a simultaneous recrystallisation of this form from the liquid melt to the more thermodynamically stable form I (under slow heating rates (5–10°C/min). Until recently, the true heat of fusion for this metastable form had never been determined. However, this type of measurement can be achieved with the use of the very fast scan rates of up to 500°C/min now available with some equipment [16]. At higher temperature scan rates, the sample does not have the time necessary to recrystallise to the higher melting form, and therefore a quantitative measurement can be made of the enthalpy of fusion for the lower melting form. Fast scan (or Hyper^{TM}) DSC methods have been used to study carbamazepine, where it was shown that as the scan rate was increased, the recrystallisation to form I (the thermodynamically stable state) was inhibited [17]. Figure 8.7 shows that even at 250°C/min the sample recrystallises to produce a higher melting form, whilst Figure 8.8 shows that this recrystallisation transition is only completely suppressed at scanning rates of 500°C/min. (For clarity, the tail end of the melting endotherm for the metastable form is shown.)

With the inhibition of the recrystallisation of form III to form I, it was shown that it is possible to measure the enthalpy of fusion directly from the calorimetric data 109.5 J/g, a value which can be estimated only from normal DSC experiments. It must be noted, however, that using fast scanning conditions in some cases cannot prevent this transition from occurring (due to the fast kinetics of the recrystallisation transition).

8.3.4.2 Measurement of small amounts of impurity

Where polymorphic impurities exist as separate crystalline phases at trace levels in drug materials, accurate quantification is an important issue. Contamination of a drug material with

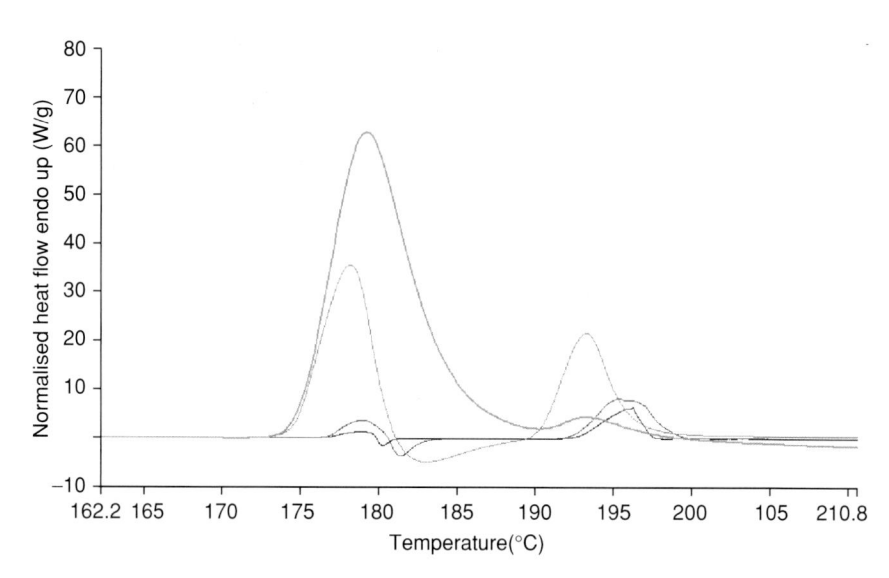

Figure 8.7 The effect of increasing scan rates from 20 to 250°C/min on the melt of carbamazepine form III. As scan rate is increased, less time is given for recrystallisation, the initial melt increases in size and the final melt proportionately reduces.

its alternative metastable polymorphic form, which can occur during uncontrolled precipitation or non-optimised crystallisation, grinding/milling or any other form of mechanical treatment, is of particular concern, since such presence of the polymorphic impurity could adversely compromise both the stability and the performance of the final products. In order

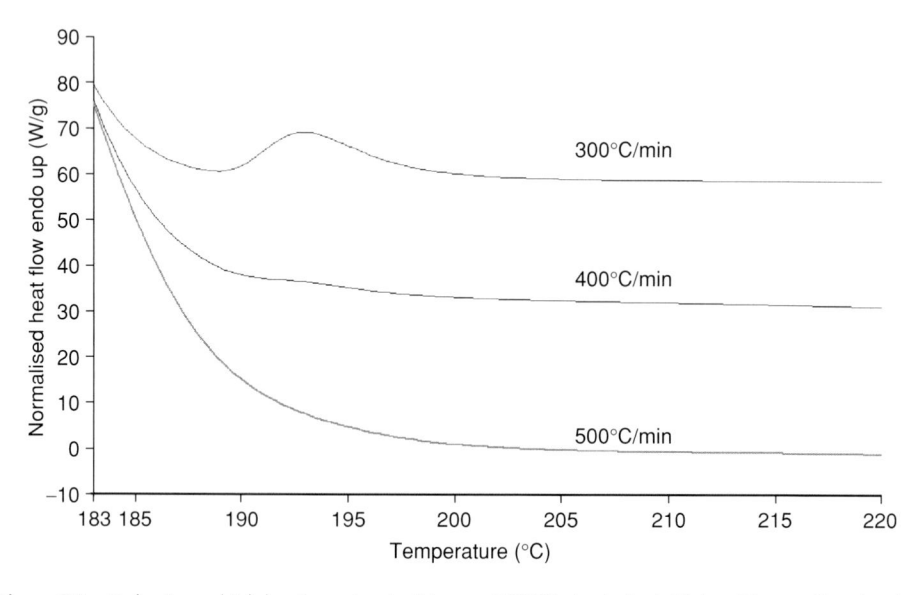

Figure 8.8 Only at very high heating rates, in this case 500°C/min, is the initial melting profile of carbamazepine free of further melting events, indicating that recrystallisation has been fully suppressed.

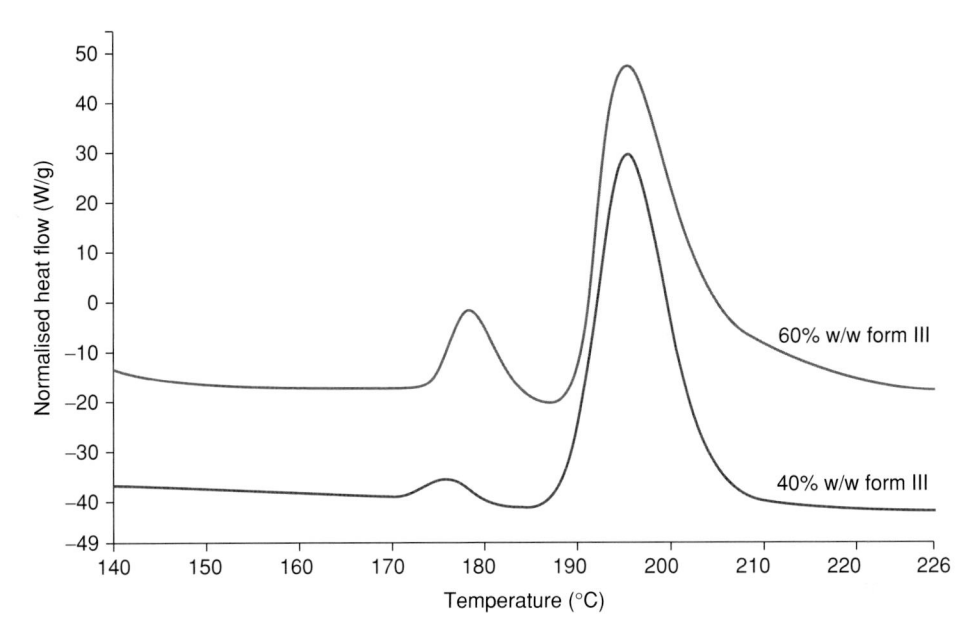

Figure 8.9 DSC traces from mixtures of form I and form III of carbamazepine scanned at 250°C/min. Small amounts of crystalline melt may be hard to detect at low rates but are more easily found at higher scan rates, where contents as low as 1% have been measured.

to limit these undesirable physical impurities in pharmaceutical materials, accurate quantification of trace levels of such impurities existing as separate crystalline phases has become an important issue. However, the most widely used methods for solid-phase characterisation, such as powder X-ray diffraction, FTIR spectroscopy and near-infrared spectroscopy, are normally not sufficiently sensitive for detecting relatively low levels (<5%) of polymorphic impurity. The measurement of polymorphic purity by DSC using slow scan rates is not without its difficulties either, for the reasons described earlier.

McGregor et al. [17] give an example of the application of high scan rates, using DSC for the determination of polymorphic purity of mixed systems containing a quantity of a metastable form which is known to undergo simultaneous melting and recrystallisation at slow scan rates (typically 10°C/min). The mixtures containing known ratios of carbamazepine forms I and III (0–100%, w/w) were analysed with a heating rate of 250°C/min, and the enthalpy of the melting endotherms was calculated for each mixture.

Figure 8.9 shows a typical thermal profile obtained for mixtures containing 40 and 60% (w/w) form III. The endothermic transitions due to melting of both forms I and III are clearly detected, and more importantly complete resolution between the two transitions was achieved. An endotherm due to melting of form III was also detected in mixtures containing as little as 1% (w/w) of this polymorph. When the same sample was analysed at a heating rate of 10°C/min, this melting endotherm was not detected, indicating the utility of high-speed DSC for detecting low levels of polymorphic impurities, which might otherwise be missed.

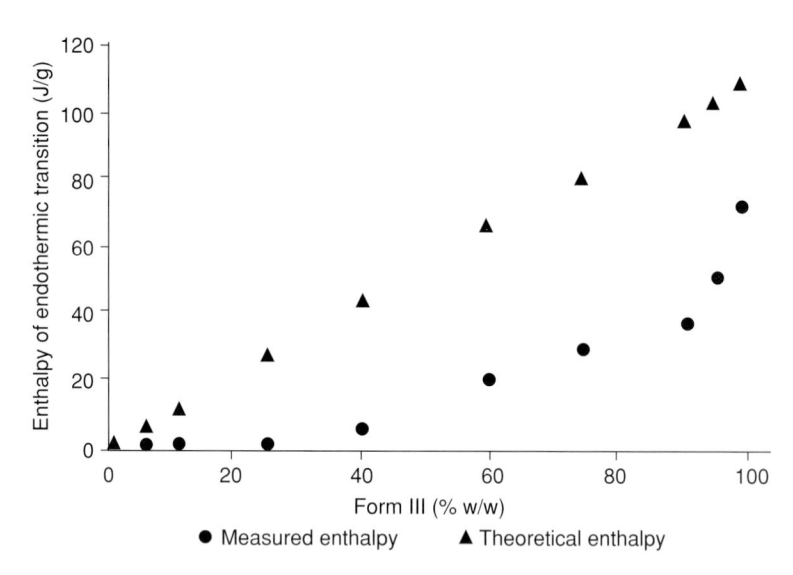

Figure 8.10 The measured enthalpies of the melting of caramazepine form III are compared with the theoretical values for a range of percentage mixtures. The measured values are significantly lower than expected, leading to the suggestion of interaction (seeding) between the polymorphs.

A graph of the measured enthalpy of the endothermic transition versus percentage of form III in the mixture is shown in Figure 8.10. For comparison, the theoretical enthalpy of the endothermic transition versus percentage of form III is included. The theoretical enthalpies were calculated from the enthalpy of the endothermic transition for pure form III and the amount of form III in the mixture.

The measured enthalpies of the melting endotherm for form III in mixtures of the two crystal forms were found to be considerably lower than expected from the calculated values across the entire range from 1 to 99% (w/w) for form III. It has already been shown that at a heating rate of 250°C/min, there is essentially total inhibition of recrystallisation of form I from the melt of form III. It was, however, postulated that the presence of form I in the mixture prior to analysis resulted in the crystal seeding of form I and partial recrystallisation of form III to form I on melting. The measured enthalpies of the melting endotherm for form I support this. As shown in Figure 8.11, recrystallisation of the lower melting form resulted in an increase in the enthalpy of endothermic transition for form I relative to the theoretical values. For example, the enthalpy at a level of 99% (w/w) was found to be 162.5 J/g, compared with a calculated value of 107.7 J/g. The enthalpy of the endothermic transition for pure form I measured at a heating rate of 250°C/min was, as expected, comparable to that measured at 10°C/min. The theoretical enthalpies were calculated from this value and the amount of form I in the mixture.

When the different polymorphs were physically separated in the pan, so that they could not interact, the data shown in Figure 8.12 were obtained. This shows a graph comparing the measured and theoretical enthalpy for the endothermic transition as a function of content of form III.

Thermal Analysis of Pharmaceuticals   **301**

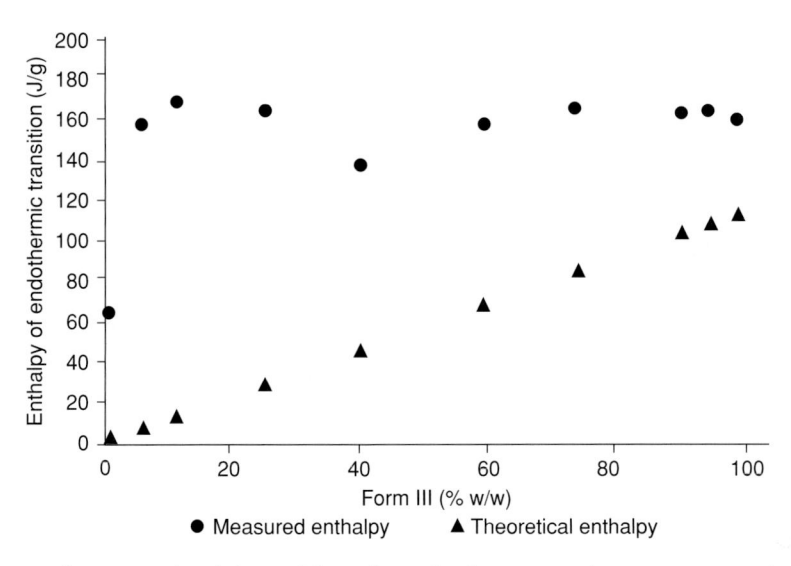

Figure 8.11 The measured enthalpies of the melting of carbamazepine form I are compared with theoretical values. These are higher than expected and correlate with the lower values found for form III (Figure 8.10) on the basis of crystal seeding.

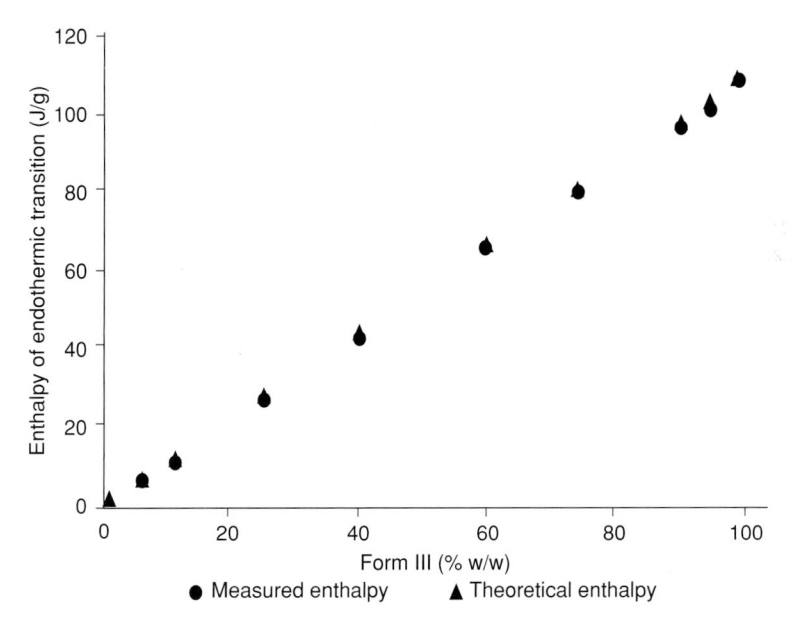

Figure 8.12 The melting enthalpies of carbamazepine form III compared with theoretical values for mixtures of carbamazepine form III and carbamazepine form I. In this experiment the individual polymorphs were not physically mixed so that no interaction could take place. The fact that measured and theoretical values are the same (when separated) indicates that, as found in previous experiments, interaction does take place where the materials are in intimate contact.

From Figure 8.12 it can be seen that there were no significant differences between the measured enthalpy of the melting endotherms and the calculated theoretical values, indicating no seeding or partial crystallisation of form I in the melt of form III.

This work demonstrates that using DSC with a heating rate (500°C/min), the recrystallisation of form III of carbamazepine is inhibited, and as a result a single melting endotherm is observed for this polymorph, thus allowing the determination of thermodynamic parameters, such as the enthalpy of fusion of the metastable melting endotherm. The melting endotherm associated with form III was detected even at a level of 1% (w/w); however, care needs to be taken when trying to quantify mixtures since interaction (in this case possible seeding) between the forms may affect the results.

8.3.5 Interpretation of DSC thermograms of samples exhibiting polymorphism

This section briefly deals with some types of DSC curves that may be obtained when investigating polymorphic tendencies of drug compounds and also explains some of the limitations of using DSC to study polymorphic behaviour.

When making the measurement it is important to make sure that there is no weight-loss profile over the temperature range of the transition, as found from the corresponding TGA curve, so ensuring that the response is not due to a desolvation/dehydration process but due to the melting of a metastable polymorph. For this reason TGA scans of materials should be routinely run. These also indicate the temperature of decomposition of the material, and the upper temperature limit of the DSC scan should be set below this value to avoid contamination.

8.3.5.1 Type 1: solid–solid transition

Figure 8.13 shows a DSC trace of a sample undergoing a low-temperature, endothermic, solid-state transformation (at ca. 25°C).

From the DSC trace shown in Figure 8.13, a low-temperature, solid–solid transition occurs prior to the main endotherm corresponding to the melting form. This transition can be distinguished from a low-temperature desolvation process since no mass loss is detected by TGA and the transition is reversible. (That is, the reversing transition will also be detected upon cooling.) For solid–solid transformations, this transition is exothermic for monotropy and endothermic for enantiotropy.

8.3.5.2 Type 2: liquid–melt recrystallisation

This refers to materials like carbamazepine (already considered) which recrystallise upon melting into a more stable form which then melts at a higher temperature. This is classic polymorphic behaviour as observed by DSC (see Figure 8.5).

Such a DSC scan can correspond to both monotropy and enantiotropy, with the sample being either a pure form or a mixture. Fast scanning rates can kinetically hinder this transformation, giving us detailed information about the actual composition of the sample.

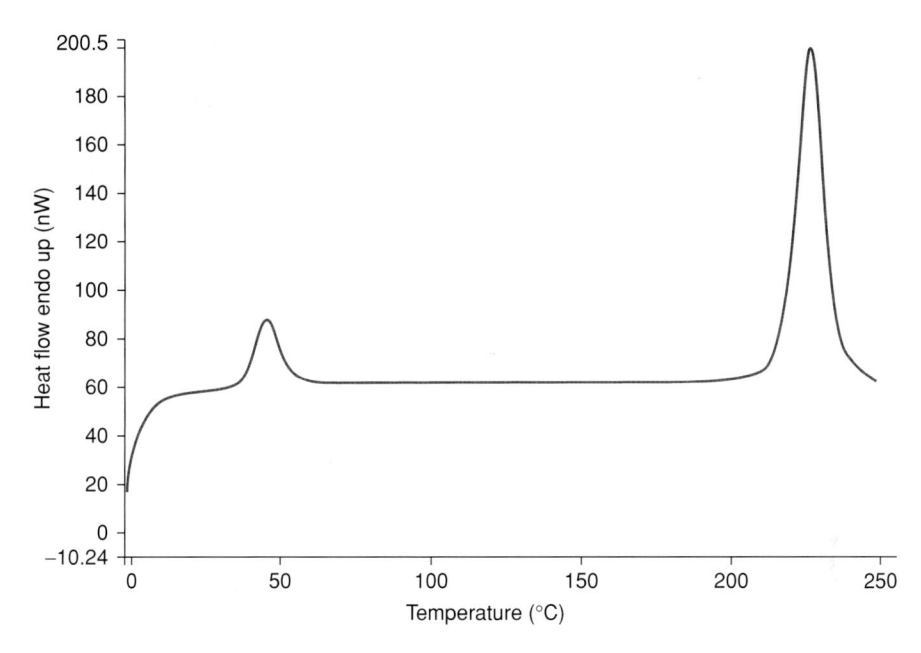

Figure 8.13 DSC trace of a sample undergoing an endothermic solid-state transition at about 25°C. There should be no corresponding weight loss on the TGA trace, and the event should be reversible. For solid–solid transformations, this transition is exothermic for monotropy and endothermic for enantiotropy.

Experiments should be performed on a known pure, lower melting substance to see if this inhibition of recrystallisation at fast heating rates is possible.

8.3.5.3 Type 3: single melting point determination

Each crystalline modification has a melting peak and no conversion between each form is seen. No conclusion can be made concerning the thermodynamic stability at room temperature; however, information can be obtained regarding the purity of samples containing a mixture of forms through calculation of the enthalpies of fusions of each transition and relating this to the pure form components.

8.4 Solvates and hydrates (pseudo-polymorphism)

In 1965, Walter C. McCrone introduced the term 'pseudo-polymorphism' [18]. By his definition, pseudo-polymorphic effects included desolvation/dehydration products, second-order transitions and dynamic isomerism; however, today the term is generally limited to all phenomena connected with solvates and hydrates. Any substance used in the pharmaceutical industry has the ability to form so-called crystalline hydrates or solvates. In these structures, volatiles (either water or solvent) are not only physisorbed at the solid–air interface but also incorporated into the crystal lattice structure (chemisorbed) as a guest molecule, either in stoichiometric or in non-stoichiometric amounts. Careful consideration must be given when progressing a solvated state for pharmaceutical development, as the crystalline solvent present is classed as an impurity within the sample and the daily dosing levels of that specific solvent should not exceed those set out in the FDA guidelines (ICH Topic Q3C (R3) Impurities: Residual Solvents).

8.4.1 Factors influencing DSC curves of hydrates and solvates

It is important to choose the correct type of DSC pan. Note that it is the loss of volatiles (mass) from a pan that gives rise to the large endothermic peak observed when heating solvates in an open pan, reflecting the fact that evaporation of a material requires significant energy input. Therefore when examining pseudo-polymorphic behaviour using DSC, be aware that the choice of pan can dramatically influence the curve that is obtained.

With hermetically sealed pans, the sample is sealed within a closed system such that the volatiles cannot escape and remain in the head space of the pan even after ejection from the sample. In this pan type, the melting peak of the solvate may be observed, but not the loss of volatiles.

If a small pinhole is made in the cover of the pan, or if a crimped or open pan is used, then volatiles can escape, but the shape of the DSC curve and the temperatures involved will depend upon the rate at which the volatiles can escape, which in turn depends upon the extent of crimping or the size of a hole made in the pan together with experimental conditions, such as heating rate and sample size. As a result, data obtained may not be very reproducible. In some cases, both melting of the solvate and desolvation in the solid state may happen, which gives rise to two endothermic peaks.

Figure 8.14 demonstrates the influence of pan type on a scan of copper sulphate. If the pan is open or crimped so that volatiles can easily escape, two broad, poorly resolved peaks are observed, as the water is easily lost from the bulk of the sample and escapes from the pan. Using a crimped pan can cause variable results since the extent of crimping may differ from pan to pan and affect the rate of volatile loss. However, if a cover with a 50-μm hole is used with the pan, a series of sharper, reproducible and better resolved events are observed, shifted slightly to higher temperature. In this case water is not lost from the pan until sufficient vapour pressure has built up to eliminate the water molecules through the very small hole in the lid. In the case of a pure solvent this occurs when the partial pressure of the vapour exceeds atmospheric pressure, which is the definition of boiling point, so this approach allows the boiling point of liquids to be determined. If a hermetically sealed sample pan is used (closed system), the solvate cannot be eliminated and only phase changes are observed.

When a sealed pan has been chosen, make sure it is capable of withstanding the internal pressure generated and take care to seal properly. If the pan cannot withstand the pressure build-up, the pan may rupture and can cause serious contamination to the instrument. Also note that the presence of solvent or water as vapour around the crystal in the atmosphere of the pan may induce metastable solvate-free forms (due to vapour diffusion of solvent into the solid phase and a resulting solvent-induced, solid–solid phase transformation) and also the amorphous state.

When performing DSC experiments on solvated/hydrated compounds, the DSC thermogram can often become rather complex due to the multiple events that can take place. To help with interpretation it is essential to perform a corresponding TGA experiment, so that

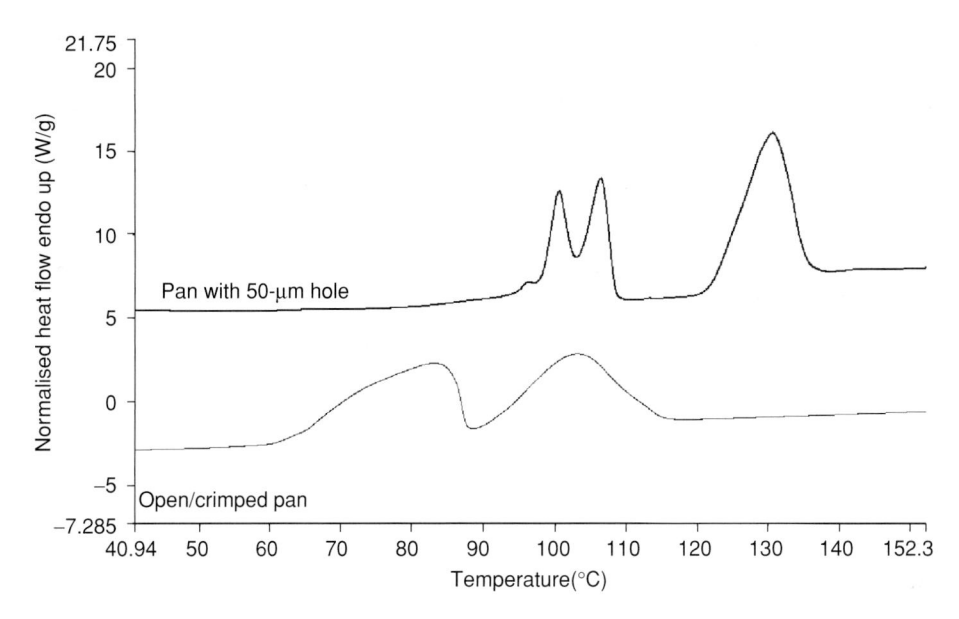

Figure 8.14 Influence of the sample pan type on DSC curves copper sulphate. The lower curve shows a typical trace from a crimped pan where volatiles may easily escape. The upper trace shows a pan with only a 50-μm hole for volatiles to escape. This gives more detailed and reproducible information than from an open or crimped pan.

the temperature range of volatile loss can be confirmed. It may also be important to increase purge gas flow rates to remove volatiles from the system. Large amounts of volatile loss can alter the atmosphere surrounding a sample or furnace and cause artefacts due to changing thermal conductivities.

8.4.2 Types of desolvation/dehydration

This section briefly deals with some types of DSC curves that may be obtained when examining solvates and hydrates and also explains some of the factors that can be responsible for misinterpretation of the data.

8.4.2.1 Type 1: dehydration/desolvation with no recrystallisation

This type of DSC curve is common in samples which exist either as a channel hydrate/solvate structure (i.e. the volatiles are condensed in capillaries or channels within the bulk structure and do not constitute any part of the unit cell of the crystal lattice structure) or as samples which undergo dehydration/desolvation with the resulting desolvated/dehydrated lattice structure remaining thermodynamically stable (i.e. does not undergo spontaneous recrystallisation to a more thermodynamically favourable anhydrous lattice arrangement). Figure 8.15 shows a typical DSC curve obtained for a hydrated compound exhibiting this kind of behaviour.

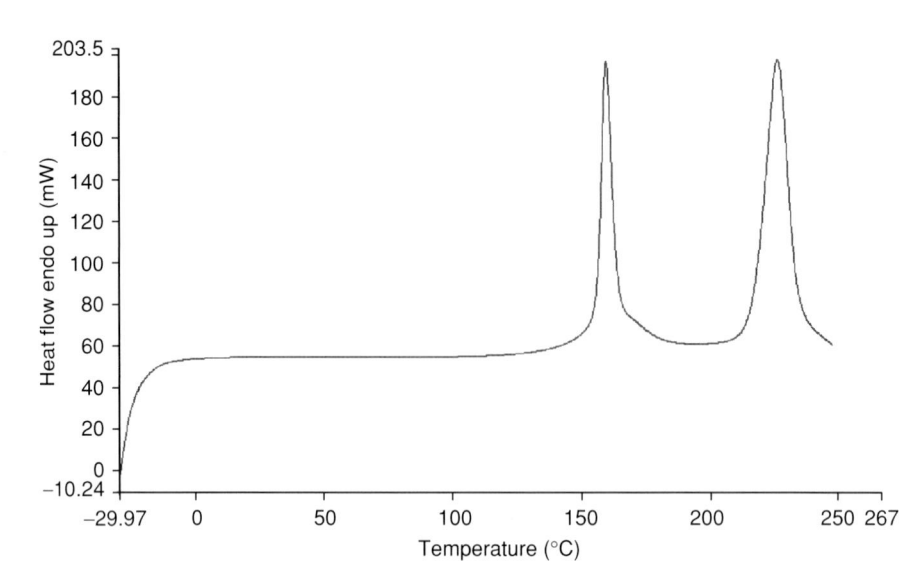

Figure 8.15 In this example the first peak at 150°C is due to loss of volatiles (hydrate) and the higher temperature peak due to melting of the crystal structure. This is confirmed by TGA (Figure 8.16).

From the DSC trace shown in Figure 8.15, the first endothermic peak corresponds to the thermal desorption of the water molecules constituting the crystalline hydrated form. In a corresponding TGA curve shown in Figure 8.16, a weight-loss profile is seen over this temperature range as the volatiles are desorbed from the sample and subsequently evaporate from the pan. The position and energy of this endothermic peak depend on the phase diagram of two components, the drug substance and the solvent present, and the stability of the component formed. Therefore, different hydrates or solvates of the same compound or compound series can lose their respective volatiles at different temperatures (i.e. the more thermodynamically stable the species, the higher the desolvation/dehydration temperature). In general if a sample is merely wet (damp) then a broad endotherm typically around 60–70°C is obtained, which becomes sharper and displaced to higher temperatures if volatile loss is restricted due to association with the sample.

8.4.2.2 Type 2: dehydration/desolvation and concurrent recrystallisation

For many hydrates and solvates, the guest molecule is incorporated into the unit cell of the lattice arrangement and has a stabilising effect on the crystalline structure. The dehydration or desolvation process then occurs during the melt or shortly after the melting of the hydrated/solvated state (Figure 8.17). In such cases, the hydrate/solvate melts first and the solvent is eliminated from the liquid phase. The subsequent exothermic transition is due to the crystallisation of the solvent-free form from the melt into a more stable anhydrous arrangement. In this case, the melting of the solvate and desolvation of the solid-phase overlap and the subsequent higher endothermic transition is the result of the melting of the more thermodynamically stable anhydrous form. It is also possible for a molecule to be

Thermal Analysis of Pharmaceuticals 307

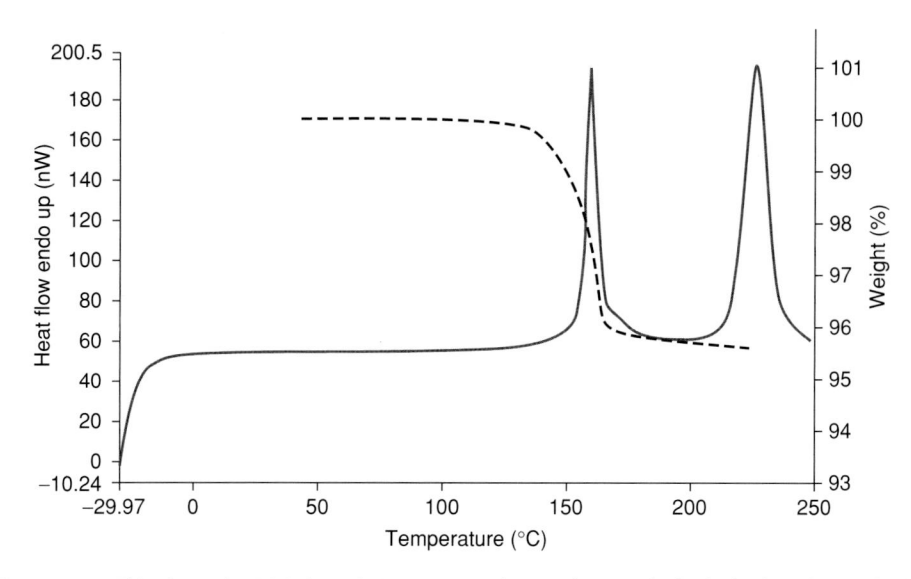

Figure 8.16 This shows the DSC data of Figure 8.15 with TGA data overlaid. The hydrate loss is clearly identified.

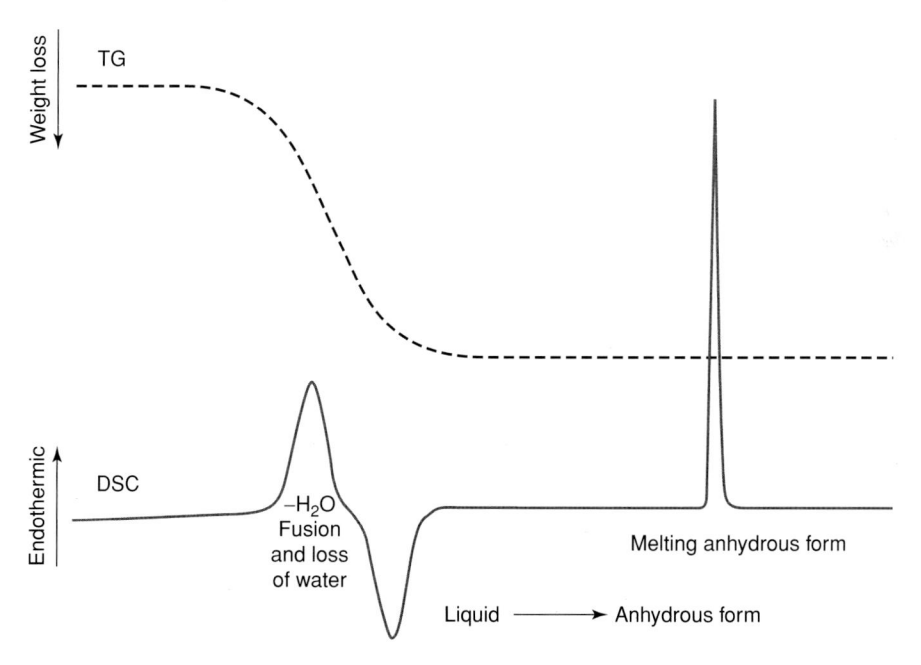

Figure 8.17 DSC and TG scans of a substance in which dehydration and melting events overlap and the anhydrous form immediately crystallises from the melt.

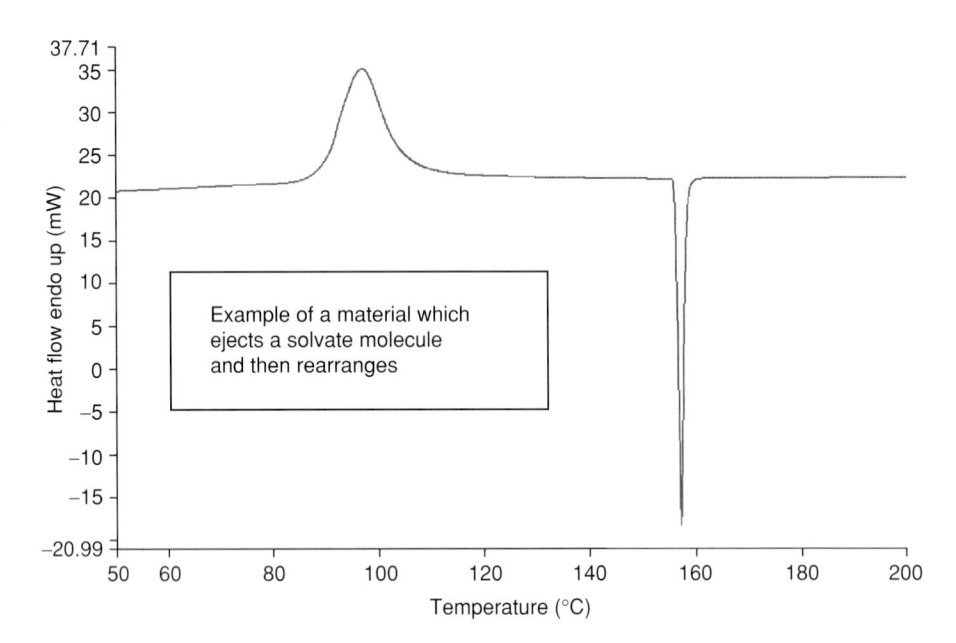

Figure 8.18 Example of dehydration without melting leading to a subsequent extremely rapid solid-state recrystallisation.

ejected from a crystal structure so de-stabilising it. The resulting de-stabilised structure is then likely to rearrange in a very rapid process (see Figure 8.18).

8.5 Evolved gas analysis

During a standard TGA experiment, the weight of a sample is recorded as a function of temperature or time under defined atmospheric condition and can be compared with that obtained for a corresponding DSC experiment. This enables us to perform quantitative compositional analysis, assuming that the evolved species which are thermally desorbed from the sample are known prior to analysis. However, since TGA is a quantitative and not a qualitative technique, if the volatile species are not known, TGA cannot be used to identify them.

Evolved gas analysis (EGA) is an approach which gives qualitative information, allowing the analyst to identify the volatiles produced. This is often of significance since many times the operator assumes that moisture or some other expected volatile is being released, but in practice this is simply an assumption or a guess. On other occasions it can help identify simultaneous reactions [19, 20]. An example of a TGA connection to the heated line for connection to a mass spectrometer (MS) is shown in Figure 8.19.

Two different approaches can be used for EGA. In the most popular approach, two analytical technologies are coupled to form a hybrid instrument and the materials investigated in real time. For example, thermogravimetric analyser coupled to a fourier transform infrared spectrophotometer (TG-FTIR) or thermogravimetric analyser coupled to a mass spectrometer (TG-MS) analysis in which volatilised products that evolve during heating

Figure 8.19 Picture of a TG-MS heated line connection.

can be monitored simultaneously by the second analyser. A second (less used) approach is a combined analysis technique, where volatiles are absorbed onto a suitable medium (typically a chromatographic tube) and then desorbed onto an analyser of interest, giving the technique of thermogravimetric analyser where the evolved gases are adsorbed onto a chromatography column with subsequent desorption through a Fourier transform infrared spectrophotometer for analysis (TG-GC-FTIR) or similar. The GC step can be chosen to be selective and so focus the analysis of interest. A classic example of the simultaneous approach is shown with calcium oxalate in Figure 3.25.

The choice of evolved gas system may be influenced by the experience of the user. An experienced FTIR spectroscopist will normally find FTIR spectra easier to deal with, and an experienced mass spectrometrist will find MS data easier to deal with. However, FTIR can be more useful when very complex molecules are produced, such as decomposition products from a polymer, and MS when simpler molecules are produced, such as a solvent from a pharmaceutical, which means this type of system predominates in the pharmaceutical industry.

Most TG-MS systems benefit from the use of a helium purge allowing mass 28 to be measured (see Section 3.11). When beginning a run in helium, a period of time needs to be given for air to be replaced by helium in the furnace and a stable atmosphere produced. Note however that any volatiles lost from the sample during this period will not be monitored, so it can be useful to analyse a sample immediately upon loading, before stabilisation, and then repeat it after stabilising in helium. Take note of the differing buoyancy effects of air and helium, so a drift in the recorded weight may also be expected during the initial equilibration period.

An example of a weight-loss curve overlaid with multiple ion-detection traces is shown in Figure 8.20, indicating that a range of different molecules may be lost together.

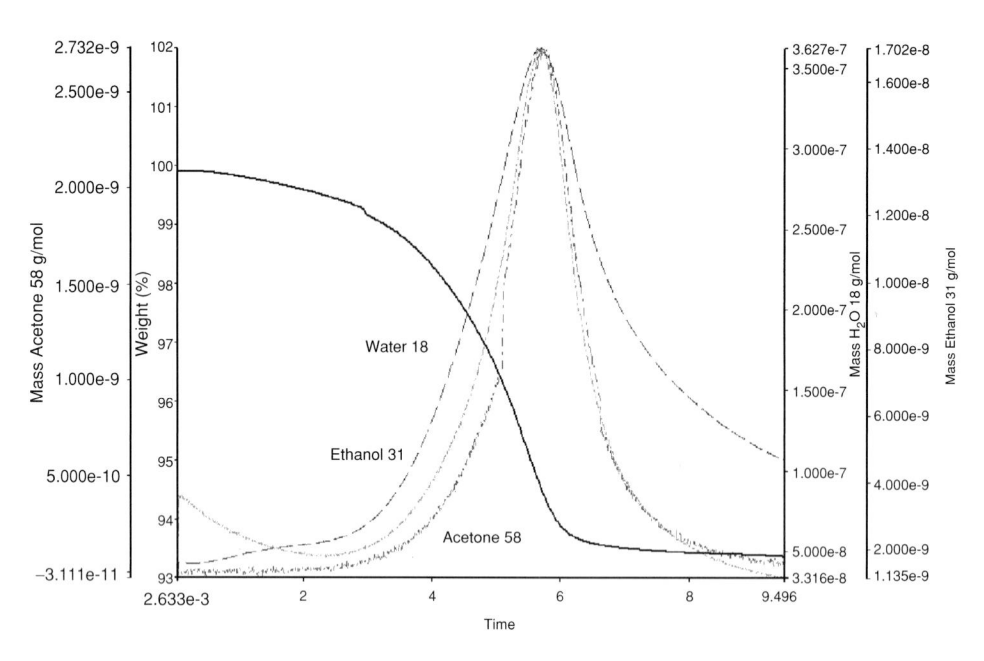

Figure 8.20 An example of EGA analysis using TG-MS. A number of different solvents are identified during a single weight-loss event.

8.6 Amorphous content

8.6.1 Introduction

Apart from perfect crystals, all crystalline solids contain some area of disorder or low-crystallinity regions. When these disordered regions make up the bulk of the material, it is said to exist in an amorphous form; below T_g this is defined as being a glass (see Section 1.5.3). Amorphous solids can be distinguished from their crystalline counterpart by their lack of macroscopic and microscopic properties, such as particle shape, birefringence [21] and fracture mechanism. When examined using XRPD methods, they exhibit a broad, 'halo' effect with no noticeable diffraction [22]. In parallel, their solid-state nuclear magnetic resonance spectrum is also broad or non-distinct [23]. The reason for this is that amorphous materials do not possess the long-range order of crystals, and yet do possess a degree of short-range order, typically over a few Angstroms.

Amorphous forms of materials are usually prepared by freeze drying [24], spray drying [25] or by rapidly cooling from the liquid melt [26]. Quench cooling from the liquid melt is useful for inorganic materials but less favourable for pharmaceuticals, since many organic compounds decompose at or near their melting points. Amorphous drugs have also been prepared by co-lyophilisation with polymers [27], such as polyvinylpyrolidone or polyethylene glycol.

With respect to stability, amorphous forms are thermodynamically less stable (metastable) than the corresponding crystalline forms. In theory, the amorphous form can be regarded as an extension of the liquid state below the melting point of the solid. Therefore, an amorphous form of a drug compound may eventually transform to the crystalline state by means of nucleation and growth of crystals. This process is independent of compound type, and the speed of conversion will depend on the nucleation and growth rate, both of which are related to the molecular mobility of the molecules in the amorphous environment.

8.6.2 Characterisation of amorphous solids: the glass transition temperature

As already noted, amorphous solids do not possess any long-range order and hence have no crystal lattice arrangement. When using thermal methods such as DSC to study such materials, they cannot be characterised by the distinctive endothermic melt transitions that are commonly observed for the corresponding crystalline structures. However, one transition that is unique to amorphous solids is that of the 'glass transition' often abbreviated to T_g, the temperature of the glass transition. The T_g (characteristic for each system) is the temperature below which the molecules are configurationally frozen in the *glassy state* and hence lack the motions of the molecules in a liquid. Above the T_g, the amorphous material is said to be a rubber or in the *rubbery state* and will exhibit a certain degree of flow (see Section 1.5.3).

The T_g of an amorphous material is not a single point; it occurs over a temperature range and can vary depending upon how it is defined and measured (see Sections 1.5.3–1.5.6). However, for a dry and pure amorphous solid, the glass transition should occur in a defined region and should not change as a function of time and pressure, provided it is stored in a suitably dry environment and well below T_g. Eihei et al. [28–30] prepared a number of amorphous pharmaceuticals by quench cooling from the liquid melt. The subsequent dry T_g was measured from the characteristic step in heat capacity and the anomalous endotherm in the DSC curve. The results of their research are shown in Table 8.3. It is clear that the ratio T_g/T_m (in degrees Kelvin) determined for a pure, dry, amorphous solid is between 0.7 and 0.85. This apparent constancy of the ratio T_g/T_m indicates that the T_g can be estimated from melting point data. Knowing the T_g allows us to predict the storage temperature needed to ensure stability of the glass, to prevent recrystallisation or transformation (both physical and chemical).

Although the T_g temperatures represented in Table 8.3 correspond to those of pure, dry amorphous solids, the T_g can be significantly decreased by addition of plasticisers or guest molecule into the matrix (e.g. absorption of water molecules during storage at elevated relative humidities). This behaviour has been known for many years and was described by Zografi and co-workers, who suggested that these smaller molecules of *plasticiser* act as an impurity by embedding themselves between the molecules of the amorphous solid [31]. This effectively increases the spacing and free volume of the sample and results in an increase in the degree of molecular mobility. For example, the 'new-car smell' is due to the initial outgassing of volatile small-molecule plasticisers used to modify interior plastics (e.g. dashboards) to keep them from cracking in the cold, winter weather. In pharmaceuticals, this is of extreme importance, as storage above the glass transition temperature increases the molecular mobility and increases the likelihood of crystallisation. For example, amorphous cephalexin absorbs large amounts of water when equilibrated at elevated humidities, and

Table 8.3 The measured T_g (K), T_m (K) and T_g/T_m values for a range of pharmaceutical compounds [29, 30]

Compound	T_g (K)	T_m (K)	T_g/T_m
Acetaminophen	302	441	0.69
Antipyrine	256	380	0.67
Aspirin	243	408	0.59
Atropine	281	379	0.74
Cholecalciferol	296	352	0.84
Cholic acid	393	473	0.83
Dehydrocholic acid	348	502	0.69
Deoxycholic acid	377	447	0.84
Ergocalciferol	290	376	0.77
Ethacrynic acid	282	398	0.71
Flufenamic acid	290	406	0.71
Griseofulvin	370	497	0.74
Methyltestosterone	270	421	0.64
Phenylbutazone	277	377	0.73
Progesterone	279	399	0.70
Quinidine	326	445	0.73
Quinidine ethylcarbonate	278	362	0.77
Salicin	333	466	0.71
Santonin	290	434	0.67
Stilbestrol	308	439	0.70
Sulfadimethoxine	339	465	0.73
Sulfathiazole	334	471	0.71
Sulfisoxazole	306	460	0.67
Tartaric acid	289	430	0.67

storage above the critical relative humidity (RH) point (75%) results in sufficient plasticisation of the glass transition such that the amorphous form becomes rubbery and is able to crystallise [32]. Therefore, if there is a chance that amorphous material is introduced into the samples as a by-product of physical processing or manufacture, it is imperative that a detailed understanding of the relative stability of the amorphous fraction introduced, as well as the extent of production, is understood.

Measurement of T_g can be performed by DSC, as described in the next section, and also by mechanical means. If a dynamic mechanical analysis (DMA) instrument has a good compressional mode capability then the T_g of a powder may be determined in compressional mode. Typically, the powder is placed in a small container such as a straight-sided DSC pan and a flat lid placed on top. The storage modulus should show a decrease as the material goes through the glass transition region and a corresponding peak in loss modulus and tan δ. Whilst the temperature of T_g can be determined in this way, the approach is non-quantitative for modulus, because this will vary depending upon how the material is compacted.

A more recent approach using DMA makes use of a materials pocket for powders, which is mounted in a flexural mode. This provides an easier method of determining the T_g of a powder, and whilst modulus values are non-quantitative, measurement of amorphous

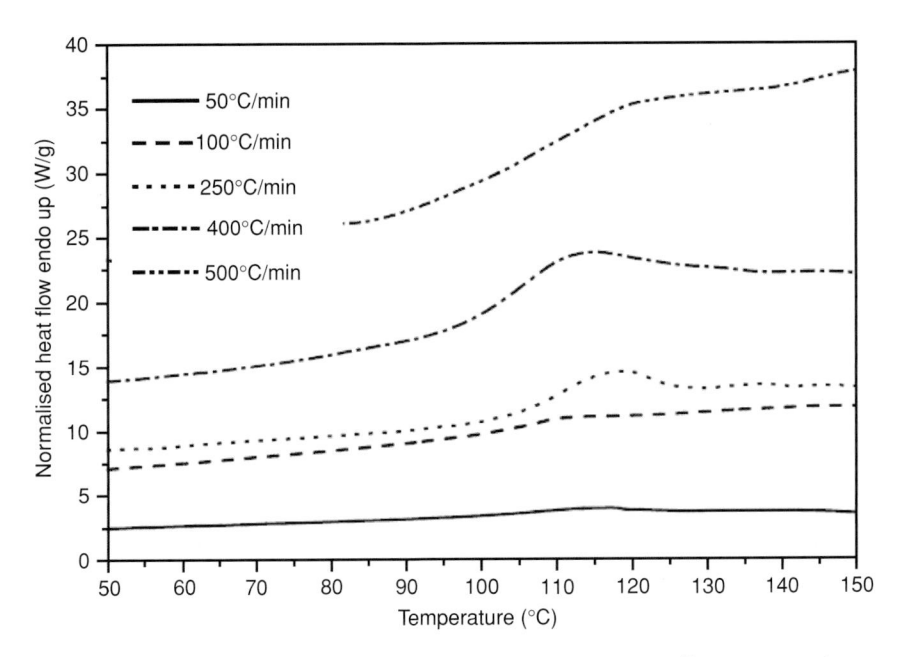

Figure 8.21 The T_g of amorphous lactose measured by DSC using a range of heating rates. The greatest sensitivity is found at the higher heating rates.

content can be estimated from the peak height of the tan δ curve (see Section 4.5.5 for further details).

8.6.3 Quantification of amorphous content using DSC

DSC, although used frequently for investigations of phase behaviour, compatibility and polymorphism, is not frequently reported in the field of determination of amorphous content. It has often been difficult to quantify very low levels of amorphous content using DSC (below 10% w/w) because of the small energy changes associated with the measurement of the glass transition at these low levels [33]. However, high-speed DSC (HyperDSC™) has been successfully used in the quantification of small levels of amorphous content through identification and quantification of the energy change associated with the glass transition [34]. In this study, α-lactose monohydrate was used as the test compound. The DSC scans of amorphous lactose revealed a glass transition, a crystallisation and a melt region (showing both alpha and beta melts), but for this study the area of interest was the glass transition region. The T_g responses of amorphous lactose have been discussed in Section 2.5.3, and in Figure 2.33 the effect of increasingly fast scan rates is shown. Similar data were obtained in this study (Figure 8.21). It was observed that the size of the DSC response increased substantially as the scan rate was increased; the change in heat capacity (W/g) was ca. 1 at 100°C/min, 3 at 250°C/min, 5 at 400°C/min and 10 at 500°C/min. Recrystallisation was also offset to higher temperatures in general, resulting in clearer data at the higher rates.

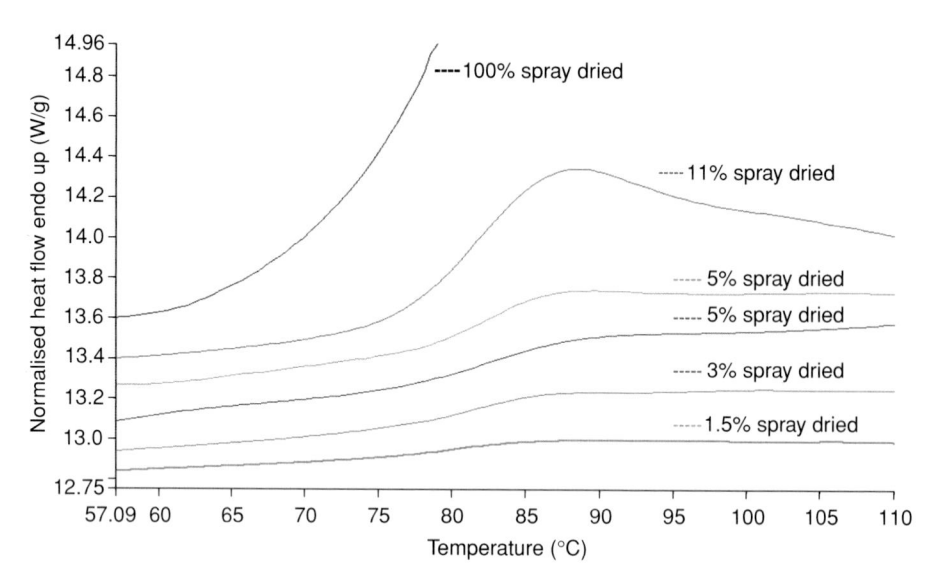

Figure 8.22 The T_g region of mixtures of amorphous and crystalline lactose as measured by DSC at 500°C/min. (The percentage of spray-dried content is proportional to the amorphous content.)

For the amorphous sample, the dry T_g was easily observed close to the literature value of 116°C [35]. Despite the fast scan rates up to 500°C/min, crystallisation exotherms were still observed, indicating the rapid mobility and kinetics of crystallisation of small molecules compared with polymers [36]. Investigating the relationship between crystallisation and heating rate may offer an opportunity to characterise, or at least rank order, the mobility within devitrified amorphous pharmaceuticals.

The T_g responses for mixtures of amorphous and crystalline lactose are shown Figure 8.22 (using a y-axis scale that shows most of the T_g values clearly). The T_g response for the 100% amorphous sample is too large to be recorded fully on this axis scale; furthermore, it is easy to rescale to show the responses more clearly for the lower amorphous content (e.g. 1.5%). The onset of T_g was not affected by scan rate (80°C at all rates) but was much lower than the accepted value of lactose T_g, probably due to the moisture content.

For scan rates of 500°C/min, it was very easy to see the detail of the T_g for the sample with 1% amorphous content. The fact that at high scan rates it is easy to detect the T_g of these samples is a significant improvement on traditional approaches. Saleki-Gerhardt et al. [37] demonstrated that a sample had to have around 10% amorphous content in order to be detected by conventional DSC (using slow heating rates), and this figure has not been challenged to any meaningful extent since that time. It has been shown in one study that modulated-temperature DSC has been used to detect approximately 1% w/w amorphous content [38]. However, the data presented here show that fast scan rates can rapidly (much more rapidly than with slow modulated-temperature experiments) detect T_g for samples with very low amorphous content, without giving time for annealing effects during measurement. There is great advantage in being able to detect small amorphous contents, and given that the amorphous form is not thermodynamically stable, there is further great advantage to being able to detect it quickly and so minimise the chance of the amorphous form recrystallising during the experiment. (Even though crystallisation is most rapid above T_g, it is also possible for materials to crystallise, albeit more slowly, near to T_g, and to suffer relaxation if low underlying heating rates are applied.)

Subsequently, the change in the heat flow signal at T_g for these lactose samples was measured as the step height change from the onset to the maximum height for the sample, giving an indication of the change of the specific heat of transition for the T_g. These data produced a linear relationship from 0 to 100% amorphous content, as shown in Figure 2.34, and it is clear that detection is possible to less than 1% amorphous content. Following the method described in the United States Pharmacopoeia and by Miller and Miller [39], it is possible to determine a theoretical limit for detection and quantification using this method for the quantification of amorphous lactose. Based on these data, the theoretical limit of detection using this method is 0.57%, and the limit of quantification is 1.89% amorphous content, though even lower limits have been claimed for sucrose (Chapter 2, [14]).

8.7 Purity determination using DSC

8.7.1 Types of impurities

Impurities in pharmaceuticals are the unwanted chemicals that remain with the active pharmaceutical ingredients (APIs) after synthesis or which may arise during the manufacturing process and/or storage of the drug substance and final drug product formulation. Generally, impurities can be classified into the following categories:

- Organic impurities (process and drug related)
- Inorganic impurities
- Residual solvents.

Organic impurities can arise during the manufacturing process and/or storage of the new drug substance. They can be identified or unidentified and volatile or non-volatile, and include:

- starting materials
- by-products
- intermediates
- degradation products
- reagents, ligands and catalysts.

Inorganic impurities can result from the manufacturing process. They are normally known and identified, and include:

- reagents, ligands and catalysts
- heavy metals or other residual metals
- inorganic salts
- other materials (e.g. filter aids and charcoal)

Solvents are inorganic or organic liquids used as vehicles for the preparation of solutions or suspensions in the synthesis of a new drug substance. Since these are generally of known toxicity, the selection of appropriate controls is easily accomplished (ICH Q3C (R3) Impurities: Guideline on Residual Solvents).

The presence of these unwanted chemicals, even in small amounts, may influence the efficacy and safety of the pharmaceutical products. Therefore impurity profiling (i.e. establishing the identity as well as the quantity of impurity in the product) is now receiving important, critical attention from regulatory authorities. The different pharmacopoeias, such as the British Pharmacopoeia and the United States Pharmacopoeia, are slowly incorporating limits to allowable levels of impurities present in the APIs or formulations. Also, the ICH has published guidelines on impurities in new drug substances, products and residual solvents (Q3A Impurities in New Drug Substances, Q3B(R) Impurities in New Drug Products, Q3C Impurities: Residual Solvents and Q6A Specifications: Test Procedures and Acceptance Criteria for New Drug Substances and New Drug Products: Chemical Substances).

In general, according to ICH guidelines on impurities in new drug products, identification of impurities below the 0.1% level is not considered to be necessary unless the potential impurities are expected to be unusually potent or toxic. (In all cases, impurities should be qualified.) If data are not available to qualify the proposed specification level of an impurity, studies to obtain such data may be needed.

8.7.2 DSC purity method

In principle, when a material is slowly heated through its melting region in a DSC, the shape of the melting profile can be used to determine the purity of the substance with respect to those organic impurities which form a eutectic mix with the substance. It requires knowledge of the molecular weight of the material and can be applied to single crystalline materials of 96–98% purity and above. In practice, this is often the case for precursors and by-products which result from the manufacturing process and which may still remain in small quantities. This simple and rapid method for estimating purity is therefore a potentially significant application, but it is an empirical approach and its use is restricted by interferences from other events that occur simultaneously with melting and affect the peak shape and the subsequent purity calculation. For example, these include solid–liquid interactions and polymorphic transformations. Other impurities which do not form a eutectic mix and interact in no way with the substance will not be taken account of by this calculation and need to be assessed separately, though it is possible for non-eutectic impurities to interact and affect the melting point. Provided any interactions do not affect peak shape, the purity calculation should remain unaffected, so it is possible for the individual purity of two mutually incompatible materials in a mix to be assessed from one scan. The nature of these restrictions means that great care is needed when the technique is applied to a range of unknown samples, but it can work well when applied as a quality control method to a material which is known to have no interferences. Plato and Glasgow have found that of 95 crystalline organic compounds they analysed, this method could be successfully applied to over 75% if sufficiently pure [40].

The basic theory behind the method is based on the observation that the presence of small degrees of impurities in an organic compound depresses the melting point [41] (Figure 8.23);

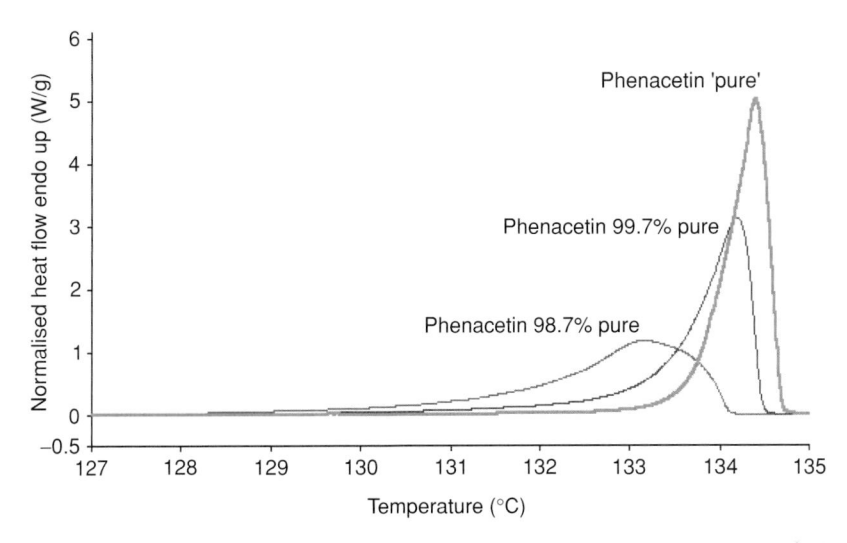

Figure 8.23 DSC traces showing the effect of increasing impurity levels on the melting peak shape of phenacetin.

that is, the melting point is lowered with increasing amounts of impurity (which is why water freezes at a temperature below 0°C with the addition of salt (NaCl)). The correlation between the melting point depression and the degree of impurity for a dilute system is defined by the van't Hoff equation:

$$T_0 - T_m = RT_0^2 X_2/\Delta H_f \times 1/F \qquad (8.2)$$

where T_0 and T_m are absolute temperatures of fusion of the pure and impure materials, ΔH_f is the molar enthalpy of fusion, F is the fraction molten corresponding to T_m and R is the gas constant. A plot of T_m against $1/F$ should yield a straight line, with the mole fraction of impurity (X_2) obtained from the slope of the plot. For further details of the development of the equations and how they are used, see the study by Gray who first applied the technique to DSC and described in detail the thermodynamic theory on which it is based [42, 43].

8.7.3 Practical issues and potential interferences

In practice, instrument companies provide software to perform purity calculations, but there is still a lot of care needed in the method. First of all it has to be noted that thermal gradients across a sample will affect the rate of melting and the resulting peak shape. For this reason sample sizes must be small, typically about 1 mg, and the scan rates slow, typically 1°C/min. Even so the rate of transfer of heat to the sample, expressed as the thermal resistance constant R_0, will influence the rate of melt, and this varies from instrument to instrument and with pan type used. R_0 must therefore be determined under the conditions of the test and used in the calculations. See Section 1.5.2 for a description of how this is measured. It is very important that this is done correctly; whilst in one sense the method is not based on the comparison of a pure material against an impure material, in reality the rate of melting of the test material is being compared with the rate of melting of a 100% pure material (normally indium) and if this value (the R_0 value) is not correct then the answer will be in error. For example, on some occasions, a calculation may give a purity value in excess of 100% (if the software does not have a limiting value). This indicates a possible error in the R_0 value.

It is probable that the use of helium as a purge gas could improve the thermal transfer and allow potentially larger samples and more rapid conditions as found with fast scanning techniques; however, slow rates are also required to allow a reasonable number of data points to be taken during the equilibrium melt region. If melting is too rapid (sharp) then the partial integration is compromised. Partial integration is performed at a series of temperatures during the melting range to obtain values of the fraction melted as a function of temperature. It is condition of the theory that this is obtained in the equilibrium melt region (where solid and liquid coexist (see Section 1.5.2)), so ideally melting should occur slowly. Methods have been produced where purity has been obtained from a series of isothermal steps in place of the slow scan, but most softwares use the slow scan method.

Sample should be well pressed down in the pan to give good thermal contact. If possible, pans should be sealed to prevent loss of volatiles or possible sublimation during melting. An air space above a sample can result in a microclimate in the pan, and possible 'snowing' effects so is best avoided. The melting range is best determined via a more rapid survey scan which may also be used to show up other unrelated impurity events, and the purity measurement should begin well below the expected melt range and continue until flat baseline appears after the trace. It is surprising how much material may melt at lower than expected temperatures due to impurity effects, so the trace should be well expanded on the y-axis to show this information so that integration limits can be properly selected (see Figure 8.24).

Once data has been obtained, it should be carefully inspected. Most interferences can be detected from inspection of the trace. Any irregularities in the peak shape or possible shoulders indicate potential interferences, and the data should be discarded. An obvious step under the melting profile is another indication. The second derivative trace should be used to give a further check. Sometimes irregularities may be due to sample movement during melting, leading to the suggestion that purity determination might be done on re-heating; however, the chances of most materials cooling into the same state without any change are remote, so reheating should not be done unless this approach has been previously verified. If sample movement is suspected then repeat the measurement on a separate sample.

If the trace looks to be acceptable then begin the purity determination by setting the limits for the partial integration. In some systems these may be fixed by the software or in others chosen by the operator. The early part of melt below 5% is unlikely to be useful, and limits are often set between 5 and 60% of the peak area which corresponds to the equilibrium melt region (leading edge of melt) for the majority of materials [40]. If the 60% limit falls on the tail of the peak, this is evidence of excessive energy under the tail, which could be due to solid–liquid interactions which tend to result in an unduly symmetrical peak shape.

If the partial area integration looks acceptable then the plot of $1/F$ against T_m can be obtained. In reality, a curve is almost always obtained, though a straight line is expected (Figure 8.25). The cause is thought to be an underestimate of the total heat of fusion which

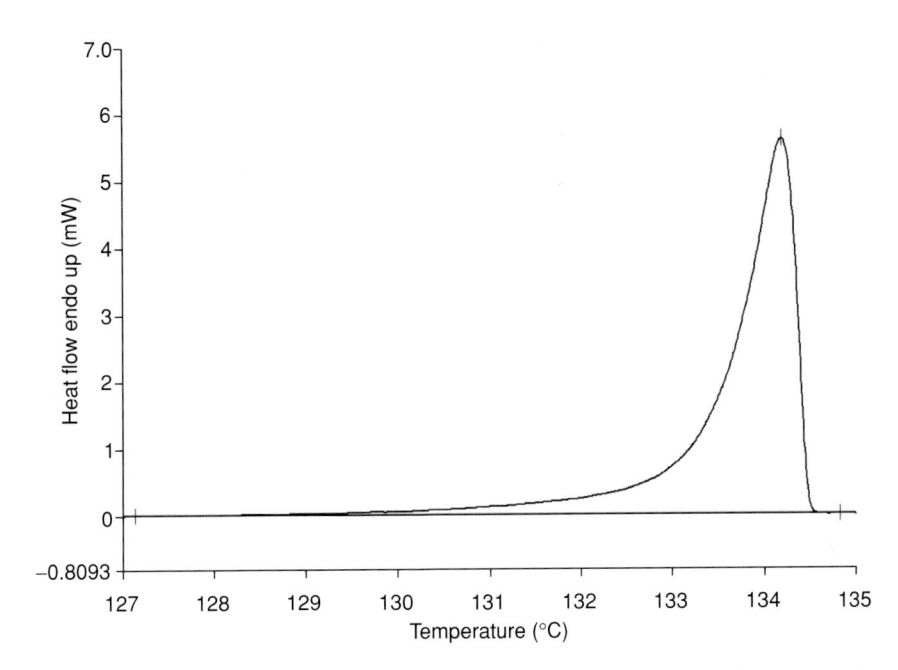

Figure 8.24 Selection of integration limits for the peak area integration for purity calculation. Make sure the initial limit allows for low melting material.

would give rise to a curve. This may be rationalised by noting that any impurity will melt with a small amount of the main component at the eutectic point, which is below the main melt, and remain unmeasured. In addition, the measurement accuracy of 1 mg of material heated at 1°C/min may be called into question since the heat flows generated will be very

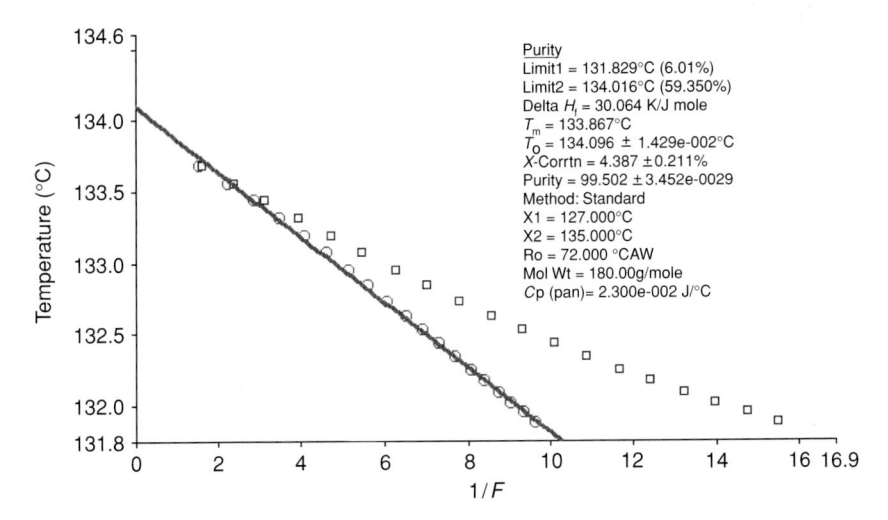

Figure 8.25 The van't Hoff plot showing a DSC purity calculation. The curved line shows the original data before use of the x-correction algorithm.

small. As a result an algorithm is introduced to alter the heat of fusion value employed in the calculation until a straight line fit is obtained. The correction involved is termed the x-correction (or similar) and represents the estimated error in the original heat of fusion calculation. This correction should not be excessively large. Values of 5% are not uncommon; anything over 10% in a modern analyser should cause concern; the size should also vary with the amount of impurity.

It is because of this correction that the purity calculation can be used despite decomposition occurring during melting. Providing the leading edge of the melt is unaffected by decomposition (i.e. looks straight with no imperfections) then even though an accurate heat of fusion measurement cannot be made, it may be obtained by the x-correction algorithm (although a more rigorous algorithm may be selected for this particular purpose). In principle, the same argument may apply to other interferences, but if there is doubt, it is wise not to use the data.

In general results should be reproduced to give confidence, and as with all extrapolated data some variability can be expected. To some extent this will be operator dependent but typically should be to within a few tenths of a percent. The method may be optimised for a given substance with the general guideline that the purer the material, the slower the scan rate that should be used and that the more impure the material, the faster the scan rate that can be used (since peaks will be broader). We have found 0.5°C/min to give most accurate and reproducible data for very pure materials.

8.8 Excipient compatibility

During the development of any drug compound into a successful pharmaceutical product, the final formulation is required to show acceptable chemical stability during the distribution, storage conditions and shelf life. All drugs are formulated with a range of excipients, such as binders, disintegrants, fillers, lubricants, etc. It is important that the drug does not interact with any of the excipients in a way that is likely to reduce its efficacy; therefore, excipient compatibility is important when considering drug stability.

Over the years, various excipient compatibility methods have been developed to guide the selection of excipients. Excipient compatibility screening is generally regarded as an essential part of the development process; however, since real-time data are often not available during the initial stages of development, formulation scientists must employ accelerated stability studies on model formulations to estimate and predict long-term ambient stability. Such studies are costly and time consuming, and it is therefore desirable to minimise the number of studies performed.

There are numerous ways to conduct excipient compatibility screening. In all cases, however, the basic method is the same – mix two or more materials together and monitor any subsequent reactions. In one type of study, drug-excipient mixtures are stored under accelerated stability conditions, such as binary blends, mini-formulations or statistically designed mixtures, and then assayed over time using liquid chromatography or spectrophotometry. A disadvantage of this technique is that it must be monitored for several weeks. Also, since the quality of the results depends on the precision of the assays, well-developed and sufficiently validated methods are required.

8.8.1 Excipient compatibility screening using DSC

Use of DSC has been proposed as a rapid method for evaluating the physicochemical interaction between two components and can provide fast and reliable information about potential physical or chemical incompatibilities between the formulation components through the appearance, shift or disappearance of endotherms or exotherms and/or variations in the relevant enthalpy values [44, 45]. However, interpretation of DSC results is not always easy, and thoughtful evaluation is necessary to avoid misinterpretation and erratic conclusion [46].

The basic approach is that of mixing two components (drug and excipient) typically in a 50/50 mix and then performing a DSC scan. The melting profile of the two individual components is then compared with the scan of the mixture. Ideally, if no interaction occurs then the mixture should show equivalent transitions to that of the individual components. If it does not, some interaction is indicated.

Issues arise because physical interactions occur that have nothing to do with the chemical interactions, which are the basic cause of concern, such as solution of one component in the melt of another. This means that many 'false' results will be obtained that are difficult to distinguish from results which are of true concern. Moreover, the fact that an interaction has been observed at elevated temperature is not necessarily an issue of significance if the temperature is beyond testing parameters, and the validity of using a 50/50 mix can also be questioned. Further, the presence of a solid–solid interaction does not necessarily indicate pharmaceutical incompatibility [47] but it might instead be advantageous, e.g. as a more desirable form of drug delivery system [48]. Therefore, other analytical techniques often have to be used in conjunction with DSC to adequately substantiate the findings, such as hot-stage microscopy, mass spectroscopy and chemical purity by HPLC.

Having said this, experienced analysts have reported that useful information has been obtained from DSC interaction studies, and it should be pointed out that data indicating non-interaction provide significant evidence that no interaction is occurring. In fact this may be the most useful information that can be gained from such studies. An alternative approach yet to be investigated is that of using fast scan DSC methods. Mixtures that have been previously prepared and aged might be usefully heated using high rates where physical interactions are of less significance, because they do not have time to occur.

8.8.2 Excipient compatibility analysis using isothermal calorimetry

Isothermal microcalorimetry offers ultimate instrument sensitivity in the nanowatt region, which is 1000-fold greater than traditional DSC. This allows for the detection of reactions at lower, more relevant temperatures, thereby improving the likelihood of valid extrapolations. See Section 8.9.2 for more details of isothermal microcalorimetry.

When using calorimetry for compatibility testing, it is imperative to realise the non-specific nature of the technique. Several physicochemical processes can give rise to a calorimetric response, including wetting, evaporation, dissolution, crystallisation and chemical reaction. Isothermal microcalorimery has been previously employed to study the degradation kinetics of drugs in solution [49] and in the solid state [50]. Recently [51], isothermal

Table 8.4 Time-averaged integration power (P) values for 16 excipients tested for compatibility

Functional class	Excipient	Interaction power (µW)
Diluents	Dibasic calcium phosphate	5.23
	Microcrystalline cellulose	2.92
	Lactose monohydrate	−0.345
	Pregelatinised starch	−1.31
Lubricants	Calcium stearate	12.15
	Sodium stearyl fumarate	10.66
	Magnesium stearate	6.51
	Zinc stearate	2.08
	Hydrogenated cottonseed oil	0.516
	Stearic acid	0.512
	Colloidal silicon dioxide	0.32
Binders and disintegrants	Sodium starch glycolate	7.12
	Povidone K30	2.59
	Crospovidone	2.37
	Hydroxypropyl methylcellulose	1.78
	Hydroxypropyl cellulose	1.66

The excipients are grouped according to functional class for comparison reasons [51].

microcalorimetry has been used for the study of excipient compatibility of the test compound ABT-627. This screen was conducted at 50°C using binary blends containing 20% (w/w) water. Sixteen excipients from various functional classes were evaluated, with the power–time curves collected for drug+water, excipient+water and drug+excipient+water. The separate drug and excipient curves were used to construct a theoretical non-interaction curve (by simple addition) which was then subtracted from the actual mixtures to give the interaction curve. (The results are expressed in terms of time-averaged power values (P) calculated over an 8-h period.)

Figure 8.25 shows the power–time curves for the mixture of ABT-627 with calcium phosphate, which illustrates typical results. The three curves were then used to calculate the time-averaged interaction power (P) as described earlier.

Table 8.4 lists the (P) values for the 16 excipients tested. (The excipients are grouped according to functional class for comparison reasons. Obviously, a 10-µW output signal for a diluent and a lubricant where the drug-to-excipient ratio could differ by orders of magnitude are not equivalent.) Table 8.5 gives the results from parallel HPLC-based compatibility study.

Comparing the results of Tables 8.4 and 8.5 show that the conclusions from the two methods are in good agreement and that for the drug molecule ABT-627, isothermal microcalorimery was suitable for predicting the stability of the drug with various excipients. The other main advantage of microcalorimery over the HPLC-based method was that the latter required significantly more resources for method development, sample preparation and data analysis. Moreover, the HPLC method required nearly 6 weeks to complete as compared with 3 days for the calorimetric method.

Table 8.5 Results from an HPLC-based compatibility (parallel) study for ABT-627 [51]

Excipient	Initial	After 3 weeks	After 5.2 weeks
Dibasic calcium phosphate	0	0.006	0.22
Microcrystalline cellulose	0	0.056	0.108
Pregelatinised starch	0.05	0.048	0.102
Lactose monohydrate	0	0.051	0.076
Magnesium stearate	0.546	1.23	2.04
Stearic acid	0	0.05	0.03
Povidone K30	0.059	0.10	0.54
Sodium starch glycolate	0	0.21	0.51
Crospovidone	0.052	0.25	0.39
Drug only (control)	0	0.07	0.05

Area percent impurities

8.9 Microcalorimetry

8.9.1 Introduction

In pharmaceutical sciences, the term 'calorimetry' is often misinterpreted to include only those experiments that are carried out using DSC. In practice, calorimetry is the universal term given to any experimental technique which involves the measurement of heat (power, q, in W) with time (t) or temperature (T). In turn, there are many different types of calorimeters that can be used to assess materials. Isothermal microcalorimeters are a type of calorimeter which measures power as a function of time. Scanning calorimeters (such as DSC) measure power as a function of temperature. The main advantage of isothermal microcalorimery over scanning calorimetry is that the instrument has the potential for much greater sensitivity. This is due to the increased stability of the background temperature against which heat flow measurements are made, coupled with the potential for a significant increase in sample size. This type of instrument is well suited for studying systems in which the heat flow rate is very small, as is expected to occur, for example, in oxidative degradation or hydrolysis of many synthetic biomaterial components during shelf storage or *in vivo*. In scanning calorimeters, it is also possible that some irreversible phase transitions may be induced by subjecting the sample to elevated temperatures, such as crystallisation of amorphous components or thermally induced degradation. However, a benefit of both approaches is that they are not dependent on the physical nature of the subject under study; the sample can be a solid, liquid, gas or any combination of the three. This allows the direct investigation of a wide range of systems, the only constraint being sample size.

During the measurement of any sample using microcalorimetry, the heat flows associated with all the reactions that are occurring simultaneously within the sample are recorded. Although this allows many complex reactions which are outside the scope of other analytical tools, to be studied, it also means that poor sample preparation can lead to erroneous heat flow signals and it may be that the heat flow signal is influenced greatly by an effect other than that which is the intended subject of the study. Consequently, the use of microcalorimetry offers much potential, but there is a need to balance good experimental design with careful data interpretation.

In general, calorimetry is highly suited to the study of pharmaceutical systems because the technique is very sensitive to changes induced by formulation or processing. This section deals with some potential applications of microcalorimetry to the characterisation of pharmaceutical systems that are undergoing physical changes.

8.9.2 Principles of isothermal microcalorimetry

Isothermal microcalorimetry involves the differential measurement of the changes in enthalpy (in J) between a test material and a reference material as the sample is undergoing some physical or chemical process. The time integral of this record, called the heat flow (Q (in W) or q (in W per test sample mass)), is directly proportional to the exothermic or endothermic heat resulting from process(es) occurring in the test material at any one point. In turn, the rate of heat flow with time is directly proportional to the rate of the process(es) occurring. Because changes in enthalpy accompany all chemical and physical reactions, in principle, the progress of all processes can be monitored with the use of isothermal microcalorimetry.

In a typical experiment, the test sample and a suitable reference material are contained in two separate, identical ampoules kept at constant temperature in separate, identically constructed wells of the calorimeter. Ideally, the reference material is identical or very similar to the test sample in mass, heat capacity and thermal conductivity, but, unlike the test sample, it is thermally inert (i.e. the reference material will not undergo changes that result in heat production or absorption under the conditions of the experiment). One example is a small quantity of ordinary glass beads in air at room temperature used as reference for the same amount of a hydrated ceramic material which is expected to lose water under the same conditions. Consequently, most of the noise arising from temperature fluctuations is removed when the reference data are subtracted. A feedback temperature control system between the wells (a) serves to ensure that the temperature difference between the wells is zero and (b) provides an output that measures any difference in electric power requirement of one well relative to the other, needed to keep the temperature of both wells the same. This power difference, as a function of time, is the output from the calorimeter, which is recorded continuously or intermittently over the duration of the test.

8.9.3 High-sensitivity DSC

In DSC the power supplied to a sample to raise it in temperature at a constant rate is measured and is compared with that supplied to an inert reference material undergoing the same temperature programme. DSC is a technique used throughout the pharmaceutical industry, typically being employed for physical characterisation of materials and, occasionally, for excipient compatibility testing or stability screening. Modern DSC instruments can operate at scan rates of up to 500°C/min and with an ultimate sensitivity, in practice, of around ± 1 μW. However, the small sample size of typically a few milligrams results in a lack of sensitivity that has precluded the use of DSC for studying reactions that occur with a small change in reaction enthalpy, ΔH, for example reactions in dilute solution.

Over the last 20 years the need to study biological molecules in their native state has led to the commercial availability of a range of DSC instruments with a much greater calorimetric sensitivity [52–55]. Such instruments (notionally referred to as high-sensitivity DSC, although the terms used by individual manufacturers vary) have a similar calorimetric sensitivity to that of traditional DSC (and sometimes less), but because they hold a much greater amount of sample (0.5–1 mL), they can measure heat flows resulting from low-energy reactions. Relatively slow scan rates (0–2°C/min) are normally used to allow the large sample sizes to follow the temperature programme. Typical applications include the denaturation of proteins, phase changes in lipid bilayers, phase transitions in dilute polymer solutions, changes in structure of creams and emulsions, and testing excipient compatibility.

8.9.4 Pharmaceutical applications of isothermal microcalorimetry

8.9.4.1 Dynamic physicochemical stability

One of the biggest challenges in the use of microcalorimetry to investigate pharmaceutical stability is the difficulty in analysing the data quantitatively. This problem is particularly acute for drug stability, which often involves solid phases, although microcalorimetry can play an important role in the development of solid-state drugs [56] and in determining long-term stability [57]. Nevertheless, many drugs have been subjected to microcalorimetric investigation including, for instance, aspirin [58], cephalosporins [58], lovastatin [59], meclofenoxate hydrochloride [60] and ascorbic acid [61].

Isothermal microcalorimetry has been used in the investigation of the oxidation of L-ascorbic acid in aqueous solution [62]. Ascorbic acid oxidises reversibly in aqueous solution forming dehydroascorbic acid, which is subsequently irreversibly hydrolysed to give diketogluonic acid. The rate of the reaction is affected by a number of factors including pH, oxygen concentration, ascorbic acid concentration, the presence of metal ions and the presence of antioxidants. Using the microcalorimeter it was possible to study this oxidation reaction under varying conditions and determine the effects of altering each of the factors.

8.9.4.2 Solution microcalorimetry: polymorphs and amorphous content

Isothermal and adiabatic calorimeters used in flow modes or in individual ampoules can allow the precise determination of the heat of solution. When using an ampoule, a two-phase system typically comprising 100 mL of solvent and a known amount of solute (solid or liquid) are housed together in the reaction vessel and equilibrated at the required temperature. The solute is sealed in a small, fragile glass ampoule immersed in the solvent. The glass ampoule prevents dissolution until thermal equilibrium has been established. Following equilibration the glass ampoule may be broken, initiating the reaction and allowing the enthalpy of solution to be measured. If a compound exists in two or more different crystalline or amorphous configurations with different lattice energies (e.g. polymorphs of the same compound), the heat of solution in any given solvent will differ. The difference in the heats of solution will be equal to the difference in lattice energies of the solids, provided that the solid compounds are chemically identical and readily dissolve in the chosen solvent system without undergoing

Table 8.6 Heats of solution of sodium sulfathiazole

Solvent	ΔH_s form I (KJ/mol, 25°C)	ΔH_s form II (KJ/mol, 25°C)	ΔH_{trans} (KJ/mol, 25°C)
Acetone	11.94	5.144	6.798
Dimethylformamide	−4.659	−11.47	6.810

any spontaneous physical form change (such as recrystallisation or solvent-mediated, solid–solid transformations).

The difference in solution enthalpy between both forms is equal to the heat of transition ΔH_t, where:

$$\Delta H_t = \Delta H_S^A - \Delta H_S^B \qquad (8.7)$$

This relationship is valid for solvents which allow a rapid dissolution in the calorimeter. Whereas AH_S^A and AH_S^B depend on the solvent (if no association or complexation takes place), the difference between them is independent of the solvent. The heat of transition ΔH_t should also equal the difference of the melting enthalpies of A and B which may be determined by DSC.

Because the heat of solution is generally of low energy, modern, accurate microcalorimeters can deliver accurate determinations and are suitable for the study of both liquid–liquid and solid–liquid interactions.

Previously, Lindenbaum and McGraw have used solution microcalorimetry to study drug forms. Because different crystal forms have different structures, they will inevitably have different heats of solution. However, the difference between the heats of solution of two polymorphs in different solvents should remain the same (Table 8.6) if there is no solvate formation. This difference is the heat of transition between the forms at that temperature.

It must be noted that these comparisons apply only to solids with the same composition (i.e. when the resulting solutions are identical). Also, a hydrate and an anhydrate cannot be compared since the heat of the solution of water will be different in different solvents and thus ΔH_{trans} will vary.

Isothermal microcalorimetry has also been used to determine the crystallinity of mixtures of amorphous and crystalline antibiotics [63]. DSC could not be used for this process since the samples decomposed prior to melting and an accurate quantification of the heat of fusion could not be determined. In contrast to studies carried out by Hogan et al. [64], in this case, it was shown that the heat of solution was not dependent on residual water content. The importance of initial water content is greatest when dealing with hydratable ionic species, since sodium and quaternary ammonium salts have very high heats of hydration. Therefore, before performing any analysis one must care to identify the extent of residual solvents or water present, as well as their effects on the heats of solution in the chosen system.

However, it must be noted that utilising fast scanning conditions can delay the onset of degradation by exceeding the kinetics of the transition, which enables us to collect data which may otherwise be masked by the decomposition process (such as a concurrent melt and degradation). Therefore, before deciding to exceed this critical temperature in the DSC instrument careful consideration must be taken.

References

1. Buckton G, Beezer AE. The applications of microcalorimetry in the field of physical pharmacy. *Int J Pharm* 1991;72(3):181–191.
2. Ramos R, Gaisford S, Buckton G. Calorimetric determination of amorphous content in lactose: A note on the preparation of calibration curves. *Int J Pharm* 2005;300(1–2):13–21.
3. Giron D. Thermal analysis and calorimetric methods in the characterisation of polymorphs and solvates. *Thermochim Acta* 1995;248:1–59.
4. Briehl H, Butenuth J. Thermal studies on the six isomers of pyridinedicarboxylic acid. *Thermochim Acta* 1992;211:121–130.
5. Kobayashi Y, Ito S, Itai S, Yamamoto K. Physicochemical properties and bioavailability of carbamazepine polymorphs and dehydrate. *Int J Pharm* 2000;193(2):137–146.
6. Quintiles Phase I Services. Islandsgatan 2, S-753 18 Uppsala, Sweden, The biopharmaceutical classification system – impact for drug development. *Eur J Pharm Sci* 1998;6(1):S18.
7. Bartolomei M, Bertocchi P, Cotta Ramusino M, Santucci N, Valvo L. Physico-chemical characterisation of the modifications I and II of (R,S) propranolol hydrochloride: solubility and dissolution studies. *J Pharm Biomed Anal* 1999;21(2):299–309.
8. Pirttimäki J, Laine E. The transformation of anhydrate and hydrate forms of caffeine at 100% RH and 0% R. *Eur J Pharm Sci* 1994;1(4):203–208.
9. Spartakov A, Trusov A, Vojtylov V. Magnetooptical determination of particle shape distribution in colloids. *Colloid Surf A Physicochem Eng Asp* 2002;209(2–3):131–137.
10. Burger A, Ramburger R. On the polymorphism of pharmaceuticals and other molecular crystals. I. Theory of thermodynamic rules. *Mikrochim Acta* 1979;2:259–271.
11. Cox Philip J, Wardell James L. Studies of polymorphism in three compounds by single crystal X-ray diffraction. *Int J Pharm* 2000;194(2):147–153.
12. Pratiwi D, Fawcett JP, Gordon KC, Rades T. Quantitative analysis of polymorphic mixtures of ranitidine hydrochloride by Raman spectroscopy and principal components analysis. *Eur J Pharm Biopharm* 2002;54(3):337–341.
13. Vickery RD, Nemeth GA, Maurin MB. Solid-state carbon NMR characterization of the polymorphs of roxifiban. *J Pharm Biomed Anal* 2002;30(1):125–129.
14. Bottom R. The role of modulated temperature differential scanning calorimetry in the characterisation of a drug molecule exhibiting polymorphic and glass forming tendencies. *Int J Pharm* 1999;192(1):47–53.
15. Rustichelli C, Gamberini G, Ferioli V, Gamberini MC, Ficarra R, Tommasini S. Solid-state study of polymorphic drugs: carbamazepine. *J Pharm Biomed Anal* 2000;23(1):41–54.
16. Gabbott P, Clarke P, Mann T, Royall P, Shergill S. A High Sensitivity, High Speed DSC Technique: Measurement of Amorphous Lactose. *Am Lab* August 2003.
17. McGregor C, Saunders MH, Buckton G, Saklatvala RD. The use of high-speed differential scanning calorimetry (Hyper-DSC™) to study the thermal properties of carbamazepine polymorphs. *Thermochim Acta* 2004;417(2):231–237.
18. McCrone WC. In: Fox D, Labes MM (eds.), Polymorphism in Physics and Chemistry of the Organic Solid State, Weissberger A1965, vol. II, pp. 726–767.
19. Sorrenti M, Bettinetti GP, Negri A. Thermoanalytical characterization of pseudopolymorphs of sulphadimidine and sulphadimidine–trimethoprim molecular complexes. *Thermochim Acta* 1998;321(1–2):67–72.

20. Fang MX, Shen DK, Li YX, Yu CJ, Luo ZY, Cen KF. Kinetic study on pyrolysis and combustion of wood under different oxygen concentrations by using TG-FTIR analysis. *J Anal Appl Pyrolysis* 2006;77(1):22–27.
21. Osaki K, Inoue T, Hwang E-J, Okamoto H, Takiguchi O. Dynamic birefringence of amorphous polymers. *J Non-Cryst Solid* 1994;172–174(pt 2):838–849.
22. Murthy NS, Minor H. General procedure for evaluating amorphous scattering and crystallinity from X-ray diffraction scans of semicrystalline polymers. *Polymer* 1990;31(6):996–1002.
23. Gustafsson C, Lennholm H, Iversen T, Nyström C. Comparison of solid-state NMR and isothermal microcalorimetry in the assessment of the amorphous component of lactose. *Int J Pharm* 1998;174(1–2):243–252.
24. Craig Duncan QM, Royall Paul G, Kett Vicky L, Hopton Michelle L. The relevance of the amorphous state to pharmaceutical dosage forms: glassy drugs and freeze dried systems. *Int J Pharm* 1999;179(2):179–207.
25. Yu L. Amorphous pharmaceutical solids: preparation, characterization and stabilization. *Adv Drug Deliv Rev* 2001;48(1):27–42.
26. Forster A, Hempenstall J, Tucker I, Rades T. Selection of excipients for melt extrusion with two poorly water-soluble drugs by solubility parameter calculation and thermal analysis. *Int J Pharm* 2001;226(1–2):147–161.
27. Badwan AA, Abu-Malooh A. Some formulation aspects of terfenadine solid dispersions. *Eur J Pharm Biopharm* 1991;37(3):166–170.
28. Eihei F, Makita M, Yamamura S. Some physicochemical properties of glassy indomethacin. *Chem Pharm Bull* 1986;37:4314–4321.
29. Eihei F, Makita M, Yamamura S. Glassy state of pharmaceuticals. II. Bioinequivalence of glassy and crystalline indomethacin. *Chem Pharm Bull* 1989;37:1047–1050.
30. Eihei F, Makita M, Yamamura S. Glassy state of pharmaceuticals. IV. Studies on glassy pharmaceuticals by thermomechanical analysis. *Chem Pharm Bull* 1989;39:2087–2090.
31. Claes A, Zografi G. The molecular basis of moisture effects on the physical and chemical stability of drugs in the solid state. *Int J Pharm* 1990; 62:87–95.
32. Makoto O, Kaneniwa N. The interaction between water and cephalexin in the crystalline and noncrystalline states. *Chem Pharm Bull* 1984; 31:230–236.
33. Saklatvala R, Royall Paul G, Craig Duncan QM. The detection of amorphous material in a nominally crystalline drug using modulated temperature DSC—a case study. *Int J Pharm* 1999;192(1):55–62.
34. Saunders M, Podluii K, Shergill S, Buckton G, Royall P. The potential of high speed DSC (Hyper-DSC™) for the detection and quantification of small amounts of amorphous content in predominantly crystalline samples. *Int J Pharm* 2004;274(1–2):35–40.
35. Hill Vivienne L, Craig Duncan QM, Feely Liam C. Characterisation of spray-dried lactose using modulated differential scanning calorimetry. *Int J Pharm* 1998;161(1):95–107.
36. Pijpers FJ, Mathot VBF, Goderis B, Scherrenberg RL, van der Vegte EW. High-speed calorimetry for the study of the kinetics of (de)vitrification, crystallisation, and melting of macromolecules. *Macromolecules* 2002;35:3601–3613.
37. Saleki-Gerhardt A, Ahlneck C, Zografi G. Assessment of disorder in crystalline solids. *Int J Pharm* 1994;101:237–247.
38. Guinot S, Leveiller F. The use of MTDSC to assess the amorphous phase content of a micronised drug substance. *Int J Pharm* 1999;192:63–75.
39. Miller JC, Miller JN. Solventless collection of analytes by rapid depressurization after static supercritical fluid extraction. *Stat Anal Chem* 1993; 65:1038–1042.
40. Plato C, Glasgow AR. Differential Scanning Calorimetry as a general method for determining the Purity and Heat of Fusion of High-Purity Organic Chemicals. *Anal Chem* 1969;41:330.

41. Gustin GM. Broad range purity analysis by melting point depression using a singular feature common to all DSC purity scans. *Thermochim Acta* 1980;39(2):81–93.
42. Gray AP. Determination of Purity by Differential Scanning Calorimetry. *Therm Anal Newsl* 1966;5.
43. Brennan WP, Divito MP, Ryans RL, et al. Purity determinations by thermal methods. In: Blaine, Schoff (eds.), Purity Determinations by Thermal Methods' ASTM Special Technical Publication 838. *ASTM STP 838.* 1984:5–15.
44. Botha SA, Lotter AP. Compatibility Study Between Atenolol and Tablet Excipients Using Differential Scanning Calorimetry. *Drug Dev Ind Pharm* 1990;16:673–683.
45. Lin SY, Han RY. Differential scanning calorimetry as a screening technique to determine the compatibility of salbutanol sulfate with excipients. *Pharmazie* 1992;47:266–268.
46. Hardy MJ. *Anal Proc* 1982;19:556–557.
47. Van Dooren AA, Duphar BV. Design for drug–excipient interaction studies. *Drug Dev Ind Pharm* 1983;9:43–55.
48. Bettinetti GP, Mura P, Liguori A, Bramanti G, Giordano F. Solubilization and interaction of naproxen with polyvinylpyrrolidone in aqueous solution and in the solid state. *Farmaco* 1988;43:331–343.
49. Tan X, Meltzer N, Lindenbaum S. Solid-State Stability Studies of 13-cis-Retinoic Acid and All-trans-Retinoic Acid Using Microcalorimetry and HPLC Analysis. *Pharm Res* 1992;9:1203.
50. Phipps MA, Winneke RA. *Proc Workshop Microcalorim Energ Mater*, 1997.
51. Schmitt Eric A, Peck K, Sun Y, Geoffroy J-M. Rapid, practical and predictive excipient compatibility screening using isothermal microcalorietry. *Thermochim Acta* 2001;380(2):175–184.
52. Noble D. *Anal Chem* 1995;67:323A–327A.
53. Sturtevant JM. Biochemical applications of differential scanning calorimetry. *Ann Rev Phys Chem* 1987;38:463–488.
54. Chowdhry BZ, Cole SC. Differential scanning calorimetry – applications in biotechnology. *Trends Biotechnol* 1989;7(7):11–18.
55. Wiseman T, Williston S, Brandts J, Lin L. Rapid Measurement of Binding. Constants and Heats of Binding Using a New Titration Calorimeter. *Anal Biochem* 1989;79:131–137.
56. Phipps MA, Mackin LA. Application of isothermal. microcalorimetry in solid state drug development. *Pharm Sci Tech Today* 2000;3(1):9–17.
57. Beezer AE, Gaisford S, Hills AK, Willson RJ, Mitchell JC. Pharmaceutical microcalorimetry: applications to long-term stability studies. *Int J Pharm* 1991;179(2):39–45.
58. Angberg M, Nyström C, Cartensson S. Evaluation of. Heat-Conduction Microcalorimetry in Pharmaceutical Stability Studies. *Acta Pharm Suec* 1988;25:307–320.
59. Hansen LD, Lewis EA, Eatough DJ, Bergstrom RG, DeGraft-Johnson D. Kinetics of Drug Decomposition by Heat Conduction Calorimetry. *Pharm Res* 1989;6:20–27.
60. Otsuka T, Yoshioka S, Aso Y, Terao T. Application of microcalorimetry to stability testing. of meclofenoxate hydrochloride and dl-a-tocopherol. *Chem Pharm Bull* 1994;42(1):130–132.
61. Angberg M, Nyström C, Cartensson S. Evaluation of heat-conduction microcalorimetry in pharmaceutical stability studies VII. Oxidation of ascorbic acid in aqueous solution. *Int J Pharm* 1993;90:19–33.
62. Willson RJ, Beezer AE, Mitchell JC. A kinetic study of the oxidation of L-ascorbic acid (vitamin C) in solution using an isothermal microcalorimeter. *Thermochim Acta* 1995;264:27–40.
63. Thompson Karen C, Draper Jerome P, Kaufman Michael J, Brenner Gerald S. Characterization of the Crystallinity of Drugs: B02669, a Case Study. *Pharm Res* 1994;11:1362–1365.
64. Hogan Sarah E, Buckton G. The quantification of small degrees of disorder in lactose using solution calorimetry. *Int J Pharm* 2000;207(1–2):57–64.

Chapter 9

Thermal Methods in the Study of Foods and Food Ingredients

Bill MacNaughtan, Imad A. Farhat

Contents

9.1	Introduction	331
9.2	Starch	332
	9.2.1 Starch structure	332
	9.2.2 Order in the granule	335
	9.2.3 The glass transition	336
	9.2.4 Extrusion and expansion	337
	9.2.5 Mechanical properties	338
	9.2.6 Starch retrogradation	338
	9.2.7 Effect of sugars	339
	9.2.8 Lipid–amylose complexes	339
	9.2.9 Multiple amorphous phases: polyamorphism	339
	9.2.10 Foods	341
9.3	Sugars	343
	9.3.1 Physical properties	343
	9.3.2 Sugar glasses	345
	9.3.3 Sugar crystallisation	345
	9.3.4 Effect of ions on crystallisation	347
	9.3.5 Effect of ions on the glass transition temperature	348
	9.3.6 Ageing	349
	9.3.7 The Maillard reaction	350
	9.3.8 Mechanical properties	350
	9.3.9 Foods	350
9.4	Fats	353
	9.4.1 Solid/liquid ratio	353
	9.4.2 Phase diagrams	355
	9.4.3 Polymorphic forms and structure	356
	9.4.4 Kinetic information	356
	9.4.5 Non-isothermal methods	359
	9.4.6 DSC at high scanning rates	361
	9.4.7 Mechanical measurements	362

		9.4.8 Lipid oxidation and the oxidation induction time test	362
		9.4.9 Foods	363
9.5	Proteins		365
	9.5.1	Protein denaturation and gelation	365
	9.5.2	Differential scanning microcalorimetry	366
	9.5.3	Aggregation	369
	9.5.4	Glass transition in proteins	371
	9.5.5	Ageing in proteins	371
	9.5.6	Gelatin in a high-sugar environment	373
	9.5.7	Mechanical properties of proteins	373
	9.5.8	Foods	373
	9.5.9	TA applied to other areas	377
	9.5.10	Interactions with polysaccharides and other materials	377
9.6	Hydrocolloids		378
	9.6.1	Definitions	378
	9.6.2	Structures in solution	379
	9.6.3	Solvent effects	381
	9.6.4	The rheological T_g	381
	9.6.5	Glassy behaviour in hydrocolloid/high-sugar systems	382
	9.6.6	Foods	386
9.7	Frozen systems		387
	9.7.1	Bound water?	387
	9.7.2	State diagram	388
	9.7.3	Mechanical properties of frozen sugar solutions	390
	9.7.4	Separation of nucleation and growth components of crystallisation	391
	9.7.5	Foods	392
	9.7.6	Cryopreservation	394
9.8	Thermodynamics and reaction rates		395
	9.8.1	Studies on mixing	395
	9.8.2	Isothermal titration calorimetry	396
	9.8.3	Reaction rates	398
	9.8.4	Thermogravimetric analysis	399
	9.8.5	Sample controlled TA	401
References			402

9.1 Introduction

Current food science in relation to thermal analysis (TA) can be crudely broken down into the areas listed in Table 9.1, which will be used as a basis for the structure of this chapter.

There are many reasons as to why foods lend themselves to TA. TA is not particularly demanding as to sample preparation and a wide range of samples from ingredients in solid and liquid forms to real foods can be studied. These can be in the amorphous state, either as hard solids, such as dried extrudates, for example starch and gelatin, or as pastes and viscous liquids, such as starch solutions in sauces or sugar syrups. Amorphous solids often show

Table 9.1 Food Science, through the eyes of thermal analysis: the principal areas of study and the main techniques

Area	Principal techniques
Starch	DSC, DMA, TGA, DTGA and RVA for gels
Sugars	DSC and DVS
Fats and lipid oxidation	DSC, hot-stage microscopy and X-ray
Proteins	DSC, DMA and DSM/NC
Hydrocolloids	Micro-DSC and rheology
Frozen systems	DSC and cold-stage microscopy
Growth of micro-organisms	DSM and reaction vessel
Thermodynamics and rates of reaction	ITC, mixing cells, DSM and DSN
Real foods	Principally DSC but most other techniques

Abbreviations: DSC, differential scanning calorimetry; DMA, dynamic mechanical analysis; TGA, thermogravimetric analysis; DTGA, differential thermogravimetric analysis; RVA, rapid viscoanalyser; DVS, dynamic vapour sorption; ITC, isothermal titration calorimetry; DSM/NC, differential scanning micro/nano-calorimetry.

well-defined glass transition and structural or enthalpic relaxation and ageing behaviour which can be detected using calorimetry. Dry amorphous extrudates can also be easily formed into bars suitable for mechanical measurements. In addition, the non-equilibrium state of many amorphous materials can result in phenomena, such as cold crystallisation, observed in simple freeze-dried sugars as well as confectionery products. Complementing the amorphous state of foods, there are many materials, which are present in the crystalline form and exhibit melting behaviour. These range from the polymer-like behaviour of starches, proteins and gelling hydrocolloids to the melting and crystallisation of small molecules, such as sugars and sugar alcohols. The crystallisation and polymorphism of fats is also a very large area of research, and the subject of dedicated reviews and books.

There are other food-related areas where TA has been used, for instance the growth of bacteria, and calorimetry. The low-temperature behaviour of foods is another distinct area and is particularly relevant to ice cream, frozen desserts and ready meals. This subject is especially complicated, as these materials contain 'excess' water, that is water which behaves in a bulk manner and can crystallise (on cooling) to form ice.

The heterogeneous and complex nature of foods results in many of the above features being observed simultaneously, producing 'busy' differential scanning calorimetry (DSC) and mechanical traces. We will consider most of the constituents of food separately, describe the main properties and attempt to demonstrate that these are observed in real foods.

9.2 Starch

9.2.1 Starch structure

Starch is the most common food polymer of biological origin and comprises the dominant component of most diets. The properties of starch are intimately connected with the unique

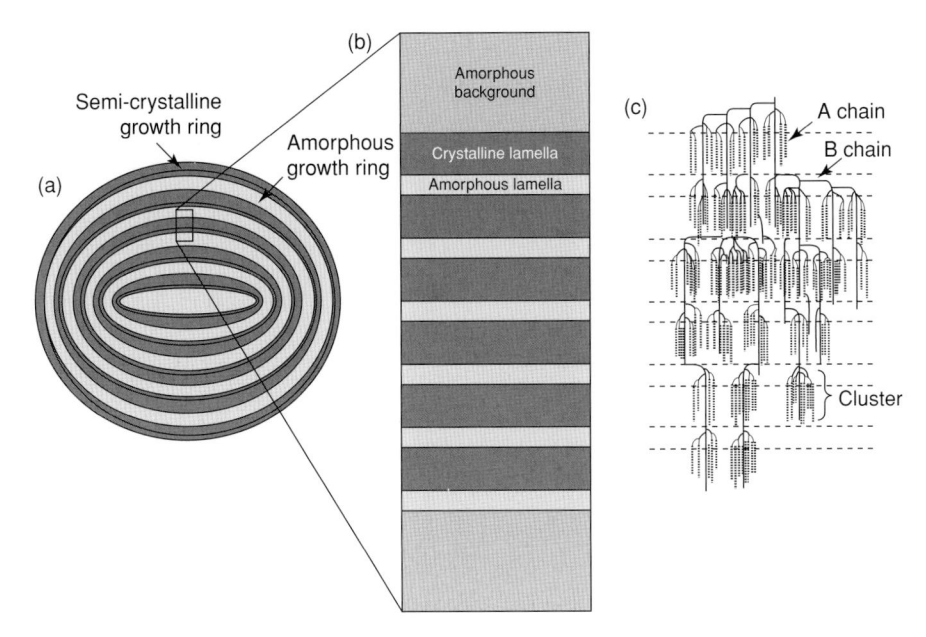

Figure 9.1 A model of starch granule structure: (a) single granule, (b) expanded view of internal structure and (c) the cluster structure of amylopectin. Amylopectin is a branched polymer, seen here extending through many crystalline and amorphous lamellae. (Taken from [3] – Reproduced by permission of The Royal Society of Chemistry.)

structure in which it is found, namely the starch granule. Starch granules are present in both cereals, such as maize and wheat (type A starches), and tubers, such as potato and tapioca (type B starches). These different forms correspond to different polymorphs with different physical arrangements of the starch-double helices and different water contents [1, 2]. Starch is a biopolymer and shows many features exhibited by conventional polymers. For instance, the morphology of the starch granule reflects the diurnal growth patterns, with a banded series of amorphous and crystalline regions (Figure 9.1) interspersed with pure amorphous regions [3]. As one molecule of amylopectin can span many of these regions, it can simultaneously be a part of both crystalline and amorphous areas, as is the case with polymers. Therefore, when the properties of the amorphous phase are examined, one cannot ignore the possibility of tie points or helical crystalline regions, which can severely limit the degrees of freedom of motion in the starch chains.

For a water-soluble polymer, the structure of starch renders it remarkably insoluble in water. One of the key observations when starch is placed in aqueous media is that the starch granules separate out quickly and show little tendency to interact with the water. It takes the application of heat together with the presence of excess water, normally of the order of two to three times the weight of water to starch, to fully bring about one of the most important characteristics of starch, namely gelatinisation [4]. Almost all food uses of starch involve gelatinisation. A series of characteristic changes are observed at this temperature. There is swelling and a loss of order in the granules, most obviously shown by the loss of the maltese cross under the optical microscope. Associated with this is an endothermic heat change and

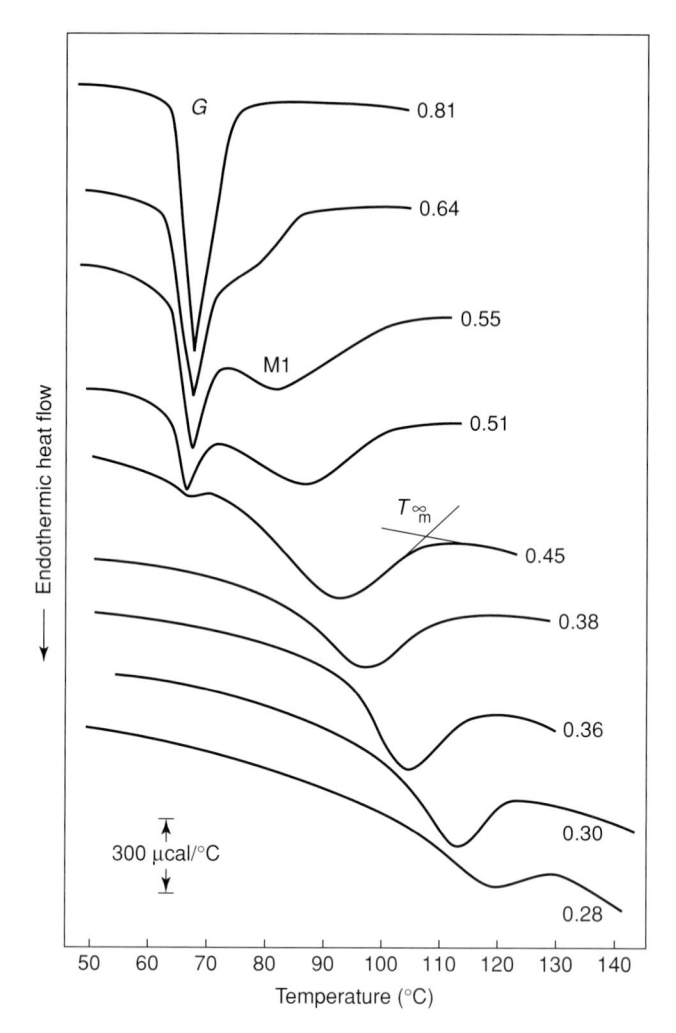

Figure 9.2 DSC profiles for potato starch at different water contents (w/total w). The gelatinisation peak occurs in the region of 60°C. As the water content is reduced, the endotherm splits eventually resulting in one high-temperature endotherm. (Reprinted from [10], Copyright (2000), with permission from Elsevier.) (See also Ref. [4].)

other changes which can be monitored by many techniques, such as relaxation nuclear magnetic resonance (NMR) [5]. Of particular importance, practically, is the increase in viscosity which can be measured at this temperature by qualitative assessments of starch 'pasting', such as by the rapid viscous analyser (RVA). This reflects the swelling of the granule and leaching of biopolymers. With continued (mechanical and thermal) work input, reductions in molecular weight of the polymers is observed. An oft-reproduced figure depicting the DSC traces for potato starch as a function of moisture content is shown in Figure 9.2 [4, 10]. A series of endotherms is observed, named the G and M1 endotherms. The main gelatinisation or G endotherm is observed distinctly at water contents above about 40–50%.

Table 9.2 Measures of order in starches using the techniques of WAXS, CPMAS NMR and DSC

	WAXS (crystalline %)	CPMAS NMR (molecular %)	DSC (enthalpy J/g)	T_{peak} (°C)
Wheat	20	39	9.7	57.7
Maize	27	43	14.3	70.2
Potato	24	40	16.2	58.3
Waxy maize	28	48	16.0	72.2

Abbreviations: WAXS, wide-angle X-ray diffraction; CPMAS NMR, cross-polarisation, magic, angle-spinning nuclear magnetic resonance; DSC, differential scanning calorimetry. (From Ref. [9].)

Below this value the endotherm splits into the G and M1 endotherms. The G endotherm was originally thought to be due to the stripping of chains from the crystallites. In limited water this process does not go to completion and the remaining crystallites melt at a higher temperature. The Flory equation [6], which treats the effect of water on the starch in a similar way to the effect of a small-molecular-weight species on a molecule composed of many subunits, has been applied successfully to the second endotherm (M1).

Other endotherms in lipid-containing starches are known as the M2 peaks and occur in the region of 100–150°C [7, 8]. These are thought to be due to (the dissociation of) lipid–amylose complexes. Unfortunately, there is some confusion over the naming of these peaks, with some references proposing that the G peak be known as the M1 peak with the other endotherms being numbered from M2 to M4.

9.2.2 Order in the granule

A particularly profitable line of enquiry involves the exact temperatures at which order, as measured by the various techniques, is lost. Gidley [9] has looked at this in detail in order to shed light on the origin of the endotherm in starch gelatinisation. Table 9.2 shows a comparison using several techniques of the overall level of order in starches.

Each technique has a characteristic distance scale. For instance, cross-polarisation magic angle spinning nuclear magnetic resonance (CPMAS NMR) is sensitive to short-range order, whilst small-angle X-ray diffraction is sensitive to stacked lamellar repeat distances of the order of 100 Å or so. Interestingly, light microscopy in the form of the measurement of the disappearance of the maltese cross from the granule has a characteristic distance of the order of micrometers. The correlation of order as measured by these techniques with enthalpy changes is achieved by heating to specific temperatures, cooling, and measuring the various order parameters on reheating and has provided considerable insight into the structure of starch. The necessity of marrying several techniques, particularly to complement TA, is a recurrent theme in this chapter.

Using a liquid crystalline approach, Waigh et al. [10] have postulated three measures of order: h is the shortest range and related to the helical content, φ is related to the orientational order of double helix and ψ is related to the lamellar order. Gelatinisation in excess water is thought to involve a slow dissociation of helices followed by a rapid helix/coil transition. We therefore see a single transition. In terms of order this implies reductions in the values of φ and h. It is interesting to propose rapid scanning as a possible means of separating these two transitions, depending on the time course of each in comparison with the scan rate. In limited water the lamellar order parameter ψ is lost initially in a smectic/nematic phase transition. These terms derive from liquid crystal studies and simply refer to the types of structure, which can be present in oriented molecules. Consequently, the SAXS peak as a measure of order disappears first, and at a lower temperature than the helix/coil transition for reasons of water effects on the amorphous region. Therefore, two endothermic transitions are observed. Interestingly, if the width of the second transition is inversely dependent on the helix length then wheat starch (type A short helices) should exhibit a wider second endotherm than potato starch (type B long helices), which is in fact observed. The origin of the endotherms is therefore related to a combination of the loss of different types of order in the granule.

9.2.3 The glass transition

If starch is completely gelatinised and then subsequently dried, a glass is formed. This has been studied extensively using DSC to monitor the glass transition temperature as a function of moisture content. Figure 9.3 shows a plot of such data [11].

These curves are well described by equations such as the Gordon–Taylor, the Couchman–Karasz and modifications thereof (see Section 1.5.8), with water taking the role of plasticiser.

Ageing phenomena are also important in starch glasses, particularly if the glass is intended for encapsulation or packaging purposes. The densification and embrittlement associated

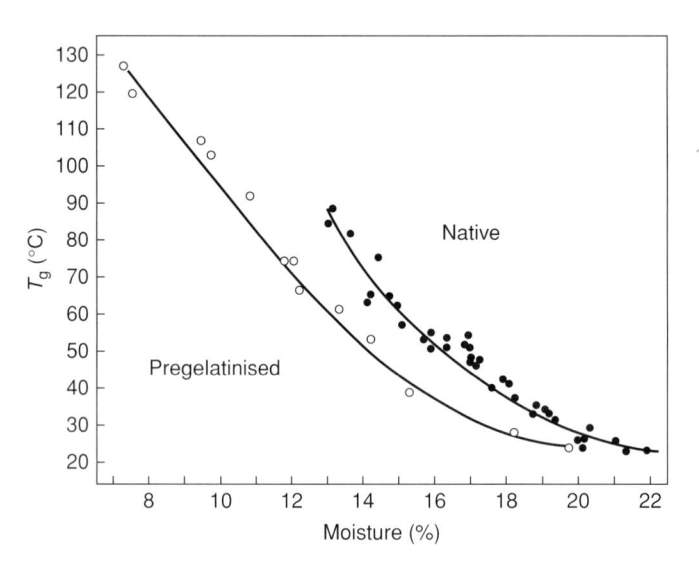

Figure 9.3 The glass transition temperature of native and pregelatinised wheat starch as a function of moisture content. (Taken from [11], with permission.)

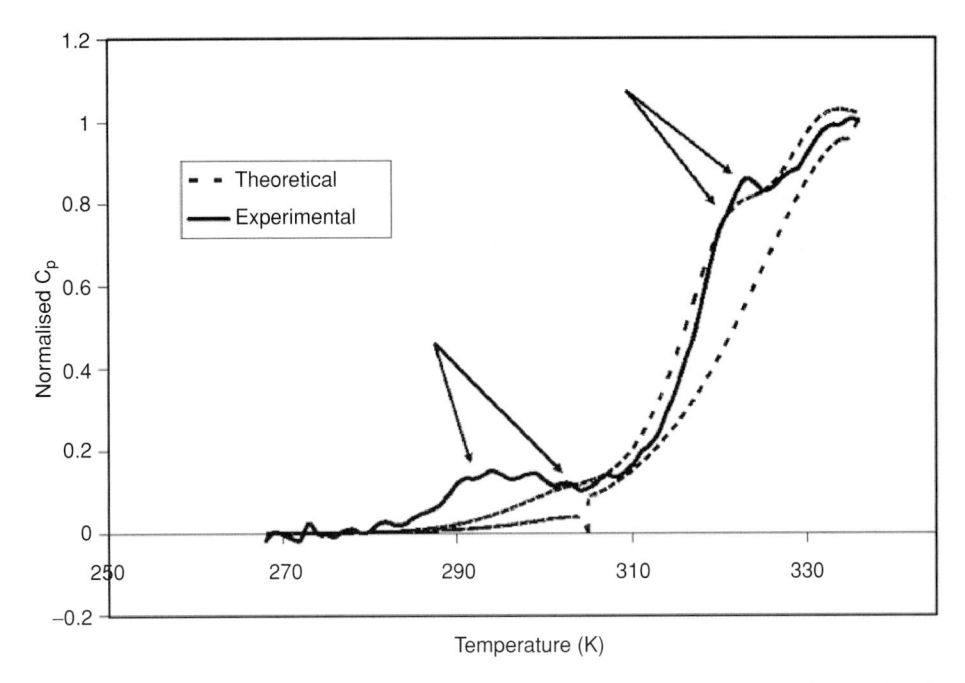

![Figure 9.4 graph showing Normalised Cp vs Temperature (K)]

Figure 9.4 Multiple ageing peaks observed in half-products. The basic TNM model did not describe the double peak well. The model did, however, predict two peaks (arrowed), but in order to show these, the time of annealing for the lower step was increased by a factor of 10. Each double-arrow pair shows the experimental and predicted peak. Theoretical cooling and heating curves are shown, whilst only the experimental heating curve is recorded.

with ageing can produce cracking, leading to loss of active component, increased rates of degradation or failure of function. Enthalpic relaxation peaks can be seen on DSC traces. Figure 9.4 shows the results of an ageing experiment on a starch-based snack precursor. These are sometimes known as half-products due to their partly being processed but requiring further heating or frying to produce the final snack.

Frequently, products such as these can experience storage at a variety of temperatures due to poor handling in the transportation process; for example shipments could be left exposed to the sun. It was therefore of some interest to look at the effect of multiple storage or annealing temperatures for set times using the formalism of the TNM (Tool Narayanaswamy Moynihan) model (see Section 1.5.6). Figure 9.4 shows that whilst the model cannot reproduce quantitatively the changes observed, a qualitative description of the multi-step ageing profile can be obtained. This raises the interesting point of a thermal history being imprinted into a product and this memory effect potentially being readable by a technique such as DSC.

9.2.4 Extrusion and expansion

One method of producing hard, amorphous, glassy, starch-based materials is to use an extruder. A starch-based melt at typically ~30% moisture can be forced through a die to produce a variety of shapes, as in pasta production, where the material is extruded and then dried under controlled conditions to produce a low-moisture glass. This is also the method of production of half-products. Under certain conditions, e.g. allowing the product to exit at temperatures well above 100°C, extrudates can be made to expand to give a porous structure. The driving force, as in the case of frying or air heating half-products, is the generation of pressure by the vapourisation of the residual water. The rheology of the material on heating determines the properties of the expanded product. Small deformation properties, such as the moduli at small strains as measured by dynamic mechanical analysis (DMA), tell only a limited story. It appears that the properties at large deformations and temperatures such as the rupture point, which are much more difficult to measure, particularly at constant moisture content, are more relevant for this type of product. These properties require different instrumentation, such as tensile pullers operating at high temperatures. So, for example, the material composing the walls of a product exhibiting a large pore size has a much larger stretch to rupture than that in the walls of a small-pore-size snack product. These properties reflect the degree of gelatinisation of the starch and give dramatically different bite properties.

9.2.5 Mechanical properties

Extruded bars or pressed starch samples of typical dimensions 10 mm wide by 2 mm deep are suitable shapes for carrying out mechanical measurements. Many studies have identified the initial drop in modulus or the peak in tan δ observed in DMA experiments [12], with the glass transition. A plot of the reciprocal of the temperature of the tan δ peak as a function of the test frequency gave activation energies in the region of 300–740 KJ/mol. These high activation energies were reported to indicate a large degree of segmental mobility.

Mechanical measurements of the glass transition are frequently preferred to those of DSC, due to the much higher sensitivity.

Reference [13] identified four DMA transitions in amylopectin/sucrose materials, which were thought to reflect the inhomogeneity of the extruded samples. These transitions were (1) a low-temperature beta transition, (2) a glass transition of a sucrose phase, (3) an ice-melting region of the sucrose phase and (4) a glass transition of an amylopectin-rich phase. Systematic variations of sugar and water content assisted with these assignments.

9.2.6 Starch retrogradation

It was already realised in the middle of the nineteenth century that the staling of breadcrumb was not solely due to drying [14]. In the 1920s Katz [15] showed, using X-ray diffraction, that staling and firming were due to a return of order, namely a crystallisation process. This can also be monitored by the return of the endothermic transition. Since then much work has been carried out on this process, particularly with a view to inhibiting the rate at which it takes place, bearing in mind the industrial and financial implications of staling in baked goods. Of particular interest is the polymer science approach to this problem, where the maximal rates of staling are found between the postulated glass transition and the melting points of the components of starch. Lipids, such as fatty acids and monoglycerides, which can complex with the amylose and enzymes which can destroy the carbohydrate structure have been found to inhibit staling and are used commercially.

9.2.7 Effect of sugars

Conventional wisdom states that the effect of sugar is to raise the temperature of the starch gelatinisation endotherm. This has profound consequences in baking sugar-rich products. The effect of sugar can be explained in terms of the associated free volume being less than that of an equivalent volume of water, and so being less effective as a plasticiser. This causes an apparent increase in glass transition temperature, which can be accurately described by a three-component Couchman–Karasz equation. The glass transition, as suggested in polymers, can control a melting event, such as starch gelatinisation, by controlling the mobility necessary for such a process to occur.

There is an issue as to how sugar concentration is expressed. If it is expressed relative to a fixed starch/water ratio, then the sugar may exert its effect by 'competing' for the water in a non-ideal way. Consequently, different plots may produce different conclusions [16]. In a thorough examination of this issue [17], it was concluded that sugars, on keeping the total solid/liquid ratio constant, increase both the onset and the peak temperatures but have little effect on the end temperature. Hence, they sharpen the transition.

9.2.8 Lipid–amylose complexes

Molecular modelling has shown that amylose–fatty acid complexes are most likely in the form of an amylose helix wrapped around a hydrocarbon chain with bulky hydrophillic groups located in the aqueous region [18]. With other complexing agents such as alcohols and α-naphthol, the existence of such structures is more equivocal with the possibility of the agents being located in the inter-helical spaces [19]. In addition the number of glucose monomers per turn of helix can increase in order to accommodate more bulky molecules in the centre of the helix. Complex behaviour can also be seen on DSC traces where there appears to be a melting of an unstable complex followed by a recrystallisation into a more stable form [20]. Whilst the complexes appear to have an effect on the pasting or viscosity properties of starches, any direct effect on baking appears to be minimal.

9.2.9 Multiple amorphous phases: polyamorphism

It seems appropriate here to introduce the current topic of multiple amorphous phases even if little work has yet been carried out. The phase rule states that at equilibrium

$$\Phi = C - P + 2$$

where, in a thermodynamic sense, Φ is the number of intensive variables required to define the system, and C and P are the number of components and phases respectively. One component, and three phases (two amorphous and one crystalline) should only be able to exist together at equilibrium at a single temperature, in a manner similar to the triple point of water. The reported existence of multiple phases over a range of temperatures is proof of profound non-equilibrium behaviour. That such behaviour is non-equilibrium can be easily demonstrated. The reaction of two species to form a product will be specified at equilibrium by the concentration of two of the species together with the equilibrium constant. However if the system is not at equilibrium, then the content of all three species or components can be varied independently as there is now no relationship between them. This leads to a higher number of degrees of freedom, the least number of intensive variables, required to specify the system, and an invalidation of the phase rule as stated above. The increase in the degrees of freedom is then expressed as a range of temperatures over which several phases can exist. Food materials tend to be formed under non-equilibrium conditions and so could form coexisting amorphous states, as well as separate amorphous phases of different properties. This is of course the case for the formation of food glasses, where the method of formation, for example, the rates of cooling used and any annealing steps present, can produce a multiplicity of states. It has been proposed that these states be known as 'pseudo' polyamorphism (see [21] for a more thorough discussion).

However the case of true polyamorphism, where there is a well-defined phase transition between amorphous phases, is more difficult to demonstrate in food materials. Polyamorphism is known in the case of several inorganic compounds and has been proposed in several pharmaceutical preparations.

9.2.9.1 The rigid amorphous fraction or phase

In addition to multiple amorphous states, Wunderlich [22] has proposed coexistent dual amorphism, where the dominant characteristics are those of a mobile amorphous phase and a so-called rigid amorphous fraction (RAF). The RAF is characteristic of semi-crystalline polymers, found in many food, pharmaceutical and polymer areas, where polymer chains, as they pass from crystalline to amorphous regions, are perturbed and have properties lying between the two extremes. A property of the RAF is that it does not contribute to the step in heat capacity at the main glass transition. In fact the method of calculating the fraction of the rigid amorphous phase φ is usually to measure the crystallinity by an independent method, such as density or possibly X-ray methods, and to estimate the mobile amorphous fraction from the difference between heat capacity steps of the sample under consideration and a 100% amorphous-generated state. The RAF is then calculated as the difference between 1 and the sum of the content of the previous phases.

$$\varphi \, RAF = 1 - (\varphi \text{ mobile amorphous} + \varphi \text{ crystalline})$$

It is obvious that such a measurement, in common with many such difference methods, will be subject to considerable error.

The demonstration of small glass transitions in foods can be difficult; however, it is possible under favourable circumstances to demonstrate the presence of the RAF directly in, for example, isotactic polystyrene. The presence of the RAF, which has a higher glass transition temperature in view of the restricted motion of the chains, is revealed by an enthalpic relaxation peak [23] and a very small shift in the baseline Cp value as the RAF transforms to the mobile amorphous phase. This demonstration however necessitates very accurate absolute C_p measurements and baselines free of artefacts, such as curvature.

9.2.9.2 Importance

It is difficult to speculate on the importance of something which has not yet been unequivocally measured in food polymers, but clearly if material was encapsulated in an amorphous matrix with different properties from the more usual mobile phase, then ageing and plasticisation could be significantly different, having both positive and deleterious effects on shelf life. Similarly, a substantial fraction of rigid material could affect the permeability to reactants, such as oxygen.

It has also been shown that in solid-state polymerisation of danol with poly(butylene terephthalate), PBT, the diol is not incorporated into the rigid fraction. The RAF was detected using DSC. The PBT chains in the rigid fraction are deemed too inflexible to take part in the transesterification reaction, in contrast to the chains in the mobile amorphous fraction [24]. The area of the RAF is one which may see significant development in food polymers, such as starch.

9.2.10 Foods

One of the principal tests for any starch-based food is the determination of the degree of integrity of the starch granules or conversely the degree of gelatinisation of the starch. This is carried out by measuring the endotherm of the remaining ungelatinised fraction. In the assessment of the extent of gelatinisation of dry materials, water must be added to generate measureable endotherms. Normally, a 2:1 ratio of water to starch is sufficient to develop maximum peak enthalpy [25]. Naturally, moist materials do not require the addition of water.

The return of the gelatinisation endotherm is indicative of true staling in starch-based products, such as bread (see Figure 9.5).

Staling is commonly confused with drying in industrial processes or with water transfer between components; however even with moisture content held constant, bread will become firmer with time [26]. Antistaling compounds such as DATA (diacyl tartaric) esters of monoglycerides and single-chain amphoteric compounds are added routinely to bread and are thought to work by binding the amylose and perhaps long amylopectin chains and preventing reassembly into crystallites which can lead to firmness and also dryness, caused by the incorporation of water into the helical structure. Ethanol appeared to slightly reduce the rate of staling/recrystallisation; however, the site of action was not identified.

In cakes which contain high levels of sugars, as well as starch, the raising of the starch gelatinisation endotherm to higher temperatures has important consequences during baking. This is because the temperature in the centre of the product can be insufficient to fully gelatinise the starch [17] and so produce an impaired texture.

Figure 9.6 shows endotherms produced during heating of a cake batter. These are thought to be due in ascending order of temperature, to the melting of fat, a combination of the coagulation of egg protein and the gelatinisation of starch and the melting of the lipid–amylose complex. The gelatinisation/coagulation endotherm is irreversible on this

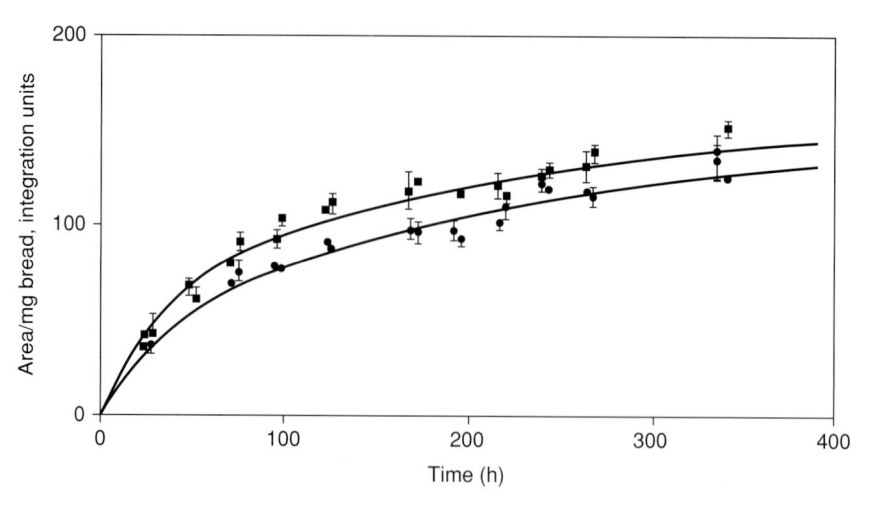

Figure 9.5 The return of the endotherm upon retrogradation or recrystallisation of starch in bread, for both control loaves and loaves painted with ethanol. The absolute limiting value was ~3.4 J/g. The upper trace was for the control loaves and showed slightly more rapid kinetics. (Taken from [26]. Copyright Society of Chemical Industry. Reproduced with permission. Permission is granted by John Wiley & Sons Ltd on behalf of the SCI.)

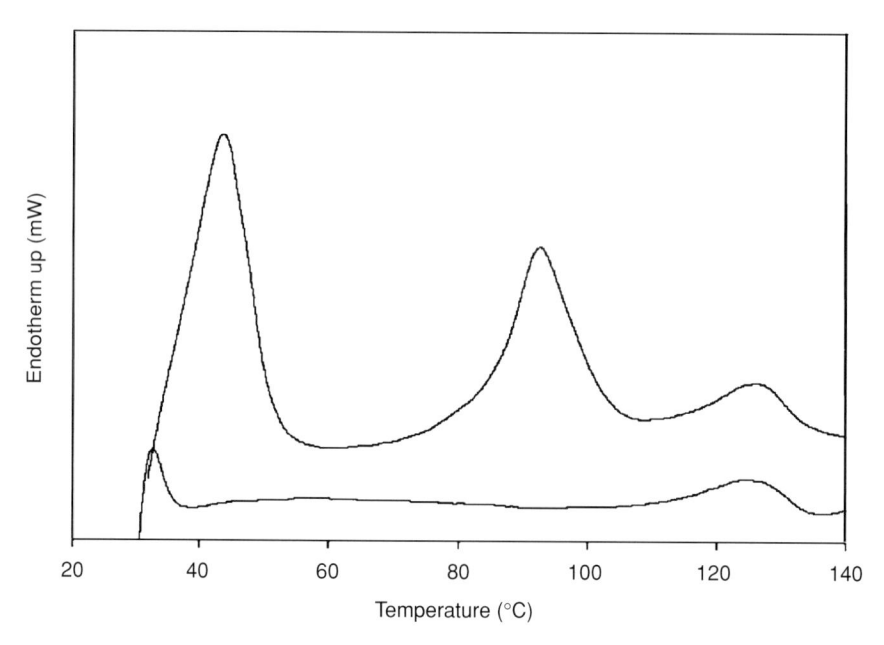

Figure 9.6 Baking á la DSC. DSC trace of a cake batter. Heating rates are 10°C/min for the first (top) and second heating runs. The low-temperature endotherm shows the melting of fat. The mid-temperature endotherm is a combination of the gelatinisation of starch and the 'coagulation' of egg protein, whilst the upper temperature endotherm is due to the melting of the amylose–lipid complex and is reversible. (Taken from [17].)

time-scale, whereas the disappearance of the fat endotherm is most likely due to a change of polymorphic form.

Impaired performance of the expansion of snack half-products has also been associated with incomplete gelatinisation during processing with resultant impaired rheological properties when the material is flowing under the vapour pressure of water. Poor expansion properties when frying half products may also be associated with ageing or densification of the carbohydrate matrix, which can be detected by ageing peaks (enthalpic overshoots) in DSC traces.

The melting of the lipid–amylose complex is completely reversible (see Figure 9.6) and can be observed on successive runs. It can be used as the basis for the assessment of the amylose content in starches [27], where the size of the melting peak in an excess quantity of lysophosphatidylcholine, a material known to form helical inclusion complexes, is dependent on the amylose level. This has been adapted to other starch-containing foods.

9.3 Sugars

9.3.1 Physical properties

Perhaps the simplest measurements, which can be made on sugars, are estimates of the melting point, the melting enthalpy and the purity. Figure 9.7 shows typical DSC traces for two samples of sucrose.

The onset temperature is 188°C and the enthalpy ranges from 129 to 138 J/g for the two samples. The broader peak and lower temperature at the start of heat absorption for the

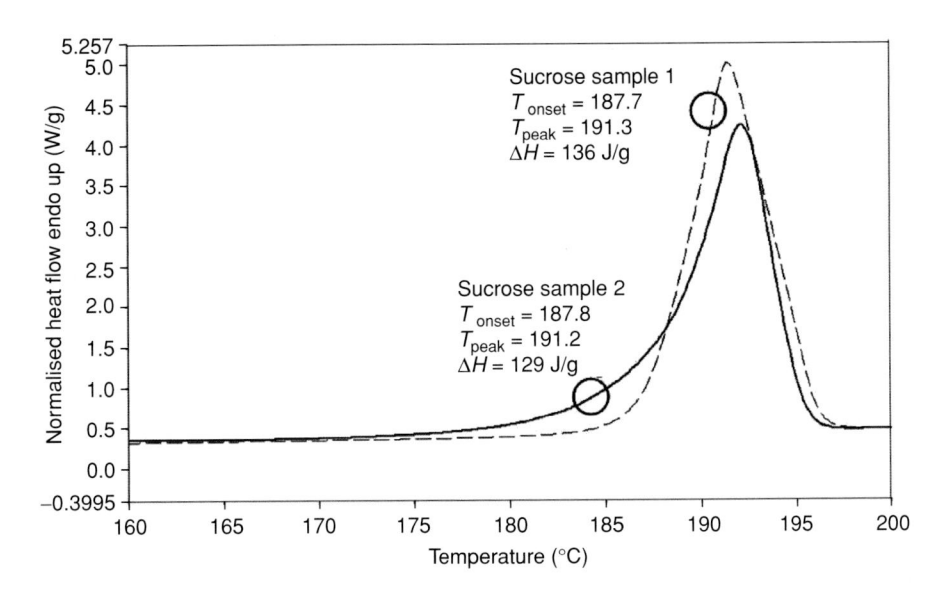

Figure 9.7 Typical DSC thermograms for crystalline sucrose. Broader peaks and lower onset temperatures indicate less pure materials but may also be caused by a higher thermal resistance in the path between furnace and sample.

Table 9.3 Some physical properties of starch, sugars, proteins and other materials for comparison

Material	T_g'	C_g'	Pure T_g (°C)	Dry T_m (°C)	n (H$_2$O)	ΔC_p (J/°C/g)
Water	—	—	−139	0	—	1.94
Glucose	−53	75	30	MH 158	1	0.88
Maltose	−37	77	98	α anh 161 β MH 102	1	0.79
Trehalose	−42	79	117	αα anh 214 DH 103	2	0.55
Sucrose	−40	81	67	192	0	0.6 → 0.77
Lactose	−36	81	101	α MH 201 α anh 222 β anh 253	1	0.52
Starch	−5 → −7	70 → 75	227	>250	A 10 B 42	0.47
Gluten proteins	−6.5 → −8.5	71 → 77	138 → 147	—	—	0.29 → 0.47
Polystyrene	—	—	92	Complex 200 → 220	—	0.4
NaAlSi$_3$O$_8$ mineral glass	—	—	823	1100	—	0.1

Abbreviations: T_g' and C_g', the glass transition temperature and concentration (in %w/w) of the maximally freeze-concentrated solution; T_g and T_m, the glass transition (usually a mid-point determination) and melting temperatures of the dry material; n, number of water molecules in the unit cell; ΔC_p, the change in heat capacity at the glass transition; MH, monohydrate; DH, dehydrate; anh, anhydrous.

These values have been taken from references cited throughout this article and are illustrative only. Some values are subject to debate, notably T_g' and C_g', whilst others (T_g) have a tendency to increase in value in more recent references as the remaining water is more effectively removed. There can be an alarming variation in values. For instance, values for the extrapolated T_g for anhydrous starch have been reported from 150 to 330°C.

second sample indicate differences in either the purity or perhaps the crystalline perfection or even the thermal resistance of the path between pan and sample.

Lactose exhibits more interesting thermograms. For the α-lactose polymorph, which has one water of crystallisation associated with the unit cell (Table 9.3), the structure alters, as evidenced by an endotherm in the region of 150°C, by rearrangements of the water of crystallisation in the crystal. As this is a closed system, the enthalpy change cannot be due to the water being lost. Lactose is a complex sugar with, in addition to the basic α and β forms, other polymorphs of 4/1 and 5/3 mixed alpha/beta crystals. These are thought to represent distinct crystals rather than physical mixtures as evidenced by the peak shifts detected in solid-state NMR.

DSC tends to be insensitive to the detection of separate polymorphs of similar properties due to the inherent large peak widths; however, α and β polymorphs of lactose with their different melting points can be distinguished. Melting can also be close to degradation temperatures producing uncertain results, particularly in the case of sealed pans, where the water is held in intimate contact with the organic sugar. Water tends to react under these conditions and promotes degradation, and hence for this type of work simple crimped or

Table 9.4 The degradation of sugars after a crude melt-quench preparation where the sample is melted on a hot plate and rapidly cooled

{PRIVATE} Sugar	T_m °C (dry)	T_g °C (dry)	Colour	Percentage degrade
Fructose	124	12	Clear	0–5
Glucose	158	30	Light brown	20–30
Sucrose	192	67	Brown	75+
Maltose	161	98	Severely brown	—

An HPLC method was used to estimate the surviving sugar.

even open pans are used which will result in perfectly dry materials when heating to high temperatures.

9.3.2 Sugar glasses

The issue of degradation is also important when producing sugar glasses. The simplest glass-forming technique, which involves melting the sugar often with poor temperature control, followed by quenching to a low temperature, will frequently produce glasses, which are coloured to varying degrees. Table 9.4 shows that these can be considerably degraded.

Controlled melting in sealed ampoules in temperature-controlled baths ameliorates this problem to a degree, for instance in the production of maltose glasses [28]; however, studies on repeated heating and cooling of glucose [29] have shown a small but significant lowering of T_g. Consequently, a favoured method to produce sugar glasses is by freeze drying a dilute solution of sugar (see Section 1.5.8). Careful attention to rates of cooling and progressive step heating during the drying stage, preferably by controlled shelf temperatures [30], will produce a very pure and high-surface-area product free from all signs of collapse. Collapse is often manifest as clear sticky patches having a high residual moisture content. Table 9.3 shows the glass and melting transitions together with the change in heat capacity at the glass transition for a range of materials.

9.3.3 Sugar crystallisation

One useful property of the freeze-dried product, with its large surface area, is the propensity to crystallise. The type of trace produced by heating this material can be seen in Figure 9.8.

This looks similar to thermograms for polyethylene terephthalate (PET) and several other polymers where the exothermic peak is characteristic of cold crystallisation. However, there are some important differences. A successive quench and reheating will not reproduce the crystallisation in the case of amorphous sucrose. This is due to the collapse and loss of porous structure of the material on the first heating run. The energy required to form a crystal inside the amorphous centre of a uniform material is large. Therefore during the second heating, only a small amount of crystallisation will occur at the surface, which goes effectively undetected. Some authors would also argue that breakdown impurities produced

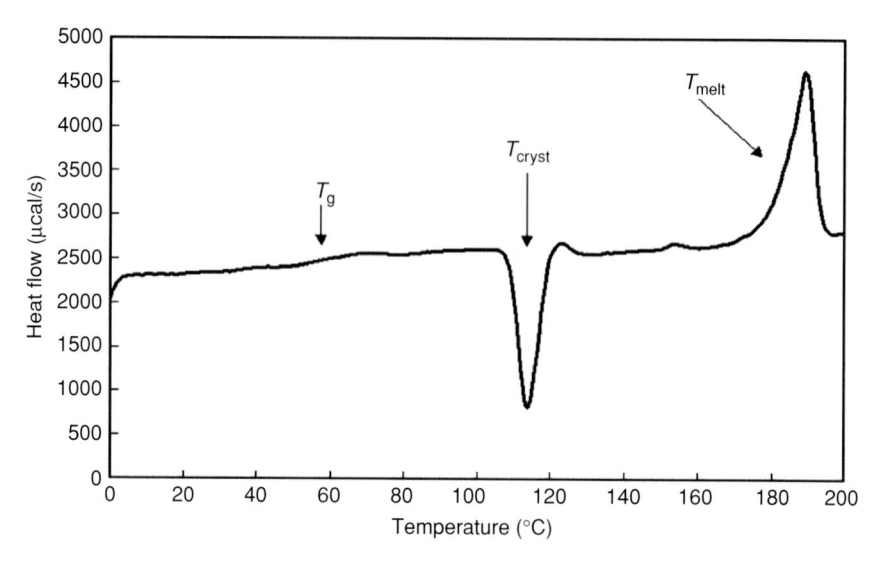

Figure 9.8 DSC trace for amorphous sucrose at a moisture content of a few percent, showing a glass transition, cold crystallisation and melting.

by heating these labile organic materials effectively 'poison' any crystals, which might form on subsequent heating.

However, providing the materials that can be made reproducibly, these problems do not prevent the application of the principles of polymer physics to the analysis of crystallisation in these systems. It is interesting to note that this type of crystallisation, observable on the DSC, will also occur in commercial products, such as confectionery chew-like materials with a porous aerated structure. With this industrial aspect in mind, it is relevant to develop methods to measure crystallisation rates. Methods for isothermal and non-isothermal analysis exist to do this. These methods are considerably easier in fats where there are no 'difficult-to-control' variables, such as the level of plasticiser/water. Nevertheless this has been attempted in sucrose using the approach of Chan et al. [31, 32]. Rates were obtained for experiments on isothermal crystallisation of amorphous sucrose [33] and crystallisation exotherms were measured from DSC curves scanned at different rates. Shift factors were then calculated for both isothermal and non-isothermal measurements and are plotted in Figure 9.9.

Only one limb of the crystallisation curve could be accessed with this measurement, but acceptable agreement was obtained between the isothermal and non-isothermal results; however, the extreme sensitivity of crystallisation rate to moisture content and surface area contributes to the errors. Similar experiments with lactose did not produce satisfactory results [33]. Irregular peaks were produced on some of the isothermal traces for lactose which were felt to represent concurrent transformations between polymorphs. It is known from work on fats that the range of scanning rates has to be carefully chosen in order to obtain a consistent set of data showing the crystallisation of only one polymorphic form. The presence of multiple forms or transformations between forms invalidates the approach.

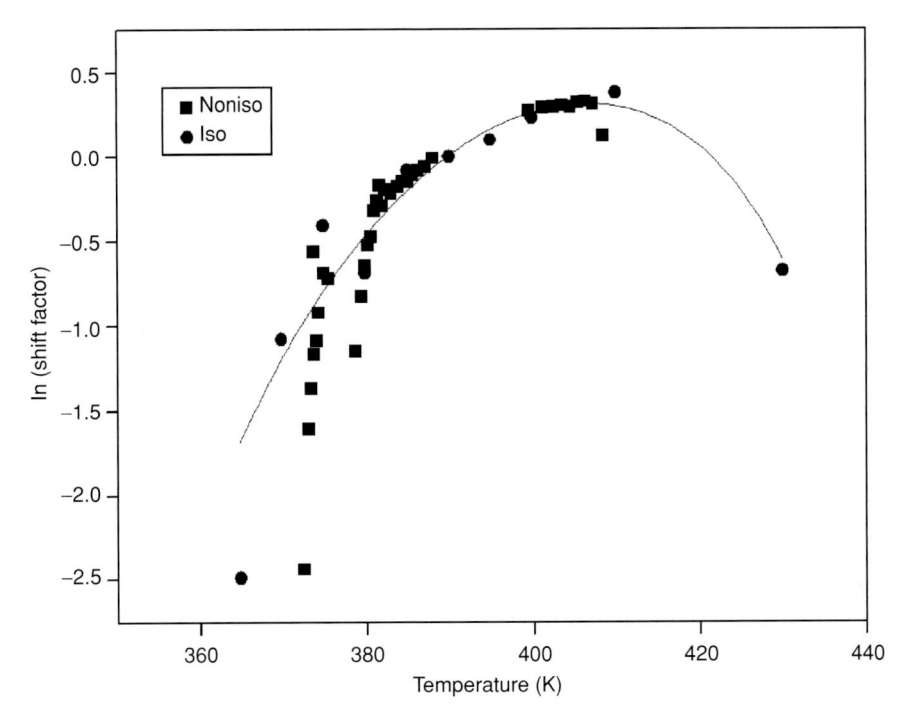

Figure 9.9 The agreement between isothermal and non-isothermal crystallisation data for sucrose. (Reprinted from [33], Copyright (2000), with permission from Elsevier.)

9.3.4 Effect of ions on crystallisation

Methods for investigating the rate of crystallisation by observing DSC traces can also be used qualitatively. For instance a crystallisation peak which is pushed to a higher temperature or has a lower absolute value of enthalpy can be taken as evidence for some form of inhibition of crystallisation. Bearing in mind that the *x*-axis of a DSC trace also represents time as well as temperature, a higher temperature on the trace implies that the event has taken a longer time to occur for the same scanning rate. When the effects of hydrocolloids, such as carrageenans and agar, on crystallisation were examined [34], part of the study involved control experiments to examine the effects of the ions necessary for the formation of a gel network throughout the amorphous structure. It can be seen in Figure 9.10 that KCl (potassium is a required co-ion for the gelation of carrageenan) had a large effect, showing a large reduction in the value of ΔH, in fact much larger than the carrageenan itself, and a shift to a higher temperature of the remaining small crystallisation peak.

The potassium was originally present in the solution before freeze drying, at a concentration of 1 g/L; however, the freeze-drying process concentrated the ions to very high levels. Ionic effects are well known when sugars crystallise directly from relatively dilute solution. In this work there appeared to be a minimal effect on T_g of these particular ions.

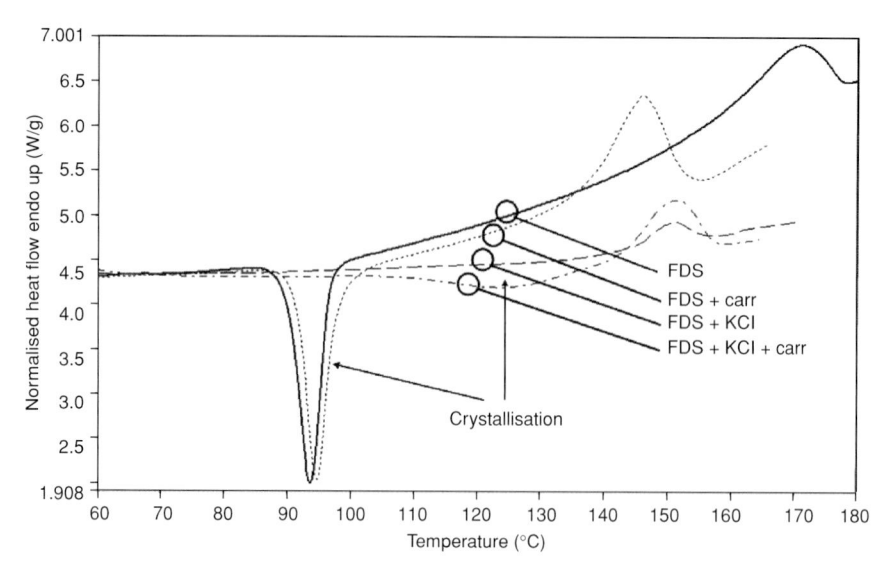

Figure 9.10 The effect of potassium chloride on the crystallisation of freeze-dried sucrose. It appeared to be the ions which had the inhibiting effect on crystallisation. FDS, freeze-dried sucrose; carr, κ-carrageenan; KCl, potassium chloride.

9.3.5 Effect of ions on the glass transition temperature

There is a body of work in the literature on the effects of ions on the T_g's of sugars at both high and low moisture contents [35, 36]. It is important to express concentration in consistent units for comparison purposes. For instance in a study of the effects of borax on the glass transition temperature of fairly dilute solutions of sucrose and trehalose, borax concentration was expressed according to the equation

$$\Delta T_g = T_g(x_T, x_s) - T_g(x_T, 0)$$

where x_T and x_S are the mole fractions of trehalose and salt respectively. This means that as the salt was added, the ratio of sugar to water changed. The question of ideal behaviour of solutions should be borne in mind when considering the large increases in transition temperature reported for borax on these sugars [37]. Polyhydroxy ions such as borax possibly act by forming a network of links due to condensation reactions between hydroxy ions on the borax and the carbohydrate. Formation of complexes has been demonstrated using boron NMR. A direct ionic effect on T_g due to the interaction of ions with a dipolar matrix has also been proposed [38]; however, this probably applies more to uniform single-component glasses, such as those composed entirely of water or alcohols. The poisonous nature of borax has led some workers to propose phosphates [39] as a possible substitute which may act by a similar mechanism. Simple salts such as sodium and potassium chloride were observed to have minimal effects on T_g but substantial effects on the crystallisation of both sugars and water in sugar solutions.

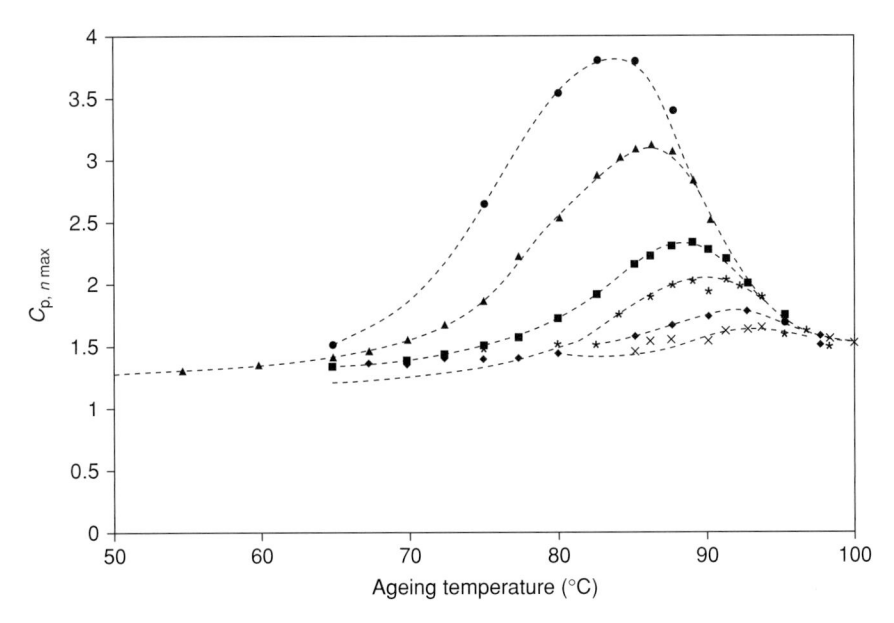

Figure 9.11 Normalised heat capacity overshoots as a function of temperature and time in aged maltose. × 1, ♦ 3, * 7, ■ 15, ▲ 60, ● 160 min. The sample is dry, glassy maltose with a T_g of 98°C. (Reprinted from [42], Copyright (1999), with permission from Elsevier.)

It is interesting to speculate on possible industrial effects of ions. For instance, in commercially supplied water with significant amounts of dissolved species, processing involving the removal of water and the concentration of ions could be significant.

9.3.6 Ageing

Much work has been carried out on the ageing of sugar matrices, from both a theoretical viewpoint and the industrial perspective of the use of sugars as encapsulation matrices. For instance, maltose and higher molecular weight maltodextrins are used to encapsulate mint- and citrus-flavour compounds. Ageing is thought to produce densification and consequently stress build-up in the glass, ultimately leading to cracking and leakage of the encapsulated material. The most thorough work, including sound recordings of cracking from aged maltose bars [40], has been carried out by Noel et al. on maltose and other carbohydrates [41]. Of particular relevance are the size of the heat capacity overshoots in maltose as a function of temperature and time (Figure 9.11) [42]. These indicate the degree of ageing in the glass. If stored at high temperatures the system is close to equilibrium and so little ageing is observed. At low temperatures little ageing can occur due to the high viscosity. A maximum can therefore be observed in plots of the overshoot against ageing temperature.

In this case, the predictions of the TNM model seem to be accurate (see Section 1.5.6). The ageing of sugars in general seems to be reasonably well described by TNM kinetics; in particular, the properties of maltose, trehalose, glucose and fructose have now been investigated [43, 44]. This work should facilitate the prediction of stability and shelf life in products whose properties are dominated by these components.

9.3.7 The Maillard reaction

Dogma dictates that for reactions, both enzymatic and non-enzymatic, which are diffusion controlled, that is, the limiting step in the reaction lies in one reactant diffusing to the active site, the rate of reaction should fall to very low values in the glass. There should therefore be dramatic temperature dependence for such reactions, effectively ceasing below T_g. Obviously, the time-scale is important here and storage trials on these systems must take account of the effective lowering of T_g over long time-scales. Nevertheless a non-zero reaction rate has been observed in many systems where the Maillard reaction has been measured in a glass [45].

A convenient model for studying this behaviour is the reaction between glucose and lysine prepared in a non-reacting glass-forming matrix, for example a mixture of sucrose and trehalose. There is still controversy as to whether a reaction which can be detected in the glass is due to imperfections in the matrix, such as cracking, where reactants can diffuse rapidly, or even local pockets of high moisture, which consequently have a depressed T_g.

In a particularly inventive experiment on reactions in the glassy state, the effect of ageing on the rate of Maillard reaction has been studied [46].

With regard to enzymatic activity in glasses, it was found that sucrose hydrolysis, as monitored by levels of glucose, fructose and sucrose, only occurred significantly above the T_g measured by DSC [47]. It appeared that hydrolysis was detected concomitantly with crystallisation of the sucrose, and neither was likely in glassy materials.

9.3.8 Mechanical properties

Three-point bending tests have been carried out on maltose bars formed in moulds from boiling maltose–water mixtures [48]. These methods determined the tensile modulus by a deformation method. Samples were allowed to age in oil baths, which controlled the water loss from the sample, for times up to 320 min, after which the decay in modulus was followed. By fitting the decays to a stretched exponential function, the time constant of the decay τ_0 and the non-exponential parameter β could be obtained.

$$\Phi(t) = \exp(-(t/\tau_0)^\beta)$$

It was found that the relaxation behaviour was dependent on the degree of undercooling from T_g and that the mechanical relaxation times and calorimetetric behaviour were correlated.

9.3.9 Foods

The level of crystallisation of sugars in many products can be an important factor determining quality. In the case of chews and candy doughs, crystallisation must reach a certain

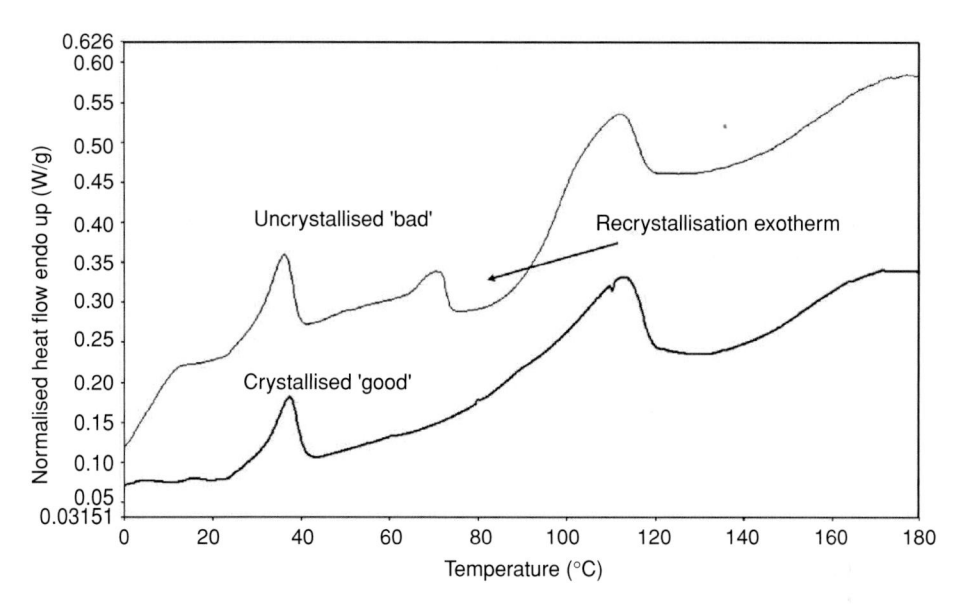

Figure 9.12 Typical DSC thermograms showing the sugar recrystallisation exotherm for soft confectionery (chews) which have not yet reached the correct level of crystallisation.

level in order to give the appropriate softness to the product. This will occur over a period of weeks but can be detected in the DSC during a heating scan, as the process occurs much more rapidly at higher temperatures; becoming comparable to the scanning rate and so producing a well-defined exotherm. Figure 9.12 shows the comparatively small exotherms detected in a complex chew material. They are not seen in materials which have already crystallised. This method can thus detect the extent of crystallisation in the product and whether it is ready to be released for consumption.

Candyfloss, being essentially a mixture of sugars in the glassy state, will exhibit crystallisation as a defect [49]. Crystallisation will lower the volume (Figure 9.13) and increase the stickiness, and it will occur under conditions of increased relative humidity and temperature.

It can be combated by the incorporation of other sugars. Ten per cent fructose raises the temperature and reduces the enthalpy of the sucrose exotherm, as shown in Figure 9.14. The mechanisms involve the competition of molecules for sites on the growing crystal.

The T_g of glucose syrups can be an important indicator of functionality. Glucose syrups are complex mixtures of glucose polymers ranging from glucose through maltose to long-chain dextrins with variable degrees of branching. A glucose syrup will be chosen due to certain properties of handling. For instance, high-maltose syrups generate lower viscosities and may have higher T_g's despite having a higher dextrose equivalent which would lead one to believe that they should have lower T_g's [34]. A lower viscosity material would be chosen, for example, when making a viscous hydrocolloid mixture in a confectionery recipe.

It has been claimed that the adulteration of honeys by glucose syrups can be detected by calorimetric methods. Relationships were found between the percentage of added syrup and T_g [50, 51].

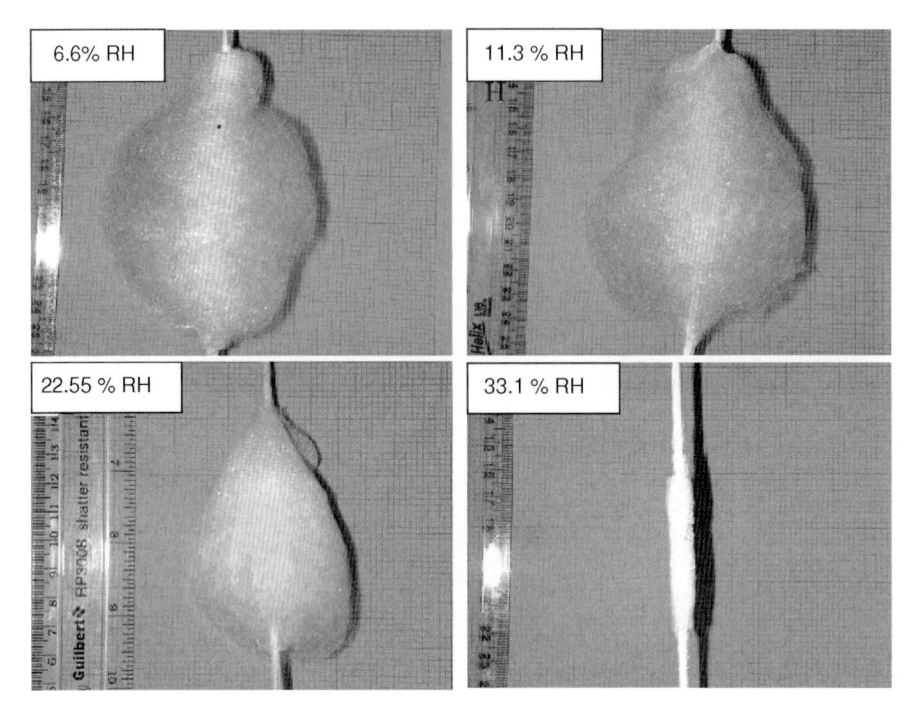

Figure 9.13 Pure sucrose candyfloss after storage at the specified RH and ambient temperature for 4 days. Collapse (associated with crystallisation) can be seen at the higher RHs. (Taken from [49].)

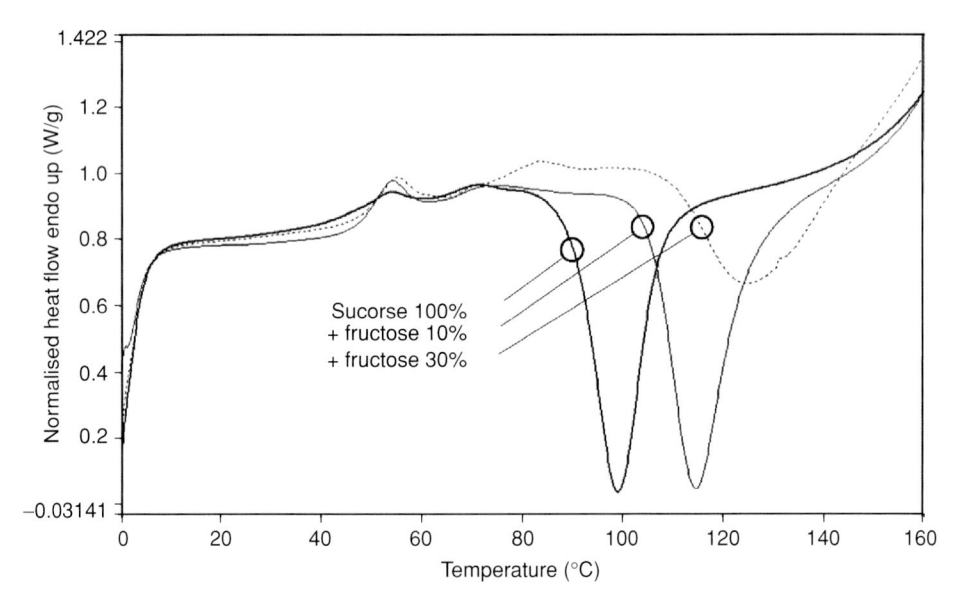

Figure 9.14 DSC of candyfloss showing the effect of sugar additives in shifting the exotherm to higher temperatures. The additives have a less dramatic delaying effect on the visually assessed collapse at higher RH. (Taken from [49].)

Milk powders have high levels of amorphous lactose and will show well-defined apparent recrystallisation peaks [52, 53]. However, if such bulk powders are left at high temperatures, for example during transit, then browning can take place, demonstrating a Maillard interaction between the proteins and the sugars present in the powder. On the DSC this produces an exotherm at a similar temperature to the recrystallisation peak of lactose, and disentangling the two events can be difficult.

It has been reported [54] that the production of water during the Maillard reaction, which is not produced during crystallisation, can be used to determine the kinetics of non-enzymatic browning and distinguish between the two processes.

9.4 Fats

There is an enormous literature [55, 56] on even comparatively small aspects of the study of fats. What we seek to show here is how thermal methods, principally DSC, can be used to understand the complex behaviour of fats and indeed show where thermal methods are insufficient and require complementary techniques, mainly X-ray diffraction. However, we will start by considering one of the simplest DSC measurements on one of the most complex systems, namely the solid/liquid ratio of a complex mixture of triglycerides.

9.4.1 Solid/liquid ratio

Many functional properties of fats, for instance how well a fat will perform in the pastry-making process, are to a large extent dependent on one simple parameter, the solids content or solid/liquid ratio. When we consider a mixture of triglycerides there is a whole series of materials with different melting points, so we will have very complex behaviour and a broad melting envelope. A simple method for determining the solids content is to measure this melting envelope and calculate the area to the right and left of a particular temperature and express this as the solid/liquid ratio of the mixture at that temperature (see Figure 9.15; [57]). The area to the left of the temperature can be expressed as the fraction of fat which has melted, the liquid, whilst the area to the right is the fraction which has yet to melt, the solid. Margarine and butter usually have around 20% water associated with the fat component. This could appear in the endotherm so this method may not be appropriate for these materials. Although Figure 9.15 gives the essence of the method, the original calculation of solids content was

$$\% \text{ solids} = \{\text{heat/unit area} \times \text{solid area/wt of sample} \times \Delta H_f\} \times 100$$

where the heat/unit area was based on the heat of fusion of indium and ΔH_f was an average value for the heat of fusion of fats (typically 35 cal/g). The calculation allows for the fact that all the materials present may not participate in the endotherm. This method was a great improvement in terms of the time taken, over the original dilatometric method [58], which involved looking for discontinuities in the volume of material as the temperature was increased. This could take up to a day to complete and would involve only discrete points. The DSC method would be complete in several hours at most, even at the lowest scanning rates, and would give effectively continuous data. Both methods have now largely been superseded by NMR-based methods [59], which can be carried out in seconds, although tests involving the

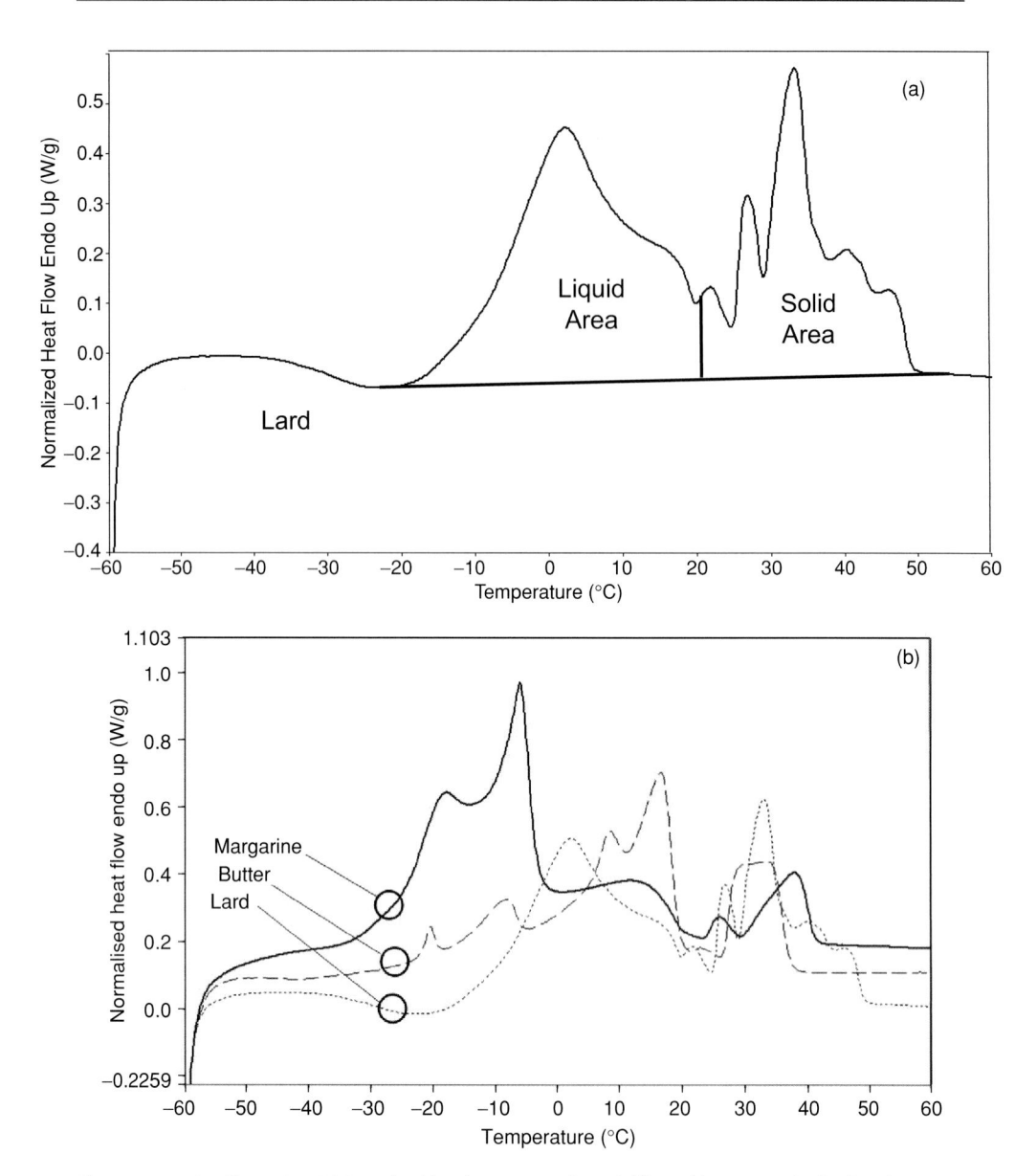

Figure 9.15 An illustrative DSC method for determining the solid/liquid fat ratio in lard. Also shown are melting profiles for butter and margarine.

comparison of solid/liquid ratios measured by the various methods at different laboratories are still occasionally organised. One of the problems of the DSC-based method is the perennial difficulty of not perturbing the sample while measurements are being made. As the temperature is increased, polymorphic transformations (see below) between forms can occur, although in the original description of the method, careful tempering of the sample was specified. Transformation is particularly obvious in single triglycerides and creates

Table 9.5 Melting temperature and enthalpy of cocoa butter as a function of polymorphic form

Form	Melting temperature (°C)	Heat of crystallisation (kJ/kg)
I	17.3	—
II	23.3	86.2
III	25.5	112.6
IV	27.5	117.6
V	33.9	136.9
VI	36.3	148.2

Taken from [60].

serious difficulties for the DSC method of solid/liquid ratio determination, as any recrystallisation will dramatically alter the property being measured. It is possible that polymorphic transformation is less likely to be so dramatic in more uniform complex mixtures. The other difficulty was the possible variation of melting enthalpy with temperature as a function of either composition, temperature or polymorphic form [60] (see Table 9.5). The latter meant that the same area in different regions of an endotherm would represent different weights of material.

9.4.2 Phase diagrams

Phase diagrams describe how composition varies with temperature and at which temperatures major transitions occur. The oils and fats literature is brimming with phase diagrams of various kinds, and particularly simple phase diagrams have already been described (see Section 1.5.8).

One additional complexity which arises on phase diagrams such as those in Chapter 1 is when one material has a limited solubility in the other, that is, it ceases to form well-mixed solutions at the extreme edges of the phase diagram (see, e.g., points (a) and (b) on Figure 9.16). These are the points at which components A and B begin to separate from the mixture. These are also referred to as solubility limits. There are many more complicated types of behaviour, including higher dimensional phase diagrams describing multi-component behaviour, but for now an experimentally determined phase diagram of a mixture of two fat molecules, tripalmitin and tristearin, will be examined (Figure 9.16).

The complete phase diagram is effectively a combination of the phase diagrams shown in Chapter 1 and shows the differing behaviour for the polymorphs. Polymorphs are different crystalline forms of the same material. For instance, the β form shows eutectic behaviour (upper line of Figure 9.16), whilst the α and β′ forms approximate to miscible liquid and solid behaviour. This is common behaviour for mixtures of similar triglycerides, although careful examination of crystallisation thermograms for the α phase demonstrates a double peak at lower scan rates, suggesting a more complicated α phase boundary than that shown in Figure 9.16.

There are certain limitations in such phase diagrams. Firstly, this type of equilibrium phase diagram gives no indication as to how quickly the phases will crystallise; that is, it

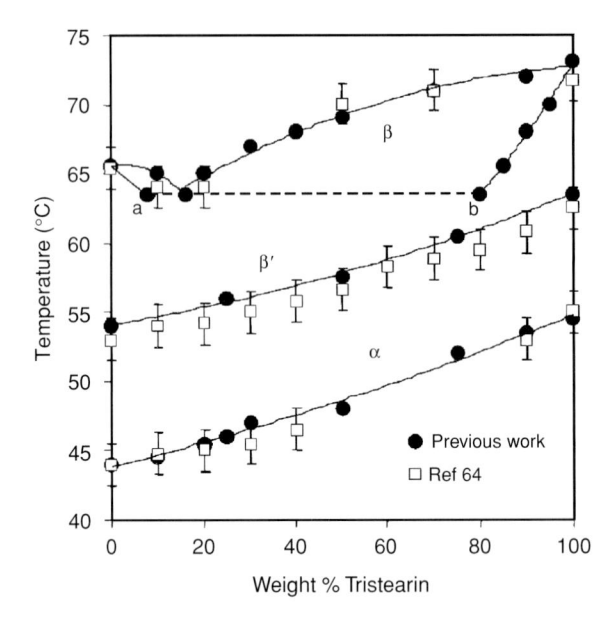

Figure 9.16 The phase diagram of a tripalmitin–tristearin binary mixture. Points (a) and (b) show limited solubility regions. (Taken from [64], Copyright (2006), with kind permission of Springer Science and Business Media.)

shows equilibrium rather than kinetic data. Secondly, due to the difficulty of forming certain polymorphic forms, there is considerable disagreement over the temperature range in which a form can exist.

9.4.3 Polymorphic forms and structure

The observation of multiple melting forms in fats originally caused much confusion in the literature (see [61] for a discussion); however, it soon became clear, principally by using X-ray diffraction, that polymorphic forms were responsible [62]. Fats tend to form lamellar structures and so a typical X-ray diffractogram will reveal details of structure via wide-angle diffraction maxima, which reflect the lateral packing of chains: the so-called subgroups and low-angle diffraction peaks, which reflect longitudinal order in the more widely spaced lamellar sheets.

9.4.4 Kinetic information

The presence of these different forms and the fact that they crystallise at different rates and have different levels of supersaturation at any one temperature affords the opportunity for some very complex behaviour. However, some simple principles can give a reasonable understanding. The kinetics of crystallisation tend to be studied less frequently than that of the equilibrium phase behaviour, but is often more important industrially [63]. Here we reproduce work carried out in our laboratory [64]. It can be seen in Figure 9.17 that there

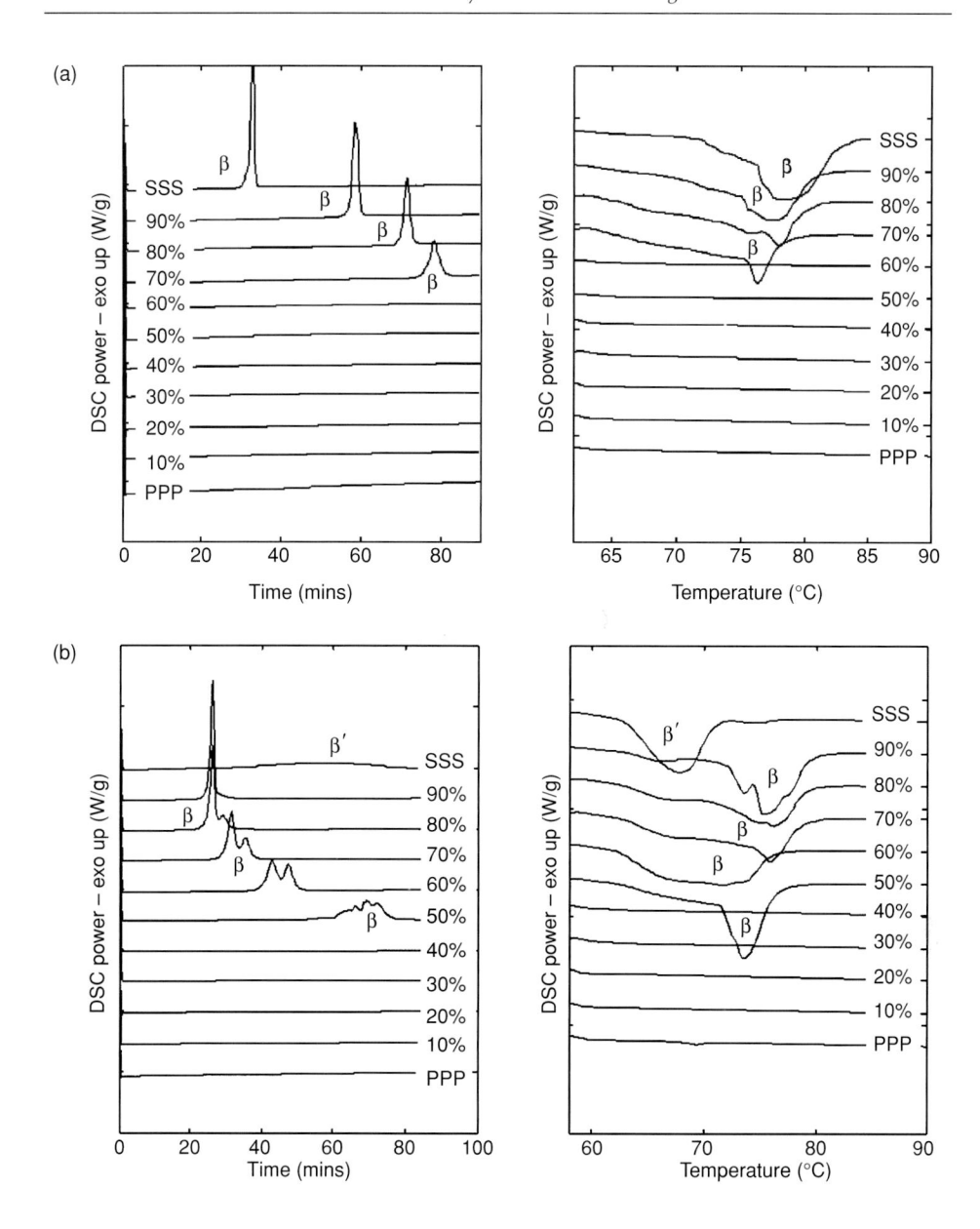

Figure 9.17 DSC thermograms (endotherms up) showing isothermal crystallisation (left) and melting (right) of tripalmitin–tristearin blends (labelled as wt % tristearin) at (a) 62.5°C, (b) 58.5°C, (c) 54.5°C and (d) 40.5°C. The y-axis scale divisions are 1 and 10 W/g for crystallisation and remelting thermograms, respectively. This demonstrates the use of DSC in identifying the polymorphic forms of fats. (Taken from [64], Copyright (2006), with kind permission of Springer Science and Business Media.)

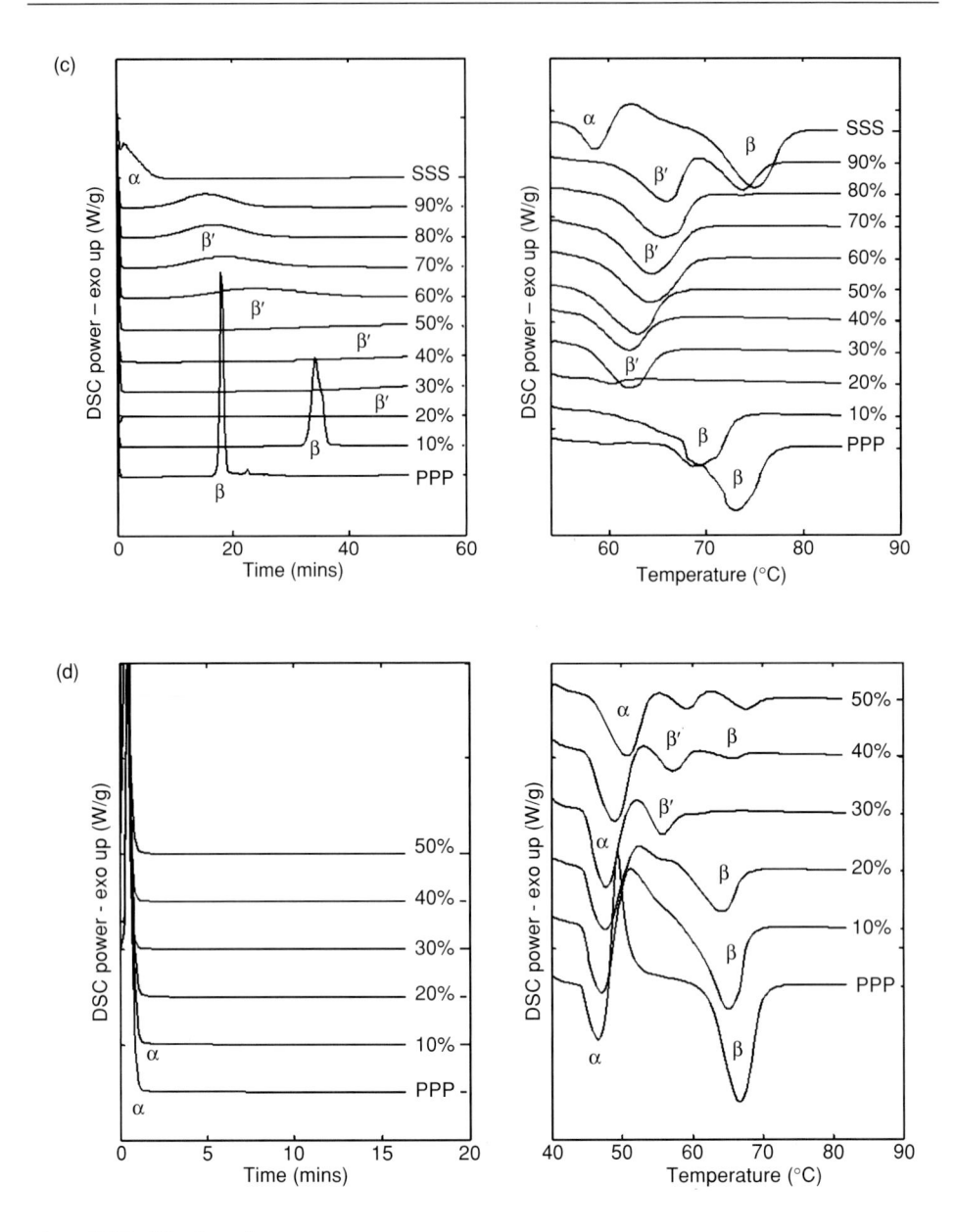

Figure 9.17 (Continued)

are fairly dramatic differences in the isothermal crystallisation rates of the polymorphic forms. Crystallisation peaks for the β form are very sharp with sometimes extremely long induction times (upwards of 1 h). Those for the α and β' forms are broader with much shorter, possibly in the case of α, negligible, induction times. The forms can be crudely identified on thermograms by the onset temperatures of subsequent melting peaks. The overall crystallisation rate is made up of two contributions, an induction time t_{ind} and a time to half crystallisation $t_{1/2}$. According to the Avrami equation (see Section 1.5.9)

$$X_{solid}(t) = 1 - \exp\{k(t - t_{ind})^n\}$$

a time to half crystallisation can be defined in terms of the parameters k and n after the induction time has been subtracted.

$$t_{1/2} = \left(\frac{\ln 2}{k}\right)^{1/n}$$

The induction time is a subject of some controversy. It in turn is made up of two components. Firstly, there is the true induction time of nucleation, the time to the appearance of the nucleus; secondly, there is the time of growth until a detectable fraction has been produced. This latter time is technique dependent, with TA methods being deemed, in general, to be insensitive. If a more sensitive technique such as microscopy or particle scattering is used then this latter time is less. There is some debate as to whether a true induction time of nucleation really exists, with some authors proposing that a sufficiently sensitive technique would detect nuclei perhaps of only a few molecules at a very early stage. The overall time is referred to as the induction time of crystallisation.

For a comprehensive list of crystallisation parameters for the complete range of concentrations and polymorphs of PPP/SSS, the reader is referred to reference [64]. The induction time for nucleation is thought to relate to the 'difficulty' or free energy of activation required for the formation of stable nuclei. However, whereas the induction time gives an indication of the free energy required to form a stable nucleus, the sharpness of the peak is indicative of the rate of crystallisation, in turn dependent on the supersaturation or supercooling below the equilibrium melting point of the particular polymorph.

Using the Turnbull–Fischer expression for crystallisation rate [65], which is inversely related to induction time, plots can be made of $\ln(T t_{ind})$ against $1/T(\Delta T^2)$. Figure 9.18 demonstrates the dramatic difference in the apparent free energy of formation of β and β' forms. Despite the difference in the sharpness of the peaks, it is found somewhat surprisingly that the half times for crystallisation of β and β' forms as a function of driving force lie on one continuous curve.

Figure 9.17(d) shows that it is difficult to observe the α crystallisation by direct cooling to the isothermal crystallisation temperature, the kinetics being so rapid, effectively a zero induction time, that the crystallisation step is merged in the cooling stage of the trace.

9.4.5 Non-isothermal methods

In order to measure rapid kinetics, it is possible to use non-isothermal methods, that is, to observe the crystallisation at different rates of cooling and then use one of the many procedures available to analyse this type of data. This is a developing area and the reader is referred to references [66, 67]. It is possible using non-isothermal data to set up a matrix of rates at constant conversions and temperatures, and extract the temperature and kinetic

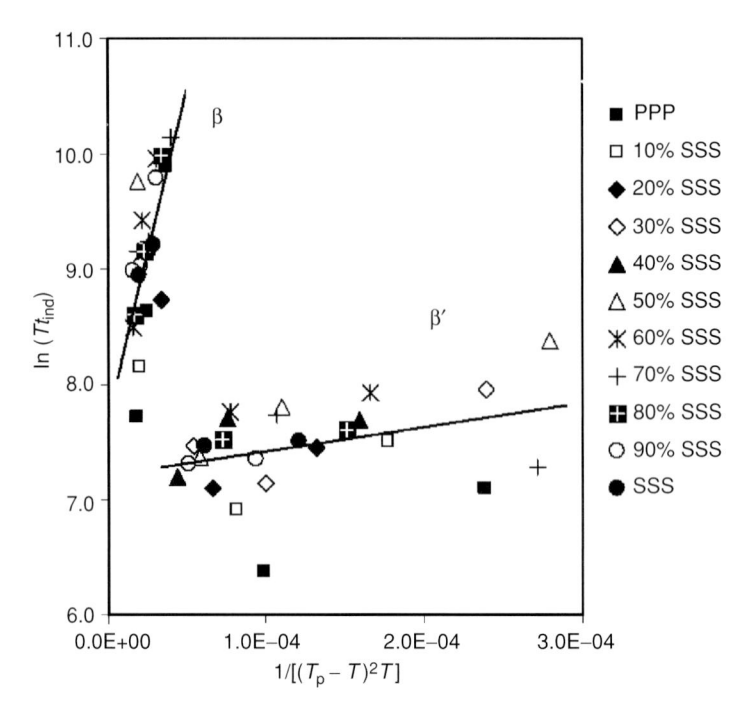

Figure 9.18 A plot of $\ln(Tt_{ind})$ against $1/[(T_p - T)^2 T]$ for all mixtures showing a higher activation energy for formation of the β form (56 500 K^{-3}) compared with the β′ form (2090 K^{-3}). The longer induction time is a consequence of the greater difficulty in forming the β polymorph. (Taken from [64], Copyright (2006), with kind permission of Springer Science and Business Media.)

functions directly. This is known as the non-parametric kinetic (npk) method and has the advantage of separating the data into independent functions before fitting with models. The method only really works for simple single-stage reactions. It is an interesting proposal as to whether the imposition of deconvolution procedures to the initial data, despite being apparently against the spirit of the method, would enable it to apply to more complex data, for example those containing multiple peaks. Sample-controlled TA adopts a diametrically opposed procedure and seeks to minimise artefacts produced by thermal gradients in the sample by slowing the heating program based on some property of an occurring transformation. This approach has not yet been applied to DSC but has been applied to dilatometry and may be an area for future work.

Whilst the previous discussion gives an idea of the type of information available, often from simple DSC experiments, much work has been directed towards predicting crystallisation behaviour using only the triglyceride composition of the mixture. For example, using the nucleation and growth behaviour, crystal-size distribution and phase diagrams for the components, it may be possible to predict the functional properties of industrial fat mixtures.

Interestingly, it is found that certain rates of cooling will produce particular polymorphic forms. For instance, in the PPP/SSS system above, rates greater than about 1 K/min will produce the α form, whereas rates between about 0.2 and 0.5 K/min will produce β′ forms. β forms are produced by still lower rates. For many fat mixtures, it is simply not possible to

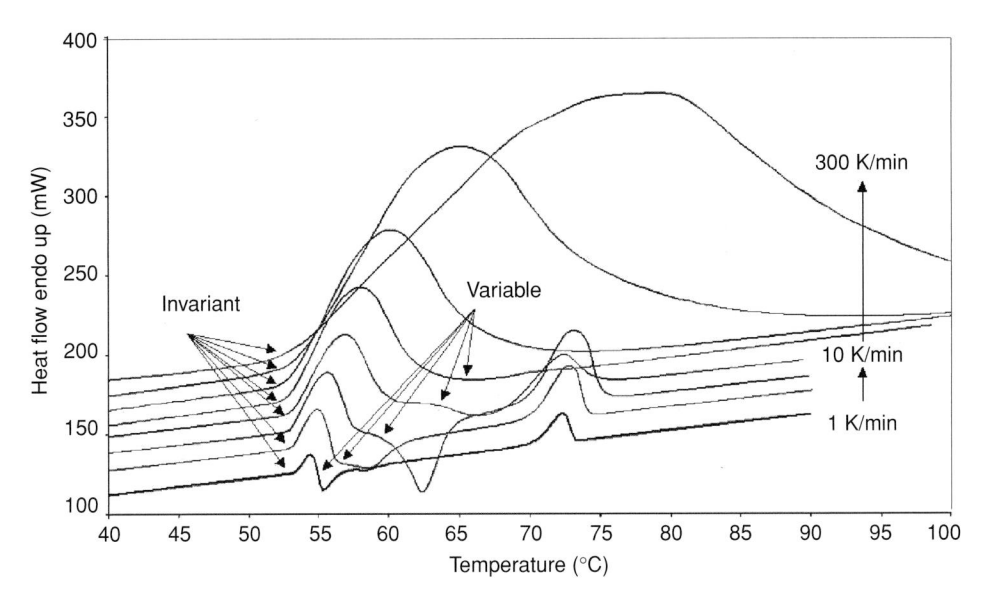

Thermal Methods in the Study of Foods and Food Ingredients 361

Figure 9.19 Tristearin scanned at rates of 1, 2.5, 5, 10, 20, 40, 100 and 300 K/min. Features which are invariant and those which change are arrowed. Sample weight is 10 mg. The exotherm decreases in enthalpy and increases in temperature, whereas the melting points are invariant.

produce the higher, more stable β-type forms directly from the melt. In this case, the sample has to be tempered, that is, have stable precursor nuclei induced by previous temperature or temperature/shear treatments, in order to crystallise in the desired form. This is particularly important in chocolate.

9.4.6 DSC at high scanning rates

DSC on its own is obviously incapable of discerning structure; however, it can give an indication of multiple melting points. In addition, where a kinetic transformation occurs, say, for example, the melting of one form followed by a recrystallisation into another polymorph, investigation of this transition can take place by varying the DSC scan rate. Figure 9.19 shows the effect of varying the scanning rate systematically into the realms of so-called hyper-DSC for a sample of tristearin. It can be seen that the recrystallisation peak (exothermic) between the two endothermic peaks at first increases in temperature and then progressively disappears. In this system the rapidity of the scan presumably prevents recrystallisation into the other polymorph; however, the widening of the peaks and smearing of the response can clearly be seen and have not been corrected for. The weight for this sample is of the order of 10 mg, which is also larger than the recommended value of a few milligrams or less. Thermal gradients will thus be larger as will the smearing function for the sample.

Moreover, the purge gas used was nitrogen rather than helium, which further impairs the resolution of the traces (see Section 2.7).

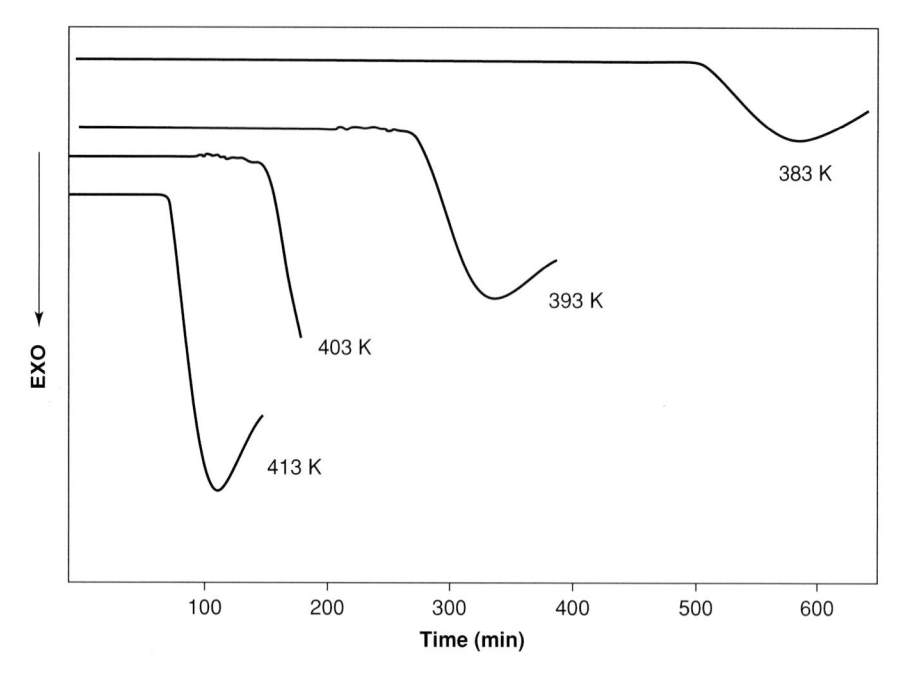

Figure 9.20 Lipid oxidation curve of refined, bleached, deodorised palm olein at 383, 393, 403 and 413 K. Oxygen was passed over the furnaces. (Taken from [70], Copyright (2001), with kind permission of Springer Science and Business Media.)

9.4.7 Mechanical measurements

Mechanical, as opposed to viscosity, measurements on fats are comparatively rare. Often this is due to the brittle nature of the materials, for example chocolate. However, one approach is to observe the major transitions using the Triton materials pocket or constrained layer damped beam method. At present, however, absolute values of moduli cannot be obtained. Nevertheless, transitions between polymorphs in cocoa butter were measurable.

9.4.8 Lipid oxidation and the oxidation induction time test

The oxidation of oils and fats, normally of the unsaturated bonds, has important consequences for human health and indeed the quality of products themselves [68, 69]. By passing oxygen over the furnaces of a normal DSC, the accuracy of temperature control of the machine gives a precise method for measuring the progress of these reactions [70] (see also Section 1.5.9). Figure 9.20 shows the exotherms produced for refined, bleached, deodorised palm olein subjected to a range of temperatures from 383 to 413 K. The reaction proceeds more rapidly at higher temperatures, as evidenced by a reduction in the onset time of the reaction and a sharpening of the peak. If the logarithm of the onset time is plotted against the reciprocal of the absolute temperature then activation parameters such as the

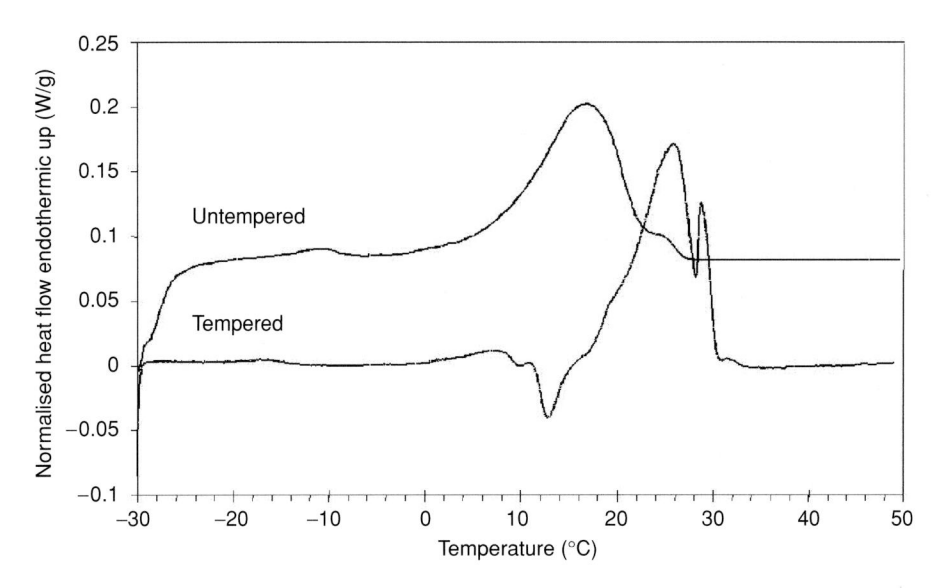

Figure 9.21 The difference in melting point of the fat component in tempered and untempered chocolate. Notice also the exothermic recrystallisation events on the tempered trace. These kinetic events can be crudely separated from the melting events using stepscan. Tempering produces high-melting-point crystal precursors. It is believed that these seed the recrystallisation to a higher melting point form when the lower forms melt, so producing the recrystallisation endotherm. (Taken from [72], Copyright (2006), with kind permission of Springer Science and Businees Media.)

enthalpy and entropy can be obtained. In this way the stability of oils can be examined and compared.

9.4.9 Foods

Complex fat mixtures represent a particular challenge to TA. The melting of different polymorphic forms in the fat component of chocolate can be clearly seen (Figure 9.21).

These observations can rationalise thermal pre-treatment of materials and can justify the success of several industrial processes. So, for example, slow cooling of tempered chocolate followed by heating at 10 K/min will demonstrate the presence of stable, high-melting polymorphs (V and VI), whereas rapid cooling and the use of untempered material produce lower forms (I and II), which will subsequently melt at reduced temperatures. The temper of a material describes the degree to which the fat has been seeded with high-melting-point nuclei. A combination of temperature treatment and shearing of the material normally produces these. The overall aim in most chocolate manufacture is to make a stable chocolate, which will not convert to another form during storage, producing the characteristic and unacceptable white-surface bloom [71]. This can be investigated thoroughly using DSC. The chocolate fats can also be studied alone by the extraction of the sugar, normally by washing with an aqueous solution. However, DSC can never provide the detailed data of X-ray diffraction in terms of structures, as the melting point of a polymorphic form will vary according to composition, impurity level and crystal perfection. The X-ray diffractogram gives the polymorphic form, irrespective of changes such as these.

9.4.9.1 Modulation

As it is suspected that the conventional DSC trace for many fat-based products is composed of a series of melting and recrystallisation peaks, it seems appropriate to examine these using modulated techniques capable of separating equilibrium (melting) and reaction (recrystallisation) events. The results of Stepscan^{TM} measurements on chocolate demonstrate the limitations of this particular method, with some transformations/recrystallisations being sufficiently rapid that they are incapable of being reliably recorded using the method as currently implemented. In addition, the presence of other artefacts means that modulation methods must be treated with extreme care and appropriate measurement parameters, such as temperature amplitude and isotherm length in stepscan (see [72]), determined for each individual case.

9.4.9.2 Other foods

Foods such as butter and margarines exhibit very wide melting envelopes (Figure 9.15), which can give problems with baseline curvature and slope over such wide temperature ranges.

The fat content and profile of yoghurts have been studied (Figure 9.22).

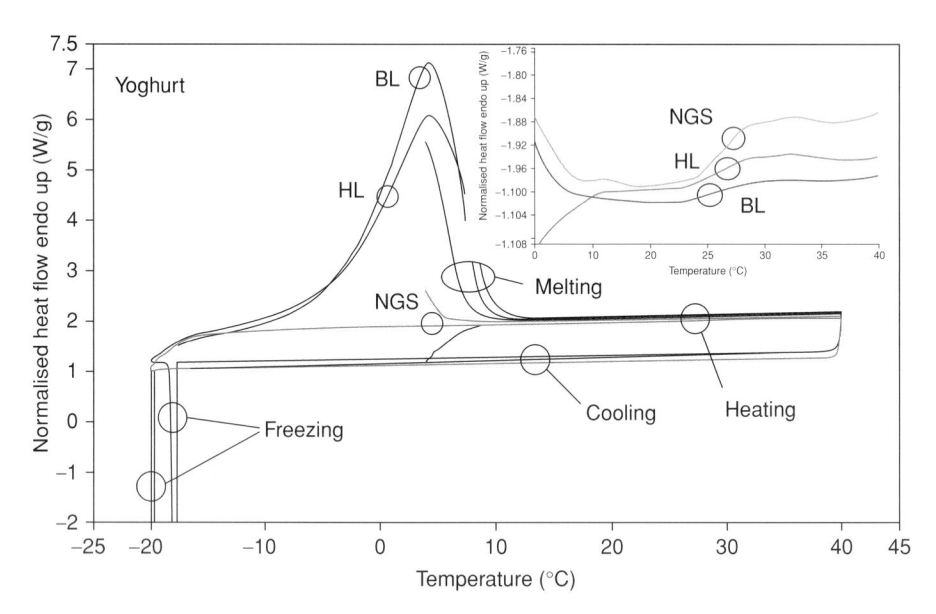

Figure 9.22 Normalised DSC traces demonstrating reduced fat content for healthy-living-style yoghurts. Interestingly, the water in the natural greek style yoghurt does not crystallise on cooling under these conditions as evidenced by the lack of an ice exotherm. NGS, natural greek style; HL, healthy living; BL, biolive. (Taken from [73].)

The endothermic peaks in high-fat yoghurts, often an indication of quality, can be clearly identified on DSC traces [73] and fat contents in different types of yoghurts can be measured (see insert trace on Figure 9.22).

9.4.9.3 The amorphous state

To date, no sustainable evidence has been offered for solid amorphous fat, which could show glassy behaviour and a glass transition. The entrapment of oils or even the presence of very low melting point species is a more likely explanation for apparently amorphous material at low temperatures. Similarly, the unequivocal demonstration of liquid crystalline forms in fats has not yet occurred.

9.5 Proteins

9.5.1 Protein denaturation and gelation

Aside from the obvious nutritional aspects, one of the most important properties of proteins, from a food processing point of view, is the ability to form gels. Protein gelation is a subtle combination of denaturation, dissociation/association reactions and aggregation, for which many different models have been proposed. So, for example, a simple general proposal for the heat-induced gel formation in small globular proteins is [74]

$$\text{Native protein} \rightarrow \text{denatured protein} \rightarrow \text{aggregated protein}$$

For the gelation of albumin [75], an additional step has been proposed where the aggregated protein is initially presumed to be a soluble aggregate, which subsequently forms a gel or coagulum.

These processes are often referred to as gel/sol transitions and are observed in many proteins and carbohydrate hydrocolloids. Gel/sol transitions range from the clotting of blood, through weak-gelatin gels formed, for example, by the cooling of meat stocks, to the clotting of casein in the presence of rennet in cheese making, and of soybean proteins in the presence of glucono δ lactone or calcium ions in tofu making [76].

DSC has been carried out on a very wide range of proteins; however, as in the case of fats, where the value of DSC-derived information was increased enormously by the complementary technique of X-ray diffraction, mainstream rheological measurements which categorise the gel behaviour are essential for even a rudimentary understanding of protein gelation. Nevertheless, the thermal properties of the processes involved in gelation can provide important thermodynamic information.

The most obvious features when scanning a solution of proteins are the denaturation peaks. For instance, if bovine serum albumin (BSA) with and without fatty acids in aqueous solution is heated, and not subjected to too high a temperature, nearly reversible endothermic peaks are obtained [77]. This is the case for many globular proteins of animal and plant origin, such as lactoglobulin and the wheat proteins gliadin and glutenin [78, 79].

One mechanism for the denaturation of small globular proteins is a two-state reversible denaturation. This is the case if the reaction is carried out far away from conditions of aggregation, that is in dilute solutions far from the isoelectric point (see 'Glossary'). If this applies then the van't Hoff law predicts the ΔH of transition, ΔH_{VH}, from the temperature dependence of the equilibrium constant describing the distribution between the two states [80]. Calorimetry in its many guises is the one general method capable of measuring these parameters, which are of use in determining the molecular nature of the interaction. For instance, hydrophobic reactions tend to have different values from those which are ionic based [81].

If we have an independent way of estimating the constant K then we can predict ΔH_{VH} and compare this with the calorimetrically determined value. These are approximately equal for dilute solutions (<0.5%) of small globular proteins [82]. The value of ΔH_{VH} can be measured as a function of temperature using DSC, by changing the stability conditions, for example by pH changes and making certain assumptions, such as there being only minor changes in ionisation enthalpies. Analysis of the thermodynamic functions can then break down the denaturation enthalpy into distinct contributions [83]. For example, there is a heat evolved with disruption of hydrophobic bonds, which decreases as the temperature is increased, and a heat absorbed, associated with the disruption of hydrogen bonds which varies little with temperature. This type of work has been carried out on several globular proteins [84].

In a method pre-empting the stepscan analysis (see Section 1.6.2), the baseline offset or heat flow can be considered to represent the heat output from a reaction and hence the reaction rate (dH/dt). Consequently, kinetic parameters for protein denaturation can be obtained [85].

9.5.2 Differential scanning microcalorimetry

In non-food areas, differential scanning microcalorimetry (DSMC) of protein and protein–ligand solutions has been studied intensively and deals with the binding of ligands to specific sites in proteins: the details of the sites often determined using X-ray diffraction and NMR structural methods. The techniques used in this area are principally differential scanning micro- or even nanocalorimetry and isothermal titration calorimetry (ITC) (see Section 9.8). These techniques are concerned with obtaining highly accurate thermodynamic data for well-defined reactions, and as such, use different instrumentation from the rest of the work presented in this chapter.

The calorimeters are extremely sensitive, capable of measuring small changes in the thermal properties of scarce and valuable materials, at levels of a few milligrams per millilitre, and designed for solutions over a limited temperature range often from around 0°C to just over 100°C. The scanning is frequently carried out at low rates and issues such as the thermal resistance to heat flow in the calorimeter, which can affect the shape of peaks, tend to be considered less. The peak shape in, say, indium melting recorded in a conventional calorimeter can look quite different from a peak due to the unfolding of a protein recorded in these high-sensitivity instruments.

The interpretation of such data is usually based around fitting a series of models and deciding which offers the best fit. Data are normally obtained under dilute non-aggregating conditions. For instance, a simple two-stage denaturation model can be fitted by the following equations:

$$\text{Native}[N] \leftrightarrow \text{Unfolded}[U]$$

$$\Delta G_{\text{unfold}} = \Delta H_{\text{unfold}}(T_m) \cdot (1 - T/T_m) + \Delta C_p \cdot (T - T_m - T \cdot \ln\{T/T_m\})$$

$$K_0 = \exp(-\Delta G_{\text{unfold}}/RT)$$

$$= [U]/[N]$$

$$\text{as} \quad C_{\text{tot}} = [U] + [N]$$

$$\text{then} \quad [U] = C_{\text{tot}}/(1 + \{1/\exp(-\Delta G_{\text{unfold}}/RT)\})$$

and the DSC signal or peak is the change of the concentration of unfolded species as a function of temperature

$$= d[U]/dT$$

In the above equations ΔG_{unfold} is the free energy of unfolding, ΔH_{unfold} is the enthalpy of unfolding, K_0 is the equilibrium constant for the unfolding reaction, C_{tot} is the total concentration of protein and T_m is the temperature of the transition, normally measured at the peak where ΔG_{unfold} is zero. By use of the equation, $\Delta G = \Delta H - T\Delta S$, so the reaction can be completely defined.

The peaks shown in Figure 9.23(a) represent the progressive unfolding of a protein whose transition temperature T_m is centred at 50°C. Also shown is the effect of changing the value of ΔH_{unfold}.

This simple model can be extended to include the effects of ligand binding firstly to the native structure, which stabilises the native state and so increases the temperature of transition. An equation describing this is

$$N + L \leftrightarrow NL$$

$$\text{and} \quad K_{N,L} = [N][L]/[NL]$$

The effective unfolding equilibrium constant is given by the ratio of the total concentration of unfolded to folded species.

$$K_{\text{unfold}} = [U]/[N] + [NL]$$

$$= K_0/(1 + [L]/K_{N,L})$$

which, if $[L] > K_{NL} \approx K_0 \, K_{N,L}/[L]$ by a similar derivation to the equation for the simple model

$$[U] = C_{\text{tot}}/(1 + \{[L]/\exp(-\Delta G_{\text{unfold}}/RT)\})$$

where ΔG_{unfold} now equals

$$\Delta G_{\text{unfold},0} + \Delta G_{\text{dis}} + RT \ln[L]$$

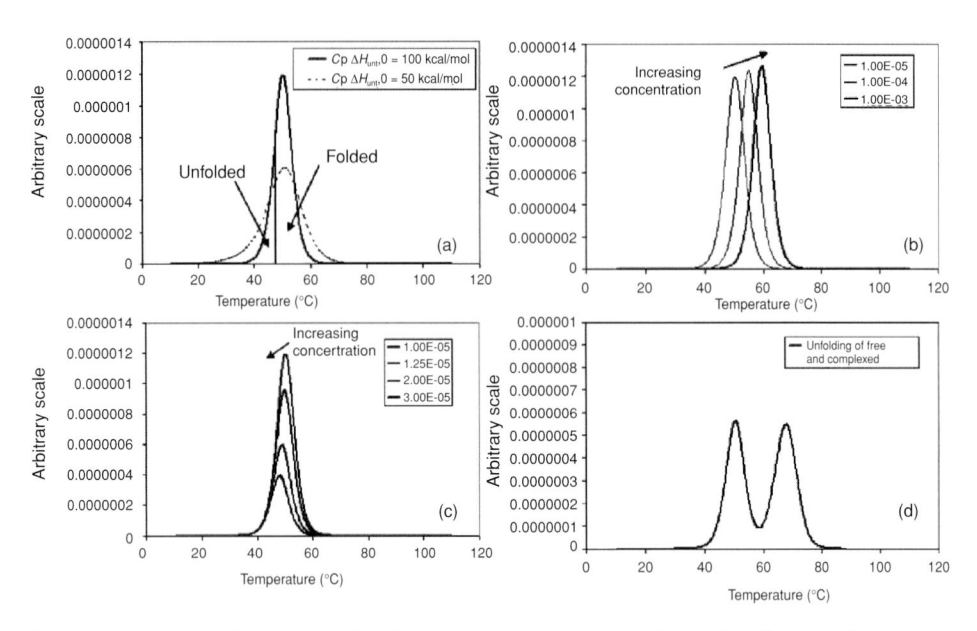

Figure 9.23 (a) An endotherm produced by a simple two-state model and the effect of a lower value for the denaturation enthalpy. The partial areas illustrate the meaning of the reaction constant in terms of the ratio of the folded-to-unfolded forms of the protein; (b) the increase in temperature of the endotherm caused by binding of a ligand to native protein; (c) the decrease in the temperature of the endotherm caused by binding of a ligand to the unfolded protein; (d) the multiple endotherm caused by the denaturation of both native- and ligand-bound protein.

showing that stabilisation results from the additional free energy to dissociate the ligand plus a contribution from the entropy of mixing. The effect of increasing the ligand concentration can be seen in Figure 9.23(b). An example of this behaviour is the binding of 2'-CMP to folded RNase.

Conversely, the binding to the unfolded state favours the destabilisation of the native state and causes a reduction in transition temperature. By similar reasoning, the binding of cyclodextrin to unfolded, egg-white lysozyme can be described (see also Figure 9.23(c)).

DSC curves are also frequently observed, which contain multiple peaks which can be deconvoluted to show individual components and which can be described by more complicated models, for instance direct unfolding of protein from the protein–ligand complex (see Figure 9.23(d)). Reactions containing sequential steps will also show broadened transitions. For an excellent review of this topic see the reference [86], from which this section has been adapted.

9.5.2.1 Van't Hoff enthalpy

As mentioned previously, in addition to the directly determined calorimetric enthalpy, it is possible to define another enthalpy based upon the shape of the peak. Using the equation

$$\Delta G = \mathrm{RT}\ \ln\{K\} = \Delta H - T\Delta S$$

It can be shown that $d(\ln K)/d(1/T) = -\Delta H_{\mathrm{VH}}/R$.

Therefore a plot of $\ln K$ vs $1/T$ gives a line whose slope is the van't Hoff enthalpy ΔH_{VH} divided by R. An estimate of K at any temperature can be obtained by the partial area ratio, as shown in Figure 9.23(a). The comparison of ΔH_{VH} with ΔH can be instructive. A broader transition due, for example, to the unfolding transition not being two state leads to ΔH_{VH} being less than the calorimetric value. Similarly if the protein unfolds cooperatively as a dimer, the transition will be sharper and ΔH_{VH} will be greater than that measured by calorimetry. Therefore this method can determine the so-called cooperativity or the cooperative unit of the system.

9.5.2.2 Why not in foods?

There are many reasons as to why this method of analysis is comparatively unusual in foods. Firstly, in general we are not dealing with well-defined proteins or indeed binding ligands. Food proteins tend to be a mix of many species with few truly molecular structures available. Secondly, the industrial interest usually lies in gelling or aggregating conditions, which are irreversible and kinetically determined unlike the previous equilibrium considerations. Aggregation models are more complicated and in any case questionable for ill-defined mixtures. This is similar to the case of isothermal titration calorimetry (see Section 9.8), which is yet more demanding in terms of the reacting species.

Nevertheless with increasing availability and purity of food materials, research in this area may see substantial development.

9.5.3 Aggregation

In addition to denaturation, which is considered to be an endothermic and reversible process, protein molecules present at higher concentrations or subject to more severe treatment, which normally corresponds to a more realistic situation, will experience aggregation after the unfolding or denaturation reaction occurs. This is generally considered to be an exothermic and irreversible process. The overall enthalpies of protein 'denaturation' therefore become more difficult to interpret, as they represent the sum of positive or endothermic denaturation and negative or exothermic aggregation phenomena. However, it is generally assumed that the energies involved in aggregation are low [87] and therefore the denaturation analysis is still applied to the treatment of DSC data.

It has also been proposed that as aggregation may be significantly slower, that is a kinetic event, whilst the denaturation event will be essentially reversible and in equilibrium, it may be possible to separate these by either using a modulated (reversing/non-reversing) approach or simply increasing the scanning rate (see Section 2.3.3). Some evidence has been obtained for this in whey protein isolate by Morris [88] who observed two events; endo- and exothermic showing a separation which increased as the scan rate was raised. There have also been reports of a similar phenomenon in the pea storage protein vicilin [89]. The separation could be controlled in the case of vicilin by changes in the salt concentration and pH value. One gram per litre of NaCl and 0.05 M NaH_2PO_4 at pH7 proved particularly effective conditions for separating the apparent denaturation and aggregation events (Figure 9.24). Solubility studies often by turbidimetry are also extremely useful in quantifying the irreversible aggregation. By varying the solvent quality so, the nature of the aggregation can be probed. For instance

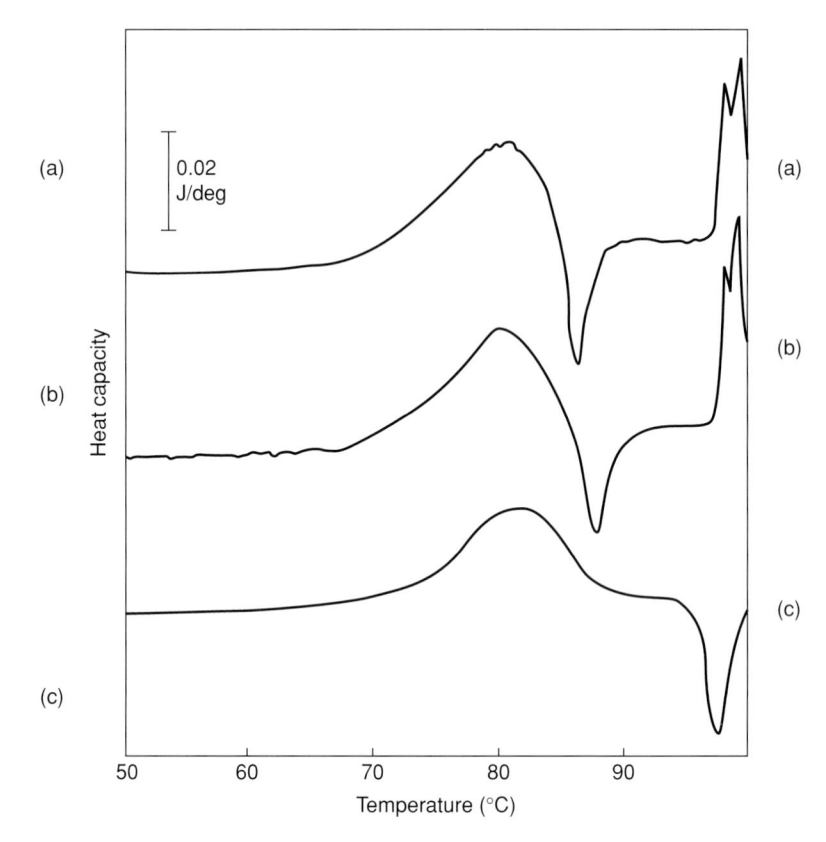

Figure 9.24 The separation of denaturation and aggregation peaks in the pea protein vicilin: 15 g/L vicilin solutions in (a) 3 g/L NaCl 0.05M NaH$_2$PO$_4$ pH 8.5, (b) 2 g/L NaCl 0.05M NaH$_2$PO$_4$ pH 8 and (c) 1 g/L NaCl 0.05M NaH$_2$PO$_4$ pH 7. (Taken from [89]. Copyright Society of Chemical Industry. Reproduced with permission. Permission is granted by John Wiley & Sons Ltd on behalf of the SCI.)

the incorporation of mercaptoethanol into the solvent mix will break disulphide bridges. If aggregates have been produced by complex chemical reactions, for example via the Maillard reaction, then the breaking of all the common bonds by appropriate solvents will still leave insoluble aggregates behind.

The gelling of some proteins, for instance when frying an egg, that is the complete aggregation of albumin, is caused by a multitude of chemical reactions: hydrogen bonding, hydrophobic association, sulphide bridge formation and entanglements. Morris has made the interesting but controversial claim that as the examples of separation of denaturation and aggregation seem to be observed more easily when using large cells, 1 g cells in the case of the Setaram micro-DSCIII, the observation could be aided by the greater time taken for the heat liberated to pass to the recording system of the calorimeter via the large mass of the sample. Microcalorimetry is often used for this type of study due to the requirements for high sensitivity. It is important to point out that aggregation does not necessarily imply that a gel will be formed.

It is interesting to note that for the pea storage protein, other thermal transitions can be induced, by changing the ionic conditions. Transitions such as these should be investigated thoroughly before deciding on exotic explanations, such as ageing peaks. However additional peaks have been observed in proteins, such as gelatin and albumin, which seem to be described well by recourse to ageing phenomena. The simple observation that the recovery of enthalpy in these peaks is linearly related to the logarithm of the ageing time is purported to be evidence for this [90].

9.5.4 Glass transition in proteins

Glass transitions in proteins are more difficult to detect than those in carbohydrates. This is principally due to the hydrophilic–hydrophobic balance of the chemical moieties in the protein and the consequence that only a fraction of the protein can mix uniformly with water [91] and so contribute to ΔCp. The small changes in ΔCp at the glass transition (see Table 9.3) also reflect the limited mobility in the protein with tight conformations and multiple tie points, which restrict the absolute degrees of freedom and the change in ΔCp.

9.5.5 Ageing in proteins

A DSC trace for gelatin, which contains both amorphous and crystalline regions, is shown in Figure 9.25. The first step G is thought to represent glassy behaviour of the amorphous part

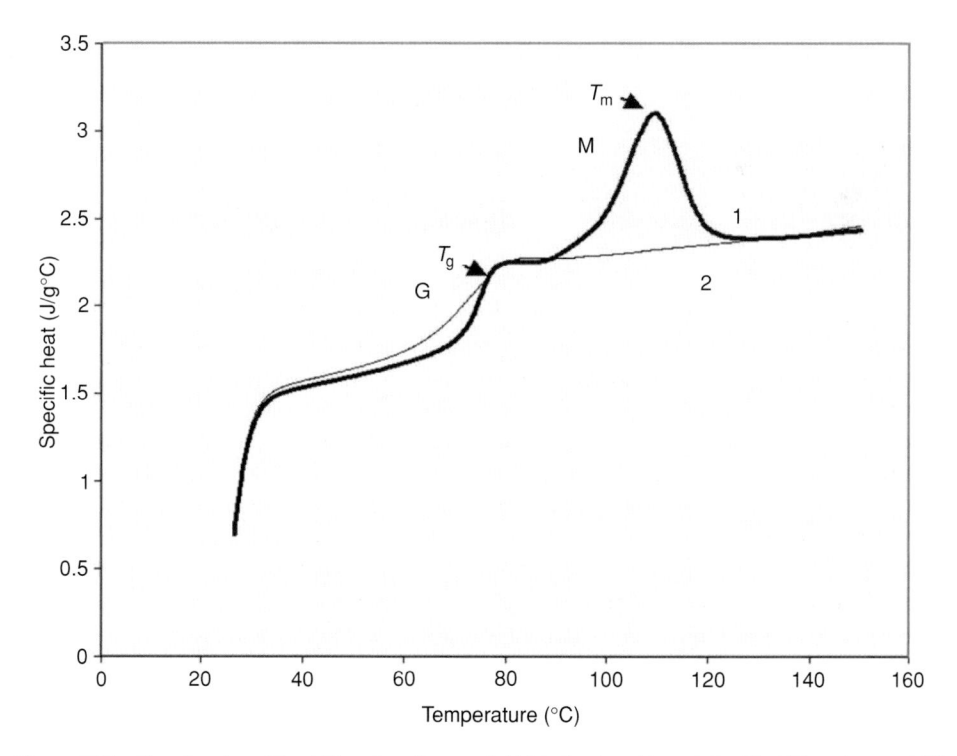

Figure 9.25 The glass transition (G) and melting peak (M) in bovine skin gelatin. First (1) and second (2) reheats are shown.

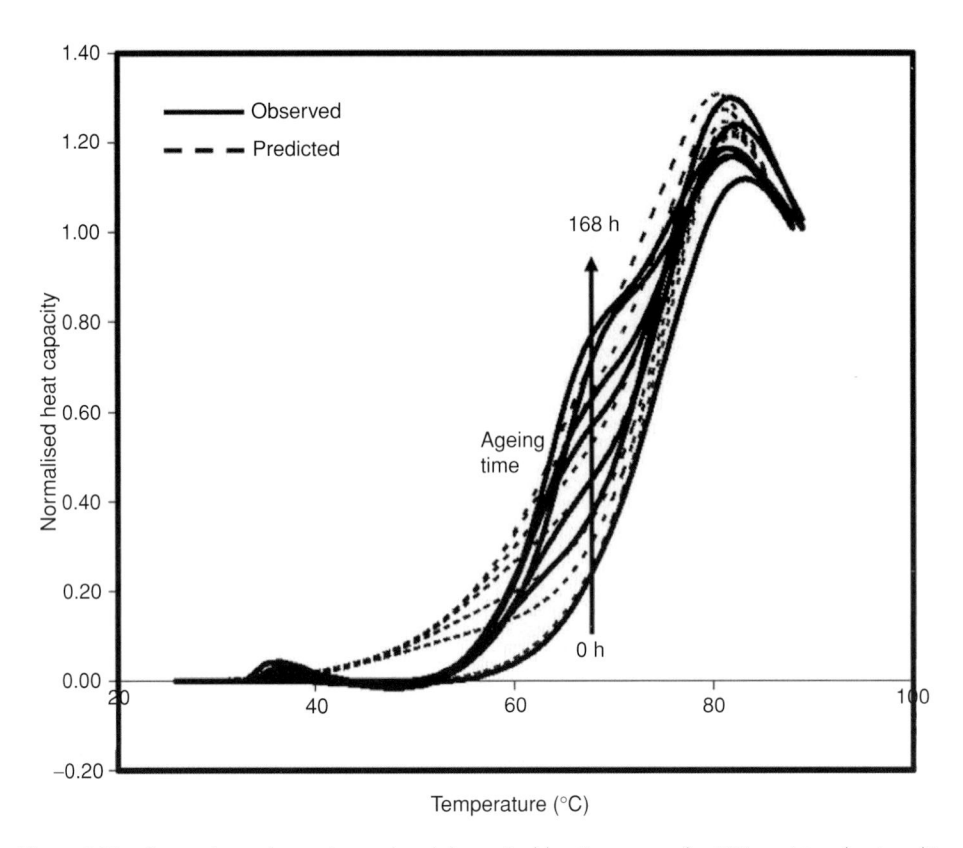

Figure 9.26 Comparison of experimental and theoretical heating curves for 11% moisture bovine skin native gelatin aged for various times. These were 0, 12, 24, 48, 72 128 and 168 h. The sample was cooled from 100 to 25°C at a rate of 10°C/min, held for the appropriate time and subsequently heated at 10°C/min back to 100°C. The heating stage was recorded. TNM model parameters were $\Delta H = 69.69$ KJ/mol, $x = 0.661$, $\beta = 0.541$ and $A = 3.50E\text{-}43$ s.

of the matrix, whilst M is due to the melting of crystalline regions. Ageing experiments can therefore be carried out on gelatin by cooling from a temperature above T_g but below T_m, which gives a partially crystalline protein, or from a temperature above T_m, which results in a completely amorphous material [90]. Figure 9.26 shows the results of ageing or annealing which simply means holding the material at a fixed temperature for times of up to 1 week.

The TNM ageing model (see Section 1.5.6) appears to describe this tolerably well. Using the optimal set of parameters (x, β, A and ΔH) for this sample (native or partially crystalline gelatin at 10% moisture), we can test the model further, by predicting the effect of ageing at different temperatures. Similarly, we can mimic the effect of moisture by changing the parameter A and holding all other parameters constant. This has the effect of moving the transition up and down on the temperature scale similar to the effect of moisture. The reader is referred to reference [92] for more details.

Whilst not quantitatively accurate, the experimental data are described qualitatively, which must be viewed as at least a partial success, in view of the complexity of the mixtures and the occasional difficulty of observing any glass transition at all in proteins. Similar results have been obtained for albumin, a globular protein [93].

9.5.6 Gelatin in a high-sugar environment

In contrast with the drop in enthalpy of the coil/helix transition in polysaccharides as the surrounding sugar concentration is increased, the enthalpy associated with the sol/gel transition in gelatin increases [94]. This is concomitant with the increase in gel strength [95] and is thought to arise due to the incompatibility and phase separation of the protein and carbohydrate components. These results are of particular interest in confectionery products, involving the use of gelatin as a gum agent in high concentrations of sugars, wine gums, gummy bears, etc.

9.5.7 Mechanical properties of proteins

The mechanical properties of gelatin have been measured by preparing films, in this case 0.7 mm thick, and using single-cantilever bending mode geometry in a DMA to determine the flexural modulus E' [96]. It was found that the modulus increased with ageing time. By extracting the shift factors required to superimpose the ageing curves on to one master curve and plotting these against the logarithm of the ageing time, a correlation with the growth in enthalpy of ageing as determined by DSC (Figure 9.27) was obtained. This is significant, because the measurement of enthalpy on the DSC is considerably easier than the measurement of mechanical properties; however, it is the mechanical properties that are usually of interest from a processing point of view. For example, the proper function of gelatin capsules is related to the stiffness and brittleness of the material.

9.5.8 Foods

It has been said that the most studied component of foods using DSC is protein [97]. The following are just a few examples from the enormous range of proteins from various sources, which have been studied. In general, denaturation studies are carried out in solutions, whereas glassy behaviour is observed in lower moisture materials.

9.5.8.1 Proteins from animal sources

Much work has been reported on dried and whole fish. T_g's in frozen fish such as cod, tuna, mackerel, sea bream and bonita have been determined using DSC [98]. This work is normally carried out with a view to determining the safe storage conditions for these products. Both dried fish muscle and extracted protein fractions showed glass transitions. The T_g's of muscle and myofibrillar proteins from red-muscle fish tended to be lower than those from white-muscle fish. There was no difference in the T_g of sarcoplasmic reticulum. The myofibrillar proteins are responsible for the contractile mechanisms in muscle. These are predominantly

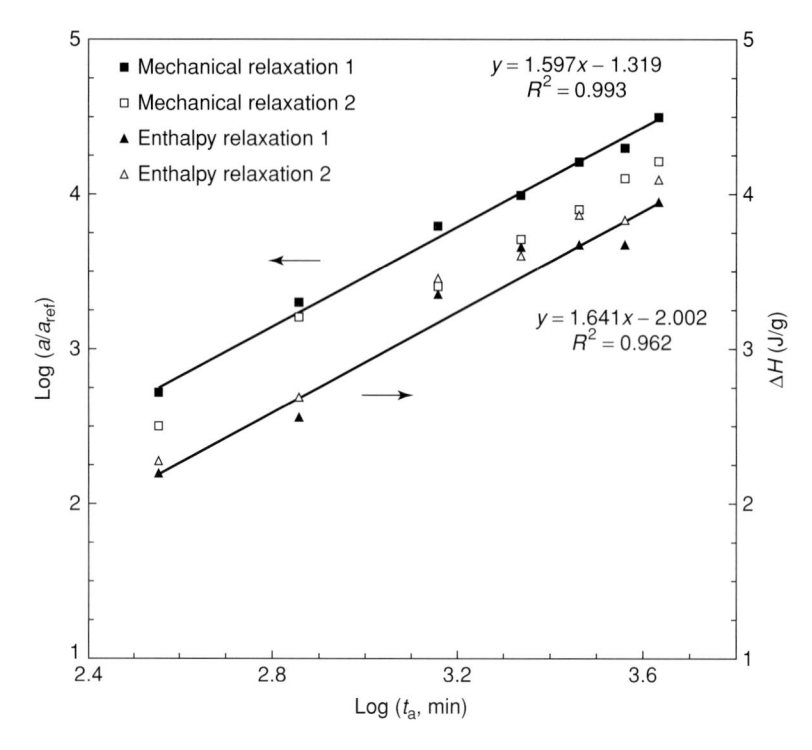

Figure 9.27 A comparison of mechanical and enthalpic relaxations for a gelatin film containing 17% moisture. Two replicate measurements (labelled (1) and (2)) were performed and are shown as solid and open symbols. Solid lines are linear fits to replicate 1. (Reprinted from [96], Copyright (2006), with permission from Elsevier.)

myosin and actin, typically 40 and 20%, respectively, of the overall myofibrillar fraction. Minor components compose the rest. The sarcoplasmic proteins are primarily enzymes and myoglobin. As expected the T_g of whole muscle was lower than extracted fractions due to the effects of low-molecular-weight materials in the muscles.

Significant thermal denaturation in cod fillets in the region of 30–80°C was found to be a function of storage conditions, with differences being found between long-term storage at −30°C and subsequent short-time storage at +2°C [99].

In a separate study [100], cod and tuna were found to have transitions at −10, −15 and −21°C, with an additional low-temperature transition at −72°C for tuna. These transitions were related to glassy behaviour; however, the storage stability at −18°C was found to be poor, suggesting that these were not the transitions governing the deterioration process.

The effect of salting herring has also been reported to increase the thermal stability of proteins but decrease the enthalpy of transition, possibly reflecting an increase in the proteolytic activity in the muscle [101]. Increased stability of proteins in this type of work is equated with an increase in transition onset temperature.

DSC endotherms of ground beef show three endotherms (Figure 9.28), which are thought to reflect the principal components: myosin, sarcoplasmic proteins and actin [102, 103].

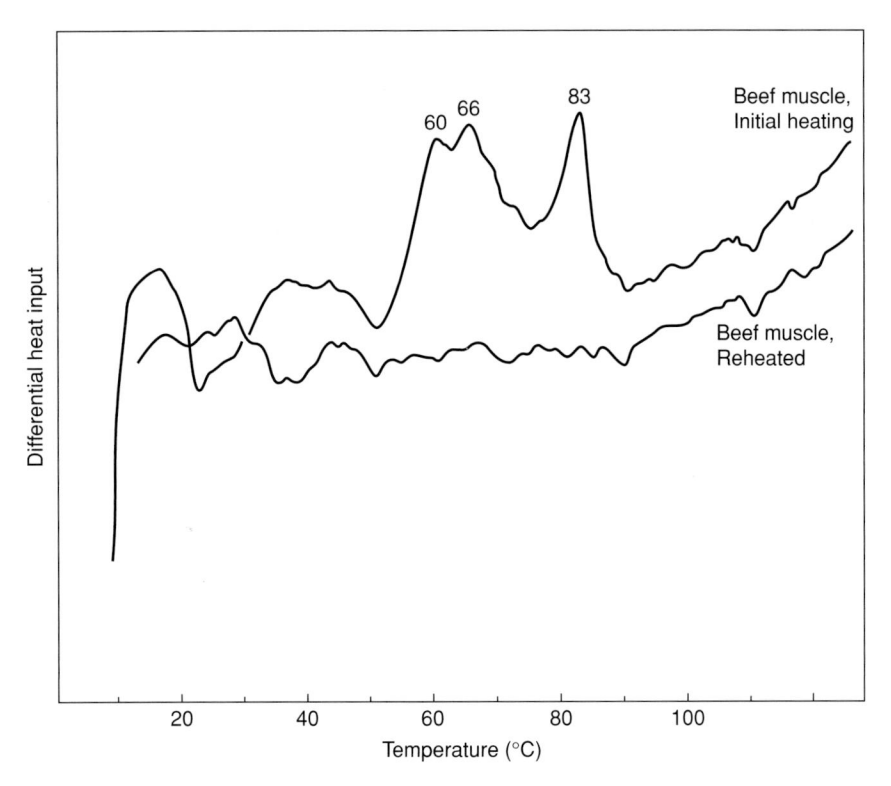

Figure 9.28 Three endotherms recorded on the first heating run of ground beef muscle. (Taken from [103]. Copyright Society of Chemical Industry. Reproduced with permission. Permission is granted by John Wiley & Sons Ltd on behalf of the SCI.)

It is also found that salt at the concentrations used in processed meats will reduce the temperature and often the enthalpy of the endotherm, indicating a decrease in the heat stability.

In the study of egg white and its components [104], it was found that thermostability was in the order ovalbumin > lysozyme > conalbumin and that lowering the pH from 9 to 7, adding Al^{3+} ions or sucrose increased the stability of these proteins as indicated by shifts of the endotherms to higher temperatures. Heat treatment of myoglobin at low moisture content suggested that only part of the protein was undergoing denaturation, resulting in apparent increases in heat stability and decrease in solubility [105].

DSC can be used to monitor the irreversible transformation of ovalbumin to S-ovalbumin [106] by monitoring the endotherm, the temperature of which increases by 8°C. This transformation is thought to be behind the poor performance of egg whites in angel cake formulations, due to decreased solubility and the lower likelihood of the baking process reaching a sufficient temperature to gel the protein.

9.5.8.2 Botanical source proteins

The thermal properties of wheat proteins have been extensively studied. DSC of wheat protein at low water content shows multiple endotherms. These have been identified as

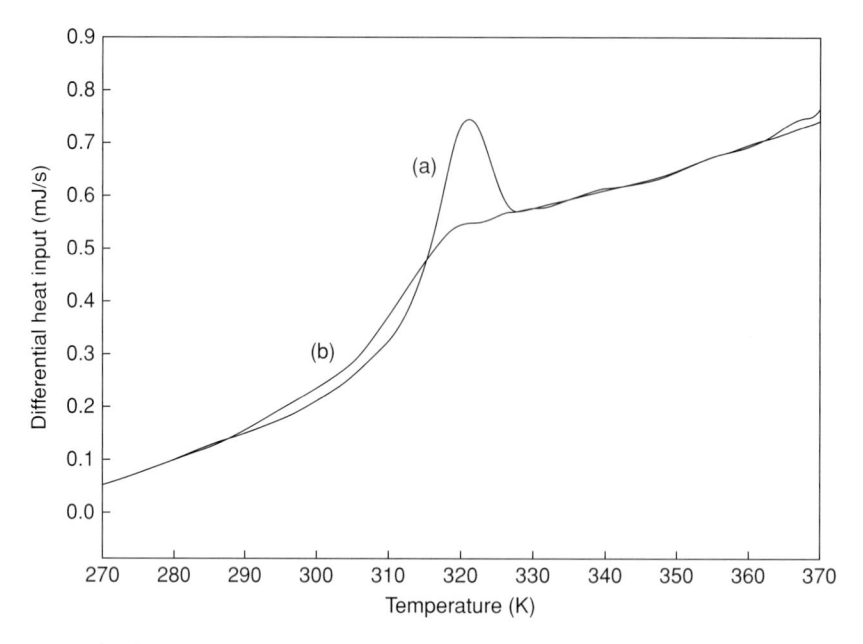

Figure 9.29 The glass transition temperature and ageing peak for a 12% moisture content ω-gliadin stored at 10 K below the glass transition temperature. (Reprinted from [108], Copyright (1999), with permission from Elsevier.)

being due to albumins and globulins at 50°C, gliadins at 58°C and glutenins at 64 and 84°C. The thermal transitions for glutenins were found to be reversible [79].

Denaturation temperatures are moisture dependent. At low moisture contents denaturation peaks will have high onset temperatures. For instance, the denaturation of sunflower globulins at 0% moisture occurs at 190°C, decreasing to 120°C as the moisture content increases to 30%, with the corresponding enthalpy of transition almost doubling [107].

Noel and co-workers [108] have analysed well-defined fractions of gliadins and glutenins and found the T_g of perfectly dry samples by extrapolation to be between 397 and 418 K (Figure 9.29).

Gluten, the protein obtained from wheat, comprises a series of protein fractions, such as the gliadins and glutenins, mentioned above. It is thought to contribute significantly to the continuous structure in bread, the strength of which reflects the interactions between the various components of the protein. Quality factors of the wheat, such as hardness, are related to protein content, which can improve the properties of the structure and increase the loaf volume by stabilising the matrix. Small-deformation DMA measurements have been carried out on a series of wheat gluten powders equilibrated with water, glycerol and sorbitol [109]. DMA is particularly suited to this work because of the increased sensitivity to the detection of glass transitions, which may have small values of ΔCp. It was found that the gluten was plasticised (i.e. the T_g reduced in a fashion well described by the Couchman–Karasz equation), with water being the most effective on a per gram basis. However, all plasticisers were equally effective on a per mole basis.

Wheat gluten has also been found to decrease the rate of bread staling as monitored by the return of the enthalpy of gelatinisation of the amylopectin fraction of starch [110]. This effect was thought to occur principally via the dilution of the starch component.

Studies on the 7S and 10S globulins, which are the two main storage proteins in soybean, have shown interesting structure-forming behaviour as the samples were heated [111]. An order of magnitude increase in the elastic shear modulus between temperatures of 70 and 150°C at moisture contents between 40 and 20% for both fractions corresponded in temperature with denaturation endotherms recorded using DSC, although the 10S fractions showed the transitions at higher temperature, thought to be related to the increased amount of energy required to break the disulphide bonds which stabilise this structure. Interestingly, for these proteins the glassy behaviour, albeit with some rather small changes in the measured heat capacity, was also recorded [91]. This enabled fairly comprehensive state diagrams to be constructed, showing glassy, rubbery, entangled polymer flow and reaction regions. This type of structure formation on heating may be a property of a class of confectionery cream-filling products, which appear to set upon heating and for which a series of changes can be detected by various physical methods including endotherms by DSC.

9.5.9 TA applied to other areas

There is interest in the use of biopolymers to replace plastics in packaging for reasons of biodegradability and that these are renewable resources [112]. Thermoplastic properties of wheat gluten have been studied by the application of DMA and DSC to gluten films. Strong, transparent and water-resistant films can be obtained by casting films from ethanol/acetic, acid/water solvents. Moreover, the water-vapour barrier properties could be improved by the incorporation of lipids. Similarly, the manipulation of the different fractions of gluten could also produce distinct characteristics.

In a model flavour-release system, 2-octanone could be released from BSA–pectin complexes by a shift in pH [113]. This was due to BSA unfolding on the oppositely charged, rigid polysaccharide matrix at a pH below the isoelectric point of BSA. DSC was used to examine the effect of flavour binding on the conformational stability of the protein.

9.5.10 Interactions with polysaccharides and other materials

An area, which has seen much activity, is the interaction between polysaccharides and proteins. Studies on the stabilisation of proteins by sugars against aggregation and chemical degradation have been carried out, designed to discriminate between stabilisation predominantly controlled by glass dynamics, and that controlled by specific interactions between sugars and proteins [114]. Calorimetric data suggested that the stabilisation by sugars such as sucrose and trehalose of the protein could not be explained solely in terms of the glass transition. The retention of native secondary structure, as monitored by Fourier transform infrared, was found to be a good indicator of storage stability; however, the determination of the dominant mechanism appeared inconclusive.

Table 9.6 Sources of polysaccharide hydrocolloids

Source		Hydrocolloid
Seaweed	Brown	Agar and carrageenan
	Red	Alginate
Botanical	Trees	Cellulose and hemicellulose
	Tree exudates	Gums arabic/karraya/tragacanth
	Plants	Starch cellulose pectin
	Seed pods	Guar/locust bean/tara gum
	Tuber	Konjac mannan
Microbial		Xanthan, gellan gum, dextran, cellulose and curdlan
Animal		Chitin, heparin, hyaluronan and glycogen

The presence of small amounts of soluble wheat proteins in a mixture with salt-soluble meat proteins (SSMP) was capable of trebling the elastic modulus, compared with a pure SSMP gel [115]. This study highlights the inadequacy of DSC as a monitor of protein gelation properties. Despite the fact that no transitions were observed over 20–120°C by DSC, a weak gel was formed. Further, strong elastic gels were formed upon more severe treatment at 120°C.

9.6 Hydrocolloids

9.6.1 Definitions

A hydrocolloid can be defined as a hydrophilic polymer, which when added to aqueous matrices controls the functional properties [116]. They are frequently but not exclusively polysaccharides. Polysaccharides appear to have more versatility in their mode of action than that of other polymers such as proteins; for instance they are more effective thickeners and arguably prevent ice and sugar crystallisation more effectively than proteins. In the pharmaceutical, personal care and food industries, hydrocolloids are used extensively and serve roughly the same functions in each area. They can be used as thickening and gelling agents, can have a direct or indirect effect on the stability of emulsions and can exhibit useful rheological properties, such as shear thinning which facilitates the application or spreading of these materials on surfaces. Examples of these behaviours are xanthan gum used as a thickener in sauces and which shows dramatic shear thinning, κ-carrageenan as a gelling agent in pet foods, and gum arabic as a stabiliser of emulsions.

Polysaccharide hydrocolloids originate from a variety of sources (Table 9.6) [117], but all have a similar method of network formation in solution, despite major differences in anhydrous structure and bond formation. Interestingly, the one protein which appears to have a similar mode of gel formation is the fibrillar protein gelatin, which has already been dealt with extensively in the previous section and which frequently acts in a similar way to polysaccharides.

9.6.2 Structures in solution

In many polysaccharide thickening and gelling agents, a prerequisite to gel formation involves the transition from a well-defined structure, in the original dry powder, to filaments finely distributed throughout the material followed by tying together at linkage points. The linkage points are associations of strands of the carbohydrate polymer and can form in a variety of ways. They can be by regions of helical intertwining as in xanthan and carrageenan. This is also the case for gelatin, which results in a similar gel structure to many polysaccharides. They can also be formed by ions either bridging structures or shielding ionic charges on the polymer itself. Frequently, the links themselves will subsequently aggregate to form large conglomerates, which can then become visible by electron microscopy. Models have been proposed [76, 118] which claim to predict properties of the linkage points from shear measurements, which in turn form the basis of the prediction of heat capacity on DSC traces. The properties are based on the number of parallel links, N, which come together at knotting points. Further, it is assumed that each link can take on G orientations when it is open, that is not tied to the other links.

As the starting material, or the conglomerates in the case of pre-formed gels, are heated, the breakdown of structure is accompanied by an endothermic melting peak. When the material is cooled, the gel will set by reformation of the links, which will exhibit an exothermic heat change (see Figure 9.30).

These heat changes (endothermic upon heating) are thought to be due to the molecular disassociation of the links, for example multiple helices to random coils coupled with ill-defined contributions from disordering of the structures.

This is observed in solutions of xanthan, carrageenan, agar, cellulose and pectin and in mixtures of hydrocolloids such as those of xanthan and konjac mannan (KM) or xanthan and locust bean gum (LBG) [119]. Figure 9.30 gives an idea of the complexity of behaviour. Xanthan tends to form viscous solutions on its own but gels in combination with KM and LBG. The effect of ions on the temperature of transition is thought to indicate the formation of different conformations of the polysaccharide chain in response to shielding of charges present on the backbone. However the double peaks observed on traces (d) and (f) (Figure 9.30) are thought to show a separation of the order/disorder transition in the xanthan and the gelling behaviour (lower temperature transition). This may also imply that KM and LBG can interact with ordered conformations of xanthan. Clearly, thermal information can be informative but is incomplete without complementary rheological data on the gels which are formed.

Certain polysaccharides, such as some substituted celluloses, will gel on heating and 'melt' on cooling, and some pectin gels are thermally irreversible. There is a bewildering variety of behaviour and it is comparatively easy to envisage how such behaviour can be put to use in many food products. In addition, celluloses can be modified in particular ways to give a range of properties, from the viscous solutions of carboxymethyl cellulose and hydroxymethyl cellulose to the gelation on heating of methyl and hydroxypropylmethyl cellulose. These gels are thermoreversible and as already mentioned 'melt' on cooling. The interaction here is thought to be primarily through hydrophobic effects.

One proposal [120] is that upon heating, initially ordered cellulosic structures will start to unravel, exposing hydrophobic groups to water and so create entropically unfavourable "water cages" around the groups. This is presumably driven by other interactions of the

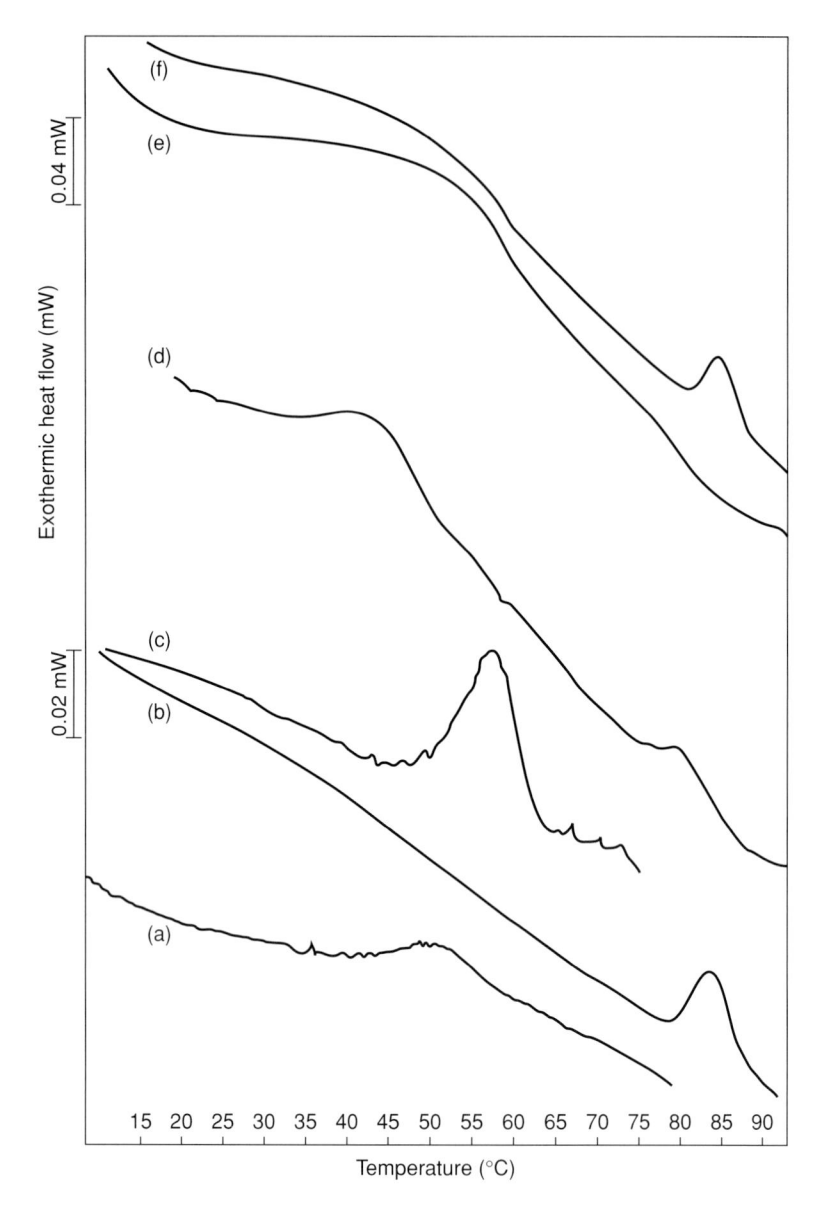

Figure 9.30 Thermal cooling traces showing complex behaviour observed in hydrocolloids and hydrocolloid mixtures: (a) 1.2% xanthan (X) in water; (b) 1.2% X in 0.04M NaCl; (c) 0.6% X/0.6% KM in water; (d) 0.6% X/0.6% KM in 0.04M NaCl; (e) 1%X/1% LBG in water and (f) 1%X/1% LBG in 0.04M NaCl. (Reprinted from [119], Copyright (1998), with permission from Elsevier.)

molecule and solvent. As heating is continued a new structure is formed involving the dissolution of the water cages and the association of hydrophobic groups, producing the gel network. The surprising fact that the enthalpy of the hydrocarbon water interaction is actually lower than the hydrocarbon hydrocarbon interaction produces, once again, an endothermic transition on heating from the solution to the gel. Of course, the complete transition will contain contributions from both these events. Interestingly, upon cooling, the events, namely the reformation of water cages and the structural re-association, may separate, producing multiple peaks on the DSC cooling trace. This complexity is in stark contrast to the simple picture given previously for conventional polysaccharide gelation where the endothermic peak represents the energy required to disrupt linkage points.

Whilst guar gum will dissolve in cold water and form viscous solutions, in combination with borate ions it will form gels. The ions are thought to interact directly or via hydrogen bonds with the OH groups on the guar backbone and so lead to cross-linking between strands.

9.6.3 Solvent effects

The heat changes accompanying these processes can be measured using DSC. Often microcalorimeters are used for increased sensitivity, as there can be a comparatively small amount of polymer, sometimes ∼1%, relative to the water in the gel. Rheology is once again an example of a technique, which is essential to the understanding of the behaviour of hydrocolloid gelation [76]. However, here we concentrate on thermal properties, incomplete as this approach undoubtedly is. The important point, from a TA perspective, is that the whole gelation process is accompanied by heat changes which relate to the structures formed in a typical gel.

If such a structure is heated then an enthalpy change can be measured at the gel/sol transition temperature, and consequently the material is said to have an enthalpic structure [121]. The enthalpy changes for agarose and carrageenan can be seen in Figure 9.31 [122].

Supporting evidence for these ideas is found in electron micrographs for gellan under conditions of gel formation [123]. For example on Figure 9.32(a), the resolution is just sufficient to show characteristic long bundles where polymer chains of gellan are grouped together. Of particular interest to the confectionery industry is the effect of high concentrations of sugars on this behaviour. At high concentrations of sugar the size of the enthalpic peak decreases (Figure 9.31). It also broadens and the peak temperature can change. This is thought to represent competition by the sugars for the 'carbohydrate' bonds between polymers. As the sugar concentration increases, characteristic changes are also seen in the rheology, as well as the enthalpy. At concentrations of about 80% the gel appears weakened, as evidenced by a drop in the elastic shear modulus G' [124]. At still higher concentrations, properties begin to be dominated by the sugar.

The result of gelation at high-sugar concentration appears to be a so-called entropic network; the description reflecting the fact that the enthalpy of transition is reduced and the carbohydrate strands are more finely distributed throughout the material and in fact cannot now be visualised easily with transmission electron microscopy (TEM) (Figure 9.32(b)). The broadening and changes in temperature of the transition can also be accommodated in this scheme. For instance there could be a distribution of stability of the links, reflected in a broader endotherm, and changes in temperature.

9.6.4 The rheological T_g

One important result from this work concerns the estimation of a glass transition temperature from conventional rheological or DMA data. It has been reported that by using

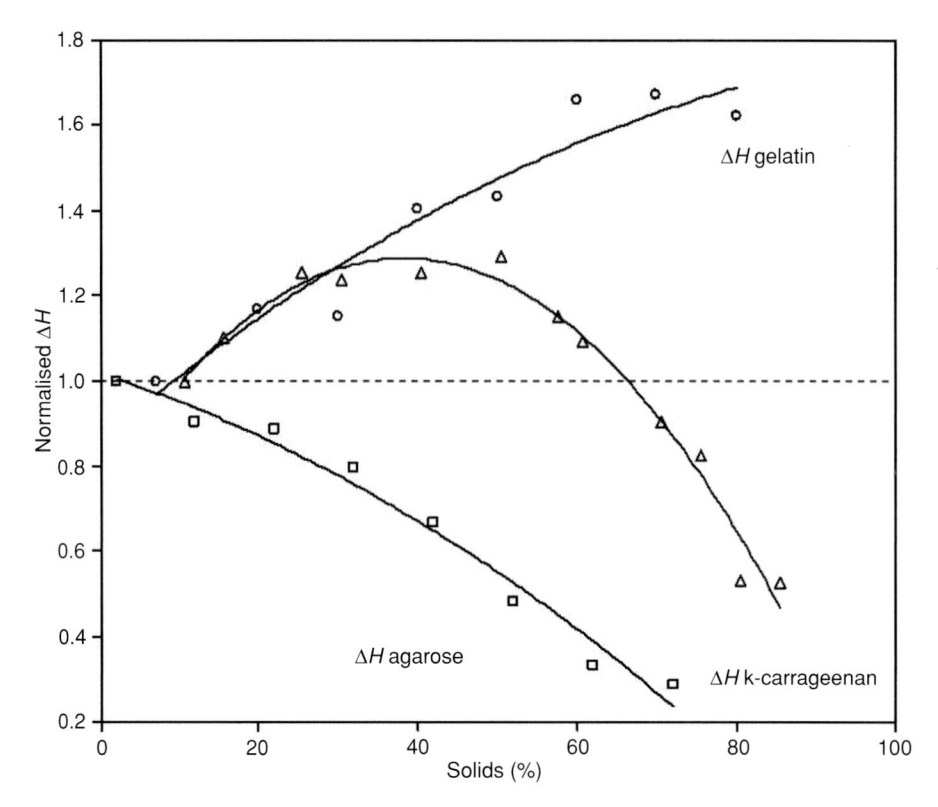

Figure 9.31 Normalised peak enthalpy values as a function of sugar content for agarose, κ-carrageenan and gelatin. The reduction in enthalpy for the polysaccharides is evidence for the ability of sugar to act as a substitute in bonding between polysaccharide units, whereas the increased enthalpy for gelatin is indicative of phase separation. (Taken from [122], with permission from American Chemical Society.)

processing previously restricted to pure polymers [125], it was possible to derive a T_g and compare this with a conventional measurement by DSC. The essence of the rheological measurement is to superimpose data obtained as a function of time (frequency) but at different temperatures, until a master curve is obtained. The master curve is so-called because in conjunction with a curve showing either the modulus against temperature or the shift factors (a_T) against $T - T_{ref}$, the response of the polymer under any conditions of time and temperature can be calculated [126]. This procedure has been carried out for a variety of hydrocolloids and gelatin in the presence of sugars. By fitting the Williams–Landau–Ferry (WLF) equation to a plot of the shift factors against $T - T_{ref}$ values, the rheological glass transition temperature, T_g, and the limiting underlying transition, T_∞, can be obtained (see Chapter 4 for more details).

9.6.5 Glassy behaviour in hydrocolloid/high-sugar systems

For certain network-forming hydrocolloids such as κ-carrageenan in high concentrations of sugar, the rheological T_g appears to be higher than the predictions of the Couchman–Karasz

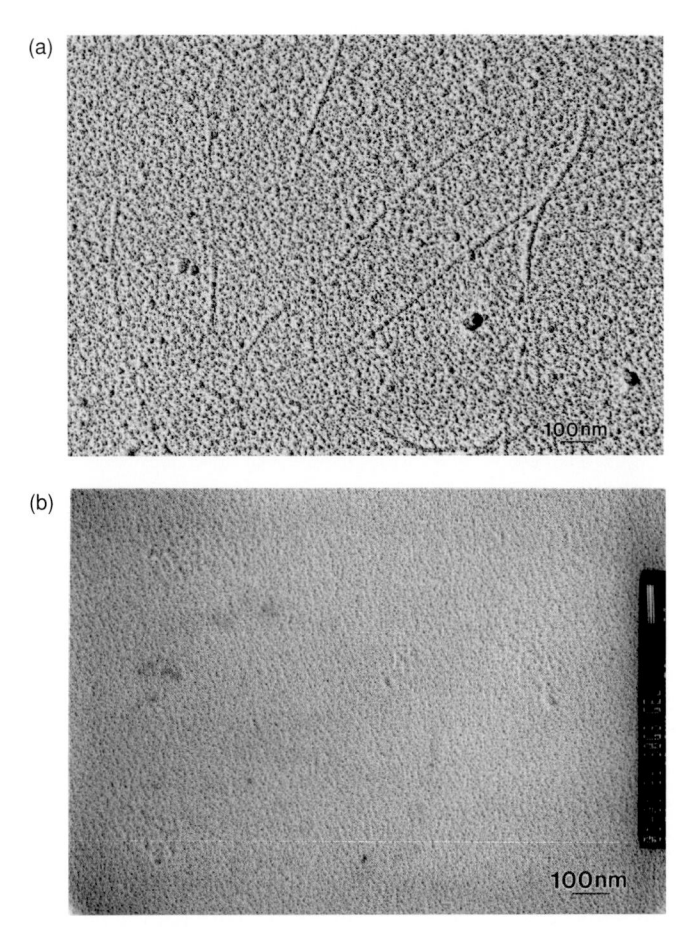

Figure 9.32 Transmission electron micrograph of (a) 0.7% gellan with 10-mN Ca^{2+} in an aqueous environment and (b) 0.7% gellan with 10-mN Ca^{2+} in the presence of 80% glucose syrup. The filamentous appearance of trace (a) is evidence for a multi-stranded or enthalpic structure. (Reprinted from [123], Copyright (2004), with permission from Elsevier.)

equation when applied to a mixture of all the components [127]. However, measurements using DSC showed no difference between the sugars and the sugars containing the hydrocolloids (Figure 9.33).

The DSC-determined T_g appears to be an average of all components, in line with the Couchman–Karasz equation, whilst the rheological T_g measures principally the properties of the polymer. It is also important to mention that DSC plots of heat capacity are normally on a linear scale, whilst rheological plots of modulus are on a logarithmic scale, which has the effect of magnifying small changes on the rheological trace and changing the shape of the curve. It can be seen in Figure 9.34 that for a glucose syrup containing pectin with a high degree of esterification, the changes in modulus occur at a much higher temperature than for either the glucose syrup alone or containing a pectin with low degree of esterification. This shows that minor changes in chemical composition as well as network formation can

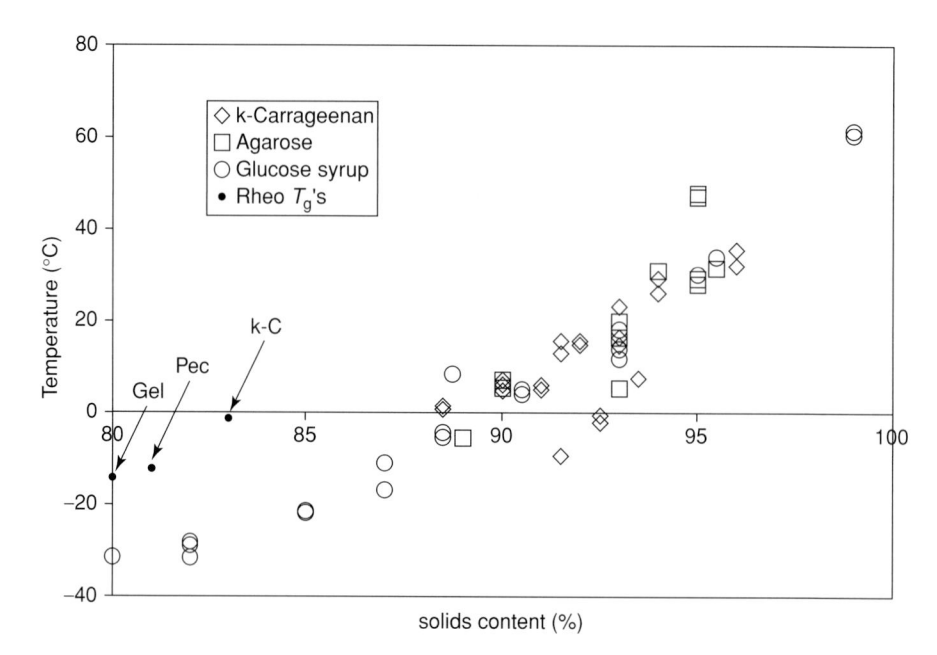

Figure 9.33 A comparison of T_g's measured by (1) DSC: ○ glucose syrup, □ agarose, ◇ κ-carrageenan and (2) Rheology: •, extreme T_g values for gelatin, pectin and κ-carrageenan are arrowed. Typical concentrations of hydrocolloids were 0.5–1.5% and for gelatin 25% the balance being made up of sugar. The rheological T_g for some hydrocolloid-containing mixtures appears to have a higher value than that determined by DSC. (Reprinted from [124], Copyright (2004), with permission from Elsevier.)

have dramatic effects on rheological properties and indeed on the apparent rheological T_g [128].

It is possible to select various temperature points on a mechanical trace as representative of a glass transition, for instance, the initial drop in modulus, the peak in loss modulus, the peak in tan δ, etc. Table 9.7 shows a range of temperature points on mechanical traces, which can be chosen as an estimate of T_g and the comparison with the DSC mid-point. The best correlation seems to be between the drop in value of the storage modulus and the DSC

Table 9.7 A comparison of DMA and DSC (mid-point) determined T_g's for various hydrocolloid/sugar mixtures

Recipe	Solids (%)	DSC	E' slope	E'' max	Tan δ
(1) i-Carrageenan/gellan	84	−33	−30	−28	−7
(2) Gelatin	80	−40	−36	−33	−12
(3) Starch	87	−33	−30	−26	−10
(4) Gellan	89	−33	−30	−26	−8
(5) Pectin	86	−33	−30	−26	−7
(6) i-Carrageenan	82	−33	−33	−30	−10
(7) i/k-Carrageenan mixture	85	−33	−34	−30	−11
(8) Starch/gelatin	84	−30	−26	−21	+3

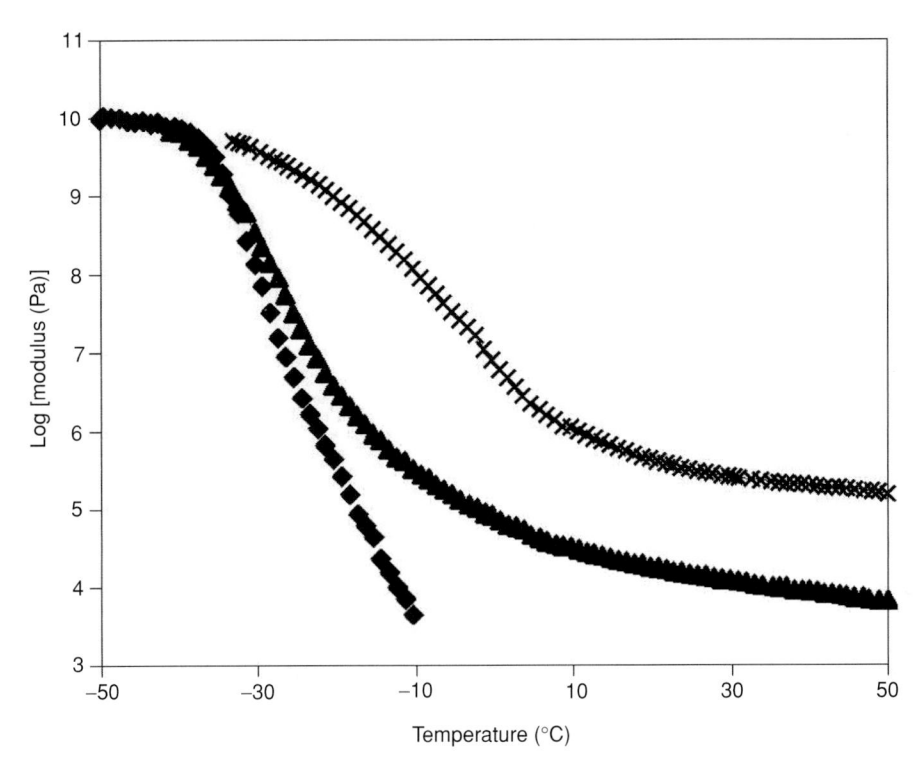

Figure 9.34 Temperature response of the storage modulus measured at 1 rad/s for high-solid pectin. Top trace: 1% pectin (DE 92)+40% sucrose+40% glucose syrup; middle trace: 1.32% pectin (DE 22)+39.8% sucrose+39.8% glucose syrup and bottom trace: 40.5% sucrose+40.5% glucose syrup. Small amounts of hydrocolloid can have large effects on the rheological properties. (Reprinted from [128], Copyright (1996), with permission from Elsevier.)

mid-point value. It is instructive here also to note the comments concerning the linear plot of heat capacity and logarithmic plot of rheological data.

In the comparison of DSC and rheological data, time-scales are crucial. The underlying rheological glass transition T_∞ is thought to be about 50°C below a T_g measured by a slow technique, such as dilatometry [126], and supposedly represents a true T_g at effectively zero frequency or infinite time of measurement. The Donth equation [129] gives a frequency to a DSC measurement based on the width of transition and the rate of the scan.

$$f = a\,\beta_0/2\pi\Delta T$$

where a is a constant (≈ 1), β_0 the scanning rate and ΔT the half width of transition.

In the mode coupling theory, the increase in T_g with increased scanning rate is due to available energy modes being filled at different rates. A crude rule of thumb gives an equivalent frequency of ~5 mHz for a scanning rate of 10 K/min for a transition of 5 K width. It is of considerable interest to examine the glass transition simultaneously under both oscillatory (equilibrium glass transition) and linear (vitrifcation or non-equilibrium glass transition) scanning rates using modulated DSC techniques [130].

9.6.6 Foods

The use of hydrocolloids as thickening agents in food materials is widespread. This can be from the replacement of fats in low-fat spreads [131] to the use of polysaccharides (<1%) and proteins (gelatin at <3%) at low concentrations as thickening agents in yoghurts [132].

There has also been a substantial effort to find replacements for gelatin [133]; for example in confectionery products where the textural characteristics of gelatin, bite, elasticity, etc., can be extremely difficult to accurately reproduce. Tests by sensory panels of trained testers and mechanical measurements on high-sugar confectionery products have shown carrageenan-based products to be closer to the rheological properties, including those monitored by DMA, of gelatin (Figure 9.35) than other simple hydrocolloids. However this is an active area, with ever more complex hydrocolloid mixtures and modifications appearing, which can mimic particular properties of gelatin closely.

The measurement of glass transition has also been carried out using DSC on confectionery products, such as commercial wine gums and found to be dominated by the sugar component, contrary to reports of large and unexpected increases in T_g produced by small amounts of hydrocolloids.

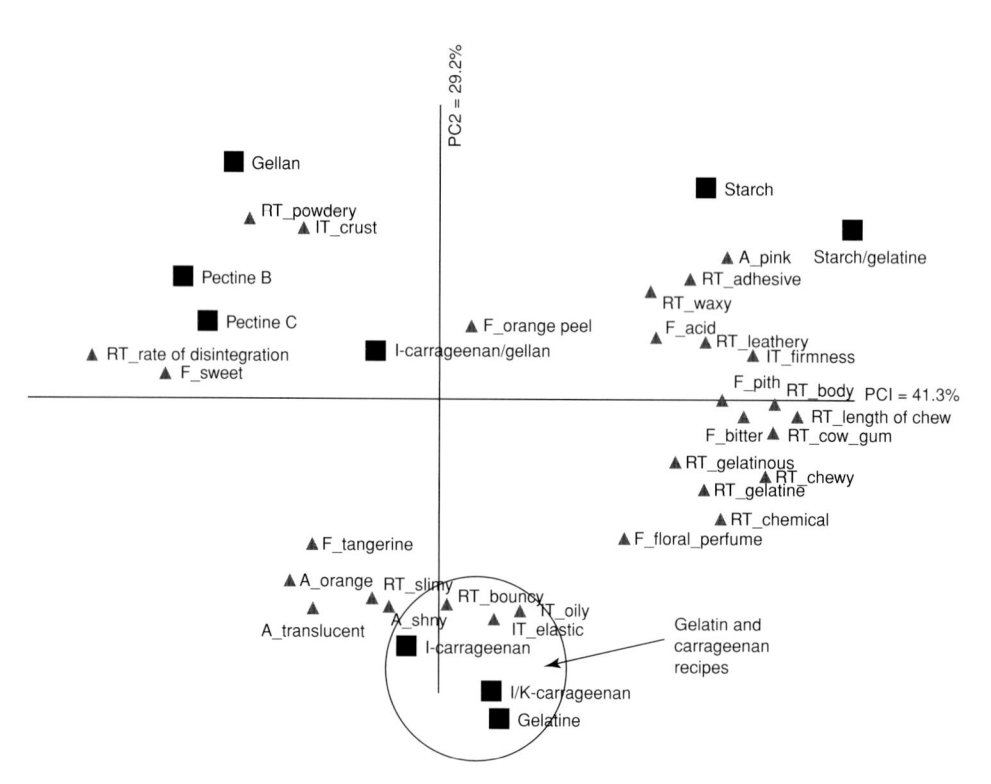

Figure 9.35 The properties of gelatin are similar to those of carrageenans. The descriptors (▲) show properties which are significant in producing the position on the chart. The properties described by the taste panel are shown here. Similar results are produced using DMA-measured parameters.

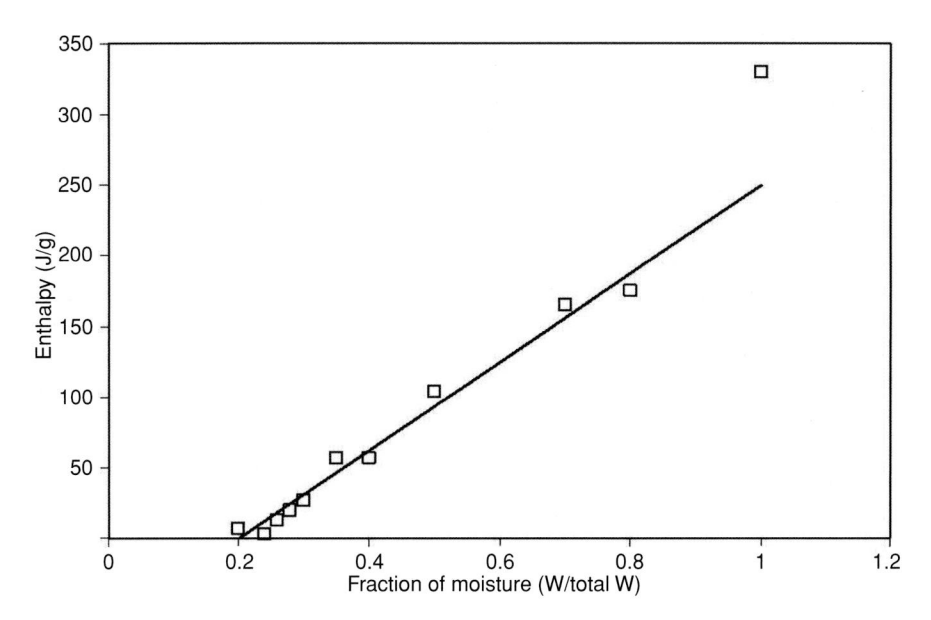

Figure 9.36 The intercept on the x-axis, i.e. the value where the enthalpy falls to zero, shows the value for the bound water in rehydrated freeze-dried garlic powder. That this plot is curved is shown by the inclusion of the 100% value, i.e. the enthalpy for pure ice. (Taken from [139].)

9.7 Frozen systems

9.7.1 Bound water?

The most common TA measurement to be undertaken on frozen materials is the assessment of frozen and unfrozen water, which is obtained by integrating the ice/water melting peak and comparing this value with the known moisture content. A typical example for hydrated freeze-dried garlic powder is shown in Figure 9.36.

This demonstrates that up to around 20% of water will not crystallise as ice. Attempts have been made to demonstrate the importance of the so-called bound fraction with regard to all manner of food properties, for example the water-binding capacity and stickiness in doughs, all without much success. The unfreezable fraction is supposed to represent the water associated with the glassy matrix. NMR relaxation measurements also appear to show different fractions of water or more strictly speaking protons of different mobility (different values of T_2), which in the past have been identified as fractions tightly bound in some way to the solid matrix. However, closer examination of this behaviour demonstrates the presence of exchange of protons between water which appears not to be significantly immobilised in any way and protons on perhaps a relatively immobile polymer, which results in an average decay time value between these two extremes, erroneously implying the presence of an immobilised or bound fraction. The 'averaged' decay value can be described according to the equations of the form

$$1/T_{2\,\text{average}} = 1/T_{2\,\text{water}} + 1/T_{2\,\text{polymer}}$$

By setting up an appropriate model including possible chemical shifts on the polymer site, an equilibrium constant between sites, and even including the possibility of diffusional effects, Belton and Hills [134, 135] have shown highly dynamic behaviour of protons. Therefore we conclude that the DSC demonstration of immobilised water indicates a fraction whose properties are perturbed and cannot form ice but nevertheless whose individual protons and entire molecules can be in exchange with other fractions in other environments. This dynamic behaviour also explains the ability to freeze dry glassy systems, in terms of the dynamic exchange of water molecules with the gaseous phase of the freeze drier (see Section 1.5.8) coupled with the anomalous movement of small molecules through the glass. This is in stark contrast to the complete immobilisation of water in the glassy phase which might have been expected.

One possible resolution of the apparent paradox of the presence of a substantial fraction of unfreezable water (DSC), which nevertheless has the properties of bulk water (NMR), has been given in the arguments of Belton [136]. Upon cooling a solution of carbohydrate, at least part of the carbohydrate, will not be able to crystallise due to increased viscosity/entanglement considerations. In order to maintain equilibrium between this carbohydrate solution and the ice as the temperature is lowered, the water activity in each phase must be the same. The ratio of the vapour pressure of ice to supercooled water at 'low' temperatures (approx $-30°C$) has been suggested to be 0.8. Therefore, for a fraction of liquid (unfreezable) water to be in equilibrium with ice, dissolved solute must lower the activity of the water. It turns out that the concentration of this solution lies in the correct range, that is 0.63–0.83-g solute/total gram solution, and as can be seen no special binding mechanism of water to solute is invoked.

9.7.2 State diagram

The C_g' value, the concentration of the maximally freeze concentrated glass (see Glossary), for most carbohydrate solutions has a similar value (see Table 9.3). Highly viscous solutions do not reach equilibrium, and a cooling curve will follow the lower trajectory on graph 1.14 (see Chapter 1). This means that the concentration of water in the glass is actually above the equilibrium content, and subsequent heating will increase mobility such that the excess water can recrystallise to form ice. Both T_g' and C_g' will thus be difficult to obtain from simple measurements.

The meaning of the features observed on DSC traces of frozen carbohydrate (maltooligomers and sucrose) solutions has been clarified by the work of Ablett et al. [137, 138].

Figure 9.37 demonstrates the complexity that can arise on these traces. These features have been ascribed in the past to glass transitions, including T_g', devitrification, ice melting and even additional transitions [139]. Izzard demonstrated that annealing at temperatures close to an estimate of T_g' could simplify the appearance and aid the understanding of these traces substantially. The annealing brought these systems closer to equilibrium and showed that the features causing confusion were the ice devitrification exotherm and the initial ice dissolution. This work also cast doubt on the previous values of T_g', the glass transition temperature of the maximally concentrated phase, suggesting that true values were somewhat lower.

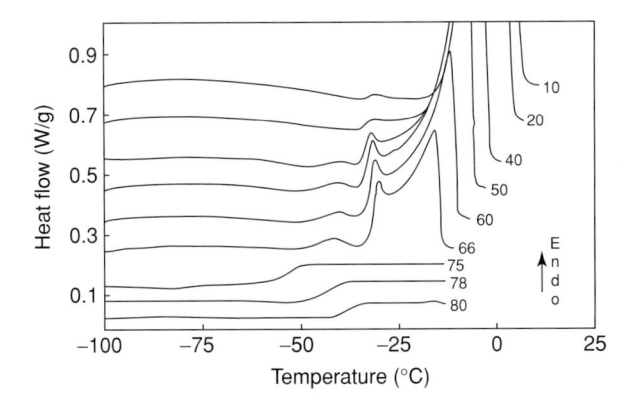

Figure 9.37 The complex nature of transitions which are observed in 10–80% sucrose/water (w/w) solutions. (Taken from [137]. Reproduced by permission of The Royal Society of Chemistry.)

As shown in Figure 9.38, the main features on DSC traces of carbohydrates were a T_g of a non-equilibrium glass, the devitrification exotherm as ice recrystallised from the non-equilibrium matrix and the beginning of ice melting coupled with some ageing of the ice phase. By annealing at temperatures spanning the T_g', it was demonstrated that the true value of T_g' was close to the maximum of the curve (Figure 9.39). Annealing at temperatures below the T_g' preserved the original non-equilibrium value for T_g (lower than T_g'), whereas annealing above T_g' produced a dilute glass which once again on cooling and rerunning had a reduced non-equilibrium value.

This work explained discrepancies, such as the apparent variations in C_g' as a function of the sucrose concentration. It was shown that the main sources of error were the assumptions of a temperature-independent latent heat of fusion of ice and a linear background for calculation of the enthalpy and hence ice content. Moreover, an error analysis of the

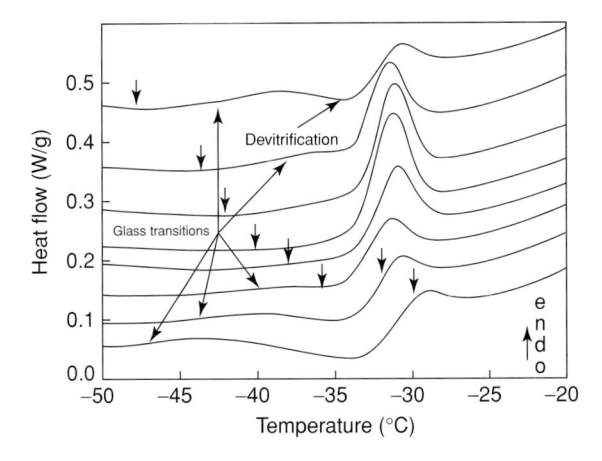

Figure 9.38 The arrowed annealing temperatures used for a 40% sucrose solution subsequently cooled and reheated. The measured glass transitions on the reheating scan are also indicated. (Taken from [137]. Reproduced by permission of The Royal Society of Chemistry.)

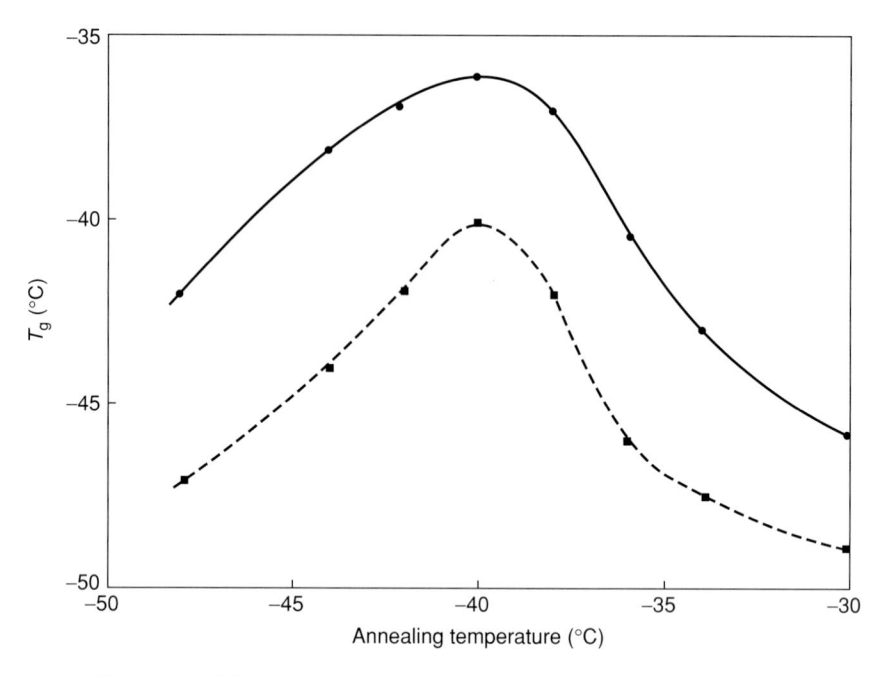

Figure 9.39 The estimated glass transition temperature as a function of annealing temperature. The lower curve shows the onset temperature. The maximum of the onset temperature curve corresponds to T_g'. (Taken from [137]. Reproduced by permission of The Royal Society of Chemistry.)

expression for C_g' showed that this parameter would be systematically lowered with errors in the apparent ice content and the error would be greatest for low sucrose contents.

By using the reversing/non-reversing approach of modulated DSC [140], the above ideas were substantiated, at least qualitatively, by the appearance of the devitrification exotherm and enthalpic relaxation features in the non-reversing trace and the glass transition in the reversing trace, in accordance with prediction (Figure 9.40).

9.7.3 Mechanical properties of frozen sugar solutions

The mechanical properties of frozen materials can be difficult to study due to the enormous stiffness and moduli involved and the fact that changes can occur rapidly over small temperature ranges. The Ares rheometer (Rheometrics Corporation/TA Instrument, Delaware, USA) has a wide range of modulus measuring capability and has been used in the oscillating plate mode in studies of sugars at comparatively low moisture contents (15% ~ −10°C) with and without hydrocolloids [141].

One particularly interesting approach involves the use of a circular cell containing a film of solution sandwiched between two plastic sheets and used in a conventional DMA apparatus [142, 143]). Aqueous solutions of fructose, glucose, sucrose, maltose and lactose were examined. The major features on the DMA traces could be matched to the supplemented state diagram. Glass transition temperatures were determined from the peaks in the loss modulus. The T_g's for an average vibration time of 100 s, the reciprocal of the measuring frequency

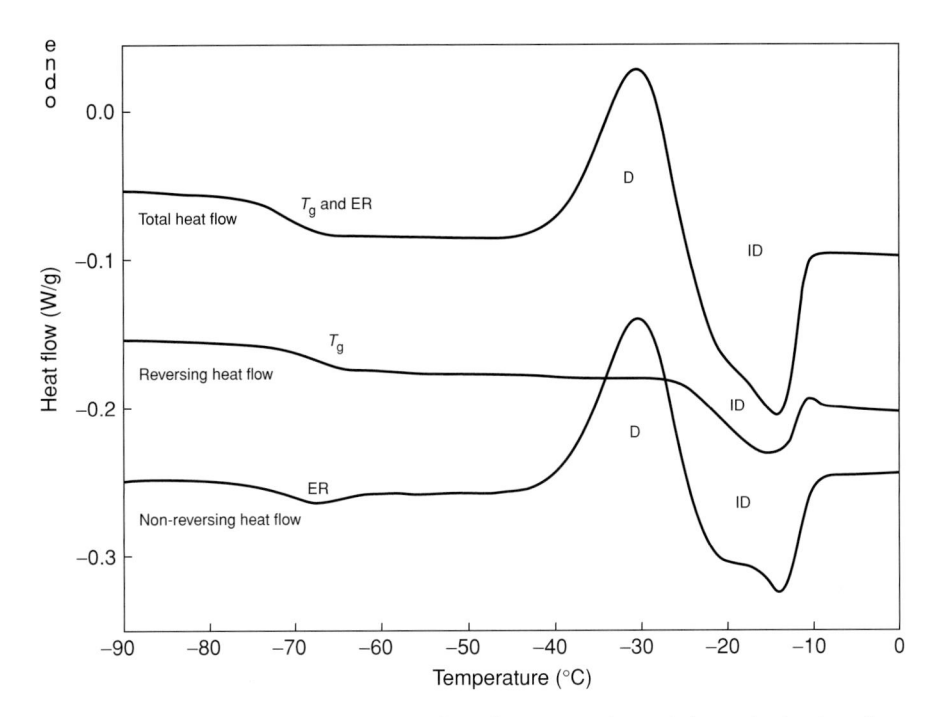

Figure 9.40 Total, reversing and non-reversing heat flow curves obtained during the heating of a non-annealed 65% w/w sucrose solution. Endothermic changes are in the negative direction. T_g, glass transition; ER, enthalpic relaxation; D, devitrification; ID, ice dissolution. (Taken from [140], Copyright (2001), with kind permission of Springer Science and Business Media.)

used in the DMA tests, were similar to published mid-point temperatures determined for maximally freeze-concentrated solutions by DSC.

9.7.4 Separation of nucleation and growth components of crystallisation

DSC will provide information on the overall volume of ice and the heat changes involved in ice formation; however, it will provide only limited information on the nucleation and growth of ice crystals. Attempts to use the DSC to give information on heterogeneous nucleation (see Glossary) have been thwarted by the unreliability of cooling curves; for example the inability to extract heat rapidly enough from a sample undergoing crystallization can lead to distortions of the trace. In addition there are problems of supercooling of solutions and temperature calibration on cooling. DSC can however be used to give information on homogeneous nucleation using an emulsion technique. The impurities responsible for heterogeneous nucleation (see Glossary) can be restricted to a statistically insignificant number of droplets in the emulsion and hence the cooling curve will be dominated by the homogeneous component of nucleation [144]. Cold-stage microscopy is normally used in the study of nucleation and crystal growth, being capable of counting the number of nuclei per unit volume and visualising approximate crystal growth.

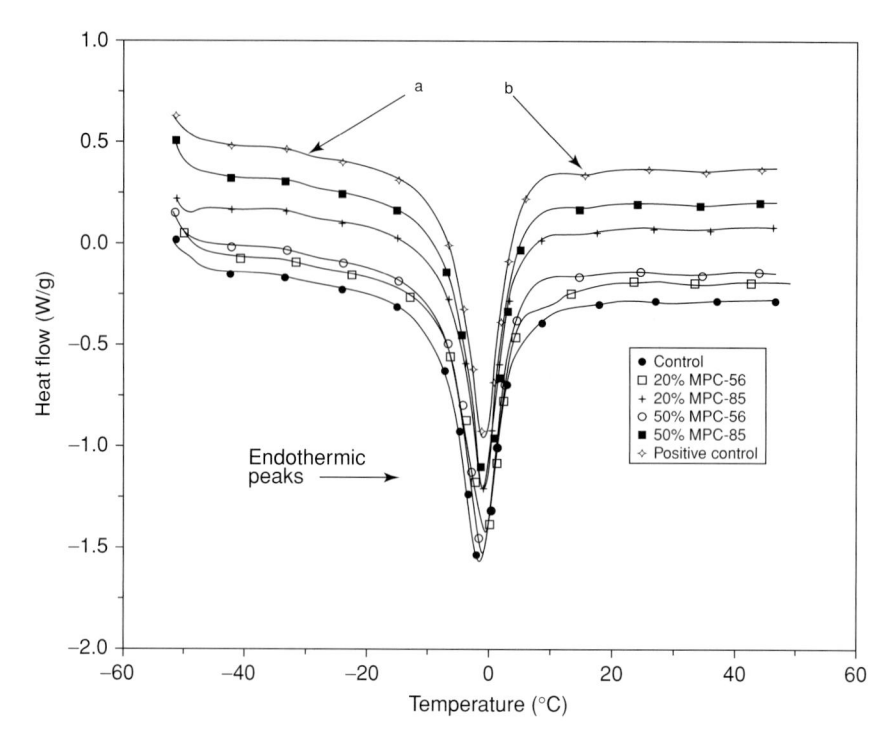

Figure 9.41 A series of thermograms from ice cream with two levels of milk protein concentrates replacing the normal protein in ice cream. Arrows show positions of (a) glass transitions and (b) fat-melting endotherms. (Taken from [148], with permission.)

In studies of recrystallisation in fructose solutions at temperatures of -10--$-30\,°C$, when crystals were small and close together accretion, that is the joining together of crystals to form networks occurred, however, migration was the dominant growth mechanism when crystals were well separated [145, 146]. Rates could be modelled by plotting r^3 against time where r is the radius of the crystal. A primitive Williams–Landau–Ferry (WLF) type of fit was also carried out, by examining relative crystalline rates with respect to a function of reference and limiting temperatures T_0 and T_∞ (see Chapter 4).

9.7.5 Foods

In the area of frozen desserts, water ices (ice lollies) and ice cream, the properties of the ice phase in terms of number and size of crystals is an important textural property of the material. The control of ice crystallisation by stabilisers or hydrocolloids such as locust bean, xanthan and guar gum is also therefore the subject of some interest and not a little controversy [147]. It has been proposed in the past that hydrocolloids can affect the amount of ice formed, affect the thermal conductivity and increase ice nucleation. However, the most likely modes of action are via the increase in viscosity by hydrocolloids on the freeze-concentrated component. For instance, convective transport in ice lollies is reduced, as is the growth rate of ice by mechanical interference of gel fibres for those materials which form gels and also by the accumulation of hydrocolloids at the interface of a growing crystal. This latter mechanism can produce a lowered diffusion rate and a lowering of freezing point. The net result is a slowing of the coarsening of ice crystals in water ices and perhaps in ice cream, although this is more difficult to discern in ice cream due to the increased complexity of the physical structure: the presence of air bubbles and the large amount of fat. Ice melting endotherms in ice cream (Figure 9.41) samples prepared with milk protein concentrates as a replacement for normal protein were similar, with a suggestion of less unfreezable water in the control as compared with the other samples [148]. It was also suggested that a narrow ice endotherm indicated a more homogeneous distribution of ice crystals. This was found for the mixes, with the control having the greatest width. Close examination of Figure 9.41 reveals glass transitions in the region of −30°C and fat transitions in the region of 5°C upwards, although the massive ice endotherms tend to obscure these [149].

DSC is useful in studies of this type both for determining the amount of unfrozen water and more rarely for estimating the amount of homogenous nucleation. Alginate was found to increase the heterogeneous nucleation of ice in water ices but probably by the introduction of impurities [150].

An interesting theory of the interference of ice formation and growth rate and conversely the damage to porous media produced by ice formation has been described in reference [150]. By use of expressions for the forces across interfaces related to the surface tension between the various components and the tension generated in a circular gel pore by an advancing curved ice front, the reduction in melting temperature of the ice can be derived from the Clausius–Clapeyron equation according to

$$\Delta T = T^*/R^2 \Delta Sv$$

where ΔSv is the entropy upon formation of a unit volume of ice. If pressures are produced sufficient to generate a tension in the gel fibres of greater than T^*, the fibre tensile strength, rupture of the fibres can result. The crucial parameters for a gel to interfere with ice formation are small gel mesh size R, high tensile strength of gel fibres T^* and a high value of the gel-fibre–ice interfacial tension. It can be seen that the temperature must be lowered for penetration into the gel. However, whether this has any substantial effect on crystal size distribution in most foods is a controversial point.

It has also been proposed that hydrocolloids, which increased the viscosity of the unfrozen matrix in gelatinised cornstarch, inhibited the additional small amount of ice formation by devitrification on heating. However, there was a minimal effect on the T_g' [151].

Due to extensive production of frozen doughs for convenience use, work has been carried out on the possibility of impairment of performance upon storage. By comparing the ice endotherms for gluten and dough, it was proposed that there is a small (1%) loss of water from the gluten phase on storage at −15°C but not at −25°C. A slightly larger effect was observed in dough, possibly due to the lower T_g [152]. Generally, however, few changes are observed in the 'bound' fraction of water or C_g'. Most large changes reported in these parameters are normally due to complications in interpretation of the traces. According to reference [153] dough freezing had no effect on starch gelatinisation but increased the retrogradation rate, as measured by DSC, when storage was prolonged over 30 days.

DSC has been used to measure T_g', the freezing point and the enthalpy of crystallisation in dehydrated fruits [154] and generate state diagrams in dates [155].

Combined mechanical (DMA) and DSC studies on moist bread [156] showed that the tan δ and E' curves spanned a temperature range of 50°C in the ice-melting region. DSC results showed that freezable water was present at >33% moisture content. At lower moisture contents the transitions began to move to higher temperatures and broaden.

9.7.6 Cryopreservation

The cryopreservation of seeds is closely allied to the study of storage of foods at low temperatures in terms of the arrestment of enzymatic and general chemical deteriorative processes. This has been given a fillip recently, with the widespread interest in the preservation of the genetic diversity of the botanical resource. The crucial element in the preservation of many seeds and plant tissue, and indeed in animals with a strategy of cold survival, appears to be the prevention of the destructive capability of ice formation in cells and surrounding tissue. This can be carried out in two ways. Firstly, desiccation, such that the remaining water is unfreezable, is important. As already discussed, this can be measured effectively using the DSC. A second strategy appears to be the replacement of water by potent hydrogen-bonding compounds such as sugars, although other materials such as oligosaccharides and proteins have been reported to be involved. Trehalose and sucrose are particularly effective compounds and there are many examples of plants and animals with desiccative strategies, which synthesise large quantities of these sugars. However, there are seeds with apparently similar mechanisms of protection and similar levels of sucrose which are intolerant of dessication [157], so called recalcitrant seeds.

The ability to form a uniform glass of high transition temperature, however, appears to be important. Trehalose with its high glass transition temperature is well suited to this function. In other complex materials, several glass transitions can frequently be detected and it can be difficult to decide on the value of temperature below which storage will be effective. Seeds which are resistant to desiccation and become metabolically quiescent upon dehydration are said to show orthodox behaviour in contrast with recalcitrant seeds which are intolerant to dessication, that is show a loss of viability. It appears that desiccation tolerance arises in the embryological development and can be influenced by factors such as the rate of drying and the temperatures to which the seed is exposed before freezing. Some seeds are also sensitive to storage at very low temperatures such as those of liquid nitrogen [158] particularly if a certain amount of water, supposedly necessary for structural integrity, is not maintained [159]. There is also the possibility of limited molecular mobility at cryogenic temperatures [160]. This demonstrates the complexity of behaviour and the difficulty of finding unitary mechanisms in a living system.

As already stated, despite there being no difference in the apparent glassy properties, some seeds will show differences in terms of longevity when stored at low temperatures, typically at −20°C. In these materials it has been proposed that lipid transitions can play an important role, with a correlation between impaired longevity and lipid transitions observed on DSC traces. For instance, in several *Cuphea* species [161], peak lipid-melting temperatures of >27°C were associated with sensitivity to storage at −18°C, whereas those with melting temperatures <27°C tolerated low-temperature exposure. This was further reinforced by sensitive species having high concentrations of lauric C12 and myristic C14 acids, while species with capric C8 or caprylic C10 or high concentrations of unsaturated fatty acids

Table 9.8 Heats of solution for crystalline sugar alcohols

Polyol	Lactitol	Maltitol	Sorbitol	Mannitol	Xylitol
Heat of solution	−52	−79	−110	−121	−153

Endothermic values are given a negative sign.

resisted low-temperature exposure. It was also postulated that the heat pulses which seem to prevent low-temperature damage being expressed in some seeds could melt solidified lipids before the rehydration stage prior to germination.

Similarly, it has been proposed that chilling sensitivity in plants was related to phase transitions in cell membranes, for example in chilling sensitive tomatoes. The membranes of those plants which were chill sensitive appeared to show a transition, below which, the lipids were in a rigid state and presumably less able to resist mechanical deformation intact. Membranes, which show continuous changes in fluidity, that is no sharp phase transitions, appeared more resistant.

Clearly, a living system is more complex than most foods; however, the postulation that mechanisms for chill sensitivity could reside in hydrophobic regions would be novel in the study of low-temperature storage in foods.

9.8 Thermodynamics and reaction rates

Whilst potentially all of this vast area could be of use in the investigation of foods and food ingredients, as is the case with TA generally, only some of the areas will be highlighted which are being used and it is felt will become of increasing importance in the near future.

9.8.1 Studies on mixing

Of particular interest in the areas of mixing calorimetry are the cooling effects of the dissolution of polyols, such as maltitol, xylitol and sorbitol in water. The sensation of cooling is a positive attribute for certain confectionery products, hence the interest in the accurate measurement of enthalpies of mixing or heats of solution. Mixing of powders and liquids can be carried out on machines such as the Setaram micro DSCIII using a manual rotating paddle system. However, more appropriate models, such as the C80, are produced by the same manufacturer (Setaram Caluire France). Table 9.8 shows typical endothermic reaction enthalpies. Contrary to conventional representation of endo- and exothermic heat changes, these endothermic changes are occasionally presented with a negative sign.

The thermal activity monitor (TAM; TA Instruments, Delaware, USA) is also used for this type of measurement and employs the principle of breaking a glass ampoule in a solution under continuous mixing to measure the enthalpy change.

This method can also provide independent measurement of the amorphous content of sugars, fundamentally different from the more usual X-ray or NMR methods. A crystalline sugar has a lattice energy due to the regularity of structure, which is manifest as an endothermic enthalpy change on dissolving. An amorphous material, not possessing the long-range

Table 9.9 Comparison of heats of solution (J/g) for crystalline and amorphous forms of sugars

Sugar	Crystalline α-lactose	Amorphous lactose	Crystalline sucrose	Amorphous sucrose
Heat of solution	+56	−56	+17	−44

Endothermic values are given a positive sign.
Taken from [162], reproduced with permission of TA instruments.

structural order responsible for the heat change when the structure breaks up, can exhibit completely different heat changes upon dissolving. Some standard values for heats of solution of lactose and sucrose are presented in Table 9.9 [162].

By assuming linear additivity of the enthalpy values of a completely amorphous and a completely crystalline sample, the amorphous content can be calculated for any sugar sample (Figure 9.42). This is a very sensitive technique with a claimed accuracy of more than 1%.

9.8.2 Isothermal titration calorimetry

This is a powerful method of obtaining thermodynamic data such as ΔH and K, and hence ΔG and ΔS, and in addition the stoichiometry n of the reaction. The stoichiometry is

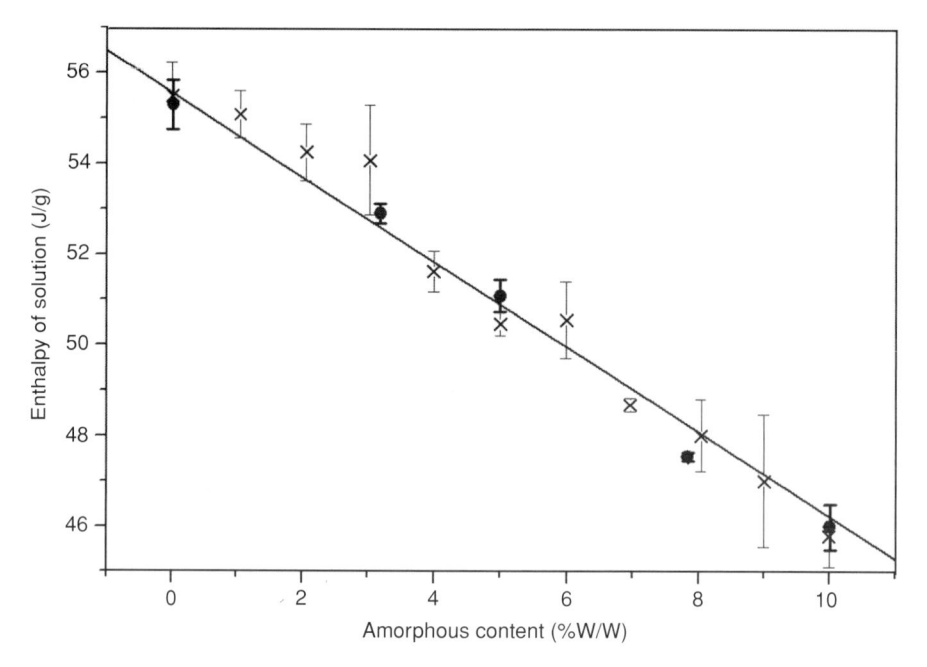

Figure 9.42 A calorimetric method for obtaining amorphous content of lactose samples. •, sample stored under vacuum at 50°C; x, sample stored over phosphorous pentoxide. (Taken from [174], reproduced with permission of TA Instruments.)

normally obtained for well-defined chemical reactions where comparatively small numbers of molecules are involved. This is carried out routinely for ligand binding, for example the binding to ribonuclease of 2'CMP, which is a strong competitive inhibitor of the substrate 2'3'CMP [163]. It can be shown that a fit of the equation

$$Q = nA_t \Delta H(V/2) \cdot \{X - [X^2 - (4B_t/nA_t)]^{1/2}\}$$

where

$$X = 1 + B_t/nA_t + 1/nKA_t$$

to a plot of heat produced per mole of injectant, against molar ratio of A/B, after correction for the heat of dilution, will return values of K, n and ΔH [164] (see Figure 9.43). A_t and B_t are the total concentrations free and complexed of A and B respectively.

In the interpretation of thermodynamic parameters of binding, it is important to realise that it is the net changes in the entire system which are being measured. So, for example, if a net increase in entropy upon binding of a ligand to a protein site is measured, this is composed of several components. As well as the reduction in entropy due to the restriction of motion of the ligand and parts of the protein site, there will be a release of solvent molecules into free solution, those, for example, displaced from the active site by the binding of the ligand. This latter component can more than compensate for the entropy reduction at the active site by the increase in disorder of the solvent molecules. Similar reasoning applies to other thermodynamic variables such as enthalpy of binding and free volume effects. This is dealt with extensively in references [164] and [165].

Frequently in foods, these reactions are not well defined. In haze production in beverages, such as beer and fruit juices, a relatively small phenol-containing molecule reacts with a large molecule, gelatin, which contains a large number of binding sites [166]. It is a more difficult exercise to obtain meaningful data from such reactions. Firstly, there is the issue of the exact molecular weight of components in food, such as proteins of indeterminate structure, being unknown. Secondly, the statistical robustness of parameters and parameter combinations, when they lie outside particular ranges, is limited. This can apply to values of the product of substrate and association constant or large values of n. Thirdly, there is the possibility of interactive binding events on a large molecule such as a protein.

Nevertheless these measurements are important in foods. For haze production in beverages, which is produced by proline-rich proteins binding to polyphenols such as tannic acid, it is reported that the maximum amount of haze production occurs when the number of polyphenol-binding ends and the number of protein-binding sites are nearly equal. Hence ITC determined binding information and stoichiometry is relevant. In some simple theories of bitterness perception, the concentration of free phenolics is a determining factor, hence the importance of knowing the binding constant K. This determines, in conjunction with partition into hydrophobic phases, the free aqueous concentration.

Studies of the binding of olive oil phenolics to food proteins have been carried out [167]. These showed small heats of interaction, which precluded the calculation of the above thermodynamic parameters. The heat changes upon interaction of proteins with tannic acid were much larger and allowed the reliable calculation of K, n and ΔH (Table 9.10).

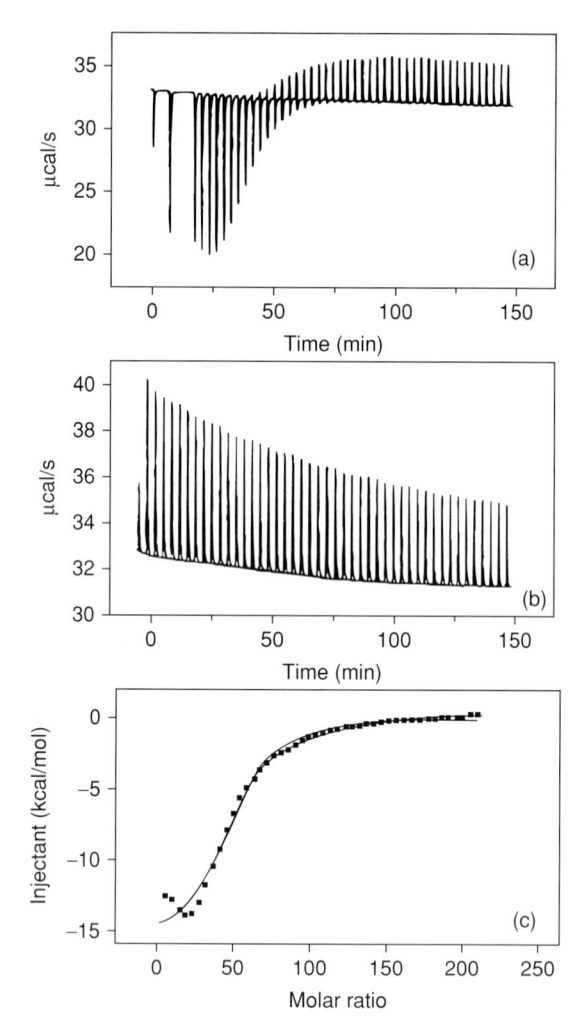

Figure 9.43 Isothermal titration calorimetry: (a) titration of tannic acid to buffered gelatin solution at pH 6.75 and $T = 35°C$, (b) titration of tannic acid to aqueous buffer solution and (c) peak areas after subtraction of heat of dilution as a function of tannic acid/protein molar ratio. (Taken from [167]. Copyright Society of Chemical Industry. Reproduced with permission. Permission is granted by John Wiley & Sons Ltd on behalf of the SCI.)

9.8.3 Reaction rates

Sensitive calorimeters such as the TAM offer the possibility of studying the chemical deterioration and hence shelf life of food materials. At present this approach is mainly confined to the study of pharmaceutical compounds. By examining the first part of the reaction curve it may be possible to predict the time to completion of the reaction and hence the stability. This is a controversial area with some workers claiming that alternative methods with powerful software and different algorithms offer better prediction.

Table 9.10 Values of K, n and ΔH for interactions between tannic acid and proteins using Isothermal calorimetry ITC

Protein	BSA	Gelatin	B-lactoglobulin	Caseinate
K (M^{-1})	2.7×10^4	3.6×10^4	5.9×10^3	2.6×10^5
N	14	48.5	6.4	5.3
ΔH (kJ/mol)	−21.7	−66	−32.4	−31.5

Taken from [167]. Copyright Society of Chemical Industry. Reproduced with permission. Permission is granted by John Wiley & Sons Ltd on behalf of the SCI.

The Maillard reaction is relevant to many foods. Using new developments in fitting software, which enable complex data to be fitted according to the optimum number of reaction steps, it was found that initially (<4 h), the Maillard reaction occurring in sugar solutions could be fitted to a three-to-four step reaction, whereas at greater times, more complex fitting was required [168]. Once again one of the major differences between food systems and others lies in the complexity of the reactions.

Degradation and pyrolysis are also interesting areas with regard to food materials. Clearly, organic materials held in the region of 200°C would be expected to react/degrade. This tendency is exacerbated if DSC pans are sealed and water is contained in the local environment, or if reactive gases are passed over unsealed systems. For instance, lactose will degrade in sealed pans and produce artefacts on the exothermic peaks due to bubbling/mechanical disruption, negating attempts to measure the melting point. Crimped pans which allow the water to escape permit the determination of melting point enthalpy as well as distinction between the melting points of α and β forms. Using DSC, dough also shows extensive degradation above about 210°C. The isothermal trace at 210°C shows an exothermic reaction peak at about 20 min (Figure 9.44).

Reactions between proteins and carbohydrates are of interest, particularly with regard to minimising protein nutritional value decreases in cooking, and also in relation to the toxic effects of thermal browning. In a comprehensive DSC study conducted with an atmosphere of oxygen over unsealed pans, two major exothermic peaks in the regions of 200–400 and 400–650°C were observed upon the reaction between gluten and carbohydrates, such as starch, cellulose and sugars [169]. The magnitude of the heats of combustion was listed for the various mixtures.

9.8.4 Thermogravimetric analysis

Thermogravimetric analysis (TGA) is a method that can examine the course of a reaction by monitoring the weight change as volatiles or gases of decomposition are released. These gases can subsequently be analysed by standard chemical methods such as coupled mass spectrometry techniques. By switching atmospheres from an inert gas such as pure nitrogen to a reactive gas such as oxygen, the contrast between the release of volatiles such as water and flavours with the products of reactivedegradation of the food matrix can be made.

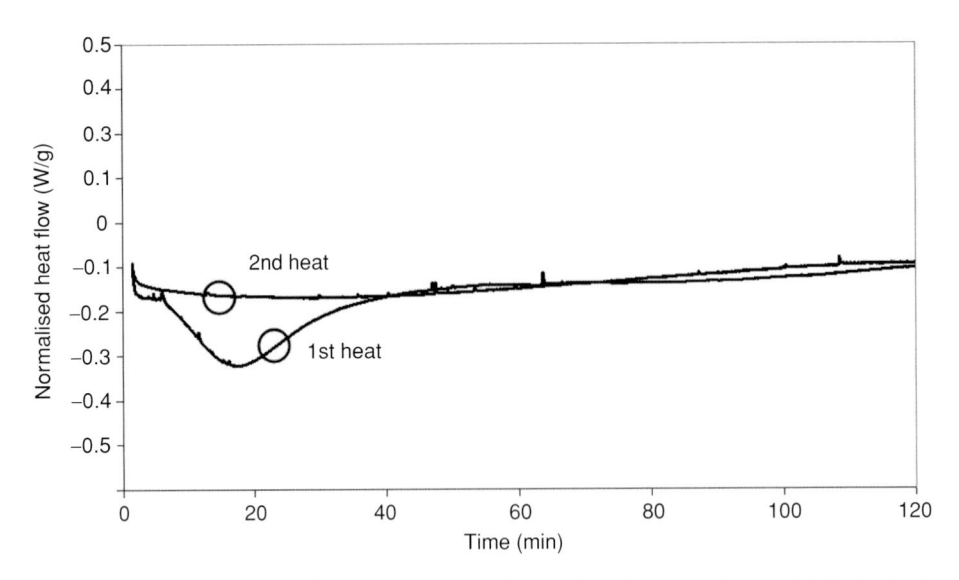

Figure 9.44 A non-yeasted dough undergoing an exothermic degradation reaction at 210°C after having been ramped to this temperature at a rate of 100°C/min. A reheat is also shown demonstrating that the reaction has gone to completion.

This type of approach does not appear to have been used extensively in food research. The decomposition observed in TGA measurements is invariably in the solid state and consequently reactions can be extremely complex, leading to many different equations being put forward to describe the loss of weight (see [170] for a discussion on this point). For example, a reaction can proceed by a mechanism similar to crystallisation, that is points of initiation followed by circular growth fronts. As might be expected, a variant of the Avrami equation (see Section 1.5.9) has been proposed to describe such reactions.

The combined use of TGA and SPME-GC/MS (solid-phase microextraction gas chromatography/mass spectrometry) is informative in identifying the major components released from dryer-feed (DF) and distillers-dried grains (DDG) [171]. These co-products formed in the production of fuel and beverage ethanol are a potential animal and human food source rich in protein, digestible fibre and fat. Three peaks in the differential of the TGA trace, which is often referred to as differential thermogravimetric analysis (DTG), were observed for both DF and DDG (Figure 9.45) using air as a purge gas. The first of these in the region of 60–70°C is due to the loss of water+some volatiles.

The second peak at 250–300°C appears to be due to the degradation of protein amino acids, whilst the third peak shows the majority of degradation products, due to the breakdown of carbohydrates and proteins, and interactions between them. Similarly, the thermal stability of coal tar dyes for food was investigated and weight-loss temperature regions identified with decompositions [172]. For example, the colours of all the dyes changed during the first stage and this was associated with the decomposition of the azo group and the release of H from the OH bond in the region of 320–450°C for the dye sunset yellow FCF.

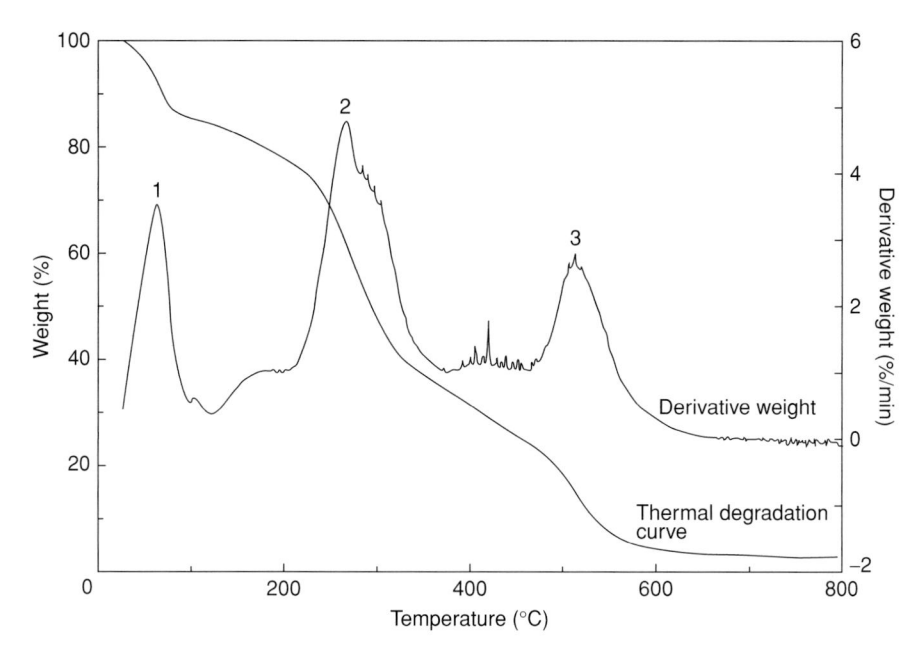

Figure 9.45 TGA and DTGA traces for drier-feed grains from a distillery. (Reprinted from [171], Copyright (2001), with permission from Elsevier.)

TGA has also been used in the analysis of moisture and ash content of flours, and the moisture content of pasta, coffee and powdered milk.

9.8.5 Sample controlled TA

One technique which to the authors' knowledge has not yet been applied in the foods area is sample controlled TA [173]. When applied to TGA, for example, this controls the applied thermal profile, that is slows the heating rate, when a loss of mass is detected. The loss of mass will then effectively occur isothermally if this is the desired method of control. This has the potential to simplify the analysis of kinetics and provide an alternative approach to the non-isothermal methods discussed earlier. Furthermore, there are potential controlling variables other than simple isothermal criteria. For example, the partial pressure of a degradation volatile could be used to control the applied temperature profile, as could the expansion of a material measured by dilatometry or, particularly interestingly, features in the non-reversing signal of a suitably adapted DSC. These latter two possibilities could be relevant with regard to transitions in fats.

9.8.5.1 Concluding remarks

This can only ever be a superficial view of TA in the foods area. We have concentrated mainly on calorimetry but hopefully have given an indication of the type of work which has been carried out and the areas which may see activity in the near future.

References

1. Wu H-CH, Sarko A. Packing analysis of carbohydrates and polysaccharides. VIII: the double-helical molecular structure of crystalline B-amylose. *Carbohydr Res* 1978;61:7–25.
2. Wu H-CH, Sarko A. Packing analysis of carbohydrates and polysaccharides. IX: the double-helical molecular structure of crystalline A-amylose. *Carbohydr Res* 1978;61:27–40.
3. Waigh TA, Jenkins PJ, Donald AM. Quantification of water in carbohydrate lamellae using SANS. *Faraday Discuss* 1996;103:325–337.
4. Donovan JW. Phase transitions of the starch-water system. *Biopolymers* 1979;18(2):263–275.
5. Lelievre J, Mitchell J. Pulsed NMR study of some aspects of starch gelatinization. *Staerke* 1975;27(4):113–115.
6. Flory PJ. *Principles of Polymer Chemistry*. Cornell University Press, 1953.
7. Biliaderis CG, Page CM, Maurice TJ. On the multiple melting transitions of starch/monoglyceride systems. *Food Chem* 1986;22(4):279–295.
8. Burt DJ, Russell PL. Gelatinization of low water content wheat starch-water mixtures: a combined study by differential scanning calorimetry and light microscopy. *Starch/Staerke* 1983;35(10):354–360.
9. Cooke D, Gidley MJ. Loss of crystalline and molecular order during starch gelatinisation: origin of the enthalpic transition. *Carbohydr Res* 1992;227:103–112.
10. Waigh TA, Gidley MJ, Komanshek BU, Donald AM. The phase transformations in starch during gelation: a liquid crystalline approach. *Carbohydr Res* 2000;328(2):165–176.
11. Zeleznak KJ, Hoseney RC. The glass transition in starch. *Cereal Chem* 1987;64(2):121–124.
12. Kalichevsky MT, Jaroszkiewicz EM, Ablett S, Blanshard JMV, Lillford PJ. The glass transition of amylopectin measured by DSC, DMTA and NMR. *Carbohydr Polym* 1992;18(2):77–88.
13. Farhat IA, Mousia Z, Mitchell JR. Structure and thermomechanical properties of extruded amylopectin-sucrose systems. *Carbohydr Polym* 2003;52(1):29–37.
14. Boussingault JB. Experiments to determine the transformation of fresh bread into stal bread. *Ann Chimie Physique* 1852;36:490.
15. Katz JR. The physical chemistry of starch and bread making. I: the changes in the Rontgen spectrum of starch during baking and staling of bread [abstract A]. *Z physik Chem* 1930;150:37–59.
16. Farhat I, Hill S, Mitchell J, Blanshard J, JF B. In: Roos Y, Leslie J, Lillford PJ (eds.), *Water Management in the Design and Distribution of Foods (ISOPOW 7)*. Pennsylvania: Technomic Publishing Co, Inc, 1999; pp. 411–428.
17. Marie V. *Starch Phase Transitions in Relation to Baking*, PhD Thesis. University of Nottingham, Nottingham, 2001.
18. Godet M, Tran V, Delage M, Buleon A. Molecular modeling of the specific interactions involved in the amylose complexation by fatty acids. *Int J Biol Macromol* 1993;15:11–16.
19. Nuessli J, Sigg B, Conde-Petit B, Escher F. Characterization of amylose-flavor complexes by DSC and X-ray diffraction. *Food Hydrocolloid* 1997;11(1):27–34.
20. Biliaderis CG, Page CM, Slade L, Sirett RR. Thermal behavior of amylose-lipid complexes. *Carbohydr Polym* 1985;5(5):367–389.
21. Hancock BC, Shalaev EY, Shamblin SL. Polyamorphism: a pharmaceutical science perspective. *J Pharm Pharmacol* 2002;54(8):1151–1152.
22. Wunderlich B. Reversible crystallization and the rigid-amorphous phase in semicrystalline macromolecules. *Prog Polym Sci* 2003;28(3):383–450.
23. Xu H, Cebe P. Transitions from solid to liquid in isotactic polystyrene studied by thermal analysis and X-ray scattering. *Polymer* 2005;46(20):8734–8744.
24. Jansen MAG, Goossens JGP, de Wit G, Bailly C, Schick C, Koning CE. Poly(butylene terephthalate) copolymers obtained via solid-state polymerization and melt polymerization: a study

on the microstructure via 13C NMR sequence distribution. *Macromolecules* 2005;38(26):10658–10666.
25. Muenzing K. DSC studies of starch in cereal and cereal products. *Thermochim Acta* 1991;193:441–448.
26. Russell PL. A kinetic study of bread staling by differential scanning calorimetry: the effect of painting loaves with ethanol. *Starch/Staerke* 1983;35(8):277–281.
27. Mestres C, Matencio F, Pons B, Yajid M, Fliedel G. A rapid method for the determination of amylose content by using differential scanning calorimetry. *Starch/Staerke* 1996;48(1):2–6.
28. Van den Dries IJ, Van Dusschoten D, Hemminga MA. Mobility in maltose-water glasses studied with 1H NMR. *J Phys Chem B* 1998;102(51):10483–10489.
29. Noel T. Molecular mobility in foods. European contract ERBFAIRCT961085, 1996–1999.
30. Saleki-Gerhardt A, Zografi G. Non-isothermal and isothermal crystallization of sucrose from the amorphous state. *Pharm Res* 1994;11(8):1166–1173.
31. Chan TW, Shyu GD, Isayev AI. Reduced time approach to curing kinetics. Part I: dynamic rate and master curve from isothermal data. *Rubber Chem Technol* 1993;66(5):849–864.
32. Chan TW, Shyu GD, Isayev AI. Master curve approach to polymer crystallization kinetics. In: *Annual Technical Conference*, Vol. 52, issue 2. Society of Plastics Engineers, 1994; pp. 1480–1484.
33. Kedward CJ, MacNaughtan W, Mitchell JR. Isothermal and non-isothermal crystallization in amorphous sucrose and lactose at low moisture contents. *Carbohydr Res* 2000;329(2):423–430.
34. MacNaughtan W. MAFF Food Link Program. Project no FQS06 High sugar polysaccharide systems, 1999–2002.
35. Kilmartin PA, Reid DS, Samson I. The measurement of the glass transition temperature of sucrose and maltose solutions with added NaCl. *J Sci Food Agric* 2000;80(15):2196–2202.
36. Longinotti MP, Mazzobre MF, Buera MP, Corti HR. Effect of salts on the properties of aqueous sugar systems in relation to biomaterial stabilization. Part 2: sugar crystallization rate and electrical conductivity behavior. *Phys Chem Chem Phys* 2002;4(3):533–540.
37. Miller DP, De Pablo JJ, Corti HR. Viscosity and glass transition temperature of aqueous mixtures of trehalose with borax and sodium chloride. *J Phys Chem B* 1999;103(46):10243–10249.
38. Dakhnovskii Y, Lubchenko V. The effect of charged impurities on a glass transition in a polar medium. *J Chem Phys* 1996;104(2):664–668.
39. Ohtake S, Schebor C, Palecek Sean P, de Pablo Juan J. Effect of pH, counter ion, and phosphate concentration on the glass transition temperature of freeze-dried sugar-phosphate mixtures. *Pharm Res* 2004;21(9):1615–1621.
40. Noel TR. *Thermal Analysis Congress*. Thermal Methods Group of the Royal Society of Chemistry, John Moores University Liverpool, 2004.
41. Noel TR, Parker R, Brownsey GJ, Farhat IA, MacNaughtan W, Ring SG. Physical aging of starch, maltodextrin, and maltose. *J Agric Food Chem* 2005;53(22):8580–8585.
42. Noel TR, Parker R, Ring SM, Ring SG. A calorimetric study of structural relaxation in a maltose glass. *Carbohydr Res* 1999;319(1–4):166–171.
43. Wungtanagorn R, Schmidt SJ. Phenomenological study of enthalpy relaxation of amorphous glucose, fructose, and their mixture. *Thermochim Acta* 2001;369(1–2):95–116.
44. Surana R, Pyne A, Suryanarayanan R. Effect of aging on the physical properties of amorphous trehalose. *Pharm Res* 2004;21(5):867–874.
45. Craig ID, Parker R, Rigby NM, Cairns P, Ring SG. Maillard reaction kinetics in model preservation systems in the vicinity of the glass transition: experiment and theory. *J Agric Food Chem* 2001;49(10):4706–4712.
46. Hill Sandra A, Macnaughtan W, Farhat Imad A, et al. The effect of thermal history on the Maillard reaction in a glassy matrix. *J Agric Food Chem* 2005;53(26):10213–10218.
47. Kouassi K, Roos YH. Glass transition and water effects on sucrose inversion by invertase in a lactose-sucrose system. *J Agric Food Chem* 2000;48(6):2461–2466.

48. Noel TR. The stability of foods and food ingredients in the glassy state. Link Project no. FQS05 Drystore, 2001–2004.
49. Alimkuu M. *Evaluation of the Effectiveness of Sugar Additives in Controlling Crystallization of Candy Floss*. MSc Thesis. Nottingham University, Nottingham, 2004.
50. Cordella C, Faucon JP, Cabrol-Bass D, Sbirrazzuoli N. Application of DSC as a tool for honey floral species characterization and adulteration detection. *J Therm Anal Calorimetry* 2003;71(1):279–290.
51. Cordella C, Antinelli J-F, Aurieres C, Faucon J-P, Cabrol-Bass D, Sbirrazzuoli N. Use of differential scanning calorimetry (DSC) as a new technique for detection of adulteration in honeys. 1. Study of adulteration effect on honey thermal behavior. *J Agric Food Chem* 2002;50(1):203–208.
52. Jouppila K, Kansikas J, Roos YH. Water induced crystallization of amorphous lactose: dehydrated lactose and lactose in skim milk powder. *Crystal Growth of Organic Materials 4*, 4th International Workshop, Bremen, 17–19 September 1997; pp. 317–324.
53. Jouppila K, Kansikas J, Roos YH. Glass transition, water plasticization, and lactose crystallization in skim milk powder. *J Dairy Sci* 1997;80(12):3152–3160.
54. Roos YH, Jouppila K, Zielasko B. Non-enzymic browning-induced water plasticization: glass transition temperature depression and reaction kinetics determination using DSC. *J Therm Anal* 1996;47(5):1437–1450.
55. Garti N, Sato K (eds.). *Crystallization Processes in Fats and Lipid Systems*. New York: Marcel Dekker, 2001.
56. Timms RE. Phase behaviour of fats and their mixtures. *Prog Lipid Res* 1984;23(1):1–38.
57. Bentz AP, Breidenbach BG. Evaluation of the differential scanning calorimetric method for fat solids. *J Am Oil Chem Soc* 1969;46(2):60–63.
58. A.O.C.S. Tentative Method CD 10–57, Solids Fat Index. AOCS Champaign Illinois.
59. Leung HK, Anderson GR, Norr PJ. Rapid determination of total and solid fat contents in chocolate products by pulsed nuclear magnetic resonance. *J Food Sci* 1985;50(4):942–945, 950.
60. Rudnicki WR, Niezgodka M. Modeling phase transitions in cocoa butter. In: *4th International Symposium on Confectionery Science*, Hershey, Pennsylvania, 2002.
61. Chapman D. Polymorphism of glycerides. *Chem Rev* 1962;62:433–456.
62. Clarkson CE, Malkin T. Alternation in long-chain compounds. II: an x-ray and thermal investigation of the triglycerides [abstracts]. *J Chem Soc* 1934:666–671.
63. Kellens M, Meeussen W, Reynaers H. Study of the polymorphism and the crystallization kinetics of tripalmitin: a microscopic approach. *J Am Oil Chem Soc* 1992;69(9):906–911.
64. MacNaughtan W, Farhat IA, Himawan C, Starov VM, Stapley AGF. A differential scanning calorimetry study of the crystallization kinetics of tristearin–tripalmitin mixtures. *J Am Oil Chem Soc* 2006;83(1):1–9.
65. Turnbull D, Fisher JC. Rate of nucleation in condensed systems. *J Chem Phys* 1949;17:71–73.
66. Smith Kevin W, Cain Fred W, Talbot G. Kinetic analysis of nonisothermal differential scanning calorimetry of 1,3-dipalmitoyl-2-oleoylglycerol. *J Agric Food Chem* 2005;53(8):3031–3040.
67. Serra R, Sempere J, Nomen R. A new method for the kinetic study of thermoanalytical data: the non-parametric kinetics method. *Thermochim Acta* 1998;316(1):37–45.
68. Smouse TH. Factors affecting oil quality and stability. In: *Methods to Assess Quality and Stability of Oils and Fat-Containing Foods*, Warner K, Eskui NAM (eds.), Illinois: AOCS Press, 1995: pp. 17–36.
69. Paul S, Mittal GS. Regulating the use of degraded oil/fat in deep-fat/oil food frying. *Crit Rev Food Sci Nutr* 1997;37(7):635–662.
70. Tan CP, Man YBC, Selamat J, Yusoff MSA. Application of Arrhenius kinetics to evaluate oxidative stability in vegetable oils by isothermal differential scanning calorimetry. *J Am Oil Chem Soc* 2001;78(11):1133–1138.

71. Schlichter-Aronhime J, Garti N. Solidification and polymorphism in cocoa butter and the blooming problems. In: *Crystallisation and Polymorphism Fats Fatty Acids (Surfactant Science Series 31)*. New York: Marcel Dekker, 1988; pp. 363–393.
72. Baichoo N, MacNaughtan W, Mitchell JR, Farhat IA. A stepscan differential scanning calorimetry study of the thermal behaviour of chocolate. *Food Biophysics*, 2006;1(4):169–177.
73. Ma KM. *Studies on Yoghurt*, MSc Thesis. University of Nottingham, Nottingham, 2004.
74. Ferry JD. Protein gels. *Adv Protein Chem* 1948;4:1–78.
75. Ma C-Y, Holme J. Effect of chemical modifications on some physicochemical properties and heat coagulation of egg albumen. *J Food Sci* 1982;47:1454.
76. Nishinari K. Rheological and DSC study of sol-gel transition in aqueous dispersions of industrially important polymers and colloids. *Colloid Polym Sci* 1997;275(12):1093–1107.
77. Michnik A. Thermal stability of bovine serum albumin: DSC study. *J Therm Anal Calorimetry* 2003;71(2):509–519.
78. Wang B, Tan F. DSC study of cold and heat denaturation of b-lactoglobulin A with urea. *Chin Sci Bull* 1997;42(2):123–127.
79. Leon A, Rosell CM, Benedito de Barber C. A differential scanning calorimetry study of wheat proteins. *Eur Food Res Technol* 2003;217(1):13–16.
80. Biliaderis CG. Differential scanning calorimetry in food research – a review. *Food Chem* 1983;10(4):239–265.
81. Tanford C. Protein denaturation. *Adv Protein Chem* 1968;23:121–282.
82. Biltonen R, Schwartz AT, Wadso I. A calorimetric study of the chymotrypsinogen family of proteins. *Biochemistry* 1971;10(18):3417–3423.
83. Privalov PL. Stability of proteins: small globular proteins. *Adv Protein Chem* 1979;33:167–241.
84. Privalov PL. Thermal investigations of biopolymer solutions and scanning microcalorimetry. *FEBS Lett* 1974;40(suppl):S140–S153.
85. Borchardt HJ, Daniels F. Application of differential thermal analysis to the study of reaction kinetics. *J Am Chem Soc* 1957;79:41–46.
86. Cooper A, Nutley MA, Wadwood A. In: *Differential Scanning Microcalorimetry in Protein-Ligand Interactions: Hydrodynamics and Calorimetry*, Harding SE, Chowdhry BZ (eds.), Oxford/New York: Oxford University Press, 2000; pp. 287–318.
87. Donovan JW, Ross KD. Increase in the stability of avidin produced by binding of biotin: a differential scanning calorimetric study of denaturation by heat. *Biochemistry* 1973;12(3):512–517.
88. Fitzsimons SM, Mulvihill DM, Morris ER. *Thermal Analysis Congress*. Liverpool: John Moore's University, 2004.
89. Bacon JR, Noel TR, Wright DJ. Studies on the thermal behavior of pea (*Pisum sativum*) vicilin. *J Sci Food Agric* 1989;49(3):335–345.
90. Badii F, MacNaughtan W, Farhat IA. Enthalpy relaxation of gelatin in the glassy state. *Int J Biol Macromol* 2005;36(4):263–269.
91. Morales A, Kokini JL. Glass transition of soy globulins using differential scanning calorimetry and mechanical spectrometry. *Biotechnol Prog* 1997;13(5):624–629.
92. Badii F, MacNaughtan W, Mitchell JR, Farhat IA, Gregson K. *Modelling of Ageing in Gelatin*, paper in preparation, 2006.
93. Farahnaky A, Badii F, Farhat IA, Mitchell JR, Hill SE. Enthalpy relaxation of bovine serum albumin and implications for its storage in the glassy state. *Biopolymers* 2005;78(2):69–77.
94. Kasapis S, Al-Marhoobi IMA, Deszczynski M, Mitchell JR, Abeysekera R. Structural properties of gelatin in mixture with sugar. *Gums and Stabilisers for the Food Industry 12*, Vol. 294. Cambridge: Royal Society of Chemistry, 2004; pp. 437–449. (Special Publication)
95. Nagatsuka N, Matsushita K, Nishina M, Okawa Y, Ohno T, Nagao K. Factors affecting the gelation of a gelatin solution in the presence of sugar. *Nippon Kasei Gakkaishi* 2004;55(2):159–166.

96. Badii F, Martinet C, Mitchell JR, Farhat IA. Enthalpy and mechanical relaxation of glassy gelatin films. *Food Hydrocolloid* 2006;20(6):879–884.
97. Farkas J. Applications of DSC thermoanalysis in food science. *Elelmiszervizsgalati Kozlemenyek* 1994;40(3):180–189.
98. Hashimoto T, Suzuki T, Hagiwara T, Takai R. Study on the glass transition for several processed fish muscles and its protein fractions using differential scanning calorimetry. *Fish Sci* (Carlton, Australia) 2004;70(6):1144–1152.
99. Jensen KN, Jorgensen BM. Effect of storage conditions on differential scanning calorimetry profiles from thawed cod muscle. *Lebensmittel Wiss Technol* 2003;36(8):807–812.
100. Jensen KN, Jorgensen BM, Nielsen J. Low-temperature transitions in cod and tuna determined by differential scanning calorimetry. *Lebensmittel Wiss Technol* 2003;36(3):369–374.
101. Schubring R. Differential scanning calorimetric investigations on pyloric caeca during ripening of salted herring products. *J Therm Anal Calorimetry* 1999;57(1):283–291.
102. Wright DJ, Leach IB, Wilding P. Differential scanning calorimetric studies of muscle and its constituent proteins. *J Sci Food Agric* 1977;28(6):557–564.
103. Quinn JR, Raymond DP, Harwalkar VR. Differential scanning calorimetry of meat proteins as affected by processing treatment. Food Sci 1980;45(5):1146–1149.
104. Donovan JW, Mapes CJ, Davis JG, Garibaldi JA. Differential scanning calorimetric study of the stability of egg white to heat denaturation. *J Sci Food Agric* 1975;26(1):73–83.
105. Hagerdal B, Martens H. Influence of water content on the stability of myoglobin to heat treatment. *J Food Sci* 1976;41(4):933–937.
106. Donovan JW, Mapes CJ. A differential scanning calorimetric study of conversion of ovalbumin to S-ovalbumin in eggs. *J Sci Food Agric* 1976;27(2):197–204.
107. Rouilly A, Orliac O, Silvestre F, Rigal L. Thermal denaturation of sunflower globulins in low moisture conditions. *Thermochim Acta* 2003;398(1–2):195–201.
108. Noel TR, Parker R, Ring SG, Tatham AS. The glass-transition behaviour of wheat gluten proteins. *Int J Biol Macromol* 1995;17(2):81–85.
109. Pouplin M, Redl A, Gontard N. Glass transition of wheat gluten plasticized with water, glycerol, or sorbitol. *J Agric Food Chem* 1999;47(2):538–543.
110. Xie F, Dowell FE, Sun XS. Using visible and near-infrared reflectance spectroscopy and differential scanning calorimetry to study starch, protein, and temperature effects on bread staling. *Cereal Chem* 2004;81(2):249–254.
111. Morales A, Kokini JL. State diagrams of soy globulins. *J Rheol* 1999;43(2):315–325.
112. Gontard N. Edible and biodegradable films: study of wheat gluten film-forming properties. *Comptes Rendus de l'Academie d'Agriculture de France* 1994;80(4):109–117.
113. Burova TV, Grinberg NV, Golubeva IA, Mashkevich AY, Grinberg VY, Tolstoguzov VB. Flavor release in model bovine serum albumin/pectin/2-octanone systems. *Food Hydrocolloid* 1998;13(1):7–14.
114. Chang L, Shepherd D, Sun J, et al. Mechanism of protein stabilization by sugars during freeze-drying and storage: native structure preservation, specific interaction, and/or immobilization in a glassy matrix? *J Pharm Sci* 2005;94(7):1427–1444.
115. Comfort S, Howell NK. Gelation properties of salt soluble meat protein and soluble wheat protein mixtures. *Food Hydrocolloid* 2003;17(2):149–159.
116. Mitchell JR. Proteins as hydocolloids. Hydrocolloid course given at Department of Food Sciences, Nottingham University, 2005.
117. Williams PA. Polysaccharide structure and properties. Hydrocolloid course given at Department of Food Sciences, Nottingham University, 2005.
118. Oakenfull D. A method for using measurements of shear modulus to estimate the size and thermodynamic stability of junction zones in noncovalently crosslinked gels. *J Food Sci* 1984;49(4):1103–1104, 1110.

119. Williams PA, Day DH, Langdon MJ, Phillips GO, Nishinari K. Synergistic interaction of xanthan gum with glucomannans and galactomannans. *Food Hydrocolloid* 1991;4(6):489–493.
120. Haque A, Morris ER. Thermogelation of methylcellulose. Part I: Molecular structures and processes. *Carbohydr Poly* 1993;22(3):161–173.
121. Evageliou V, Kasapis S, Sworn G. The transformation from enthalpic gels to entropic rubbery and glass-like states in high-sugar biopolymer systems. *Gums and Stabilisers for the Food Industry 9*, Vol. 218. Cambridge: Royal Society of Chemistry, 1998; pp. 333–344. (Special Publication)
122. Kasapis S, Al-Marhoobi IM, Deszczynski M, Mitchell JR, Abeysekera R. Gelatin vs polysaccharide in mixture with sugar. *Biomacromolecules* 2003;4(5):1142–1149.
123. Kasapis S, Abeysekera R, Atkin N, Deszczynski M, Mitchell JR. Tangible evidence of the transformation from enthalpic to entropic gellan networks at high levels of co-solute. *Carbohydr Polym* 2002;50(3):259–262.
124. Kasapis S, Mitchell J, Abeysekera R, MacNaughtan W. Rubber-to-glass transitions in high sugar/biopolymer mixtures. *Trends Food Sci Technol* 2004;15(6):298–304.
125. Kasapis S. Definition of a mechanical glass transition temperature for dehydrated foods. *J Agric Food Chem* 2004;52(8):2262–2268.
126. Aklonis JJ, McKnight WJ. *Introduction to Polymer Viscoelasticity*. New York: Wiley, 1983.
127. Kasapis S, Al-Marhoobi IM, Mitchell JR. Testing the validity of comparisons between the rheological and the calorimetric glass transition temperatures. *Carbohydr Res* 2003;338(8):787–794.
128. Kasapis S, Al-Alawi A, Guizani N, Khan AJ, Mitchell JR. Viscoelastic properties of pectin-co-solute mixtures at iso-free-volume states. *Carbohydr Res* 2000;329(2):399–407.
129. Schawe JEK. Investigations of the glass transitions of organic and inorganic substances: DSC and temperature-modulated DSC. *J Therm Anal* 1996;47(2):475–484.
130. Weyer S, Merzlyakov M, Schick C. Application of an extended Tool-Narayanaswamy-Moynihan model. Part 1: description of vitrification and complex heat capacity measured by temperature-modulated DSC. *Thermochim Acta* 2001;377(1–2):85–96.
131. Brummel SE, Lee K. Soluble hydrocolloids enable fat reduction in process cheese spreads. *J Food Sci* 1990;55(5):1290–1292, 1307.
132. Snoeren THM. Use and action of hydrocolloids in milk products. *LWT Edition [Gelier Verdickungsm Lebensm]* 1980;5:263–274.
133. Morrison NA, Sworn G, Clark RC, Chen YL, Talashek T. Gelatin alternatives for the food industry. *Prog Colloid Polym Sci, Physical Chemistry and Industrial Application of Gellan Gum* 1999;114:127–131.
134. Hills BP, Takacs SF, Belton PS. A new interpretation of proton NMR relaxation time measurements of water in food. *Food Chem* 1990;37(2):95–111.
135. Belton PS, Hills BP, Raimbaud ER. The effects of morphology and exchange on proton NMR relaxation in agarose gels. *Mol Phys* 1988;63(5):825–842.
136. Belton PS. NMR and the mobility of water in polysaccharide gels. *Int J Biol Macromol* 1997;21(1–2):81–88.
137. Ablett S, Izzard MJ, Lillford PJ. Differential scanning calorimetric study of frozen sucrose and glycerol solutions. *J Chem Soc Faraday Trans* 1992;88(6):789–794.
138. Ablett S, Clark AH, Izzard MJ, Lillford PJ. Modeling of heat capacity-temperature data for sucrose-water systems. *J Chem Soc Faraday Trans* 1992;88(6):795–802.
139. Rahman MS, Sablani SS, Al-Habsi N, Al-Maskri S, Al-Belushi R. State diagram of freeze-dried garlic powder by differential scanning calorimetry and cooling curve methods. *J Food Sci* 2005;70(2):E135–E141.
140. Izzard MJ, Ablett S, Lillford PJ, Hill VL, Groves IF. A modulated differential scanning calorimetric study: glass transitions occurring in sucrose solutions. *J Therm Anal* 1996;47(5):1407–1418.
141. Kasapis S, Mitchell JR. Definition of the rheological glass transition temperature in association with the concept of iso-free-volume. *Int J Biol Macromol* 2001;29(4–5):315–321.

142. Braga Da Cruz I, MacInnes WM, Oliveira JC, Malcata FX. Supplemented state diagram for sucrose from dynamic mechanical thermal analysis. *Amorphous Food and Pharmaceutical Systems*, Vol. 281. Cambridge: Royal Society of Chemistry, 2002; pp. 59–70. (Special Publication)
143. Cruz IB, Oliveira JC, MacInnes WM. Dynamic mechanical thermal analysis of aqueous sugar solutions containing fructose, glucose, sucrose, maltose and lactose. *Int J Food Sci Technol* 2001;36(5):539–550.
144. Rasmussen DH, MacKenzie AP. Effect of solute on ice-solution interfacial free energy: calculation from measured homogeneous nucleation temperatures. *Water Struct Water Polymer Interface, Proc Symp* 1972:126–145.
145. Sutton RL, Lips A, Piccirillo G. Recrystallization in aqueous fructose solutions as affected by locust bean gum. *J Food Sci* 1996;61(4):746–748.
146. Sutton RL, Lips A, Piccirillo G, Sztehlo A. Kinetics of ice recrystallization in aqueous fructose solutions. *J Food Sci* 1996;61(4):741–745.
147. Muhr AH, Blanshard JMV. Effect of polysaccharide stabilizers on the rate of growth of ice. *J Food Technol* 1986;21(6):683–710.
148. Alvarez VB, Wolters CL, Vodovotz Y, Ji T. Physical properties of ice cream containing milk protein concentrates. *J Dairy Sci* 2005;88(3):862–871.
149. Granger C, Schoeppe A, Leger A, Barey P, Cansell M. Influence of formulation on the thermal behavior of ice cream mix and ice cream. *J Am Oil Chem Soc* 2005;82(6):427–431.
150. Muhr AH. *The Influence of Polysaccharides on Ice Formation in Sucrose Solutions*, PhD Thesis. Nottingham University, Nottingham, 1983.
151. Ferrero C, Martino M, Zaritzky N. Influence of xanthan gum addition on frozen starch paste properties. *Proc Int Conf Ind Exhib, Food Hydrocolloid* 1993:461–466.
152. Bot A. Differential scanning calorimetric study on the effects of frozen storage on gluten and dough. *Cereal Chem* 2003;80(4):366–370.
153. Leon A, Duran E, Benedito de Barber C. A new approach to study starch changes occurring in the dough-baking process and during bread storage. Zeitschrift fuer Lebensmittel-Untersuchung und Forschung A. *Food Res Technol* 1997;204(4):316–320.
154. Cornillon P. Characterization of osmotic dehydrated apple by NMR and DSC. *Lebensmittel Wiss Technol* 2000;33(4):261–267.
155. Kasapis S, Rahman MS, Guizani N, Al-Aamri M. State diagram of temperature vs date solids obtained from the mature fruit. *J Agric Food Chem* 2000;48(9):3779–3784.
156. Vodovotz Y, Hallberg L, Chinachoti P. Effect of aging and drying on thermomechanical properties of white bread as characterized by dynamic mechanical analysis (DMA) and differential scanning calorimetry (DSC). *Cereal Chem* 1996;73(2):264–270.
157. Farrant JM, Walters C. Ultrastructural and biophysical changes in developing embryos of Aesculus hippocastanum in relation to the acquisition of tolerance to drying. *Physiol Plant* 1998;104(4):513–524.
158. Dussert S, Charbrillange N, Rocquelin G, Engelmann F, Lopez M, Hamon S. Tolerance of coffee (*Coffea* spp.) seeds to ultra-low temperature exposure in relation to calorimetric properties of tissue water, lipid composition, and cooling procedure. *Physiol Plant* 2001;112(4):495–504.
159. Buitink J, Walters C, Hoekstra FA, Crane J. Storage behavior of typha latifolia pollen at low water contents: interpretation on the basis of water activity and glass concepts. *Physiol Plant* 1998;103(2):145–153.
160. Walters C. Temperature dependency of molecular mobility in preserved seeds. *Biophys J* 2004;86(2):1253–1258.
161. Crane J, Miller AL, Van Roekel JW, Walters C. Triacylglycerols determine the unusual storage physiology of Cuphea seed. *Planta* 2003;217(5):699–708.
162. TA Instruments. Characterisation of processing induced changes in morphology by solution calorimetry. Application Note 2225-02 2001.

163. Wiseman T, Williston S, Brandts JF, Lin LN. Rapid measurement of binding constants and heats of binding using a new titration calorimeter. *Anal Biochem* 1989;179(1):131–137.
164. Ladbury JE, Chowdhry BZ. Sensing the heat. *Chem Biol* 1996;3:791–801.
165. Handout to Applications of BioCalorimetry (abc 5) organised by Microcal., Zaragoza, Spain, July, 2006.
166. Siebert KJ. Effects of protein-polyphenol interactions on beverage haze, stabilization, and analysis. *J Agric Food Chem* 1999;47(2):353–362.
167. Pripp AH, Vreeker R, van Duynhoven J. Binding of olive oil phenolics to food proteins. *J Sci Food Agric* 2005;85(3):354–362.
168. O'neill MAA, Beezer AE, Tetteh J, et al. Determination of kinetic and thermodynamic parameters from complex reaction systems. *Thermal Methods Group Meeting*, Kings College, London, November, 2003.
169. Ziderman II, Gregorski KS, Friedman M. Thermal analysis of protein-carbohydrate mixtures in oxygen. *Thermochim Acta* 1987;114(1):109–114.
170. Haines P. *Thermal Analysis*, Chapter 2. Cambridge UK: Royal Society of Chemistry, 2002.
171. Biswas S, Staff C. Analysis of headspace compounds of distillers grains using SPME in conjunction with GC/MS and TGA. *J Cereal Sci* 2001;33(2):223–229.
172. Taru Y, Takaoka K. Thermal stability and mechanism of the thermal degradation of the coal-tar dyes for food additive. *Shikizai Kyokaishi* 1982;55(8):537–545.
173. Sorensen OT, Rouquerol J (eds.). *Sample Controlled Thermal Analysis*, Netherlands: Kluwer Academic Publishers Group, 2003.
174. TA Instruments. The application of solution calorimetry to quantify amorphous lactose occurring at 10% w/w or less. Application Note 2225-01 2001.

Chapter 10
Thermal Analysis of Inorganic Compound Glasses and Glass-Ceramics

David Furniss, Angela B. Seddon

Contents

10.1 Introduction	411
10.2 Background glass science	411
10.2.1 Nature of glasses	411
10.2.2 Crystallisation of glasses	413
10.2.3 Liquid–liquid phase separation	416
10.2.4 Viscosity of the supercooled, glass-forming liquid	416
10.3 Differential thermal analysis	418
10.3.1 General comments	418
10.3.2 Experimental issues	420
10.3.3 DTA case studies	421
10.4 Differential scanning calorimetry	426
10.4.1 General comments	426
10.4.2 Experimental issues	427
10.4.3 DSC case studies	427
10.4.4 Modulated Differential scanning calorimetry case studies	432
10.5 Thermomechanical analysis	432
10.5.1 General comments	432
10.5.2 Experimental issues	433
10.5.3 Linear thermal expansion coefficient (α) and dilatometric softening point (M_g)	433
10.5.4 Temperature coefficient of viscosity: introduction	436
10.5.5 TMA indentation viscometry	438
10.5.6 TMA parallel-plate viscometry	443
10.6 Final comments	447
References	448

10.1 Introduction

The International Commission on Thermal Analysis (ICTA),* when it was set up in 1965, put together rigorous guidelines to bring conformity to the reporting of thermoanalytical results. Since then TA has matured into a key analytical technique. There has been continuous development of convenient modular TA equipment incorporating small responsive furnaces, software control and integral data handling. TA has become a reliable, fast, standard technique for the characterisation of many materials, leading to even greater need for rigour in reporting TA results, which will be addressed here.

This chapter is primarily concerned with how TA can facilitate greater insight into inorganic compound glasses, not only their technology and processing, but also increasing fundamental understanding of their nature. Examples will be drawn from glass systems based on silicates, heavy metal oxides and heavy metal fluorides, oxyfluorides and chalcogenides (*glasses based on Group VIb of the Periodic Table (S, Se and Te) usually together with Groups IVb and Vb (Ge, As, etc.)*). It so happens that the examples taken are mainly of glasses intended for applications in photonics (*communication by the manipulation of light*) but the subject matter is generic and transferable to inorganic glasses intended for most end applications.

The chapter is aimed at materials undergraduates and physical science, and engineering, undergraduates taking materials options, as well as graduate students newly specialising in inorganic compound glasses and glass-ceramics. We very much hope that the examples given here will stimulate analogous investigations into new glass systems. The chapter begins with the background glass science and practice necessary to make the chapter self-supporting. It is then organised into sections according to TA technique covering differential thermal analysis (DTA), differential scanning calorimetry (DSC), including modulated differential scanning calorimetry (MDSC), and thermomechanical analysis (TMA). For each TA technique we point out generalities, practicalities and describe pertinent case studies.

10.2 Background glass science

10.2.1 Nature of glasses

Glasses have been defined as X-ray amorphous materials which exhibit the glass transition (T_g), this being a more or less abrupt change in the derivative thermodynamic properties with increasing temperature, such as change in heat capacity at constant pressure and thermal expansion coefficient, from crystal-like values to liquid-like values (see Figure 10.1) [1].

Figure 10.1 is probably the most important diagram in glass science and illustrates one way of making glasses – that is by quenching the glass-forming liquid from above the equilibrium melting point (T_m) (or liquidus (T_L) for a multicomponent system) fast enough 'to cheat' on the crystallisation process (abcX), which normally occurs on cooling through the liquidus. Thus, heat energy is extracted rapidly from the system so that the atoms (or ions) do not have sufficient time to diffuse and rearrange to form an ordered lattice. As cooling

---

* Since 1992 it has become ICTAC: International Confederation for Thermal Analysis and Calorimetry.

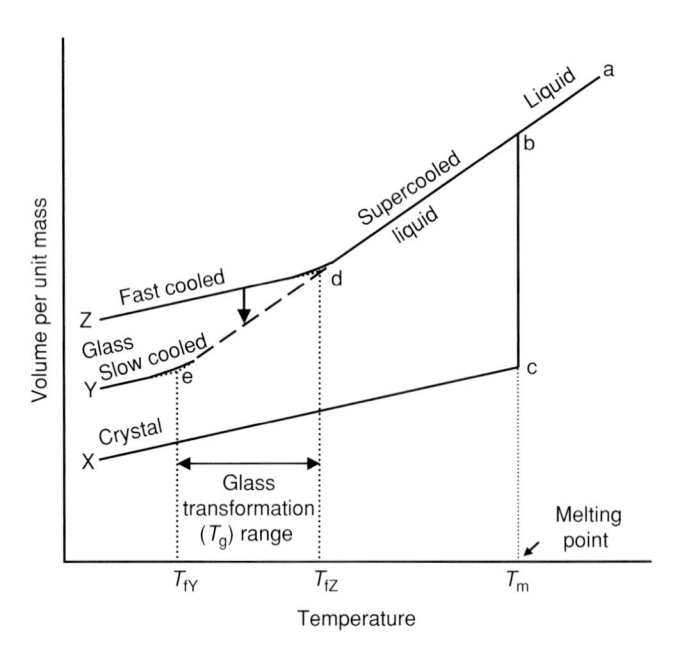

Figure 10.1 Plot of volume per unit mass versus temperature at constant pressure, showing the contrasting behaviour of fast quenched (high fictive temperature T_{fZ}) and slow quenched (low fictive temperature T_{fY}) glasses and equilibrium freezing to the crystalline state.

proceeds (abde), the supercooled liquid increases in viscosity (macroscopic resistance to flow), eventually reaching $\sim 10^{12.5}$ Pa s in the glass transition temperature range (T_g). Here the supercooled liquid freezes to a glass, i.e. a disordered solid exhibiting the frozen-in structure of the supercooled liquid and a lack of atomic periodicity. Only below T_g is it correct to describe the material as a glass. Glass should be considered as a state of matter rather than a particular material, and in theory any liquid may be quenched from above its liquidus to form a glass so long as the cooling rate is sufficiently fast to prevent crystallisation.

Silicate melts exhibit high viscosities at the liquidus ($10 - 10^2$ Pa s); this resistance to flow reflects the lack of mobility of molecular units relative to one another which inhibits crystallisation on melt-quenching. On the other hand, many of the more novel inorganic glasses have low viscosities at the liquidus; for example, fluorozirconate glasses have a viscosity similar to that displayed by water at standard temperature and pressure ($\sim 10^{-2}$ Pa s) [2]. For glasses such as these, rather it is considered to be competition for growth between several crystalline phases that accounts for the low rate of crystallisation manifested on cooling below the liquidus [3]. This last observation is embodied in the so-called 'confusion principle' of glass-making: that incorporating more components in the glass melt usually, but not always, enhances glass-forming ability during melt-cooling.

Glasses, being metastable, have no place on equilibrium phase diagrams, but it should be noted that glass formation is likely to occur most readily close to a eutectic [4], being a low-lying discrete liquidus encompassing intimate competition for solid phase growth.

Figure 10.1 shows that the supercooled liquid decreases in volume with decrease in temperature along bd. This is partly due to a decrease in the mean position of each atomic bond vibration and partly due to a configurational volume contraction of the disordered structure. When the fast-supercooled liquid is held isothermally in the high-viscosity glass transformation region, its volume tends to decrease (occurs *slowly*) to the dashed curve 'de' in Figure 10.1, which is an extrapolation of the equilibrium (more correctly *metastable* equilibrium) contraction curve of the supercooled liquid, to approach a volume configuration characteristic of the supercooled liquid at that temperature.

In Figure 10.1, density decreases along the ordinate (*vertical axis*) as molar volume increases. On cooling to ambient, the density of the glass obtained is affected by the rate of cooling through T_g (compare abdZ with abdeY). Such density difference, primarily caused by the actual configurational structural volume frozen into the glass in the glass transformation region, manifests as differences in properties of the glass so-obtained, for instance refractive index. The potential for controlling glass density, and density-dependent properties, by controlling the cooling behaviour through the glass transformation region has been usefully embodied in the concept of the fictive temperature of a glass.

Fictive temperature is that temperature in the glass transformation region at which the cooling curve achieved empirically during a melt-quench departs from the equilibrium contraction curve of the supercooled liquid (e.g. at d or e in Figure 10.1). The fictive temperature therefore notionally indicates the extent of configurational volume frozen into the glass at T_g and the fictive temperature of a glass could usefully be a performance indicator of glass properties below T_g. More complicated cooling schedules through the T_g region may be obtained if the supercooled liquid lingers in the T_g region. Thus, suppose the melt quenching of the supercooled liquid along dZ (Figure 10.1) is arrested and there is an isothermal hold such that the configurational volume contracts partially, but not fully, down the arrow in Figure 10.1 before fast cooling is resumed. The fictive temperature in that case would be the temperature at which the extrapolated-back, resumed-fast-cooling curve intersects the equilibrium contraction curve of the supercooled melt.

It is very important to avoid temperature variation through the glass article during cooling in the glass transformation region as local fictive temperature variation across the article can result in permanent strain 'frozen into' the finished glass sufficient to cause fracture. It is for this reason that annealing of glasses is carried out. Unlike in the field of metallurgy where the term annealing has more general application and means a heat treatment, glass annealing is a specific type of heat treatment designed to bring the whole article to the same temperature at T_g and then to cool it slowly through the glass transformation region to achieve the same configurational volume structure throughout the article so as to avoid any permanent strain.

It is worth noting that above T_g, supercooled inorganic compound glass melts generally exhibit Newtonian viscosity (see Section 10.5.4). In the T_g region itself they are viscoelastic supercooled liquids, transforming below T_g to glasses which commonly behave elastically up to fracture under normal loads.

10.2.2 Crystallisation of glasses

Devitrification (i.e. crystallisation) of the supercooled liquid is free energy driven but kinetically limited. During melt quenching, the liquid above T_M (or T_L if multicomponent) has no thermodynamic tendency to crystallise (Figure 10.2). However, in the temperature range

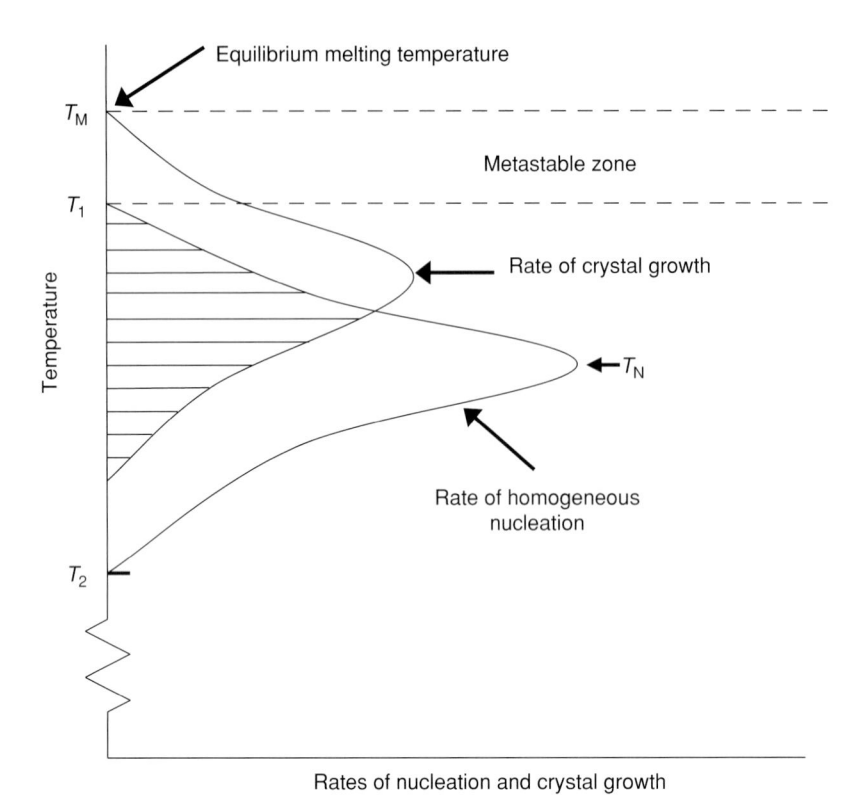

Figure 10.2 Schematic plot showing change in homogeneous nucleation and crystal growth rates with temperature [5]. On melt-cooling to just below T_M, homogeneous nucleation is absent in the small temperature range T_M to T_1 termed the metastable zone. However, heterogeneous nucleation can occur in the metastable zone if heterogeneous nuclei are present, for instance provided by the crucible wall in contact with a glass-forming melt. Below T_1 the homogeneous nucleation rate tends to increase with decreasing temperature but this tendency is offset by the lowered diffusion rates as melt viscosity increases with lowering of temperature and a zero nucleation rate will be obtained at T_2 as T_g is approached. The peak rate of crystal growth occurs at the same, or a higher, temperature than the peak homogeneous nucleation rate (T_N). The hatched region in the figure is where simultaneous nucleation and crystal growth can occur and time spent here should be limited on melt-cooling or glass reheating to avoid crystals in the finished glass.

from T_M down to T_g, a critical cooling rate (CCR) characteristic of the melt must be exceeded to avoid devitrification, for instance during a shaping process like casting into a mould.

On the other hand, the annealed glass, having been cooled to ambient, can be reheated above T_g to reaccess the supercooled liquid, again for shaping purposes, during which time again unwanted devitrification may occur.

The devitrification process, whether during melt-cooling or on glass-reheating, is composed of a nucleation step and then crystal growth step (Figure 10.2) [5]. Homogeneous nucleation involves the formation of embryos of longer range atomic order than normally present in the liquid state. These continuously form and then redissolve in the supercooled liquid due to competition between: (i) the surface energy required for embryo formation and (ii) the bulk free energy released as the disordered structure orders. With temperature decrease below T_L, embryos tend to exceed a critical diameter and become stable nuclei. Crystal growth can then occur via diffusion of atoms or ions to the nucleus surface. Homogeneous nuclei are likely to be of nanometre-scale and are of the same phase as the larger crystal. The rates of homogeneous nucleation and crystal growth tend to increase with decreasing temperature below the liquidus until T_g (T_g is at, or somewhat below, T_2 in Figure 10.2) is approached where the high viscosities encountered prevent atomic rearrangement, and diffusion, and inhibit devitrification. The temperature of the peak homogeneous nucleation rate (T_N) tends to coincide with, or be lower than, the peak temperature of crystal growth. On melt-cooling, or reheating of the glass, it is the overlap of the homogeneous nucleation and crystal growth curves (hatched area in Figure 10.2), where both homogeneous nucleation and crystal growth can occur simultaneously, that is the most problematic temperature range. Homogeneously nucleated devitrification leads to the glass containing many small crystals, rather evenly distributed throughout the bulk. Homogeneous nucleation, but not devitrification, may occur on melt cooling to produce a seemingly crystal-free glass, which in fact contains latent nuclei which may grow to crystals on subsequent heat treatment. Similarly, reheating an already successfully quenched glass from below T_g to above T_g into the temperature region of peak homogeneous nucleation (T_N) and then cooling through T_g to ambient may result in nuclei, dispersed in the glass matrix, but not necessarily crystal growth until an additional heat treatment at a temperature higher than T_N is subsequently carried out. Non-homogenised glass melts on cooling may exhibit patchy bulk homogeneous nucleation and crystal growth at locations where the glass melt itself, through lack of proper mixing, has strayed from the stable composition regime.

Heterogeneous nucleation is more commonplace than homogeneous nucleation and takes place at foreign surfaces in contact with the supercooled melt. The heterogeneous nuclei are not of the same phase as the growing crystal. Heterogeneous nucleation tends to increase with decreasing temperature below the liquidus and is finite through the so-called metastable zone (Figure 10.2), where homogeneous nucleation is inhibited primarily due to the low-free-energy difference between crystal and liquid. Heterogeneously nucleated crystals formed at higher temperatures during melt-cooling tend to be large, isolated and faceted crystals of the equilibrium phase. In contrast, crystals formed during glass-reheating to relatively low temperatures above T_g, at large undercoolings where diffusion coefficients are smaller, tend to grow as dense dendrites, preferentially located at the glass surface and/or at the surfaces of the container, seed (*bubbles*) or undissolved impurities. Also, at these lower temperatures, the supercooled liquid tends first to form intermediate, metastable crystalline phases rather than the equilibrium phase.

Glass-ceramics are made by the controlled partial, or complete, devitrification of the parent glass. A classical example is lithium disilicate glass for which the daughter glass ceramic may be made from the parent glass by means of a two-stage heat treatment: firstly, an isothermal hold at the temperature maximum of the homogeneous nucleation rate (T_N in Figure 10.2) to form nuclei efficiently followed by a second isothermal hold at the higher temperature of maximum crystal growth rate in order to grow deliberately these nuclei [6] (see Figure 10.2). Alternatively, heterogeneous nuclei can be deliberately added throughout the bulk as a dispersion of insoluble foreign particles and then the glass heat-treated to effect crystal growth. We have coined the term nano-glass-ceramics for use where the final crystallite diameter is too small to cause light scattering in the visible for optical path lengths of $\sim$10 mm and hence any visible transparency of the parent glass has been unchanged by ceramming.

10.2.3 Liquid–liquid phase separation

Some glass-forming melts exhibit liquid–liquid immiscibility, that is phase separation in the liquid state. Similar effects are readily observable by shaking oil and water together. Such immiscibility present between T_g and the liquidus is termed metastable, sub-liquidus liquid–liquid immiscibility, and as long as the melt-quenching during glass making is carried out quickly enough a single glass phase will result. However, lingering in the temperature range between T_g and the liquidus for sufficient time, either during melt-cooling or reheating from below T_g, will result in the supercooled single-phase liquid tending to phase-separate into two or more liquids. Moreover, it should be noted that metastable liquid–liquid phase separation can sometimes be a precursor state for devitrification of one or more of the liquid phases. The phase transformation of the initial single liquid into two or more liquids is accompanied by an overall lowering of free energy, albeit to a state of better metastability. The mechanism of transformation is governed by the presence or absence of an activation energy barrier to the transformation. In the absence of an activation barrier, a spinodal decomposition [7] occurs, which is spontaneous and results in an interdigitated structure of the liquids at temperatures between T_g and the liquidus; this structure is frozen-in at T_g to give similarly nano- or microstructured glasses. Conversely, a free energy barrier leads to liquid–liquid phase separation via nucleation and growth of one or more discrete droplet phases in the residual liquid, which is the continuous phase. Again this microstructure is frozen-in at T_g and is diagnostic of the mechanism of phase separation.

10.2.4 Viscosity of the supercooled, glass-forming liquid

T_g is an isoviscous point for all glasses occurring at about $10^{12.5}$ Pa s. The ratio of T_g to the liquidus, expressed in Kelvin, for most glasses is usually close to 0.67. This so-called *two-thirds rule* is a useful rule of thumb.

Above T_g, the supercooled glass-forming liquid exhibits decreasing viscosity with increasing temperature. On the one hand the glass-forming liquid of silica exhibits a relatively small and constant temperature coefficient of viscosity, e.g. decreasing from $\sim 10^{12.5}$ Pa s (T_g) at 1050°C to $\sim 10^{4.5}$ Pa s ($\sim$fibre-drawing viscosity) at 1730°C. In complete contrast to this, multicomponent heavy metal fluoride glass-forming melts have a nonlinear temperature coefficient of viscosity and experience the same order of magnitude decrease in viscosity in only 100°C. These are traditionally termed 'short' glasses (*refers to the length of the working temperature range*). Angell [8] has recognised the value of the universal plot of $\log_{10}$(viscosity) versus reduced temperature (i.e. $T_g(K)/T(K)$) on which the behaviour of all glass-forming melts may be compared (Figure 10.3). As T_g is a universal viscosity point, all glasses are coincident at the reduced T_g (=1).

As Figure 10.3 shows, the silica glass-forming liquid exhibits one extreme of behaviour, the plot being almost linear. As the temperature is raised above T_g, the silica liquid becomes more fluid; the atomic units which flow past one another under shear become smaller but the mechanism of shear remains cleavage of silica bridging chemical bonds ($\equiv$Si–O–Si$\equiv$) of covalent character, which are successively increasingly broken in number as the temperature is increased. This accounts for the linearity (Arrhenian) of the silica plot. Silica is described by Angell as a *strong* liquid, meaning a three-dimensional, mutually supporting, strongly

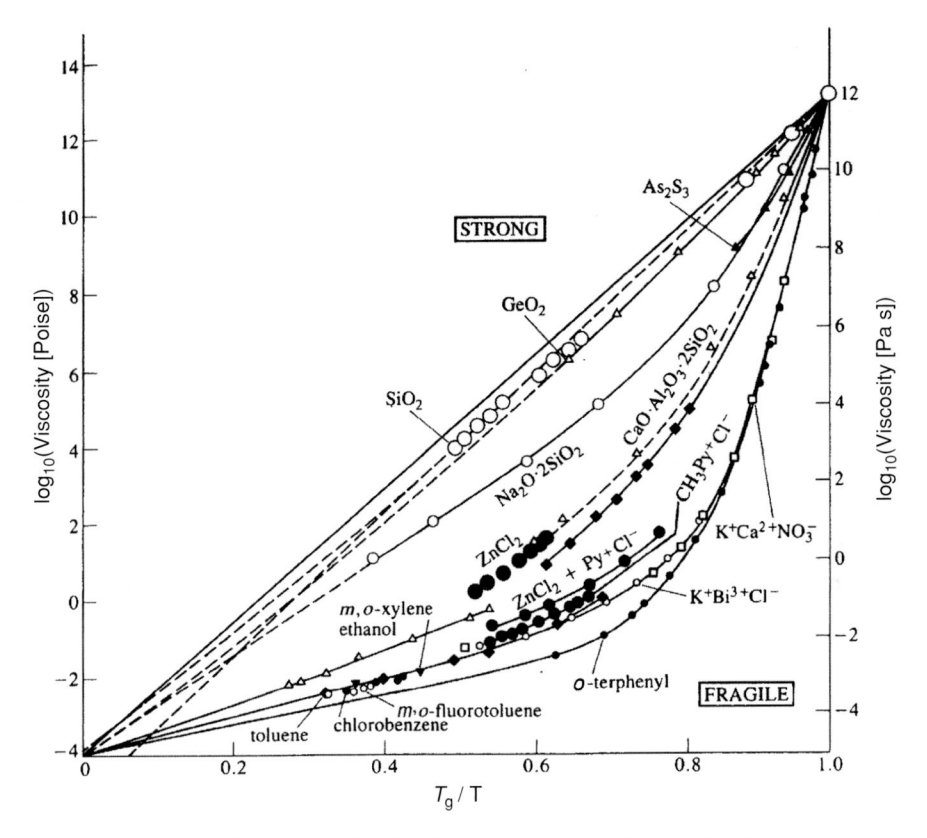

Figure 10.3 Viscosities of a variety of glass-forming systems as a function of reduced reciprocal temperature (Kelvin) T_g/T, where T_g is defined as the temperature when the viscosity $\eta = 10^{12}$ Pa s. The difference in temperature-dependent behaviour between strong and fragile glass-forming liquids is clear [8].

covalently bonded liquid. In direct contrast, multicomponent fluorozirconate ionic liquids experience a catastrophic breakdown of the lattice just above T_g and consequent large fall in viscosity; they are termed *fragile* liquids. The multicomponent fluorozirconate glass lattice is composed of a variety of relatively weak chemical bonds and as the temperature is increased above T_g, stronger chemical bonds are successively broken under shear, accounting for the nonlinear $\log_{10}$ (viscosity) versus reduced-temperature behaviour.

It is worth pointing out that many of the novel inorganic compound glasses being developed relatively recently exhibit more fragile behaviour than silica, implying that rigorous temperature control is thus a prerequisite for shaping operations.

The viscosity (η)–temperature (T) relation of strong glass-forming liquids follows the Vogel–Fulcher–Tammann model [9]:

$$\eta = A \exp\left(\frac{B}{T - T_0}\right) \tag{10.1}$$

where A, B and T_0 are material constants. Some fragile liquids have been demonstrated to follow more closely the Cohen–Grest model [10].

10.3 Differential thermal analysis

10.3.1 General comments

Differential thermal analysis (DTA) and differential scanning calorimetry (DSC; see Section 10.4) are sister techniques which provide fast, convenient analysis of the glass characteristic temperatures such as T_g and temperatures of phase changes, for instance devitrification and melting. The glass sample must neither gain nor lose mass during a DTA or DSC run. DTA operates up to higher temperatures (~1500°C) than does true DSC (~730°C). Unlike DSC, DTA does not allow accurate measurement of enthalpy changes although DTA can be calibrated to give an indication of the enthalpy changes involved.

During DTA the temperature difference (ΔT) between a small sample of the glass and a reference material is recorded, whilst each is supplied with the same rate of heat input. The reference material is chosen for its inertness over the temperature range of the DTA run.

When a new glass has been made, its DTA or DSC curve (conveniently collected by heating at 10°C/min) guides the choice of optimum temperature schedules for the future melting, annealing and shaping processes for this glass composition. Figure 10.4 depicts a typical DTA curve obtained on heating an inorganic compound glass. It shows the geometrical constructions required to specify the characteristic temperatures of a glass from its DTA or DSC curve, viz. T_g, $Tc_{1,onset}$ (onset temperature of first crystallisation on reheating above T_g), T_S and T_L (solidus and liquidus). By convention, DTA (and heat flux DSC) curves are displayed with

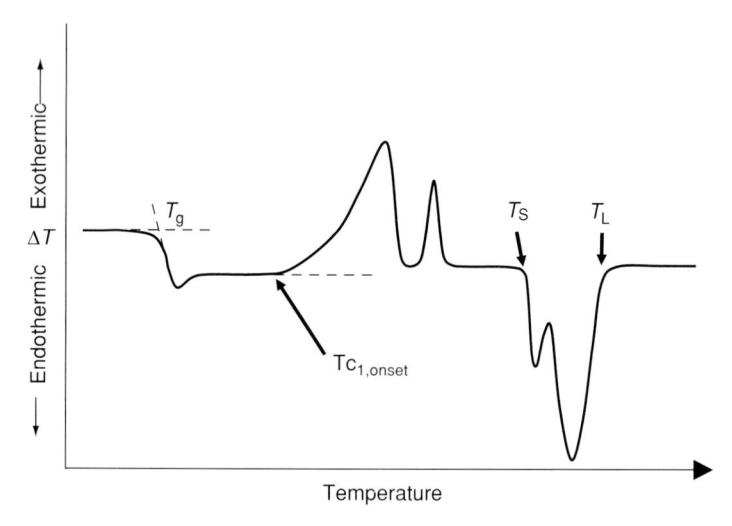

Figure 10.4 Schematic of a typical DTA curve of an inorganic compound glass, showing the geometrical constructions required to specify the characteristic temperatures of the glass, viz. the glass transformation temperature (T_g), the first onset of crystallisation above T_g ($Tc_{1,onset}$), the solidus (T_S) and liquidus (T_L).

peaks due to endothermic events (such as melting) pointing downwards and power compensation DSC curves with peaks due to endothermic events pointing upwards.

It is clear, from Figure 10.4, that at temperatures below T_g the DTA baseline is flat and there are no thermal events other than the controlled temperature rise of sample and reference according to their heat diffusivities (which are dependent on individual thermal conductivity, density and molar heat capacity). At T_g, the glass transforms to the supercooled liquid accompanied by an endothermic shift in the DTA baseline as the sample diffusivity changes from crystal-like to liquid-like values. Above T_g, the now shifted baseline is flat again and controlled temperature rise of the supercooled liquid and reference material takes place. Stepwise crystallisation of the supercooled liquid then occurs (see Section 10.2.2), and the lower temperature crystallisation often proceeds via metastable phases. The enthalpy evolved on crystallisation is manifested as exothermic peaks on the DTA curve. The crystallisation process can involve solid-state transformations and these may be endothermic and superposed on the exothermic phase changes. T_S and T_L mark the onset and end, respectively, of the melting events (crystal-to-liquid transformations). These events are manifested as endothermic DTA peaks. Above the liquidus, T_L, the equilibrium liquid (or liquids if immiscibility occurs) is 'sitting' in the DTA crucible. Thus, the DTA or DSC curve is a pictorial record of the events encountered as one heats or cools a glass isochronally (i.e. with constant change in temperature per constant time change) or isothermally.

A rather crude, but useful, measure of glass stability towards crystallisation on reheating is the temperature gap: $[Tc_{1,onset} - T_g]$; this is known as the Hruby parameter [11] of glass stability. Glass forming (i.e. shaping) may be carried out within this temperature gap, provided that the viscosity required for the forming process is accessible within the temperature range of the gap. Nevertheless, such heat treatment may 'sow the seeds' of homogeneous nuclei, leaving the potential for crystal growth during a subsequent heat treatment (see Section 10.2.2).

The extrapolated onset of T_g from a DTA or DSC curve is a guide to successful annealing for the particular glass composition. Because the thermal history of the sample can affect the measured T_g, properly annealed glass should be used. Alternatively, the glass sample should be run at, for instance, 10°C/min to T_g, cooled at the same rate in situ in the TA equipment (to give the sample a known thermal history) and then rerun at the same rate to determine T_g. The thermal history prior to the T_g measurement affects the T_g value found and should be quoted when reporting T_g values. It should also be borne in mind that very unstable glasses may not survive such deliberate pretreatment and devitrify.

For laboratory-scale glass melts ($10^{-3} - 10^{-1}$ kg) we use the following annealing schedule:

(a) isothermal hold at $[T_g + X°C]$ for about 1 h;
(b) then controlled cooling through the T_g region at 1°C/min to $[T_g - 100°C]$;
(c) finally, faster cooling ($\leq 10°C/min$) down to ambient.

X ranges from +10°C for glasses which do not devitrify during a standard DTA or DSC run (e.g. silica glass) to −20°C for those with a small Hruby parameter. For an insulated, resistance-heated annealing furnace, Step (c) is simply achieved by switching the furnace off and allowing the sample to cool in situ.

The onset temperature of crystallisation above T_g, $Tc_{1,onset}$, is best defined conservatively as that temperature at which the baseline first deviates prior to the lowest temperature DTA or DSC crystallisation peak. Typically, $Tc_{1,onset}$ is assessed by expanding this part of the DTA curve and judging the onset baseline deviation by eye.

Heterogeneous nucleation is dependent on kinetic factors such as: the thermal history of the glass; its state of subdivision (i.e. surface area to volume ratio, affecting the number of surface nucleation sites); how much time has been spent at each successive temperature to allow homogeneous nucleation and crystal growth (i.e. DTA heating rate employed) and whether heterogeneous nuclei are present, provided, for instance, by: the contacting surface of the DTA crucible and/or the mechanical surface history (grinding to particulate glass provides high-energy fracture sites which act as heterogeneous nuclei) and chemical surface history (e.g. residual polishing agents) of the glass. $Tc_{1,onset}$ is a good but not absolute guide to onset crystallisation for independent shaping operations.

Certain practices are to be discouraged. The DTA *extrapolated onset* crystallisation temperature is the temperature at the intersection of the extrapolated steepest part of the onset gradient of the DTA crystallisation peak and the extrapolated baseline. It should not be taken as the crystallisation onset because there has already been considerable evolved enthalpy of devitrification. Peak DTA crystallisation temperatures have no physical meaning.

The DTA liquidus temperature (T_L) is the minimum temperature at which the process of glass melting can be carried out. Glasses are usually melted at temperatures well above T_L and hence at lower viscosity to assist melt homogenisation. However, at the end of the glass melting, the melt temperature is lowered to just above the liquidus (guided by T_L) to remove excess heat to aid efficient quenching to T_g before annealing.

10.3.2 Experimental issues

Using optimised TA procedural parameters gives accurate and informative TA output and avoids damaging TA equipment. Points made here apply to many of the TA techniques.

Calibration of DTA equipment, with appropriate temperature material standards (In, Zn and Al melt at 156.6, 419.4 and 660.2°C, respectively), should be carried out regularly (recommended monthly) and always quoted in reported work.

It is important to ensure that there is no adverse crucible–glass chemical reaction, especially above the solidus. For instance, heavy metal oxide glasses, such as those based on TeO_2, may attack and solubilise alumina, thereby actually creating a new glass composition in situ during the DTA run. Platinum and gold crucibles are generally the most inert crucible materials but even these may be attacked and solubilised slowly by silicate, other oxide glasses and heavy metal fluoride glasses. Pt scattering centres from containment have been detected in ZBLAN glass optical fibres [12]. All crucibles should be scrupulously cleaned and dried prior to DTA runs [13]. For instance, platinum crucibles to be reused after thermal analysis of silicate glasses can be cleaned: by subjecting to a Na_2CO_3 fusion; digesting the fused product in *aqueous* HCl; followed by rinsing the crucible in copious distilled water and then high-purity propan-2-ol and finally drying to constant mass at 100°C in a clean drying oven and cooling in a desiccator. Platinum crucibles to be reused after thermal analysis of fluorozirconate glasses are best soaked overnight at room temperature in a mixture of 1 M $ZrOCl_2 \cdot 8H_2O$ and 1 M HCl (M is molarity) and then rinsed and dried in the same way. A rigorous cleaning and drying procedure should be devised and adhered to for each glass–crucible pairing.

Many non-oxide glasses are attacked by oxygen gas and most glasses are hydrolysed at high temperatures by water vapour at a greater or lesser rate. Hence, if an unlidded crucible is to be used to hold the sample then the DTA crucible head assembly must be designed to retain an inert static or dynamic atmosphere around the crucible. A dynamic atmosphere of high-quality grade argon gas (typically 0.03 dm^3/min through a $\sim 10^3$ mm^3 DTA (or DSC) cell), which may be further dried by passage through 0.4-nm-pore-diameter molecular sieves, is recommended. Take heed that fragile glass melts (see Section 10.2.4) heated to the liquidus in an open DTA crucible may creep by capillary action over the top and down the outside of the crucible and hence damage the DTA head.

Chalcogenide glass melts are often volatile above T_g, and even Ga–La–S-based glasses, unusual for their low volatility, over a period of time in open crucible DTA leave a sulfurous deposit inside the DTA equipment. Chalcogenide glasses attack noble metal crucibles and DTA should be carried out with the chalcogenide glass sample sealed under high vacuum inside a silica glass capillary against a reference of a matched sealed silica glass capillary which can be empty. Silica glass capillary tubing of suitable external dimensions to fit snugly in the DTA crucible holder may be widely sourced with (or without) one end sealed. After weighing out the required sample mass into the tube sealed at one end, the remaining end is sealed in an oxyfuel flame whilst applying a vacuum. During the DTA run, the low thermal conductivity of the silica inevitably leads to broadening of the DTA peaks and loss of accuracy. This can be partially mitigated by using as large sample mass as possible simply to create larger DTA peaks. (Using DSC, we have successfully measured the T_g of chalcogenide glasses held in a cold-sealed, lidded, aluminium crucible sealed inside a glovebox equipped with recirculating N$_2$ of <0.2 ppm (H$_2$O + O$_2$). But above T_g the cold seal will not remain hermetic and will allow egress of chalcogenide vapour. Also above T_g the aluminium tends to be solubilised into the chalcogenide melt.)

A standard useful reference material for DTA is high-purity alumina that has been recently calcined (i.e. heated in air to 1000°C, cooled in a dry environment, e.g. desiccator charged with P$_2$O$_5$). The idea is to match the thermal inertia of the reference material to that of the sample at least at the start of the DTA run. The practice of diluting particulate glass samples with calcined alumina in an attempt to match more closely the sample and reference thermal inertias is to be abhorred. Not only is there a risk of chemical reaction or solubilisation of the alumina into the glass-forming liquid, but also the relatively coarse alumina particles readily act as heterogeneous nucleation sites, provided the lattice disregistry of the alumina and ensuing crystal phase is not too great. These unwanted processes will affect the whole nature of the DTA response.

10.3.3 DTA case studies

10.3.3.1 Analysis of glass crystallisation behaviour on reheating glass

ZBLAN (ZrF$_4$–BaF$_2$–LaF$_3$–AlF$_3$–NaF) glasses have been developed to make very low optical loss waveguides (e.g. optical fibres). To achieve this, the glass composition had to be optimised for glass stability and the melt-cooling and glass-reheating for shaping had to be designed to avoid devitrification, because unwanted crystals cause light scattering out of the finished waveguides. In order to optimise glass preparation and processing, the devitrification

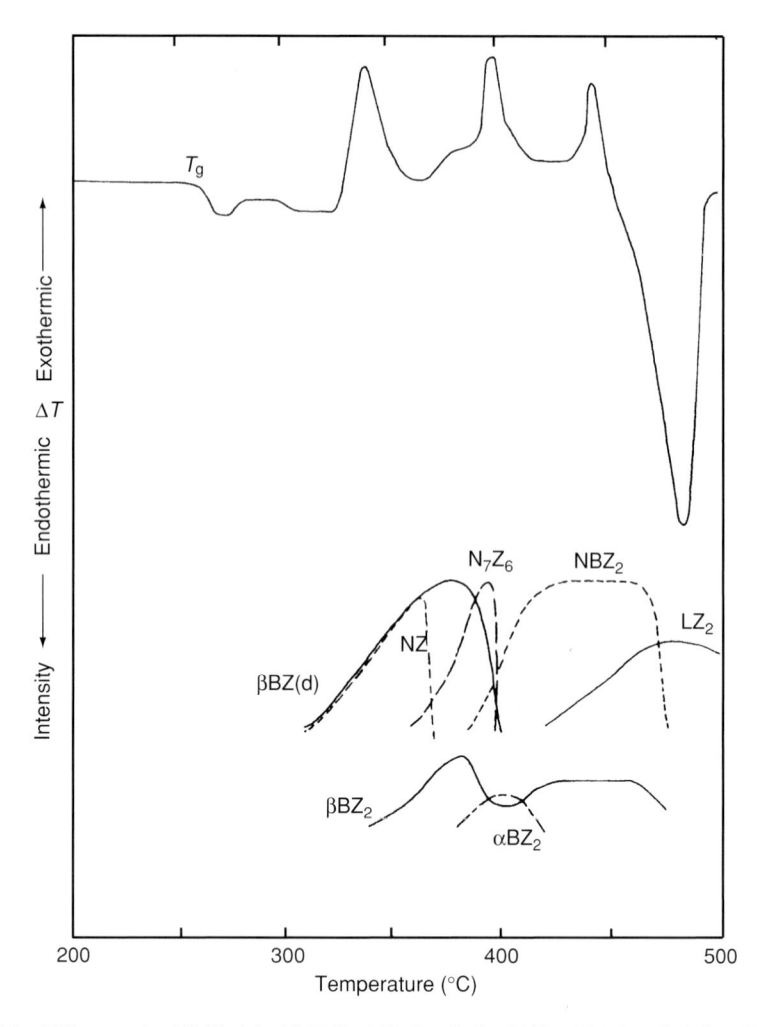

Figure 10.5 DTA curve (at 10°C/min) of 54ZrF$_4$–20BaF$_2$–4LaF$_3$–3AlF$_3$–19NaF mol % (i.e. ZBLAN) and the phase growth responsible for the DTA peaks. To elucidate the phase growth, the DTA curve was used as a guide to design glass heat treatments followed by XRD (X-ray diffractometry) of the quenched products [14].

behaviour both on melt-cooling and glass-reheating had to be understood in detail. Crystallisation on melt-cooling is discussed later in Section 10.4.3.2. Figure 10.5 is a DTA curve of ZBLAN glass [14] which exhibits at least four crystallisation exotherms on reheating above T_g. The first crystallisation peak on the DTA curve shifted to lower temperatures when powdered glass (sieved to ensure <150 μm diameter) rather than 1–2-mm-diameter glass chunks were used, indicating that the first crystallisation event was heterogeneously nucleated. Heterogeneously nucleated crystallisation almost always occurs from glass surface. The temperature at which a DSC or DTA exothermic peak occurs is highly sensitive to the state of subdivision of the glass sample and powdering the glass leads to the peak appearing at lower temperature than if the glass sample were instead present as small glass chunks.

This is because the powder has greater surface area, and probably greater mechanical surface damage, providing more heterogeneous nuclei for surface crystal growth. Hence, reporting the particle size of the glass sample during DTA is important especially when discussing the Hruby parameter [11] of the glass (see Section 10.3.1).

A ZBLAN sample was quenched in situ in the DTA crucible after the first DTA crystallisation peak and DTA rerun. The DTA rerun showed a shifted T_g, indicating the presence of residual glass after the first crystallisation event.

The ZBLAN curve was used as a guide to design a comprehensive set of isothermal heat treatments at selected temperatures and times followed by quenching and XRD (X-ray diffractometry) analysis of the product. Families of XRD peaks, characteristic of crystalline phases, were found to grow and then decrease with increasing temperature or time, allowing detailed understanding of crystallisation behaviour (see Figure 10.5) and hence glass processing to be tailored accordingly. This study was extended to the series $(ZBLA)_{100-x}N_x$ [3] which revealed an eutectic across this section of the phase diagram allowing refinement of glass composition choice. Furthermore, addition of a small amount of NaCl (a pinch of salt!) to the fluoride glass reduced the enthalpy of crystallisation on reheating according to DTA [15] and it was thought that the presence of the mixed anions inhibited diffusion, analogous to the mixed alkali effect commonly encountered in silicate glasses [1].

Interestingly, during heating or cooling a 'post-liquidus' peak (endothermic on heating) is sometimes detected by DTA for ZBLAN glasses. This has been identified as the solubilisation of LaF_3, the most refractory component of this glass. Ideal solubility calculations predicted that at 5 mol % LaF_3, its solution would not occur until $>600°C$. Indeed the true liquidus of ZBLAN should be taken as $\sim 650°C$ and certainly not $450°C$, which cursory examination of the DTA curve would suggest [3]. Further study of the crystallisation behaviour showed that there is a critical balance for ZBLAN glass stability between having sufficient LaF_3 present to prevent crystallisation of the lower temperature βBaF_2ZrF_4 phase (see Figure 10.5) on glass reheating yet not so much that LaF_3 crystallises out (precipitates) on melt-quenching.

10.3.3.2 Glass stability linked to eutectic behaviour

The stability of gallium–lanthanum–sulfide glasses towards crystallisation is critically dependent on the level of oxide content, [O]; glass formation requires a critical threshold level of oxide. $70Ga_2S_3$–$30La_2S_3$ (molar %) melts containing 0.13 wt % [O] devitrify during melt cooling. DTA of glasses showed that glass stability on reheating varies in an unusual but reproducible way as the oxide level is increased shown by the variation of $(Tc_{1,onset} - T_g)$ gap (see Section 10.3.1) with weight % [O] (Figure 10.6) [16]. Independent heat treatment of glasses, quenching and XRD analysis of products helped to explain the observed glass stability [17]. It was found that on heating $70Ga_2S_3$–$30La_2S_3$ (molar %) glasses ($T_g \sim 560°C$) to $630°C$, for batches with less than 0.49 wt % [O], an unidentified new phase A crystallised alone. In contrast, for batches with greater than 2.12 wt % [O], a phase B was the only crystalline phase to form (B is $Ga_6La_{10/3}S_{14}$, melilite). However, when [O] was between 0.49 and 2.12 wt %, both phases A and B formed and this coincided with the maximum glass stability according to the Hruby parameter from DTA. The stabilising mechanism was suggested to depend on eutectic or eutectoid crystal growth of phases A and B. This was supported by analytical and imaging transmission electron microscopy. Figure 10.7 shows a transmission

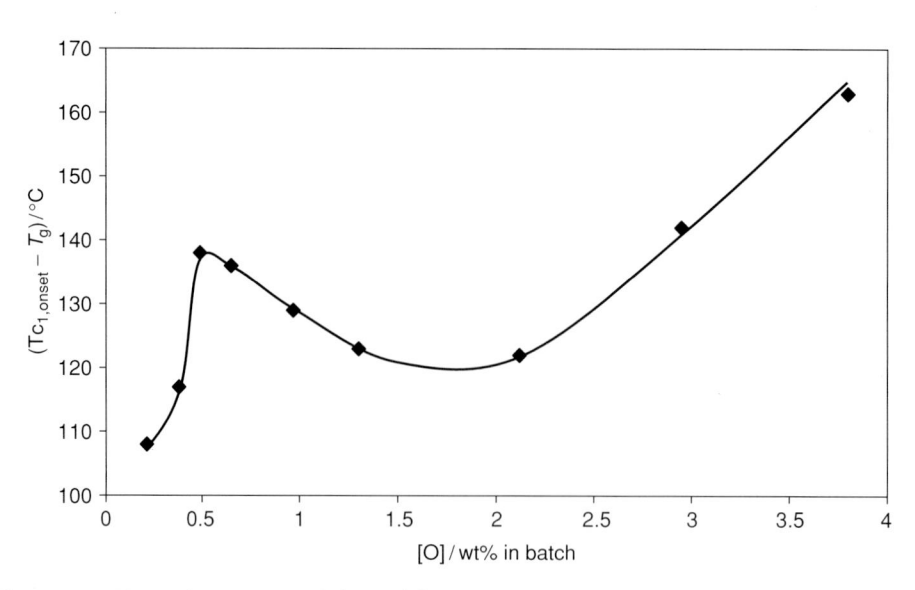

Figure 10.6 The Hruby parameter of glass stability (Tc$_{1,onset}$ − T_g) gap versus weight % [O] for 70Ga$_2$S$_3$–30La$_2$S$_3$ (molar %) glass [11, 16].

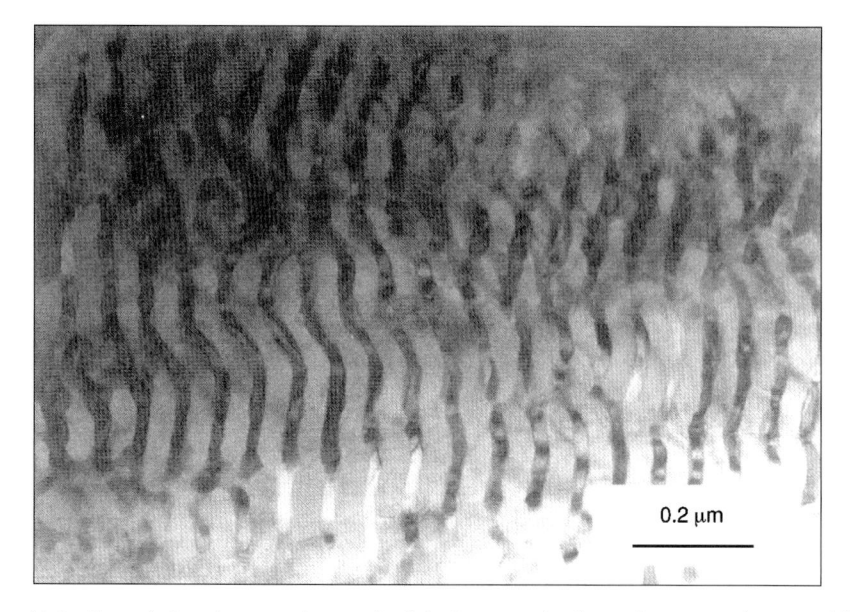

Figure 10.7 Transmission electron micrograph of the intergrowth of two phases A and B, resembling a eutectic or eutectoid microstructure obtained when 70Ga$_2$S$_3$–30La$_2$S$_3$ (molar %) glasses with [O] between 0.49 and 2.12 wt % ($T_g \sim 560°C$) were heated at 630°C. A is a new phase of unknown identity and B is Ga$_6$La$_{10/3}$S$_{14}$, melilite [17].

electron micrograph of the intergrowth of the two phases resembling a eutectic or eutectoid microstructure. At high resolution the alternating laths were shown to be phases A and B by electron diffraction. It is interesting to note that the characteristic eutectic microstructure was grown on reheating the supercooled liquid.

10.3.3.3 Multiple T_gs

Glasses which tend to undergo free-energy-driven, liquid–liquid phase separation are not common and the telltale observation of multiple T_gs on the DTA curve for a single glass sample is a rare event that gives us valuable insight into the particular glass structure (see Section 10.2.3). However, sometimes the occurrence of multiple T_gs on the DTA curve of a single glass sample is an artefact, as demonstrated by the following study.

Background
Traditional chalcogenide glass preparation comprises melting the batch, inside a sealed silica glass ampoule, for many hours in a rocking furnace using resistance heating. Rocking during melting helps to achieve homogeneity of the finished glass. As an alternative approach to resistance heating, in order to try to shorten the glass-making process, we are investigating microwave heating of the batch components. At present, the microwave heating is carried out in the absence of mechanical agitation [18].

Study
The aim of the study was to make As_2S_3 glass using microwave heating of the batch. Thus a stoichiometric batch of amorphous arsenic and sulfur, ground together and sealed into a silica glass ampoule, was melted inside the multimode cavity of a domestic microwave oven (2.45 GHz, 800 W) for 35 min. The silica ampoule containing the melt was then removed from the oven and air quenched. Part of the As–S product was indeed found to be amorphous according to both powder XRD and selected area electron diffraction (the latter acquired by means of transmission electron microscopy). This amorphous product was post-annealed. According to subsequent DTA analysis, the amorphous product exhibited at least three T_gs (at 166°C, ~194°C (broad) and 300°C; Figure 10.8a) indicating the possible presence of free-energy-driven, liquid–liquid phase separation for this product. The same DTA sample was then subjected to an identical DTA rerun but after first heating it to above the liquidus, T_L, and cooling back again to 100°C in situ inside the DTA equipment (cooling was at ~20°C/min from T_L to 400°C and at ~10°C/min from 400 to 100°C). The DTA rerun exhibited only one T_g (at 202°C; Figure 10.8b) and it was concluded that melt homogenisation of the glass must have been achieved during the heating to above T_L inside the DTA equipment in-between the two DTA runs. Thus, the originally observed triple T_g was merely an artefact caused by the lack of homogeneity of the as-prepared glass, probably due to the absence of sufficient mechanical agitation while microwave heating during the glass preparation.

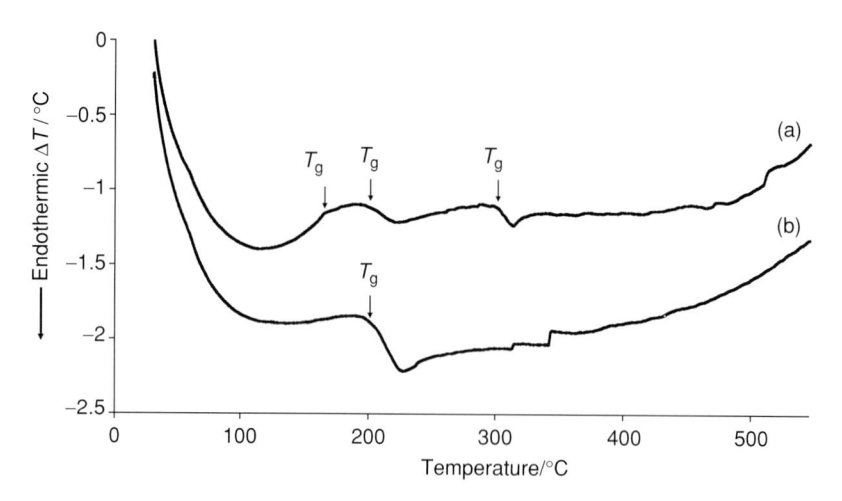

Figure 10.8 Microwave synthesis of batched components in atomic ratio 2As/3S produced a glassy product, and the DTA curve of the product (curve (a)) exhibited at least three T_gs at 166°C, ~194°C and 300°C. A rerun (b) of the same sample under the same DTA conditions as run (a) resulted in a single T_g at 202°C (curve (b)), indicating that homogenisation had occurred above the liquidus before run(b) run(a) [18].

10.4 Differential scanning calorimetry

10.4.1 General comments

True differential scanning calorimetry (DSC) gives a quantitative measure of the electrical energy, for resistance heating, which must be supplied to, or withheld from, the sample in order to maintain a zero-temperature difference between the sample and an inert reference material during: isochronal heating or cooling, or maintaining isothermal conditions. This principle of operation limits the useful range of true DSC to temperatures where conduction, not radiation, is the main heat transfer mechanism, i.e. <1000 K. DSC is therefore well suited to the study of novel inorganic compound glasses composed of weaker chemical bonding than silica and hence having rather low characteristic temperatures. These temperatures may be determined from the DSC curve in an analogous way to DTA. As mentioned above (see Section 10.3.1) by convention, DSC plots may be inverted (exotherms point downwards) to distinguish them from DTA curves (see Figures 10.4 and 10.5), though modern software allows curves to be displayed either way.

The general comments made in Section 10.3.1 for DTA also apply here to DSC. DSC is rather more versatile than DTA. DSC allows quantitative determination of heat capacity, and also the enthalpy of a phase transformation. Thus the enthalpy of devitrification may be measured during isothermal heating by integration of the energy under the DSC peak with respect to time. Compared with traditional isothermal equilibrium calorimetry, the DSC has the advantages of speed and temperature precision gained by the use of small sample mass. Such determinations are made more difficult if the DSC peak is a convolution of several peaks, each representing a separate phase transformation. Deconvolution is usually done by fitting Boltzmann statistics, but this is idealised as the DSC peaks tend to be asymmetric.

Kinetic information may be determined including isothermal and non-isothermal kinetics of crystallisation [19, 20] and measurement of CCR for glass formation (see Section 10.4.3.2) for liquidus temperatures below 1000 K [21].

An interesting point regarding DSC is that the overall enthalpy evolved as the supercooled, glass-forming liquid fully, or partially, crystallises above T_g during isochronal heating should very nearly match the overall enthalpy taken in during the melting sequence, which marks the return of the crystalline phases to the liquid state. However, if enthalpy gained by the system during melting is larger yet a T_g is present, then this signifies that the glass sample was already partially crystalline prior to the DSC run. To the authors' knowledge this approach has not been rigorously applied to date.

10.4.2 Experimental issues

All of the experimental issues discussed concerning DTA of glasses (Section 10.3.2 above) also apply to DSC. In addition, due to the quantitatively more accurate information provided by DSC, it is necessary to eliminate instrument response. This may be done by running a baseline of the empty sample crucible versus the empty reference crucible prior to an experimental DSC run. This baseline should be subtracted from the DSC curve obtained for the sample glass versus the reference using the same crucibles. However, recent DSC equipment automatically offsets instrument response.

Because of the sensitivity of the DSC technique, a relatively small sample mass can be taken, typically a few milligrams, and the reference may be an empty crucible matched in mass to the sample crucible. These steps mean that: (i) the DSC curve corresponds closely to the sample enthalpy changes and (ii) thermal lag between the sample-measuring thermocouple and the sample interior is minimised.

10.4.3 DSC case studies

10.4.3.1 Heat capacity at constant pressure (C_p)

The heat capacities, assumed at constant pressure (C_p), of several chalcogenide glasses were measured [22] using DSC employing standard software issued by the DSC supplier as follows. The measurement was checked using a sample of National Bureau Standards (NBS) 711 glass ((weight percent) 46.0SiO_2–45.3PbO–5.6K_2O–2.5Na_2O–0.6R_2O_3) and the measured value at 550°C was 666 ± 7 J/(kg °C) (NBS value: 686 ± 4 J/(kg °C)).

A baseline was run under a flowing atmosphere of nitrogen (BOC, white spot) using matched sample and reference empty, cold-sealed aluminium crucibles. Then, around 10 mg of powdered glass (<100 μm) sample was sealed (under a nitrogen atmosphere) into a new matched aluminium DSC crucible and the crucible squashed flat, without breaking the seal, in an effort to increase thermal contact with the flat-plate thermocouple on which the crucible was supported. Samples were held isothermally at 50°C for 3 min and then ramped at 5°C/min to 200°C and a second isothermal hold of 3 min was carried out at 200°C; these hold temperatures were selected to bracket T_g of the glass sample and all experiments were carried out under flowing nitrogen. Figure 10.9 indicates how T_g was determined from the heat capacity measurement for $As_{46.3}Se_{33.7}Te_{20}$ (atomic %) by extrapolation of tangents

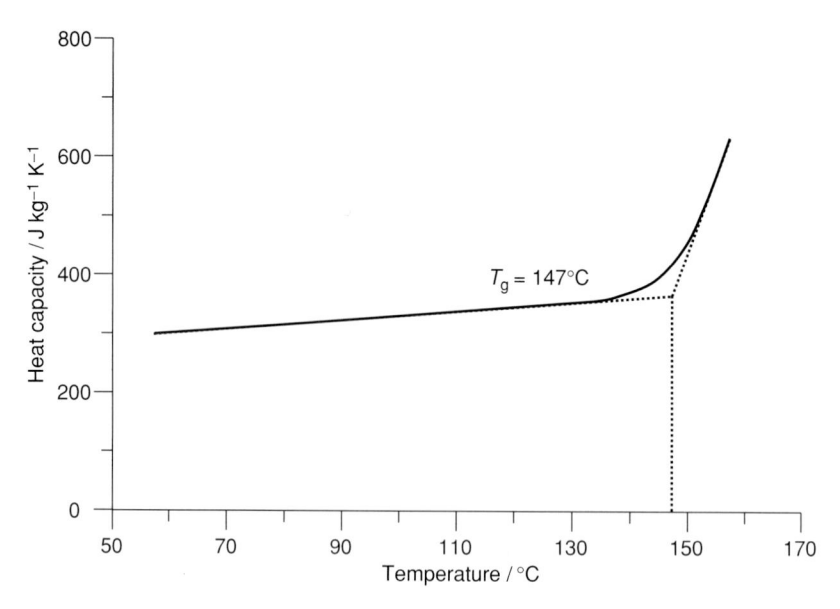

Figure 10.9 T_g was determined for $As_{46.3}Se_{33.7}Te_{20}$ (atomic %) by extrapolation of tangents drawn to the heat capacity curve before and after the increase associated with T_g [22]. T_g values calculated in this way agreed with those measured during normal DSC analysis.

drawn to the curve before and after the increase associated with T_g. T_g values calculated in this way agreed with those measured during normal DSC analysis.

C_p values are typically expressed as a function of temperature T in the form $A + BT + C/T^2$, where A, B and C are constants for a particular glass composition.

10.4.3.2 Time–temperature–transformation (TTT) curves and critical cooling rates (CCRs) for glass formation

The critical cooling rate (CCR) for obtaining glass by melt-cooling from the liquidus may be determined by constructing a time–temperature–transformation (TTT) plot via DSC. An example is the work of Busse et al. [21]. They sealed a sample of ZBLAN glass in a gold, lidded DSC crucible, by cold welding using a crimping press inside an atmosphere-controlled glovebox thereby excluding oxygen and moisture from the sample environment. A series of DSC runs was carried out (against an empty matched sealed pan reference) whereby the ZBLAN glass was heated to a temperature above the liquidus in the DSC crucible and then quenched in situ as fast as possible (~300°C/min) to a temperature intermediate between T_L and T_g and held isothermally. This allowed isothermal crystallisation to proceed, as evidenced by the ensuing DSC peak after a short induction time. The isothermal part of these DSC runs was used to measure both the induction time required for the onset of measurable crystallisation as well as the time for evolution of 50% of the total enthalpy evolved at the particular temperature. Graphical display of: (i) the isothermal temperature versus time to crystallisation onset and (ii) the isothermal temperature versus time for 50% of total enthalpy evolved, yielded a double-nosed plot in each case ((i) and (ii) Figure 10.10).

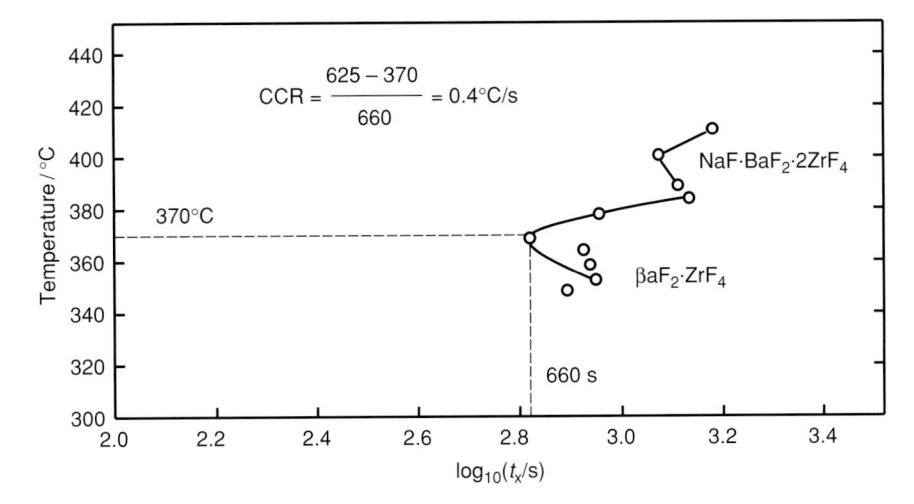

Figure 10.10 Time-temperature-transformation (TTT) plot of ZBLAN (54ZrF$_4$–20BaF$_2$–4LaF$_3$–3AlF$_3$–19NaF mol %) glass and the tangent construction required to calculate the critical cooling rate (CCR) [21].

The gradient of the tangent drawn from the nose, occurring at the shortest time, to the liquidus temperature gave CCR for the ZBLAN glass to be 0.4°C/min. The CCR of the LiF analogue was found to be over an order of magnitude greater because of its lowered stability to crystallisation during melt-cooling.

In Figure 10.10, the lower temperature nose corresponds to growth of both β BaF$_2$ZrF$_4$ and 6NaF–7ZrF$_4$ (see Figure 10.5). The higher temperature nose corresponds to growth of the ternary phase NaF–BaF$_2$–2ZrF$_4$. The ternary phase NaF–BaF$_2$–2ZrF$_4$ is stable during extended periods of subsequent heat treatment at the lower temperatures [14] and hence is the true equilibrium phase. Therefore the initial phases to form above T_g such as βBaF$_2$–ZrF$_4$ and 6NaF–7ZrF$_4$ are concluded to be metastable phases and represent an intermediate step down in free energy for the glass but not true equilibrium [14]. Clearly, concerted growth of a ternary phase is kinetically more difficult and the transformation occurs more easily via the binary metastable phases to the equilibrium ternary phase. Such stepwise approach to equilibrium is widely observed across glass science.

10.4.3.3 T_g response affected by thermal history of glass

As$_{40}$Se$_{60}$ (atomic %) glass was prepared and a sample was sealed into a lidded aluminium crucible inside a glovebox (<0.1 ppm H$_2$O and <0.1 ppm O$_2$) and run in the DSC against an empty, sealed, matched, lidded aluminium crucible as the reference.

Initially, the sample was heated in the DSC to just above T_g, next cooled slowly at 1°C/min and then reheated in the DSC back through the T_g event at 10°C/min. The sample was then cooled back through T_g at 5°C/min, reheated at 10°C/min to just above T_g followed by cooling at 70°C/min to below T_g. Finally, the same sample was reheated at 10°C/min to just above T_g. The DSC responses at T_g for the reheating runs are shown in Figure 10.11.

It may be seen from Figure 10.11 that the lowest cooling rate of 1°C/min, and proximate reheating at 10°C/min, led to a much more pronounced endothermic peak (i.e. the heat

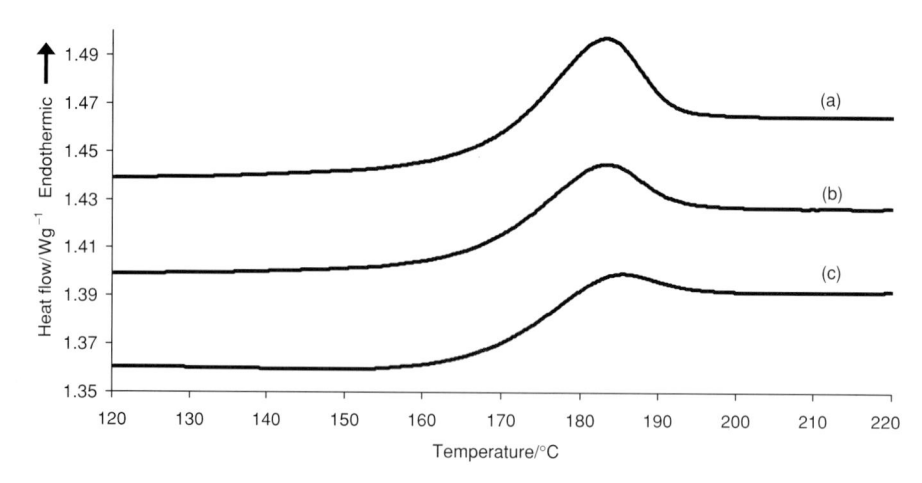

Figure 10.11 The DSC response at T_g is shown for a sample of As$_{40}$Se$_{60}$ (atomic %) glass reheated at 10°C/min having been cooled through T_g just prior to the reheating at a cooling rate of: (a) 1°C/min; (b) 5°C/min and (c) 70°C/min. The DSC curves show that at the lowest cooling rate of the glass relative to the reheating rate there is a large endothermic peak in the T_g region due to the sudden increase in configurational volume and accompanying absorption of enthalpy.

capacity rises) associated with the T_g event on reheating. This enthalpic excursion may be explained by the configurational volume of the slowly cooled, lowest fictive temperature glass (see Figure 10.1) increasing as T_g is reached accompanied by absorption of enthalpy. Thus, where the cooling rate was much lower than the reheating rate, during reheating the actual temperature of the glass increased faster than the fictive temperature of the glass in the T_g region, until the viscosity was reduced to a sufficiently low value.

10.4.3.4 Empirical constant for glass series

It has been suggested [23] that the quantity αT_g (α is the thermal expansion coefficient) may be constant for families of glasses, thus providing a means of estimating the thermal expansion coefficient from T_g and vice versa. T_g may be measured using DTA or DSC and α may be measured using thermomechanical analysis (see Section 10.5.3). Clare and Parker found that the quantity αT_g is remarkably constant for fluorozirconate glasses at 9.9×10^{-3} so long as the expansion coefficient is measured over the same range (in this case 50–250°C) (see Physical Properties in [2]).

10.4.3.5 Calculated temperature coefficient of viscosity from width of heat capacity change at T_g

A correlation exists between the width of the glass transition measured using DSC or DTA during heating and the activation enthalpy of the shear viscosity ($\Delta H_{act.\eta}$) of high-T_g (>250°C) inorganic compound glasses. Moynihan [24] has elegantly demonstrated and made use of the constancy of the dimensionless parameter $C = (\Delta H_{act\eta}/R)$

$[(1/T_g) - (1/T_g^*)]$, where T_g and T_g^* are temperatures marking the extrapolated onset and finish of the glass transition region on the DSC or DTA curve (R is gas constant). The logarithm of the shear viscosity at T_g ($\log(\eta)_{T_g}$) is also approximately constant for high-T_g glasses. Using the universal values of C and $\log(\eta)_{T_g}$, fairly reliable estimates of shear viscosities can be made from the T_g and T_g^* values obtained using DSC or DTA for the temperature range from the glass transition to the working viscosity regime.

10.4.3.6 Heterogeneous and homogeneous nucleation

Er^{3+}-doped fluoroaluminosilicate glasses have been developed which on heat treatment above T_g transform to nano-glass-ceramics [25]. Thus, efficient homogeneous nucleation in certain parent glass compositions results in the growth of spherical nanocrystals up to ~15 nm diameter (see Figure 10.12), which are Er^{3+}–$PbF_{\sim 2}$. These materials exhibit unusual Er^{3+} luminescence characteristics. DSC has shown that in the absence of the Er^{3+} dopant, crystallisation of the nanocrystals does not occur and it is therefore suggested that Er^{3+} stimulates the nucleation process [25]. Altering the state of subdivision of the glass sample does not alter the position of the DSC exotherm due to growth of the nanocrystalline phase, confirming that homogeneous nucleation dominates. However, the temperature at which a second higher temperature DSC crystallisation exothermic peak occurs decreases as the state of subdivision of the glass sample is increased [26], and hence heterogeneous nucleation is occurring for this higher temperature crystallisation of the glass matrix.

The DSC curves on heating the Er^{3+}-doped fluoroaluminosilicates sometimes show an extra, very small exotherm located between T_g and the peak associated with growth of the Er^{3+}-doped nanocrystalline phase [27]. This small exotherm has been tentatively interpreted as due to the homogeneous nucleation process itself.

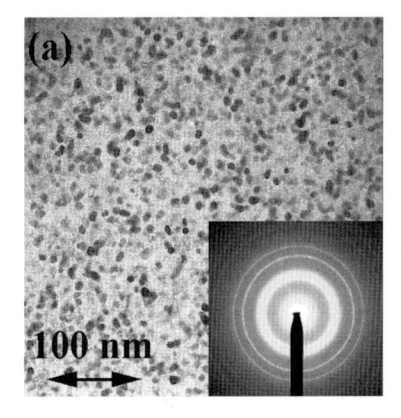

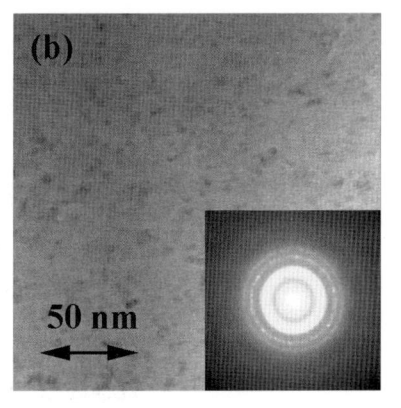

Figure 10.12 Er^{3+}-doped fluoroaluminosilicate glasses have been developed which, on heat treatment above T_g, transform to nano-glass-ceramics. Efficient homogeneous nucleation results in the growth of spherical nanocrystals which are Er^{3+}–$PbF_{\sim 2}$. The figure shows high-resolution transmission electron micrographs of: (a) a developed nano-glass-ceramic (~15 nm diameter) and (b) a *quasi*-glass-ceramic (2–3 nm diameter) [25].

10.4.4 Modulated Differential scanning calorimetry case studies

For modulated differential scanning calorimetry (MDSC), the sample temperature (T) is modulated sinusoidally about a constant temperature ramp given by:

$$T = T_0 + rt + A \sin(2\pi t/P)$$

where T_0 is the initial temperature, r is the heating or cooling rate, t is time, A is the amplitude and P is the period of the modulation. A Fourier transform of the amplitude of the instantaneous heat flow and the average heat flow, incorporating the sinusoidal temperature change, gives two quantities: reversing heat flow and non-reversing heat flow. Measurement of heat capacity at constant pressure (C_p) of glasses is more straightforward by means of MDSC than DSC and typical experimental parameters for chalcogenide glasses are use of ground glass and $r = 5°C/min$, $A = \sim \pm 1°C$ and $P = 80$ s [28].

MDSC is able to deconvolute an overlapping reversible phase transition (e.g. T_g) and a non-reversible transition such as crystallisation. Furthermore, MDSC is able to probe the nature of the glass transition itself for chalcogenide glasses by deconvoluting the total enthalpy near T_g into reversible and non-reversible components, collatable with the rigidity of the glass network [29]. The rigidity of the chalcogenide glass network is conceptually based on the average coordination number of the glass composition. MDSC results provided thermal fingerprints of the three elastic phases, namely floppy, intermediate and stressed-rigid for the range of glasses Ge_xSe_{1-x} [29]. Specifically, the non-reversible enthalpic contribution was found nearly to vanish for the intermediate phase yet be an order of magnitude larger for both the floppy and stressed-rigid phases. Glass transitions in intermediate phases are thus interpreted as being completely reversible in character and non-ageing.

10.5 Thermomechanical analysis

10.5.1 General comments

Thermomechanical analysis (TMA) gives an accurate measurement of the height of a sample (resolution is <0.1 μm for a few millimetres' high sample) under an applied load as the temperature is ramped up or down or held isothermally. This sample height output with temperature is updated at ≥1 Hz and allows measurement of a range of properties useful for characterising glasses, namely linear thermal expansion coefficient, dilatometric softening point, viscosity–temperature behaviour of the glass-forming liquid in the range T_g to 10^4 Pa s, and hence melt fragility. Properly annealed glasses are usually isotropic and hence volumetric thermal expansion may be calculated from the measured one-dimensional expansion. However, thermal history can affect both the thermal expansion coefficient and the dilatometric softening temperature. Overall, TMA yields some of the most important properties for determining the conditions for hot shaping, joining, annealing and avoidance of thermal shocking of inorganic glasses.

10.5.2 Experimental issues

The temperature and force which can be applied during a TMA run depend on the instrument design. The sample holder's hot parts may be metallic, semiconducting, ceramic or vitreous. For inorganic compound glasses analytical temperatures above 400°C are usually required and vitreous silica, for the main body of the sample holder, is recommended for inorganic compound glass analysis offering a maximum temperature of ~1000°C, maximum force of ~1 N and the advantage of a low thermal expansion. Typical commercially available TMA vitreous silica sample holders have a relatively small sample space, compared to the metallic holders, restricting sample diameter and height to <10 and <15 mm, respectively, though slightly larger designs can be found on some recent equipment.

Due to the sample size being much larger than in the case of DTA or DSC, care must be taken to limit temperature ramps to a rate that does not incur an appreciable temperature lag between the sample and sample thermocouple. This problem is exacerbated by poor thermal connection between the sample, thermocouple and furnace. If temperature ramps are required then use of helium as the purge gas helps because of its relatively high thermal conductivity. For samples reactive to water vapour and oxygen gas, the purge gas should be purified of these. If a fast temperature ramp is required, it is advisable to test the temperature lag by performing a temperature calibration with a metal calibrant such as indium, zinc or aluminium; the metal calibrant's melting point must occur below the glass sample T_g. The metal calibrant is placed on top of the glass sample with all of the same sample probe fittings as are to be used subsequently for the glass sample run and the calibration run is carried out at the ramp rate proposed for the glass sample. The onset of melting of the metal is observed and compared to the expected value. With this information the temperature lag may be assessed.

In general, a baseline run should be carried out with the same sample holder, temperature programme and applied force as for the subsequent sample run, but without the sample present. This enables the instrument response to be subtracted from the overall response of the sample (not essential when carrying out measurements not requiring absolute height or deformation rate information). It is important for all TMA measurements that the fitments are fitted correctly and tightened.

10.5.3 Linear thermal expansion coefficient (α) and dilatometric softening point (M_g)

Most inorganic compound glasses and glass-ceramics expand on heating above room temperature, exhibiting a wide range of linear thermal expansion coefficients (α) from almost zero for TiO_2–SiO_2 glass systems to $>200 \times 10^{-7}\,°C^{-1}$ for some chalcogenide glass. Negative α are possible, e.g. silica glass <150 K. Table 10.1 shows some typical linear thermal expansion coefficients for inorganic compound glasses at standard temperature and pressure.

The linear thermal expansion coefficient (α) is expressed as:

$$\alpha = \frac{\Delta l}{l_0 \Delta T} \quad (10.2)$$

Table 10.1 Typical linear thermal expansion coefficients of inorganic compound glasses at standard temperature and pressure.

Glass	Linear thermal expansion coefficient $\times 10^7 /°C^{-1}$	References
SiO$_2$	5	[1]
GeO$_2$	77	[1]
B$_2$O$_3$	150	[1]
ZBLAN*	172	[2]
GeAsSeTe system	~ 200	

*54ZrF$_4$–20BaF$_2$–4LaF$_3$–3AlF$_3$–19NaF mol %.

where Δl is the change in sample length on heating through a temperature (T) range of ΔT and l_0 is the initial sample length.

Traditionally α was measured in a dilatometer, which permitted the thermal expansion of a long rod of glass, typically 100 mm, to be compared to that of a standard, typically vitreous silica, heated slowly in a large furnace with a long and even hot zone. This method provided large absolute expansion, due to the sample length, and the comparison to a standard eliminated the effects of the instrument expansion. However, modern TMA provides very sensitive height measurement, accurate and precise temperature control (at best ±0.1°C according to the measuring thermocouple) and repeatable behaviour, allowing accurate absolute thermal expansion measurement of small samples.

The TMA sample is prepared as a rod, of round or square section, with width suitable for the sample holder in use (usually ~5 mm) and of the greatest height compatible with the sample holder and furnace hot zone (commonly ~7 mm). Rod ends are ground flat and parallel and sample length is measured with a micrometre screw gauge or similar to ensure maximum accuracy. The sample probe should be a flat silica rod of similar diameter to the sample. The temperature ramp rate should be low and isochronal, of the order of 2–3°C/min, and the applied load small ($\leq$50 mN) to prevent fast compression of the sample on reaching T_g. A baseline run is carried out under identical conditions but without the sample, usually before each measurement; baselines should be checked at least monthly.

A typical inorganic glass linear thermal expansion curve, plotted as the change in sample length divided by initial sample length ($\Delta l/l_0$) versus temperature, is schematically illustrated (solid curve) in Figure 10.13. The thermal expansion coefficient α is given by the gradient (m) and α has units of (temperature)$^{-1}$. The variation of α with temperature is polynomial but often α is assumed to be linear over a restricted temperature range and the temperature range of assumed linearity must be stated when reporting results.

It is clear from Figure 10.13 that the plot of $\Delta l/l$ versus temperature increases gradually in gradient with increasing temperature until at point (a), where an abrupt increase in gradient is observed as the transformation range is reached. The curve continues with this, typically around threefold, increase in expansion until an apparent contraction occurs due to the supercooled glass-forming liquid beginning to flow under the force applied by the probe. Figure 10.13 shows the geometrical construction on the TMA curve required to obtain the dilatometric softening point, conventionally denoted as M_g, which occurs at

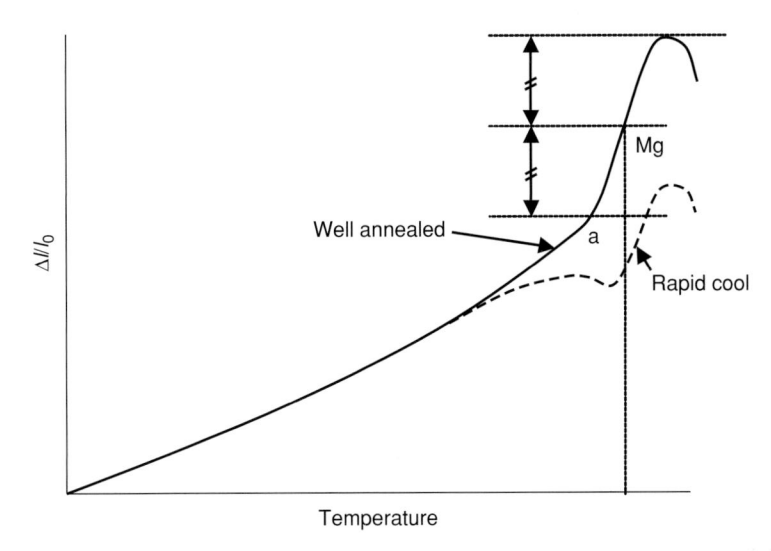

[Figure: Schematic TMA thermal expansion curve with y-axis Δl/l₀ and x-axis Temperature, showing "Well annealed" solid curve and "Rapid cool" dotted curve, with M_g marked.]

Figure 10.13 Schematic of a thermomechanical analysis (TMA) thermal expansion curve of an inorganic compound glass (solid curve). The geometrical construction to obtain the dilatometric softening point, M_g, at ~$10^{11.5}$ Pa s is given. The analogous TMA thermal expansion curve is shown (dotted curve) when a high-fictive-temperature glass sample is heated under load; relaxation to a lower-fictive-temperature configurational arrangement can produce an observable contraction as illustrated.

~$10^{11.5}$ Pa s. Thus, M_g is taken conventionally as the midway point (on the ordinate) between the abrupt increase in expansion and the onset of contraction. At M_g, the glass-forming liquid is subject to a macroscopic rearrangement as well as the microscopic or configurational rearrangement at T_g responsible for stress relaxation during annealing.

Thermal history of the sample affects the values of α and M_g. The TMA response when a relatively high fictive temperature glass sample is heated under load is schematically shown in Figure 10.13 (dotted curve). As the temperature approaches T_g the sample undergoes microscopic rearrangement and relaxation to a lower fictive temperature configurational arrangement. For some glass systems this produces an observable contraction as illustrated. It is therefore necessary to ensure a standardising heat treatment prior to TMA when comparing α and M_g of glass compositions within a particular system and to quote the prior heat treatment when reporting α and M_g of glasses, in addition to the heating rate during the measurement of expansion.

To continue the α measurement beyond the softening point requires that the supercooled glass-forming liquid be enclosed in containment such that volume change can be measured. But α above M_g is not usually relevant to the performance of a glass material in normal service since above M_g the glass melt cannot support sustained stress and undergoes viscous flow and stress relaxation.

Figure 10.14 shows the TMA dilatometric response for an annealed chalcogenide glass $As_{38}Se_{62}$ (atomic %) as sample height versus temperature. From this plot the instantaneous α is found to be the gradient divided by the initial length. α was calculated to be 22.4×10^{-6} °C^{-1} from 40 to 60°C, 27.4×10^{-6} °C^{-1} from 120 to 140°C and an average of 24.1×10^{-6} °C^{-1} from 40 to 140°C.

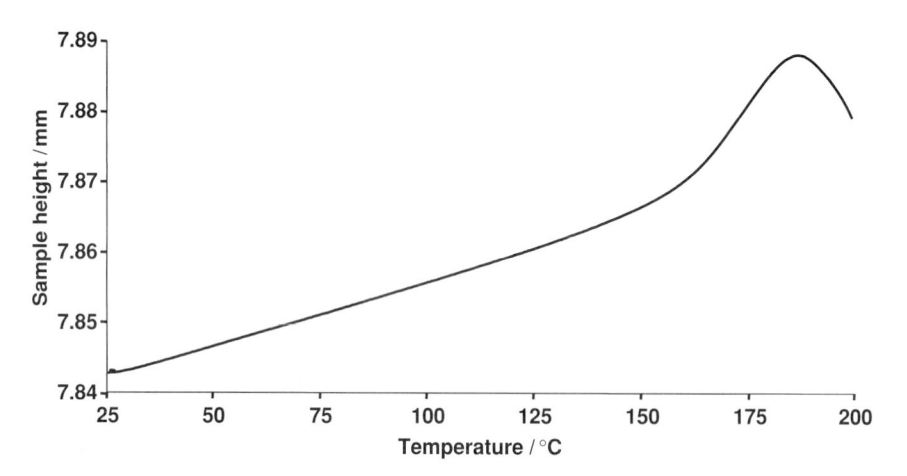

Figure 10.14 Thermomechanical analysis (TMA) dilatometric response for the chalcogenide glass As$_{38}$Se$_{62}$ (atomic %).

10.5.4 Temperature coefficient of viscosity: introduction

Hot shaping
Viscosity is one of the most important physical properties to be considered when hot forming glasses, and hot forming is by far the most common method for producing a shaped glass item, be it by blowing, pressing, blow moulding, extrusion or fibre-drawing. In order to choose a working temperature for a forming process, the temperature dependence of viscosity must be determined and considered alongside information on devitrification temperatures obtained by DSC or DTA. This is to ensure that the chosen temperature schedule allows the required deformation for shaping and yet does not produce crystals in the glass. The viscosity at $Tc_{1,\text{onset}}$ obtained from DTA or DSC data (Figure 10.4) is the minimum viscosity that can be used. Any hot working process must be achieved below this temperature. However, it should be noted that $Tc_{1,\text{onset}}$ is dependent on the timescale of observation. The supercooled liquid does not have to be held isothermally for very long at the $Tc_{1,\text{onset}}$ measured during a typical DTA run at 10°C/min for significant crystallisation to occur. If the shaping process requires some minutes or more, then a temperature somewhat lower than $Tc_{1,\text{onset}}$ must be considered the maximum. Repeating DTA or DSC runs at smaller and greater heating rates gives non-isothermal information on the rate dependence of the crystallisation [19, 20, 30]. However, reducing the shaping temperature to avoid crystallisation is accompanied by viscosity rise, requiring a greater force to shape in the same time. Care should also be taken of the nucleation temperature, even if crystallisation is avoided the supercooled liquid may have spent some time in the nucleation region (Figure 10.2) and if reheated for further processing may crystallise rapidly.

Theory
Viscosity is more correctly termed 'shear viscosity' and is closely analogous to the shear modulus [1]. Where a velocity features in the shear viscosity expressions, a displacement is found for shear modulus. To compare these terms, consider two cases:

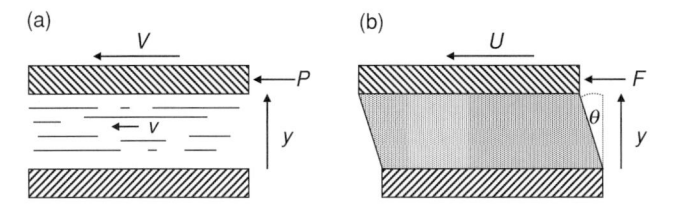

Figure 10.15 (a) An incompressible fluid is contained between two infinite, flat plates with a separation of y. The lower plate is stationary and the upper plate moves in its own plane with velocity V. For a Newtonian fluid of velocity v, at a height y above the lower plate, dv/dy = constant. The pressure to sustain this, P, is proportional to the velocity gradient, and the proportionality constant is the shear viscosity η. (b) The analogous situation is an elastic solid held between two plates of unit area and separation. The force F applied to the top plate produces a displacement U leading to an angle of shear θ. For small θ, this becomes du/dy (u is displacement at height y from the bottom plate). The shear modulus is the proportionality constant between force applied and du/dy.

1. *Shear viscosity.* In Figure 10.15a, an incompressible fluid is contained between two infinite flat plates with a separation of y. The lower plate is stationary and the upper plate moves in its own plane with a velocity V. If the velocity is not so great as to cause turbulence in the liquid, then for most types of inorganic glass-forming supercooled liquids the velocity v, at a height y above the lower plate, will be proportional to y. If $y = 0$ then $v = 0$, stationary with the lower plate, and if $y = Y$ then $v = V$, moving with the top plate; hence, dv/dy = constant. In order to sustain this movement, a force must be applied to the top plate. For unit area of the plate this is then a pressure (P) proportional to the velocity gradient. Thus:

$$P = \eta \frac{dv}{dy} \quad (10.3)$$

where η is the shear viscosity and is Newtonian.

It is worth considering the units of viscosity. In SI units, P is pressure in Pa and dv/dy is m^{-1}, and therefore the unit of η is Pa s. For many years in the glass industry, the cgs (centimetre, gramme, second) was used giving η the units of (dyn s)/cm^2, known as 'Poise'. This is still encountered in much literature and can easily be converted as 1 Poise ≡ 0.1 Pa s.

2. *Shear modulus.* In Figure 10.15b, an elastic solid is held between two plates of unit area with separation y. The bottom plate is fixed and a force F is applied to the top plate in the plane of the plates so as to produce a displacement U. This leads to an angle of shear θ. The relationship between these terms is:

$$F = G\theta \quad (10.4)$$

where G is the shear modulus. If the shear angle is small and u is the displacement at a height y then $\theta = du/dy$ and hence:

$$F = G \frac{du}{dy} \quad (10.5)$$

Compare this to Equation (10.3). For an incompressible elastic solid $E = 3G$, where E is Young's modulus. From this, many equations for deformation can be applied to viscous flow.

Table 10.2 Summary of viscometry techniques which have been verified using a modified commercial thermomechanical analysis (TMA) together with details of specimen geometry required. (η is viscosity.)

Technique	η range/Pa s	Specimen geometry	TMA cell geometry	Reference
Beam bending	10^{11} to 10^9	Rectangular beam about $10 \times 0.8 \times 2.5$ mm	Specimen supported across top of a notched holder made from silica glass tubing about 8 mm internal diameter and 10 mm deep. Short, 1-mm-diameter silica glass rod (ground flat) as knife edge between silica probe and the beam	[31]
Indentation	$10^{10.7}$ to 10^7	Cylindrical 7 mm diameter, 3 mm deep	Typically, Nimonic 105 alloy cylindrical indenter of 1 mm diameter placed between silica glass probe and sample	[32]
Indentation	10^{13} to 10^7	Flat plate $5 \times 5 \times 1$ mm^3	Hemispherical quartz indenter of radius 1.2 mm placed between silica glass probe and sample	[33]
Parallel- plate	10^8 to 10^4	Cylindrical 4–6 mm diameter, 2 mm deep	Specimen sandwiched between flat polished discs and placed under silica glass probe	[39]

TMA viscometry

The viscosity of the supercooled glass melt varies over many orders of magnitude, so it is common to express η on a log scale or by powers of 10 such that T_g is approximately $10^{12.5}$ Pa s and the viscosity at the melting temperature can be of the order of ≤ 1 Pa s. With such a large range of viscosities possible, no one measurement technique is sufficient for the whole range. Measurement methods can be broadly classified as high- or low-viscosity methods.

Low-viscosity methods, viscosities less than 10^4 Pa s, typically employ a rotating sphere, cylinder or disc immersed in the glass melt. This type of measurement is beyond the scope of the TMA and requires a dedicated instrument.

High viscosities, more than 10^4 Pa s, can be satisfactorily measured in the commercial TMA and it is in this region that the glass-forming liquid is typically shaped. Table 10.2 summarises the way in which TMA has been used for beam bending, indentation and parallel-plate viscometers to cover the viscosity range $10^4 - 10^{11}$ Pa s. The TMA indentation and parallel-plate viscometry are described in detail below (Section 10.5.5 and 10.5.6, respectively).

10.5.5 *TMA indentation viscometry*

10.5.5.1 *Indenter*

Here a probe is pressed into the surface of a flat, polished glass sample which is held isothermally at the required temperature. Many geometries can be used for the probe tip, but spherical and cylindrical are the most suited to the TMA. Although the viscosity measurement range of the cylindrical indenter is smaller than that of the spherical indenter, the cylindrical indenter has the advantage that the absolute height of the indentation is not needed due to the constant indenter cross-sectional area.

10.5.5.2 Theory

The theoretical justification of viscosity measurement based on the surface penetration of the glass-forming supercooled liquid by a flat bottom cylindrical indenter was considered by Nemilov [34]. Both the elastic and viscous elements are involved but by considering the steady-state condition the elastic terms can be neglected. This simplified model gives the viscosity as:

$$\eta = \frac{mF}{kv\sqrt{A}} \tag{10.6}$$

where η is the viscosity, F is the applied force, m is a constant depending on the indenter geometry (for a cylinder $m = 0.96$), k is a constant, A is the indenter cross-sectional area and v is the steady-state penetration rate.

10.5.5.3 Verification

Figure 10.16 shows the experimental arrangement for TMA indentation viscometry using a cylindrical indenter. The silica rod probe is fitted with a removable cylindrical pin of suitable material for the temperature and glass type to be encountered. Although vitreous silica could be used, its delicacy is not desirable as the pin can become stuck into the glass sample. A high-temperature nickel–chromium alloy is usually best suited for most glasses. The pin must be cylindrical and of smooth surface finish, with a close fit to the probe rod to ensure it remains perpendicular to the sample. If the pin is retained to the probe then its weight can be tared off. If not, the weight of the pin must be added to the applied force to give the total force. The sample of glass to be measured should be of large diameter relative to the pin. Since the TMA sample holder allows a maximum of about 8 mm diameter, a suitable pin is 1 mm in diameter and 2 mm or more in length. The glass should be about 3–5 mm or more in thickness and ground flat and parallel top and bottom. It is not important that the sample is cylindrical. The top surface should be polished to better than a 3 µm finish. It is advisable to place the sample on a thin piece of material in the TMA sample holder in order to prevent sticking to the silica holder. Platinum foil is suitable for most oxide and halide glasses, silica for chalcogenides, etc. Since the force applied is a dominant term, the TMA should be calibrated for force and checked for constancy over the working height range. The TMA should be raised to the temperature for the measurement with the probe raised just about the sample and allowed to equilibrate thermally before lowering the probe with the required force and then data taken. When carrying out the viscosity measurement, the most important point is obtaining a steady-state value for v, the deformation rate, of about 2 µm/min [32] by optimising the load and temperature range.

To verify the technique the viscosity–temperature curve of the NBS (National Bureau of Standards, USA) 711 glass was determined using a Nimonic 105 indenter. The TMA was set to ramp at 100°C/min to 40°C below the required temperature and then at 10°C/min to the set temperature. The force (F) applied to the sample was chosen by estimating the

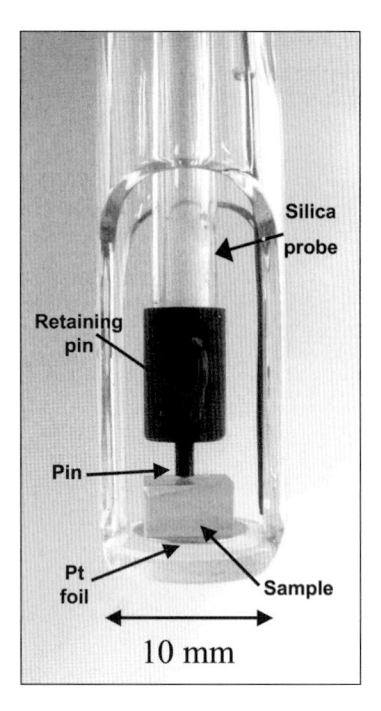

Figure 10.16 Photograph of the experimental arrangement for thermomechanical analysis (TMA) indentation viscometry using a cylindrical indenter [32].

viscosity at the required temperature and calculating back to the force which would give the recommended deformation rate of 2 μm/min [32], m is a constant dependent upon the shape of the probe tip. The isothermal penetration was measured as a function of time and measurements were carried out for between 15 and 40 min, once a linear penetration rate had been obtained. The regressed measured penetration rates had correlation coefficients of at least 0.999, so any residual viscoelastic effects were probably small. Thus, the simple physical picture of a limiting penetration rate corresponding to a purely viscous response appears justified for these practical measurements. The isothermal steady-state penetration rate allowed calculation of the viscosity using the modified Nemilov method (see Equation (10.6)) [32].

The viscosities measured for NBS 711 glass were compared to those obtained using the Vogel–Fulcher–Tammann [9] equation issued by the NBS (Figure 10.17):

$$\log_{10}(\eta[\text{Pa s}]) = -2.621 + \frac{4254.649}{T - 152.1} \tag{10.7}$$

where temperature T is in °C.

With k of Equation (10.6) adjusted to 8, the fit of the measured viscosity–temperature curve between 10^{10} and 10^6 Pa s had a mean error of 1.2%, for the particular experimental set-up.

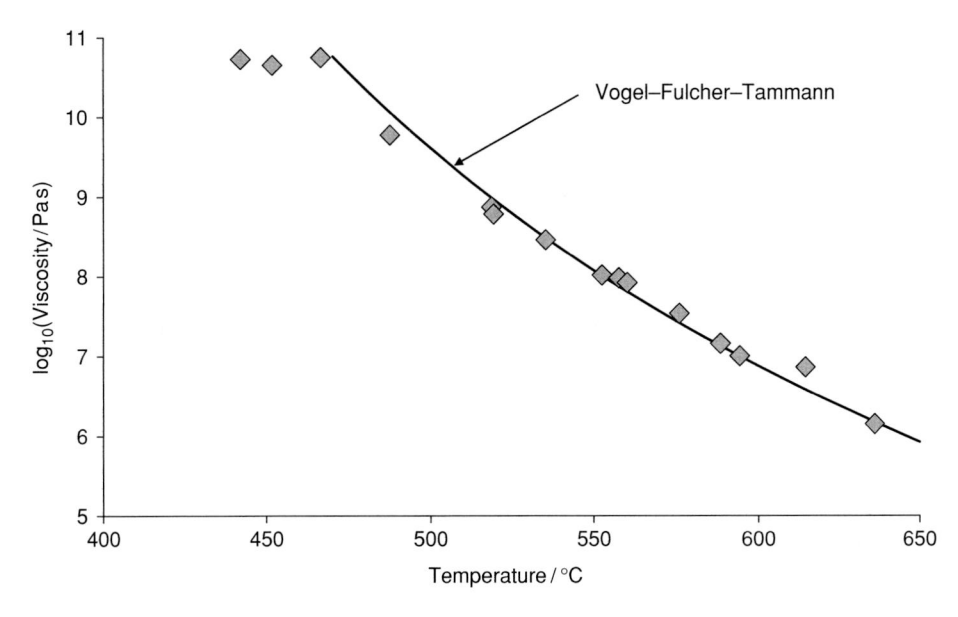

Figure 10.17 The viscosity–temperature data set measured using thermomechanical analysis (TMA) indentation viscometry of the National Bureau of Standards (NBS), USA, 711 glass is compared to that obtained using the Vogel–Fulcher–Tammann [9] model issued by the NBS [32].

10.5.5.4 Case studies

Fluoroaluminate glasses

The viscosity–temperature curve for the fluoroaluminate glass $37AlF_3$–$12BaF_2$–$15CaF_2$–$12MgF_2$–$9SrF_2$–$15YF_3$ (molar %) was measured using indentation viscometry with a cylindrical probe [35]. During some of the measurements the time spent at the selected temperature was sufficient to cause devitrification, resulting in an apparent infinite rise in viscosity. Figure 10.18 is a photograph of such a specimen after the viscosity run, showing that crystallisation has occurred in the vicinity of the indentation.

Germanium–sulfur–iodide glasses

The viscosity–temperature curves for $[Ge_{0.3}S_{0.7}]_{100-x}I_x$ glasses in the range $0 \leq x \leq 30$ were measured using indentation viscometry with a cylindrical probe [36]. Figure 10.19a shows that for the Ge–S–I supercooled liquids, isoviscous points are found to shift to a lower temperature as the iodine content is increased. As discussed earlier (Section 10.2.4), plotting log viscosity data on a reduced inverse temperature (absolute) scale enables insight into the relative relaxation time of the supercooled liquid, assuming that this is manifested in the shear viscosity of the supercooled liquid. The temperature at which the viscosity reaches T_g (here taken as 10^{12} Pa s) is used as the normalising parameter. Figure 10.19b shows that the addition of iodine tends to break down the connectivity of the germanium sulfide network in the supercooled melt, producing an increasing fragile liquid. Interestingly, there is a large increase in fragility for compositions with ≥ 15 atomic % iodine and this threshold has been

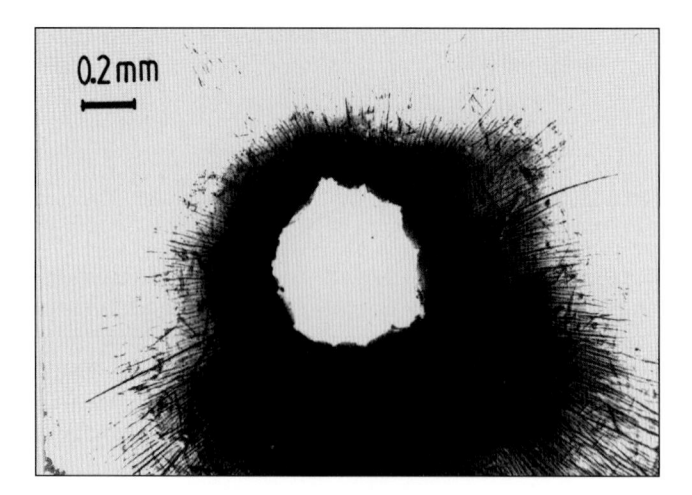

Figure 10.18 Photograph of a sample of the fluoroaluminate glass 37AlF$_3$–12BaF$_2$–15CaF$_2$–12MgF$_2$–9SrF$_2$–15YF$_3$ (molar %) that has undergone thermomechanical analysis (TMA) indentation viscometry with a cylindrical probe. The sample has devitrified during the indentation [35].

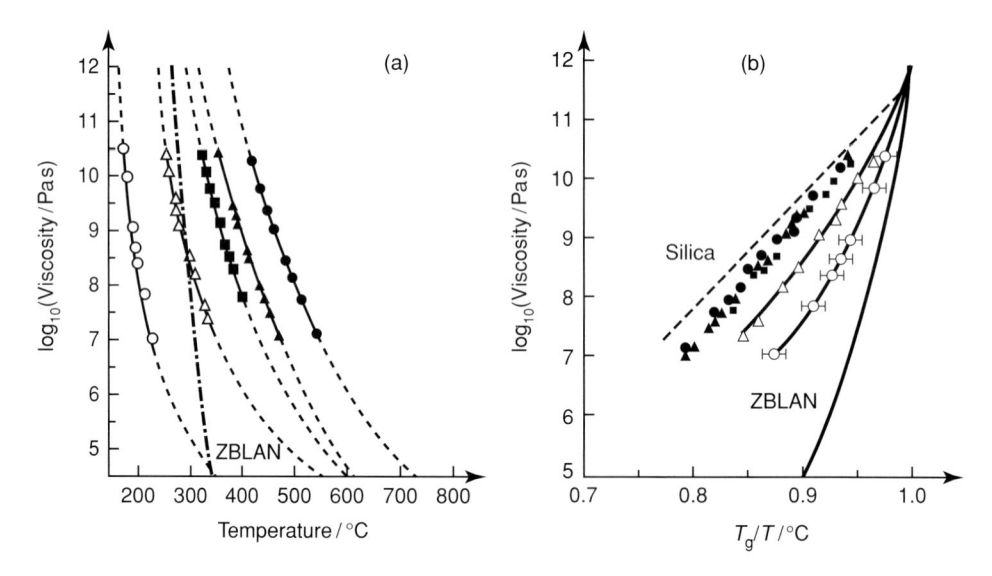

Figure 10.19 Viscosity–temperature curves for [Ge$_{0.3}$S$_{0.7}$]$_{100-x}$I$_x$ glasses in the range $0 \leq x \leq 30$ were measured using indentation viscometry with a cylindrical probe [36]: (a) isoviscous points are found to shift to a lower temperature as the iodine content is increased; (b) addition of iodine tends to break down the connectivity of the network in the supercooled melt producing an increasingly fragile liquid [8, 36 and 37]. ●, 0at% I; ▲, 5at% I; ■, 10at% I; △, 15at% I; ○, 20at% I.

correlated with the onset of the calculated rigid-to-floppy threshold of the glass network [37].

10.5.6 TMA parallel-plate viscometry

10.5.6.1 Theory

When a sample of glass is pressed between two parallel plates and heated to above T_g, the sample will exhibit viscous flow. The patterns of flow can be complex; so in order to calculate the rate of flow, a simple geometry must be considered. The case for a glass sample in the form of a disc with the height less than the diameter and for the sample sticking to the plates such that the glass melt must flow and not slide has been considered by Gent [38] and an expression for the viscosity was derived:

$$\eta = \frac{2\pi F h^5}{3 V^{ol}(2\pi h^3 + V^{ol})(dh/dt)} \tag{10.8}$$

where η is the viscosity, F is the applied force, V^{ol} is the sample volume and h is the height of the sample at time t.

10.5.6.2 Practical considerations

The technique requires linearity of the probe movement in the TMA cell. Due to the usual use of force motors, the force applied to the probe may not be constant as a function of probe height. This should be checked and corrected for. We have found that the use of a small external balance is helpful with this kind of calibration. The external balance is supported below the sample holder, and a small horizontal bridge passing through the sample holder and resting on the balance is placed such that the probe applies load to the bridge and hence to the balance. In this way the relationship between force applied and probe height can be determined for a range of forces.

The parallel plates must be arranged to remain parallel during the measurement and be flat and smooth. Plate material must be chosen so as to be compatible with the sample glass type and temperatures to be encountered. Figure 10.20 is a photograph of the experimental arrangement for TMA adapted for use with stainless steel grade 310 plates. The top plate is fitted to the silica sample probe and retained with a pin. This maintains the plates parallel and allows for the weight of the top plate to be tared off. If the plate is not attached to the probe, the weight of the plate must be added to the applied force. The flats of the plates are ground and polished to a 6 μm finish. Such plates are suitable for use up to 600°C and forces up to ~2 N. For higher temperatures, Nimonic alloys, silicon carbide or silica glass may be used. Platinum foil may be placed between the sample and plates to facilitate sample removal after the run.

Sample preparation will depend on the glass to be measured and the TMA to be used. In typical TMA equipment, the sample space is of about 10 mm diameter. In order to allow for the sample to be pressed out during the measurement, the maximum sample diameter at the start will be about 5 mm. Sample height is not critical but should be less than the

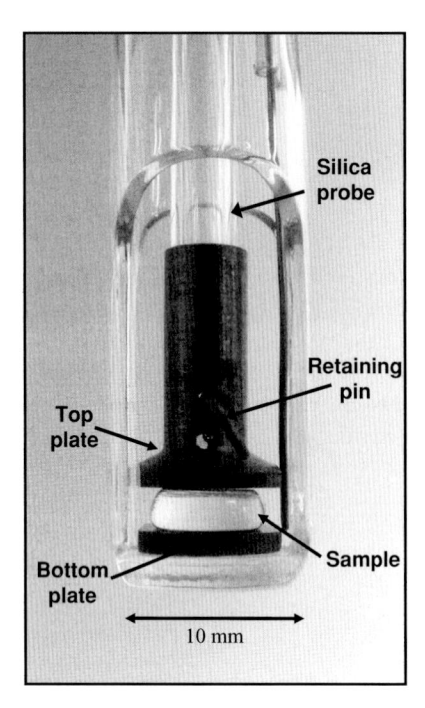

Figure 10.20 Photograph of the experimental arrangement for thermomechanical analysis (TMA) parallel-plate viscometry using stainless steel grade 310 plates.

diameter, ~2–4 mm. As the sample diameter increases during pressing, a thinner sample can notionally be pressed for longer to access lower viscosities/higher temperatures. In many cases the glass sample may be core drilled from a block and sliced to give the correct size. It is most important that the sample is accurately produced with top and bottom faces parallel, flat and ground to 1000 grit 25 μm finish or finer. The circumferential surface must be round and smooth and edges free from chips. Any defects will usually result in a volume smaller than a perfect disc and so indicate a lower viscosity than is correct.

10.5.6.3 Verification

The instrument was calibrated using the NBS 710 glass [39]. A 4.85-mm core was drilled from the glass block supplied and cut into samples of approximately 2.2 mm high, and then ground and polished flat. Diameter and height were measured with a micrometre screw gauge for the volume calculation. The sample was placed in the TMA with the load set to 250 mN. The temperature was ramped to 600°C at 20°C/min and then held isothermally for 4 min to achieve thermal equilibration and then again ramped to 750°C at 5°C/min. The resulting TMA output with the baseline subtracted is shown in Figure 10.21. After pressing the sample was 1.3 mm high and 6.7 mm in diameter; the plates used were 8 mm in diameter.

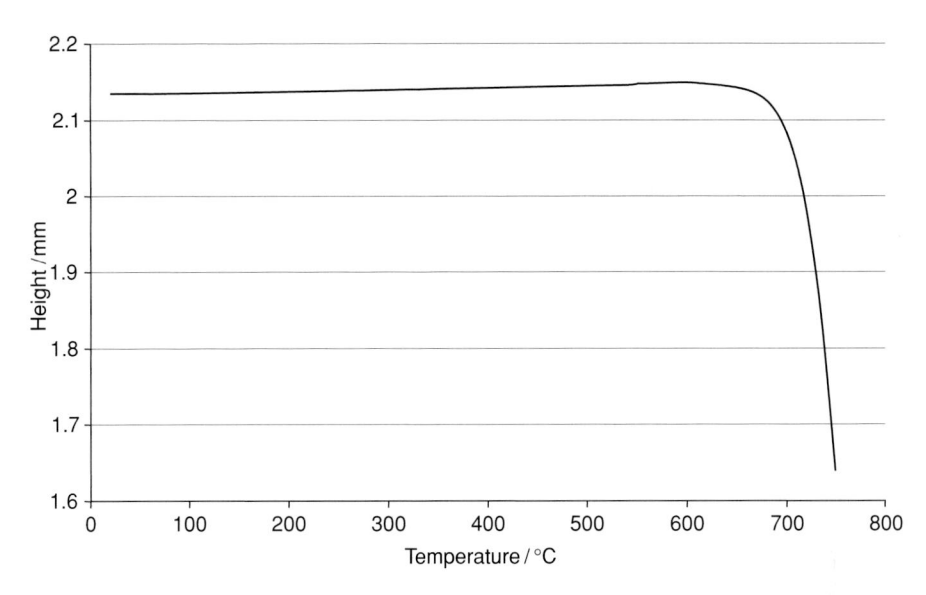

Figure 10.21 The thermomechanical analysis (TMA) output for parallel-plate viscometry of the National Bureau of Standards (NBS), USA, 710 glass after baseline subtraction.

The viscosities measured for NBS 710 glass were compared to those obtained using the Vogel–Fulcher–Tammann [9] equation issued by the NBS (Figure 10.22):

$$\log_{10}(\eta/\text{Pa s}) = -2.626 = \frac{4236.118}{T - 266} \tag{10.9}$$

where temperature T is in °C.

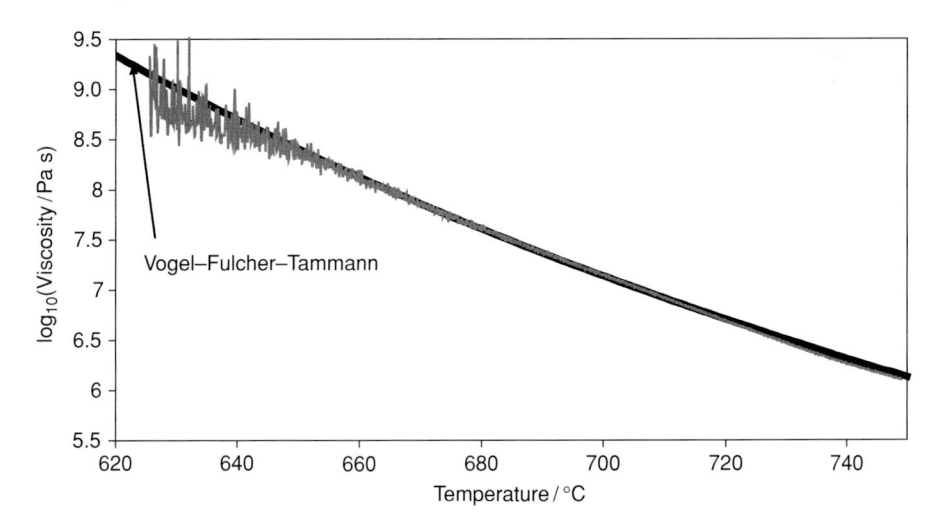

Figure 10.22 The viscosity–temperature data set measured using thermomechanical analysis (TMA) parallel-plate viscometry of the National Bureau of Standards (NBS), USA, 710 glass is compared to the Vogel–Fulcher–Tammann [9] model issued by the NBS.

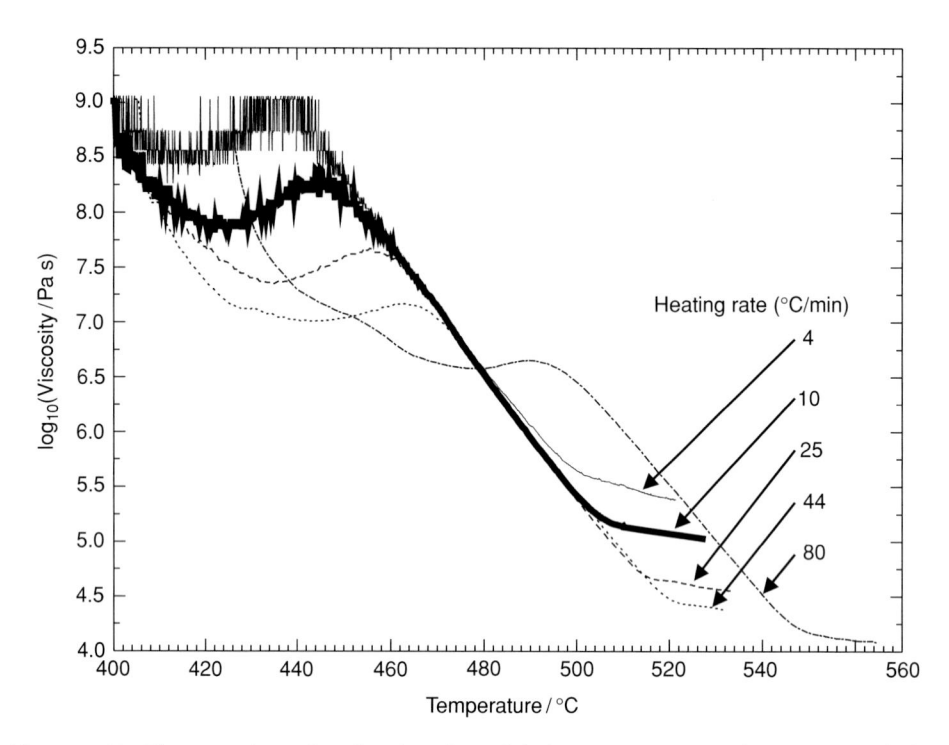

Figure 10.23 Thermomechanical analysis (TMA) parallel-plate viscometry is used to investigate the behaviour of the oxyfluoride glass 32SiO$_2$–9AlO$_{1.5}$–31.5CdF$_2$–18.5PbF$_2$–5.5ZnF$_2$–3.5ErF$_3$ (mol %) in which nanocrystals are homogeneously nucleated above T_g [39]. Description of the viscosity–temperature curves may be found in Section 10.5.6.

Agreement is excellent between $10^{8.5}$ and 10^6 Pa s. The verified range was extended to $10^{4.5}$ Pa s by using a smaller load and smaller sample height.

10.5.6.4 Case study

Oxyfluoride nano-glass-ceramics
Parallel-plate viscometry has the advantage over indentation viscometry in that a temperature ramp can be used to obtain viscosity data over a large temperature range in one run in a relatively short time. Parallel-plate viscometry was used to investigate the behaviour of the oxyfluoride glass: 32SiO$_2$–9AlO$_{1.5}$–31.5CdF$_2$–18.5PbF$_2$–5.5ZnF$_2$–3.5ErF$_3$ (mol %). This composition is amorphous as-prepared, but on heat treatment just above T_g forms a nano-glass-ceramic (see Section 10.4.3.6 and Figure 10.12). To investigate whether a fibre-drawing viscosity can be accessed to draw these materials into optical fibres, parallel-plate viscometry was carried out at different heating rates [39].

Figure 10.23 shows that at 80°C/min sample temperature lagged behind that measured by the thermocouple, invalidating the results. For heating rates up to 44°C/min, between 400 and 460°C the viscosity initially fell as expected but then increased as the fluoride nanocrystals grew and the residual liquid became more oxide-rich and less fragile in nature. As the heating rate was increased, this viscosity rise was delayed to higher temperatures.

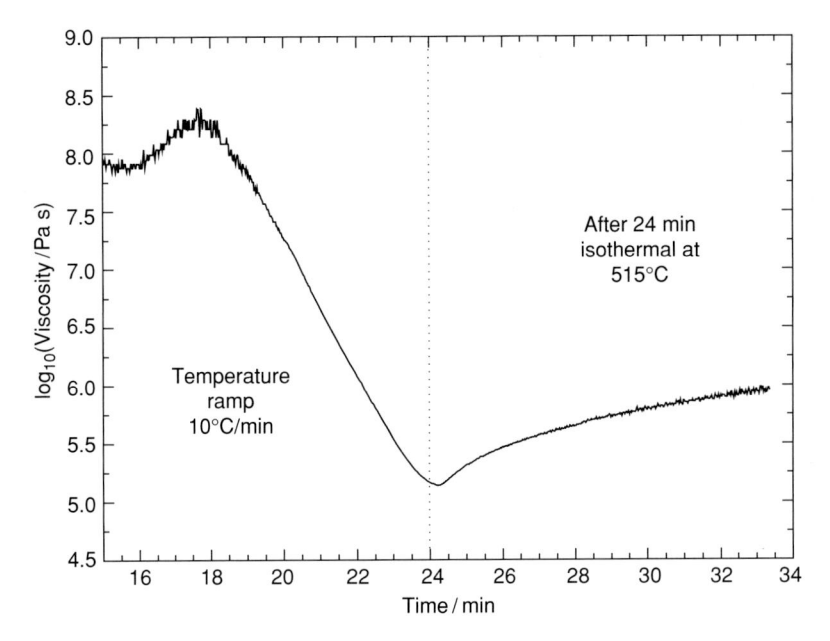

Figure 10.24 The viscosity behaviour of the oxyfluoride glass 32SiO$_2$–9AlO$_{1.5}$–31.5CdF$_2$–18.5PbF$_2$–5.5ZnF$_2$–3.5ErF$_3$ (mol %) is shown when ramped at 10°C/min up to 515°C and then held isothermally. Viscosity first rises due to the changing nature of the residual viscous liquid as the nanocrystals grow, then falls as the residual supercooled liquid is heated, and finally viscosity gradually increases as the matrix develops secondary crystals [39].

Beyond 460°C the viscosity of the residual supercooled liquid, in which the nanocrystals were dispersed, fell for all heating rates as nanocrystal growth was completed. In this temperature region the measured viscosities at heating rates of 4–44°C/min overlay, indicating thermal equilibrium had been achieved with the thermocouple and that the physical nature of the sample was similar despite the different heating rates. Above about 500°C the residual matrix tends to crystallise and viscosity plateaus (Figure 10.24); however, the higher the heating rate, the higher the onset temperature of this plateau. Moreover, an independent work has shown that this secondary crystallisation, which is to a cadmium silicate rich phase, is surface nucleated and can be inhibited by ensuring that the sample is hermetically sealed and residual oxygen is excluded [26]. Figure 10.24 shows the viscosity behaviour of this glass when ramped at 10°C/min up to 515°C and then held isothermally. Viscosity first rises due to the changing nature of the residual viscous liquid as the nanocrystals grow, then falls as the residual supercooled liquid is heated, and finally gradually increases as the matrix develops the secondary crystals. These results give valuable insight into the growth of the crystals during heat treatment and assists in development of thermal processing of the glass into fibre.

10.6 Final comments

This chapter has demonstrated the value of thermal analysis in its own right in providing quantitative, reliable data on inorganic compound glasses. The case studies selected have shown that thermal analysis is most effective when used in conjunction with other analytical techniques, for instance XRD and imaging at the micro- and nanoscale, to clarify phase behaviour.

References

1. Rawson H. *Properties and Applications of Glass*. Amsterdam: Elsevier; 1980.
2. Comyns AE (eds). *Fluoride Glasses*. Critical Reports on Applied Chemistry Volume 27. Chichester: John Wiley & Sons; 1989.
3. Seddon AB, Shah WA, Clare AG, Parker JM. The effect of NaF on the crystallisation of ZBLAN glasses. *Mater Sci Forum* 1987;19–20: 465–474.
4. Rawson H. *Inorganic Glass-Forming Systems*. London: Academic Press; 1967.
5. McMillan PW. *Glass-Ceramics*. London: Academic Press; 1964.
6. James PF. Kinetics of crystal nucleation in lithium silicate glasses. *Phys Chem Glasses* 1974;15(4):95–105.
7. Vogel W. *Glass Chemistry* (translation of 3rd edition). New York: Springer-Verlag; 1992.
8. Angell CA. Structural instability and relaxation in liquid and glassy phases near the fragile liquid limit. *J Non-Cryst Solids* 1988;102:205–221.
9. Fulcher GS. Analysis of recent measurements of the viscosity of glasses. *J Am Ceram Soc* 1925;8:339–366, 789–794.
10. Moynihan CT, Mossadegh R, Gupta PK, Drexhage MG. Viscosity temperature dependence and crystallisation of ZrF_4-based melts, *Mater Res Forum* 1985;6:655.
11. Hrubý A. Evaluation of glass-forming tendency by DTA. *Czech J Phys* 1972;22:1187–1193.
12. France PW, Drexhage MG, Parker JM, Moore MW, Carter SF, Wright JV. *Fluoride Glass Optical Fibres*. London: Blackie & Son; 1990.
13. Bassett J, Denney RC, Jeffery GH, Mendham J. *Vogel's Textbook of Quantitative Inorganic Analysis*. London: Longman; 1989.
14. Parker JM, Seddon AB, Clare AG. Crystallisation of ZrF_4–BaF_2–NaF–AlF_3–LaF_3 glasses. *Phys Chem Glasses* 1987;28(1):4–10.
15. Seddon AB, Clare AG, Parker JM. Chlorine doped ZBLAN glasses. *Mater Sci Forum* 1987;19–20:475–482.
16. Li R, Seddon AB. Gallium–lanthanum–sulphide glasses: a review of recent crystallisation studies. *J Non-Cryst Solids* 1999;256/257:17–24.
17. Reaney IM, Morgan SP, Li R, Seddon AB. TEM studies of $70Ga_2S_3$–$30La_2S_3$ glasses. *J Non-Cryst Solids* 1999;256/257:149–153.
18. Prasad N, Furniss D, Rowe H, Miller CA, Gregory D, Seddon AB. Microwave assisted synthesis of chalcogenide glasses. In: *Proceedings of the International Symposium on Non-Oxide and New Optical Glasses*, Bangalore, India, 2006.
19. Yinnon H, Uhlmann DR. Applications of thermoanalytical techniques to the study of crystallization kinetics in glass-forming liquids. Part I: Theory. *J Non-Cryst Solids* 1983;54:253–275.
20. Augis JA, Bennett JE. Calculation of Avrami parameters for heterogeneous solid state reactions using a modification of the Kissinger method. *J Therm Anal* 1978;13:283–292.
21. Busse LE, Lu G, Tran DC, Sigel GH. A combined DSC/optical microscopy study of the crystallisation in fluorozirconate glasses upon cooling. *Mater Sci Forum* 1985;5:219–228.
22. Lainé MJ. *Glasses for Acousto-Optic Devices*, PhD Thesis. University of Sheffield, UK,1996.
23. Bruce AJ. Structural relaxation in fluoride glass. In: Almeida RM (ed.), *NATO Advanced Research Workshop on 'Halide Glasses for Infrared Fibreoptics'*, Vilamoura, Portugal., Dordrecht: Martinus Nijhoff, 1987.

24. Moynihan CT. Correlation of the width of the glass transition region and the temperature dependence of the viscosity of high T_g glasses. *J Am Ceram Soc* 1993;76(5):1081–1087.
25. Tikhomirov VK, Furniss D, Seddon AB, Reaney IM, Begglora M, Ferrari M, Montagna M, Rolli R. Fabrication and characterisation of nano-scale, Er^{3+} – doped, ultra-transparent oxyfluoride glass-ceramics. *Appl Phys Lett* 2002;81:1937–1939.
26. Beggiora M, Reaney IM, Seddon AB, Furniss D, Tikhomirova SA. Phase evolution in oxyfluoride glass-ceramics. *J Non-Cryst Solids* 2003;326–327:476–483.
27. Tikhomirova SA. *Transparent Glass-Ceramics for Photonic Applications*, , PhD Thesis. University of Nottingham, UK,2006.
28. Wagner T, Frumar M, Kasap SO. Glass transformation, heat capacity and structure of $Ag_x(As_{0.4}Se_{0.6})_{100-x}$ glasses studied by temperature-modulated differential scanning calorimetry. *J Non-Cryst Solids* 1999;256/257:160–164.
29. Boolchand P. Intermediate phases, reversibility windows, stress-free and non-aging networks, and strong liquids. *Chalc Lett* 2006;3(2):29–31.
30. Seddon AB, Cardoso AV. Crystallisation studies of multicomponent AlF_3 glasses. In: *Proceedings of International Conference on New Materials and Their Applications*, Warwick, UK. Institute of Physics Conference Series No. 111. United Kingdom: Institute of Physics; 1990.
31. Jewell JM, Shelby JE. Transformation range viscosity measurements using a thermomechanical analyser. *J Am Ceram Soc* 1989;72(7):1265–1267.
32. Cardoso AV, Seddon AB. Penetration viscometry using a thermal mechanical analyser. *Glass Tech* 1991;32(5):174–176.
33. Malek J, Shanelova J. Viscosity of germanium sulfide melts. *J Non-Cryst Solids* 1999;243:116–122.
34. Nemilov SV, Petrovskii GT. A new method for measurement of the viscosity of glasses. *J Appl Chem USSR* 1963;36(1):208–210.
35. Seddon AB, Cardoso AV. Low temperature viscosity of infrared transmitting glasses. In: *Proc SPIE EC04 'Glasses for Opto-Electronics II'*, The Hague, March 1991, SPIE Vol. 1513.
36. Seddon AB, Hemingway MA. Thermal characterisation of infrared-transmitting Ge-S-I glasses. *J Non Cryst Solids* 1993;161:323–326.
37. Seddon AB. Chalcohalide glasses: glass-forming systems and progress in application of percolation theory. *J Non-Cryst Solids* 1997;213/214:22–29.
38. Gent AN. Theory of the parallel-plate viscosimeter. *Br J Appl Phys* 1960;11:85–87.
39. Seddon AB, Tikhomirov VK, Rowe H, Furniss D. Temperature dependence of viscosity of Er^{3+}– doped oxyfluoride glasses and nano-glass-ceramics. *J Mater Sci: Mater Electron (in press)* available on-line March 2007.

Appendix 1
The Williams Landau Ferry (WLF) equation and Time Temperature Superposition

The method of reduced variables, or time temperature superposition, has traditionally been applied to pure polymers, where it is used to make predictive measurements of the material. It has also found wider application in the foods area with the treatment of sugar rich hydrocolloid gels and gelatin. A detailed treatment can be found in standard texts such as Viscoelastic Properties of Polymers by Ferry (1) and some basic information is reproduced here.

The fundamental WLF equation can be derived from the Dolittle equation which describes the variation of viscosity with free volume.

$$\ln \eta = \ln A + B([V - V_f]/V)$$

Where η is the tensile viscosity, V and Vf are the total volume and free volume of the system respectively, and A and B are constants.

$$\Rightarrow \quad \ln \eta = \ln A + B(1/f - 1)$$

If it is assumed that the fractional free volume increases linearly with temperature then

$$f = f_g + \alpha_f(T - T_g)$$

where α_f is the thermal coefficient of expansion, f is the fractional free volume at T, a temperature above T_g, and f_g is the fractional free volume at T_g.

Therefore

$$\ln \eta(T) = \ln A + B(1/[f_g + \alpha_f(T - T_g)] - 1) \quad \text{at } T > T_g$$

And

$$\ln \eta(T) = \ln A + B(1/f - 1) \quad \text{at } T = T_g$$

subtracting $\Rightarrow \quad \log \eta(T)/\eta(T_g) = -B/2.303 \, f_g[T - T_g]/\{(f_g/\alpha_f) + T - T_g\}$

where the universal constants $C_1 = B/2.303 \, f_g$ and $C_2 = f_g/\alpha_f$

It can be shown that $\log \eta(T)/\eta(T_g) = \log a_T$

The method of extracting the universal constants of the WLF equation is to use the method of reduced variables (time temperature superposition). A number of DMA frequency scans are recorded at a series of isothermal temperatures, and the resulting data displayed on a screen. A reference temperature is chosen (typically the isothermal temperature from the middle section of data where properties vary most as a function of frequency, equivalent to T_g) and the factors required to shift the responses at the other temperatures calculated so that one continuous smooth mastercurve is produced. Most commercial analysers provide software to create the mastercurve and calculate the shift factors involved. Different approaches have been used to obtain this type of data including a frequency multiplexing approach where a low underlying scan rate is used and the frequency continuously varied. However this means that data collected at each individual frequency will be at a slightly different temperature, and it is best to ensure that truly isothermal conditions are employed.

The factor for the reference temperature (a_T) is 1 with those for temperatures less than the reference temperature being less than 1 and those for temperatures greater than the reference temperature being greater than 1. The mastercurve, in conjunction with either the modulus temperature curve or the shift factors (a_T) relative to a reference temperature (T_{ref}), enables the response of the polymer under any conditions of time and temperature to be obtained.

There is also a correction to be applied in the vertical direction as well as the shift along the x-axis. This is related to the absolute value of temperature, on which the moduli depend, and the change in density as the temperature changes. The entire procedure can be expressed thus

$$E(T_1, t)/\rho(T_1)T_1 = E(T_2, t/a_T)/\rho(T_2)T_2$$

This correction should only be made for the rubbery state, ie above T_g. It is invalid for the glassy state.

By fitting the WLF equation to the shift factors, (y-axis), at a particular temperature, (x-axis) we can obtain an estimate of the universal constants C_1 and C_2 and hence obtain an estimate of T_∞, the underlying true second order transition temperature, where viscosity becomes infinite. The WLF equation can also be expressed in terms of the shear moduli as

$$\log a_T = G(T)/G(T_{ref}) = -C_1(T - T_{ref})/(C_2 + T - T_{ref})$$

and can be rearranged to obtain the universal constants C_1 and C_2 from a linear graphical plot.

$$(T - T_{ref})/\log a_T = -1/C_1(T - T_{ref}) + (-C_2/C_1)$$
$$y = mx + c$$

If the term $[C_2 + T - T_0]$ went to zero the term $\log a_T$ would tend to $-\infty$ which would correspond to a shift factor of 0 and the response of the polymer at the limiting true transition temperature T_∞.
and
$$T_\infty = T_{ref} - C_2 = T_g - 50$$

if T_g is chosen as the reference temperature.

T_∞ is thought to be the true underlying 2nd order transition for these materials and only accessible for experiments carried out over very long time scales. In practice the 50°C, which is normally added to the T_∞, returns a value for T_g more in keeping with everyday experience and measured values. This procedure places the glass transition somewhere in the transition region.

It can therefore be seen that C_1 can give information on the free volume at the glass transition whilst C_2 can give information on the thermal expansion as well as the underlying glass transition.

WLF Kinetics

It is found for some diffusion limited reactions that the temperature dependence of the rate of the reaction follows a WLF type of equation rather than an Arrhenius equation in the region of the glass transition. This is also true for rheological properties such as shear moduli. Plots which linearise the data are of the form of the logarithm of reaction rate against the reciprocal of terms similar to $T - T_g$, rather than the Arrhenius form, $1/T$. It is only for reactions limited by diffusion, such as the translational motion of a reactant in a glass/rubber. Reactions which are limited by some other non diffusion based step are known as reaction limited and will obey Arrhenius kinetics.

1) Viscoelastic Properties of Polymers, Third edition, Ferry, 1980, Wiley, Pages 280–298

Glossary

Ageing Typically, this involves leaving a sample for period of time to see what effect this has on the material. In a controlled test this may well be at elevated temperature or under ultraviolet light or other conditions of interest. Decomposition or other effects may be observed. In the region of the glass transition, ageing may result in the relaxation of a material parameter (e.g. enthalpy; see 'Enthalpic relaxation') not at equilibrium towards the equilibrium value at that temperature. In the region of melting of a polymer, ageing may result in changing crystalline structure. Ageing and annealing are terms often used interchangeably.

Amorphous An amorphous material is one that has no regular crystalline structure. Liquids possess local order but no long-range order. This is the same in the case of amorphous solids, which may be considered as supercooled liquids. They exhibit a glass transition, or softening temperature, but have no melting point.

Annealing (see also 'Ageing', 'Stress relaxation' and 'Tempering') The annealing period is the time given for relaxation towards equilibrium at the specified annealing temperature. Annealing with reference to frozen solutions: a non-equilibrium frozen solution, in the sense of being at too low a concentration for that temperature, will form ice and approach the equilibrium concentration if given a sufficient annealing period in the region of the T_g'.

Argand diagram Graphical representation of complex numbers, where the real component P' is drawn along the x-axis and the imaginary component P'' is drawn along the y-axis. The vector length, drawn from the origin, has the value P^* and the angle subtended between the vector and the x-axis is δ, where $P' = P^* \cos \delta$ and $P'' = P^* \sin \delta$.

Bulk modulus (K) It is measured by the hydrostatic loading of a sample and resultant volumetric strain (hydrostatic stress/volumetric strain).

Cold crystallisation This refers to the crystallisation process that occurs when amorphous material is heated through its glass transition temperature and is able to recrystallise.

Complex compliance (D^* or J^*) A reciprocal modulus, defined as strain/stress. How easily a material responds to stress; i.e. large displacement means high compliance. Unit: Pa^{-1} or reciprocal Pascals (m^2/N).

Complex modulus ((Young's (E^*), shear (G^*) and bulk (K^*) Any material exhibiting viscoelastic behaviour will have a complex modulus. See storage and loss modulus. Unit: Pa or Pascals (N/m^2).

Composite A composite material is one made up of more than one material. The term is usually used to describe thermoset (more commonly) or thermoplastic resins reinforced with high-modulus fibres. The result is a high-modulus, tough and light material that has many transport, military and leisure applications.

Creep It is the processing of continuously changing length (or thickness for compression geometry) under an applied load. The deformation is time dependent, therefore yielding a time-dependent modulus.

Curie point The temperature at which the ferromagnetic property of a material is lost.

Damping factor It gives a measure of the energy-absorbing potential of a material. See tan δ.

Devitrification It refers to the crystallisation of a glass (e.g. quartz) that occurs when taken to a high temperature.

Elastic behaviour Exemplified by the type of deformation that a spring undergoes with an applied load. The observed extension for an applied load is proportional to that load (Hooke's law) and occurs instantaneously. The spring returns to its original length when all load is removed; i.e. all deformation is recoverable.

Elastomers These are materials that are in their rubbery phase at room temperature (see 'Glass transition').

Enthalpic relaxation Many amorphous materials below T_g are in a non-equilibrium energy state, but are able to 'relax' towards equilibrium values when in the region of the glass transition. The enthalpy (or total energy) curve is parameter used to describe the energy state of a material, hence the term 'enthalpic relaxation' when a material relaxes in the T_g region. Characteristic traces in the enthalpy and heat capacity are seen when reheating a material which has relaxed in this way.

Fictive temperature This is the temperature on the equilibrium enthalpy curve of a material obtained by extrapolation of the enthalpy of the glassy curve to the equilibrium line. It corresponds to the true temperature of T_g of a material.

Free volume It can be defined as the unoccupied space in an amorphous polymer due to the inefficient packing of polymer chains. A fully crystalline polymer would represent perfect packing and therefore has the highest density.

Freeze-concentrated glass A glassy material resulting from a lyophilisation (freeze-drying) process.

Geometry constant This parameter relates the measured stiffness to the modulus and depends upon the sample shape (see Chapter 4, Appendix 1 for different geometries).

Glass A rigid material is said to be a glass when it has no crystalline structure. Such materials are described as amorphous. All glasses behave in a similar fashion, but they may appear to be quite different, since their glass transitions (see below) may be very different. In comparison to other materials they usually have low toughness. Materials that easily form glasses often have high molecular weights and it is their lack of mobility within the liquid region that prevents the formation of a crystalline solid.

Glass transition temperature (T_g) The T_g is a fundamental property of all glass-forming materials, and a significant change in mechanical properties occurs at this temperature. Below T_g an amorphous material is a glass and above T_g it behaves more like a rubber and is defined as being in a rubbery state. Below T_g the relaxation time (see below) will be much longer than its value above T_g. This affects how the material responds to any mechanical loading or other stimulus. The T_g is a second-order process and will therefore depend upon how it is measured. Unlike a melting point, there is no unique value for the T_g process, since its value depends upon how it is measured and how it has been conditioned. One of the simplest visualisations of the glass transition process comes from measuring the thermal expansion of the sample using a TMA instrument. The T_g manifests itself as a change in the coefficient of thermal expansion, which is usually lower for the glassy state than the rubbery state (see Section 4.6). DMA is invariably more sensitive than TMA for determination of the glass transition temperature. The definition of T_g is discussed in detail in Section 4.1.3.

Heat flow The flow of energy into or out of a sample which is the primary measurement made using a DSC.

Heterogeneous nucleation Nucleus formation with the aid of a foreign body such as a dust particle.

Homogeneous nucleation Spontaneous nucleus formation as a result of reduced temperature. Normally this occurs at a lower temperature than for heterogeneous nucleation.

Linear deformation It is where a force is applied to the sample as a push or a pull, i.e. a back and forth linear motion, as used with a DMA. The motor motion is always linear in typical DMA analysers and in most cases it is sinusoidal in nature, although some instruments can apply other waveforms.

Liquidus The temperature above which a material is fully liquid, essentially the end of the melting range of a material.

Loss modulus (Young's (E''), shear (G''), bulk (K'')) It represents the viscous part of the deformation experienced in a viscoelastic material. It is commonly referred to as the out-of-phase and imaginary modulus (see 'Argand diagram representation'). It is proportional to the amount of energy lost and irrecoverable in a sample. Unit: Pascals (Pa).

Modulus It is the resistance to an applied force per unit area (stress/strain). It differs from stiffness only in that the sample geometry is taken into account. Therefore it has a constant value for a given material, irrespective of sample size. Modulus is a measure of how a material resists the application of stress; i.e. small displacement means high modulus. Unit: Pa or Pascals (N/m^2). Modulus is given a different symbol dependent upon how it is measured: Young's modulus (also Flexure or Tension deformation modes in DMA) are given the symbol E, Shear the symbol G, Bulk the symbol K.

Poisson's ratio It is defined as the ratio of lateral strain to longitudinal (lateral strain/longitudinal strain) in tension or compression geometry.

Polymers These are long-molecular-chain materials that can readily exhibit both crystalline and amorphous behaviour. Many polyolefines (e.g. PE and PP) are semi-crystalline, possessing both crystalline and amorphous regions, whilst other polymers such as acrylics and epoxies are generally completely amorphous. Their glass-forming ability typically comes from their high molecular weights.

Relaxation time It is a characteristic value relating to the time-scale of a molecular motion described above. Typically, relaxation times are very long for glassy solids (>1000s of seconds) and are almost instantaneous for rubbers (<1 ms) (see 'Glass transition'). Consequently, glasses never attain thermodynamic equilibrium, whereas rubbers will always be in equilibrium (see 'Glass transition').

Relaxations Mechanically, these refer to a deformational mechanism that occurs within a sample. The best example of this is the one that occurs in the glass transition process. A relaxation is evident when the materials' response time is sympathetic to the frequency of the applied deformation. Energy absorption is at maximum here and peaks are observed in E'' and $\tan \delta$. The activation energy for the process can be determined from the temperature shift of these peaks as a function of frequency.

Rubbers These are highly extensible materials that can sustain large strains (200–1000%) without breaking and are characterised by a low modulus ($\approx$5 MPa). They have the same local structure as glasses but behave as rubbers once glassy materials have passed through (above) the glass transition temperature.

Scragging Carbon-filled rubbers are normally conditioned or 'scragged' before testing by exposing them to a high strain. This affects filler/rubber interaction (see Section 4.3.5).

Semi-crystalline Many polymers are semi-crystalline. This means they are composed of two phases, often with quite different properties. One phase is amorphous and the other crystalline. They are chemically identical. The amorphous phase exhibits a glass transition, T_g, which is always below the melting point of the crystalline phase. The amount of softening observed at T_g will be decreased in proportion to the amount of crystalline material present.

Shear modulus (G) It is measured from a sample in simple shear (linear displacement) (shear stress/shear strain).

Solidus The temperature above which a material begins to melt, the start of the melting range.

Stiffness It is the resistance to an applied force. The stiffer a sample is, the lesser it will deform to an applied force. For linear deformation (see its definition above) it always has units of N/m. The stiffness depends on two factors, the sample size and its modulus (see below).

Storage modulus (Young's (E'), shear (G'), bulk (K')) It represents the elastic part of the deformation experienced in a viscoelastic material. It is commonly referred to as the in-phase and real modulus (see 'Argand diagram representation'). It is proportional to the amount of energy stored and recoverable in a sample. Unit: Pa (Pascals).

Stress relaxation It is often used synonymously with annealing or enthalpic relaxation.

Tangent (or tan) δ It is the ratio of loss modulus divided by storage modulus. It gives a measure of the energy-absorbing potential of a material and is also called the damping factor. It has no units.

Tempering (Foods) A combination of temperature and shear treatments designed to produce nuclei of a desired fat polymorphic form.

TG-FTIR ThermoGravimetric Analysis (or Analyser) coupled to a Fourier transform Infrared Spectrophotometer for evolved gas analysis.

TG-MS ThermoGravimetric Analysis (or Analyser) coupled to a Mass Spectrometer for evolved gas analysis.

TG-GC-FTIR ThermoGravimetric Analysis (or Analyser) where the evolved gases are adsorbed onto a chromatography column with subsequent desorption through a Fourier transform Infrared Spectrophotometer for analysis.

Thermal history This refers to the thermal conditions a sample experiences before it is analysed. Different conditions may affect a sample in different ways so that a common DSC approach to the measurement of the effects of thermal history is to heat, cool (using a known repeatable cooling rate), and then reheat a material. The initial heat is then compared with the reheat.

Thermal resistance constant *(Ro)* In a DSC this is a measure of the resistance to the flow of energy input into a sample. Usually measured from the leading edge of the melt of indium it gives a measure of the rate of flow of energy into a sample, taking into account that the instrumental pathway energy must flow before it reaches the sample.

Thermoplastic It describes polymers that can be processed by heating. The material can be repeatedly softened and formed. The chemical structure is linear (see 'Thermoset' for comparison) and can be made up from an aliphatic or aromatic backbone. There is no chemical bonding between neighbouring chains. These polymers can be amorphous or semi-crystalline. Generally lower modulus and better impact strength than thermosets.

Thermoset It describes polymers that cannot typically be reformed by heating (see 'Thermoplastic'). They are usually formed from liquids or prepreg (partially cured thermoset material). A cross-linking reaction occurs between the polymeric chains, forming a chemically bonded three-dimensional structure. Material has high modulus, especially above T_g, and is usually more brittle than thermoplastics.

Third harmonic distortion It is the result of non-linear behaviour. It can be detected as a frequency three times the applied stimulus.

Time constant (τ) In a DSC this is a measure of the time taken for the system to return to baseline after a rapid endothermic event.

Torsional deformation It is where mechanical force is applied to the sample as a torque. This occurs in rheometers, where the sample is typically sheared between parallel plates. Alternatively, a solid bar can be twisted, resulting in a shear deformation.

Transient In a DSC trace this refers to the period of instability at the start of the run before a stable baseline is obtained.

Viscoelastic behaviour It defines the response to an applied force or displacement for a material exhibiting time-dependent behaviour, as seen in a polymer for example. As the name suggests it is a combination of viscous and elastic behaviour. Deformation occurs at a certain rate, determined by the structure (molecular or physical) of the sample under test.

Viscous behaviour The type of deformational behaviour experienced by a fluid under an applied force or displacement, i.e. flow. Any deformation is permanent; the liquid will not return to its previous shape without further stimulus.

Young's modulus (E) It is measured from a sample in static tension (or compression) where the sides are free to contract (or grow) – effect of Poisson's ratio – as the sample extends (contracts). It is the applied stress/longitudinal strain.

Further Reading

Calorimetry and Thermal Analysis of Polymers, Edited by Vincent B Mathot. Hanser 1994
Differential Scanning Calorimetry, Höhne, Hemminger and Flammersheim, Second edition, Springer 2003
Modulated Temperature Differential Scanning Calorimetry: Theoretical and Practical Applications in Polymer Characterisation, Edited by Mike Reading and Douglas J. Hourston, Springer 2006
Principles of Thermal Analysis and Calorimetry, Edited by P.J. Haines, RSC paperbacks, 2002
Thermal Analysis of Pharmaceuticals, Edited by Duncan Q.M. Craig and Mike Reading, TF-CRC 2006
Thermal Methods of Analysis, Principles, Applications and Problems, P.J. Haines, Blackie 1995
J. Ferry, Viscoelastic Properties of Polymers, 3rd ed., John Wiley & Sons, New York, 1980.
K. Menard, Dynamic Mechanical Analysis – A Practical Introduction, CRC Press, Boca Raton, 1999.
Thermal Characterisation of Polymeric Materials, Second Edition, Edith Turi, Acedemic Press 1997
Thermal Analysis of Foods, edited by V.R. Harwalkar and C.-Y. MA, Elsevier Applied Science, 1990.

Web Resources

ATHAS Databank
http://athas.prz.edu.pl/

Evitherm: Virtual Institute for Thermal Metrology
www.evitherm.org/

Laboratory of the Government Chemist
http://www.lgc.co.uk/

NIST Virtual Library
http://nvl.nist.gov/index.cfm

Index

Page numbers in *italics* refer to figures; those in **bold** to tables.

Accuracy (*see* calibration)
Activation energy   36, 147–149, 206
Additives   194–199, 205–206
Ageing (*see also* annealing, oxidation, decomposition)   224–227, 238–241
  proteins   371–373
  sugars   349
Air   10, 91
Aluminium
  melting point   6
  oxide (Alumina)   94
  pans   6
Amorphous material (*see also* Glass transition)
  amorphous state   22, 31
  fats   365
  measurement by DSC   74–75, 79–81
  measurement in pharmaceuticals   310–315, 325–327
Annealing (*see also* ageing)   57–59, 69–74, 411, 417
Area calculations (*see also* calculations, heat of fusion, enthalpy, heat of reactions)   18–21
Argand diagram   128
Argon   10
Arrhenius kinetics (*see also* equations)   34, 36, 166
Artefacts   103
Aspartame   99–100
Atmosphere (*see* purge gas)
Autosamplers   97
Averaging   42
Avrami kinetics   37–39

Baseline   41, 85, 161
Balance Type   89
Beef   374–375
Beta transition   23, 124, 144, 216
Bending   130–134
Bioactive composites   264, 282
Biomaterials   257–262
Blowing agents   243–244
Bulk modulus *see* modulus
Buoyancy   90, 103

Calcium oxalate   89, 110, 116
Calculations
  peak   14, 20, 21
  Tg DSC   22, 25–26
  TGA   104–111
  molar mass   111
  Tg DMA   150–152
Calibration
  reference materials   15, 421, 433, 444
  DSC   12–17
  DMA   140–142
  effect of high scan rate (DSC)   52
  TGA   112–113
  TMA   157
Cantilever modes   130–134
Carbamazepine   78, 298–302
Carbonaceous residue   192–193, 203
Carbon black   99, 112, 185–186, 193, 196–197
Cements (Calcium Phosphate)   278–281
Chalcogenide glasses   411, 425, *426*, **434**, *436*
Chalcohalide glasses   441, *442*
Chlorpropamide   76–79
Chocolate   362

Index

Cleaning (*see also* contamination) 85
   power comp. DSC 11, 12
Clamping samples 135, 142
Coefficient of thermal expansion 156–161, 433–436, *435, 436*
Comparative studies 64, 65, 176
Complex modulus (stiffness) 125–127
Composites 152
   particulate / fibre composites 264–270
Compositional analysis 181–188, 199–200
Compression mode 130–134
Controlled rate methods (SCTA) 88
Contamination (contaminants) 8, 11, 85, 172–175
Cooling
   effect on Tg 26–30, 74–75
   into the amorphous state or other crystalline forms 31, 54
   controlled cooling 56
Copper sulphate 105
Creep 122, 137–138
Critical cooling rates 428–429
Cross linking 168–170, 176–177
Crucibles – *See* Sample Pans
Cryopreservation 394–395
Crystalline structure 57–59
Crystallisation (*see also* polymorphism) 38
   effect of fast scanning 61–64
   effect of ions 347–348
   fats 356–361
   polymers 231–235
   sugars 345–348, 351–353
   glasses 413–415, *412, 414,* 447
Curie transition 88, 102–103, 113
Curing (*see also* reactions and cross linking) 110, 218–220, 241–243

Damping factor 120
Decomposition (*see also* reactions and stability) 97, 99, 224–227
   explosive 99, 101
Defects (*see* quality control)
Dehydration 92, 100
Denaturation (*see* proteins)
Derivatives
   DSC 5, 77, 173–175
   TGA 89, 92, 100–107, 180–188
   DMA 150, 152
Detection limit
   TGA 198–199
Devitrification 413–415, *412, 414*
Dielectric analysis 247
Differential techniques (*see* derivatives)

Differential Photo Calorimetry DPC 48, 248
Drugs (*see* pharmaceuticals)
Drying (*see* evaporation) 101
DSC 1–50, 426–431
DMA (DMTA) 119–163
DTA 48, 418–426

Egg white 370, 375
Elastic behaviour 122
Elastomers 97
Empirical content 107–108
Enantiotropy 292–294
Encapsulation (*see* sample encapsulation)
Endotherms 3, 91
Enthalpic relaxation 26–30, 72–75
Enthalpy 5, 24, 27
Equations:
   arrhenius 36
   avrami 37–39
   couchman-karasz 33
   gibbs free energy 22
   gordon-taylor 33
   heat flow (DSC) 46
   tool narayanaswamy moynihan 28, 29
Equilibration time (*see* transient)
Eutectic 412, 423–425
EVA polymers 181–182, 203
Evaporation 101–102
Evolved Gas Analysis (EGA) 104, 114–117
   polymers 210–212
   pharmaceuticals 308–310
Eutectic 32
Excipient compatibility 320–323
Extrapolated onset (*see* onset calculations)
Expansion coefficient (*See* Thermal expansion coefficient)
Experimental variables
   DMA 129–140, 213
   DSC 6–12
   DTA 420–421
   TGA 93–98, 180–181, 192–193
   TMA 432
Extension mode (*see* measuring systems)
Extent of cure 170–171
Extrusion studies 178
   starch 337–338

Fast scanning applications 52, 172–175, 186–188, 245
   amorphous content 313–315
   fats 356
   polymorphism 298–302
Fats 353–365

Fictive temperature (*see also* glass transition, enthalpic relaxation) 27, 412, 413
Fillers 111, 181, 196–199, 264–270, 275
Fire Retardants 179–180, 188, 197–198, 205–206
Fish 373–374
Fluids (*see* liquids)
Fluoroaluminate glasses 441
Fluoroaluminosilicate glasses 431, 446–447
Fluorozirconate glasses 421–423
Foams
   DMA 274–275
   DSC 275
Food categories 332
   thermodynamics 395–398
Force 122, 125, 141
   static force 134
   dynamic force 134
Forensic studies 65, 66, 172–175
Fragile liquids 416, *417*
Free volume 124
Freeze concentrated glass (*see also* glossary, lyophilisation) 387
Freeze drying (*see* Lyophilisation)
Frequency scanning 138
Frozen systems 387–396
FTIR (Evolved gas analysis) 116
Furnace
   calibration 16

Gallium-lanthanum-sulfide glasses 423–425
Gamma transition 23, 124, 216
Gas (*see* purge gas, atmosphere)
Gelation (*see* curing, cross-linking, proteins, hydrocolloids)
Gelatin (*see* hydrocolloids)
Gelatinisation
   starch 341–343
Geometric constant 127, 162
Geometry
   DMA measuring systems 129–134
Geometry constants 162
Germanium-sulfur-iodide glasses 441–442
Glass (*see also* glass transition) 410, 411
   annealing 413–415
   crystallisation 413
   forming 436
   transformation temperature 413
Glass-ceramics 410, 415, 431, 446–447
Glass Transition, Tg
   DSC 22–31
   DMA 123–124, 139, 150–152, 217–218
   dilatometric 433–435
   fast scanning DSC 62–64, 67–76

frequency effects (DMA) 139
glasses 411–413
hydrocolloids 382–385
pharmaceuticals 311–315
pharmaceuticals: values for 312
polymers 236–238
proteins 371
standard DSC method (polymers) 49
starch 336–338
state diagram 388–390
sugars 348–349
Glucose 344
Gluten
   (wheat protein) 344, 375–377

Hazard 39
Heat capacity (*See also* specific heat) 3, 4, 43–45, 427–428
Heat flow
   measured curve 3, 13, 30
   effect of increasing scan rate 8–10
Heat flux DSC 47, 48
Heat rate (*see* scan rate)
Heat of fusion 19
Helium 10, 60, 83–84, 95
Heterogeneous nucleation 431
High pressure 49
High resolution TGA 208–210
Homogeneity 98
Homogeneous nucleation 431
Hooke's law 121
Hruby parameter of glass stability 419
Humidity
   TGA 97
   DMA 155–156
Hydrates 303–308
Hydrocolloids 378–386
   gels in solution 379–382
Hydrogen 95
Hydrolysis of polymers 270–271
HyperDSC (*see also* fast scanning) 54

Ice (*see* frozen systems)
Ice cream (*see also* frozen systems) 392–394
ICTAC 411
Identifying unknowns 171–172, 177, 188, 289–290
Impurities (*see* Purity)
Indium (*see* calibration)
Indomethacin 72–74
Infrared initiated reactions 49
Initial transient 17, 54, 55, 82
Integration limits 20
International Commission on Thermal Analysis 411

International Confederation for Thermal Analysis and Calorimetry (ICTAC) 411
Interpretation of data
  DSC 17, 40, 41
  TGA 98–104
Iso-kinetic curve (*see* Modulated temperature DSC)
Inorganic bone cements 278
Iso-step methods (*see* TGA)
Isothermal calorimetry (excipient compatibility) 321–325
  foods 396–399
Isothermal titration calorimetry (ITC) 396–399

Kinetic studies 36, 112

Lactose 343
  amorphous content by DSC 79–81
  DMA 153
Lag (*see* Thermal Lag)
Limestone 108
Linear viscoelastic region 134–135
Liquid-liquid immiscibility 416, 425
Liquids 7, 102, 122
Liquidus 411
Lithium disilicate glass and glass-ceramics 415
Lyophilisation (freeze Drying, *see also* frozen systems, 386) 31

Maillard reaction 350
Maltose 344
Mass spectrometry (TG-MS) 115
Material pocket 152–155
Measuring systems DMA 129–135
Methods (*see* experimental variables)
Melting
  DSC 18
  single crystal 18
  fast scanning DSC 61–64
  TGA 101
  Sucrose 343
Melting point
  DSC 19, 20, 288–290
Metastable state (*see* polymorphism)
Microthermal analysis 246
Microcalorimetry 323–327
Microwave oven 429
Modulus (*see also* stiffness) 23, 127, 142
Modulated temperature DSC 42–46, 244–245, 364, 431
Moisture effects 59–60, 82–83, 155–156
Molecular weight 25
Monotropy 292–294

Nano-glass-ceramics 415, 431, 446–447
Newtonian behavior 122
Newtonian viscosity 413, 437
Nitrogen 10, 91
Non-reversing curve (*see* modulated temperature DSC)
Nucleation 414–415, 431

OIT test (Oxidative Induction Time) polymers 40, 165–166
  Oils and fats 362
  Standard method 49
Onset temperature 14, 19, 22
Overshoot 27
Oxidation,
  DSC 40
  TGA 99, 100
Oxidative stability (*see* OIT)
Oxygen 95, 96

Particulate size 92
Peak
  area 14
  integration 19–21
  maximum 19
  separation (*see also* resolution) 21
  shape 18–20
Pharmaceuticals
  fast scanning DSC applications 76–82
Phase diagrams 31, 32
  fats 355–356
Phase lag 125–127
Phase transformations 413–415, *412*, *414*
Phosphate based glass 280
Plasticisers 24, 25, 60, 83, 195–196
  effect of solvents 276–278
Platinum 95
Poisson's ratio 163
Polyamorphism 340
Poly(carbonate)
  DMA 144, 159
Poly(ethylene) 166–169
Poly(ethylene terephthalate): PET 56, 62–64, 46, 56, 69
Poly(lactic acid) 266–269
Polymers
  DSC 20, 229–244
  fast scanning DSC 61–65
  DMA 215–216, 220–224
  TGA 201–206
Poly(methyl methacrylate): PMMA 147–149, 152, 276
Polymorphism
  DSC 37
  fast scanning DSC 76–78

fats   356
pharmaceuticals   290–303
Polymorphic purity   76–78, 297–302
Poly(propylene)   166
Polyols   395
Poly(styrene) PS
  DMA   143, 144
Poly (vinyl chloride): PVC   109, 117, 152, 184–185
Powders
  DMA   152–155
Power compensation (DSC)   46
Pressure (*see* high pressure)   6
Proteins   365–378
  denaturation and gelation   365–371, 373
  DMA   373
Proximate analysis (*see* compositional analysis)
Pseudo-polymorphism (*see* solvates and hydrates)
Purge gas
  DSC   10, 17, 60, 83
  TGA   91, 95, 98, 104
Purity
  determination   21, 315–320
  sucrose   342

Quality control   166–178, 184–185
Quasi-isothermal mode   43

Reactions   34–36
  chemical   39
  conversion   99–100
  with pan materials   7, 94
Recrystallisation (*see also* crystallisation)   34, 37
Reduced pressure (*see* vacuum)
Reference materials (*See also* calibration)   15, 16, 141, 420
Relaxations (*see* enthalpic relaxation)
Relaxation time   123
Residual oxygen   95–96
Resolution of events
  DSC (*see also* peak separation)   10, 21
  TGA   94, 208
Reversing curve (*see* Modulated Temperature DSC)
Rigid Amorphous Fraction (RAF)   24, 340–341
Rubbers (*see also* glass transition)   22
  TGA analysis   111–113, 203–206
  rubber blends using DMA   220–224
  Tg by DSC   236–238
  Curing   241–243

Sample
  Encapsulation   6–8
  Mass   10, 67, 85
  Pans   6–9, 17, 84, 94, 420–421
  Preparation TGA   92
Sample controlled TGA (SCTA)   93–94
Scaffolds   271, 273–275
Scan rate
  DSC   8–10, 17, 52–54, 68–74, 85
  TGA   93
  DMA   141–142
Semi-crystalline materials   23, 24, 145–147
Sensitivity
  due to scan rate in DSC   9, 57, 67–68
Separation of events   42–45, 59, 60
Shear mode (*see* geometry)
Simultaneous TGA-DTA   90
Single cantilever mode (*see* geometry)
Softening point   175–180
Solid / liquid ratio
  fats   353–355
Solubility
  measurement of undissolved material   81
Solvates   37, 303–308
Solvent loss   200
Solution calorimetry   325–327
  foods   395
Specific heat (*see* heat capacity)   3, 4
  standard method   49
Spinodal decomposition   415
Stability (*see also* OIT, oxidation and decomposition)   206–207, 282–283
  of pharmaceuticals (polymorphs)   291–293
Standards (*see* calibration)
Standard test methods
  DSC   49
  TGA   113–114
  DMA   140
Starch   332–343
State diagram (frozen systems)   33, 388–390
Sterilisation (gamma radiation) effects   269–270
Stiffness (*see also* modulus)   125
Storage modulus (*see also* modulus)   126
Step-Scan (*See* temperature modulated DSC)
Strain
  (applied DMA strain)   134
Stress relaxation (*see* enthalpic relaxation, creep)
Sub-ambient operation DSC   11
Sublimation   102
Sucrose   105, 106, 108
  DSC of glass   346

Sugars   343–353
　frozen sugar solutions   387–392
　ITC   396–399
　stabilising properties   377–378, 392–395
　table of properties   344
Supercooled liquid state   411, *412*

Tangent delta (tan δ)   127
Tempering (*see* annealing)
Tensile tests   130–134
Test methods (*see* standard test methods, experimental variables)
Thermal conductivity   249
Thermal contact   7
Thermal expansion coefficient (*see* coefficient of thermal expansion)
Thermal history   56, 166–168, 230, 429–430
Thermal lag   52, 82
Thermal resistance constant (Ro)   19
Thermally stimulated current (TSC)   248
Thermogravimetric analysis TGA: 87–118
Thermomechanical analysis TMA: 156–162, 432–448
Thermoplastics (*see* polymers)
Thermosets (*see* composites)
Three point bending   130–134
Time temperature superposition   138
Time-temperature-transformation curves   428–429
Tissue engineering scaffolds   271–275

Transitions (*see also* glass transition)   41
Transient (*see* initial transient)   17
Trehalose   343

Unfreezable fraction   388–389
UV initiated reactions   48, 49

Vacuum   96, 196
Validation (verification) DSC   13, 17
Viscous behaviour (*see* viscosity)
Viscoelastic behaviour   121
Viscometry   438–447
　beam bending   **438**
　indentation   438–443
　parallel-plate   443–447
Viscosity   412, 416–418, 436–447
　Cohen-Grest model   441
　from heat capacity change   430–431
　Nemilov   439
　Newtonian   413, 437
　Vogel-Fulcher-Tammann model   417, 440
Vitrification (*see* devitrification)
Volatile components Identification by EGA   115–118

Weight loss calculation   105

ZBLAN glasses   421–423
Zinc (*see* calibration)